Introduction to Differential Geometry with Tensor Applications

Scrivener Publishing
100 Cummings Center, Suite 541J
Beverly, MA 01915-6106

Publishers at Scrivener
Martin Scrivener (martin@scrivenerpublishing.com)
Phillip Carmical (pcarmical@scrivenerpublishing.com)

Introduction to Differential Geometry with Tensor Applications

Dipankar De

Scrivener Publishing

WILEY

This edition first published 2022 by John Wiley & Sons, Inc., 111 River Street, Hoboken, NJ 07030, USA and Scrivener Publishing LLC, 100 Cummings Center, Suite 541J, Beverly, MA 01915, USA

For more information about Scrivener publications please visit www.scrivenerpublishing.com.

Wiley Global Headquarters
111 River Street, Hoboken, NJ 07030, USA

For details of our global editorial offices, customer services, and more information about Wiley products visit us at www.wiley.com.

Library of Congress Cataloging-in-Publication Data

ISBN 9781119795629

Cover image: Geometry images provided by De
Background image: Abstract Spiral, Fabian Schmidt | Dreamstime.com
Cover design by Kris Hackerott

Set in size of 11pt and Minion Pro by Manila Typesetting Company, Makati, Philippines

Printed in the USA

10 9 8 7 6 5 4 3 2 1

This book is dedicated to the late Dr. G. Suseendran whose inspiration made this book possible.

Contents

Preface

Differential Geometry is the study of geometric properties of curves and surfaces and their higher dimensional analogues using the methods of tensor calculus. It has a long and rich history and, in addition to its intrinsic mathematical value and important connections with various other branches of mathematics, it has many applications in various physical sciences. Differential Geometry is a vast subject. A comprehensive introduction would require prerequisites in several related subjects. In this elementary introductory course, we discuss tensor calculus and its applications in details and many of the basic concepts of Differential Geometry in the simpler context of curves and surfaces in ordinary 3-dimensional Euclidean spaces. Our aim is to build both solid mathematical understanding of the fundamental notions of Differential Geometry and sufficient visual and geometric intuition of the subject.

We hope that this study is in the interest of researchers from a variety of mathematics, science, and engineering backgrounds and that Master level students will also be able to readily study more advanced concepts, such as properties of curves and surfaces, geometry of abstract manifolds, tensor analysis, and general relativity.

This book has three parts. The first part contains six chapters dealing with the calculus of tensors. In the first chapter, we deal with some preliminaries necessary for treatment of the materials in the succeeding chapters. In the second and third chapters, we discuss the algebra of tensors and metric tensors. Tensor calculus with Christoffel's symbols and Riemannian Geometry are dealt with in the remaining chapters of this section. Part II contains six chapters discussing the applications of tensor calculus to geometry, mainly curves and surfaces in three-dimensional Euclidean space. The remaining two chapters in Part III mainly study Analytical Mechanics with tensor notation. At the end of each chapter, a large number of problems have been completely solved and exercises containing carefully motivated examples have been incorporated.

I would like to thank Professor Dr. Hanaa Hachimi and Dr. G. Suseedhran who gave me the scope and encouragement, with technical support, to write this book. I am also thankful to Professor Arindam Bhattacherya from Jadavpur University, India for offering some valuable suggestions for the improvement of this book.

Lastly, I wish to record my appreciation to the publishers and printers for their care and effort in bringing out this study in book form.

Dr. Dipankar De
Agartala, India
20th April 2021

About the Book

An Introduction to Differential Geometry with Tensor Applications discusses the theory of tensor, curves, and surfaces and its application in Newtonian Mechanics. Since tensor analysis deals with entities and properties that are independent of choice of reference frames, it forms an ideal tool for the study of Differential Geometry and classical and celestial mechanics. This book provides a profound introduction to the basic theory of Differential Geometry: curves and surfaces. The author has tried to keep the treatment of the advanced material as lucid and comprehensive as possible mainly by including the utmost detailed calculations, numerous illustrative examples, and a wealth of complementing exercises with complete solutions making the book easily accessible even to beginners in the field. Differential Geometry, from the beginning on, characterizes the author's utmost efficient didactical approach. This book is one of the outstanding standard primers of modern Differential Geometry and a basic source for a profound introductory course or as a highly recommendable reference for effective self-study of the subject.

Introduction

Euclidean Geometry was one of the most important branches of mathematics for the last 2000 years. It was also the main tool used by scientists for the development of astronomy. In 300 B.C., Euclid treated geometry as a deductive system. In 1621, Sir Henry Savile raised some questions concerning what he called "two blemishes" in geometry: the theory of proportion and the theory of parallels. Euclid's axiom of parallels (postulate V in the first book of *Elements*) propagates that any two given lines in a plane, when produced indefinitely, will intersect if the sum of two interior angles made by a transversal with these lines is less than two right angles. In 1826, a Russian mathematician, Nicolai Lobachevski, presented a paper to the mathematician faculty of the University of Kazan based on the assumption that it is possible to draw through any point in a plane two lines parallel to a given line. Hungarian mathematician John Bolyai published some results in 1831, which, conceptually, have little difference from those of Lobachevski and perhaps it contains a deeper appreciation of the metric properties of space. However, it was only after Riemann's profound dissertation on the hypotheses that the underlying foundations of geometry appeared posthumously in 1867, showing the importance of the metric concepts in geometry.

Riemann appeared to have been unaware of the work of Lobachevski and Bolyai, although it was well known to Gauss. Later, Beltrami published his classical paper on the interpretation of non-Euclidean geometries (1868) in which he analyzed the work of Lobachevski, Bolyai, and Riemann and stressed the fact that the metric properties of space are mere definitions. It appears that three consistent geometries are possible on surfaces of constant curvatures: Lobachevskian on a surface of constant negative curvature, Riemannian on a surface of constant positive curvature, and Euclidean on a surface of zero curvature. These geometries are called hyperbolic, elliptic, and parabolic, respectively.

Tensor calculus is concerned with the study of abstract objects, called tensors, which are independent of frames of reference used to describe them. In the early part of the 17th century, geometry was developed by French mathematician Descartes. The great physicist W. Voigi discovered tensors and gave them this name in his remarkable book *A Textbook of*

Dipankar De. Introduction to Differential Geometry with Tensor Applications, (1–6) © 2022 Scrivener Publishing LLC

Crystal Physics published in 1910. Ricci and his student, Levi-cita, (1901) have been developing the subject of tensors, but it is well known that Einstein's use of tensors as a tool in his general theory of relativity (1914) was mainly responsible for the sudden emergence of tensor calculus as a popular field of mathematical activities. The application of tensors in the field of mathematics and physics was mainly accelerated after the publication of Einstein's famous paper *The General Theory of Relativity.*

Tensor is a generalization of the term vector and *Tensor Calculus* is a generalization of vector analysis. The concept of invariance of mathematical objects, under coordinate transformations, permeates the structure of tensor analysis to such an extent that it is important to get at the outset a clear notion of the particular brand of invariance we have in mind. In the given reference frame, a point P is determined by a set of coordinates, x^i. If the coordinate system is changed, point P is described by a new set of coordinates, but the transformation of the coordinates does nothing to the point itself.

Here, we discuss how the term tensor may be considered as a generalization of the term vector.

We start with a two-dimensional Euclidean space, E_2, provided with a system of rectangular Cartesian coordinates. Let P and Q be two points of E_2, with (x_1, x_2) and (y_1, y_2) as their respective coordinates. Then, the coordinates of the directed line segment $\overline{PQ}$ are $(x_1 - y_1, x_2 - y_2)$.

Denoting $x_1 - y_1$ *and* $x_2 - y_2$ by z_1 and z_2 respectively, the coordinates of $\overline{PQ}$ can be expressed as

$$z_i = x_i - y_i,\ i = 1,2. \tag{0.1}$$

Next, we consider an orthogonal transformation of coordinate axes given by

$$\begin{aligned} \bar{x}_1 &= a_{11}x_1 + a_{12}x_2 + b_1 \\ \bar{x}_2 &= a_{21}x_1 + a_{22}x_2 + b_2, \end{aligned} \tag{0.2}$$

where $\begin{pmatrix} a_{11} & a_{12} \\ a_{21} & a_{22} \end{pmatrix}$ is orthogonal, with its determinant equal to 1.

We can express (0.2) as follows:

$$\bar{x}_i = \sum_{j=1}^{2} \bar{\mathrm{x}}_{\mathrm{i}} = \sum_{j=1}^{2} a_{ij}x_j + b_i, \tag{0.3}$$

denoted by (x_1, x_2) and (y_1, y_2), the coordinates of Q and P in the new coordinates system. Then, the coordinates of vector $\overline{PQ}$ *in the new* coordinate system are $(\bar{x}_1 - \bar{y}_1, \bar{x}_2 - \bar{y}_2)_1)$ *and* are denoted by $(\bar{z}_1, \bar{z}_2)$. Then $\overline{PQ}$ can be expressed as

$$\bar{z}_i = \bar{x}_i - \bar{y}_i \tag{0.4}$$

$$\bar{x}_i - \bar{y}_i = \sum_{j=1}^{2}(a_{ij}x_j + b_i) - \sum_{j=1}^{2}(a_{ij}y_j + b_i)$$

$$= \sum_{j=1}^{2} a_{ij}(x_j - y_j) = \sum_{j=1}^{2} a_{ij}z_j$$

Hence, Equation (0.4) can be written as

$$\bar{z}_i = \sum_{j=1}^{2} a_{ij}z_j \tag{0.5}$$

$$\text{or } \bar{z}_i = \sum_{j=1}^{2} \frac{\partial \bar{x}_i}{\partial x_j} z_j \tag{0.6}$$

[from Equation (0.3) we get $a_{ij} = \dfrac{\partial \bar{x}_i}{\partial x_j}$

The above equation shows that the coordinates of a vector $\overline{PQ}$ of E_2 transform according to a certain law in Equation (0.6) when referring to a new coordinate system.

This was first pointed out by Felix Klein in 1872.

Similarly, if we consider n vectors, it can be shown that there exists an object with components $C^{j_1 j_2 .. j_n}$ in a coordinate system, according to

$$\bar{C}^{j_1 j_2 .. j_n} = \sum \frac{\partial y^{i_1}}{\partial x^{j_1}} \frac{\partial y^{i_2}}{\partial x^{j_2}} \cdots\cdots \frac{\partial y^{i_n}}{\partial x^{j_n}} C^{j_1 j_2 .. j_n} \tag{0.7}$$

Hence, we conclude that a tensor of E_2 may be regarded as generalization of a vector of E_2 defined from the transformation law. Similarly, a tensor of E_n can be obtained as a generalization of a vector of E_n. A tensor obtained from the orthogonal transformation of a rectangular Cartesian coordinate system is called a *Cartesian Tensor* and a tensor from a general transformation coordinate system is simply called a *tensor.*

The flourishing of the subjects of tensors and Differential Geometry and Mechanics is due to Einstein and Grassman. Then, many mathematicians and researchers developed Differential Geometry with tensor applications.

Sasaki and Hsu defined and studied almost all contact structures and their integrability conditions. In 1970, Yano and Okumura studied structure manifold and Walker, A.G. (1955) studied the properties of the manifold (λu, v,) with an almost product structure in which there exists a (1,1) tensor field, *f*, whose square is unity. K. Yano (1963) generalized the concept of an almost complex structure and defined f-structures as a (1,1) tensor field *f* (satisfying $f_3 + f = 0$). In 1972, K. Kenmotsu studied a certain class of an almost contact manifold. Janssen and Vanhecke (1981) named this structure a Kenmotsu structure and the differentiable manifold equipped with this structure is called a Kenmotsu manifold. Many authors have studied slant immersions in almost Hermitian manifolds. The study of Differential Geometry of tangent and cotangent bundles was started by Sasaki (1958) and then Yano and Davies and Ledger. The theory of submanifolds as a field of Differential Geometry is as old as Differential Geometry itself. A study of the submanifolds of a manifold is a very interesting field of Differential Geometry. In 1981, B.Y. Chen, D.E. Blair, A. Bejancu, M.H. Sahid (1994-95), and some others studied different properties of submanifolds. Sasaki (1960) and others studied differentiable manifolds in detail.

Differential Geometry is the study of geometric properties of curves, surfaces, and their higher dimensional analogues using the methods of Tensor Calculus. For the study of curve by this method of calculus, its parametric representation is a covariant and discuss tangent and normal and binormal, which is of fundamental importance to the theory of the curve. We will study the geometric properties of surface imbedded in the three-dimensional Euclidean space by means of Differential Geometry, termed as intrinsic properties and intrinsic geometry of surface. The study of the geometry of surfaces was carried out from the point of view of a two-dimensional being whose universe is determined by the surface parameters u^1 *and* u^2 and it was based entirely on the study of the first quadratic differential form.

Differential Geometry has a long and rich history and, in addition to its intrinsic mathematical value and important connections with various other branches of mathematics, it has many applications in various physical sciences, e.g., solid mechanics, computer tomography, and general relativity. Differential Geometry is a building block in Physics and Classical Mechanics, which was developed extensively by Newton. It deals with the motion of particles in a fixed frame of reference. Within those frames, other coordinate systems may be used so long as the metric remains Euclidean. The reference system generally used in astronomy is determined by "fixed stars". It is termed as the primary inertial system. The motion of the earth relative to its primary inertial system is so negligible that Newtonian laws which can be applied without modification to the study of motion of particles is referred to as a system of axes fixed in the earth.

Part I
TENSOR THEORY

1

Preliminaries

1.1 Introduction

Some quantities are associated with their magnitude and direction, but certain quantities are associated with two or more directions. Such a quantity is called a tensor, e.g., the stress at a point of an elastic solid is an example of a tensor which depends on two directions: one is normal and the other is that of force on the area. Tensor comes from the word tension.

In this chapter, we discuss the notation of systems of different orders, which are applied in the theory of determinants, symbols, and summation conventions. Also, results on some matrices and determinants are discussed because they will be used frequently later on.

1.2 Systems of Different Orders

Let us consider the two quantities, a_1, a_1 *or* a^1, a^2, which are represented by a_i or a^i, respectively, for $i = 1, 2$. In such cases, the expressions a_i, a^i, a_{ij}, a^{ij}, *and* a^i_j are called *systems.* In each value of a_i *and* a^i are called *systems of first order* and each value of a_{ij}, a^{ij}, *and* a^i_j is called a *double system* or *system of second order*, of which a_{12}, $a_{22}a^{23}$, a^{13}, *and* a^3_4 are called their respective *components.* Similarly, we have *systems of the third order* that depend on three indices shown as a_{ijk}, a^{ikl}, a_{ijm}, a^{ijn}, *and* a^i_{jk} and each number of their respective components are 8.

In a system of order zero, it is shown that the quantity has no index, such as a. The upper and lower indices of a system are called its indices of *contravariance* and *covariance,* respectively. For a system of A^{ij}_k, *i and j* are indices of a *contravariant* and k is of *covariance.* Accordingly, the system A^{ij} is called a contravariant system, A_{klm} is called a covariant system, and is A^i_j called a *mixed system.*

Dipankar De. Introduction to Differential Geometry with Tensor Applications, (9–24) © 2022 Scrivener Publishing LLC

1.3 Summation Convention Certain Index

If in some expressions a certain index occurs twice, this means that this expression is summed with respect to that index for all admissible values of the index.

Thus, the linear form $\sum_{i=1}^{4} a_i x_i$ has an index, i, occurring in it twice. We will omit the summation symbol Σ and write $a_i x_i$ to mean $a_1x_1 + a_2x_2 + a_3x_3 + a_4x_4$. In order to avoid Σ, we shall make use of a convention used by A. Einstein which is accordingly called the *Einstein Summation Convention or Summation Convention.*

Of course, the range of admissible values of the index, 1 to 4 in this case, must be specified. If the symbol i has a range of values from 1 to 3 and j ranges from 1 to 4, the expression

$$a_{ij}x_j,\ (i = 1,2,3 \text{ and } j = 1,2,3,4) \tag{1.1}$$

represents three linear forms:

$$a_{11}x_1 + a_{12}x_2 + a_{13}x_3 + a_{14}x_4$$

$$a_{21}x_1 + a_{22}x_2 + a_{23}x_3 + a_{24}x_4$$

$$a_{31}x_1 + a_{32}x_2 + a_{33}x_3 + a_{34}x_4 \tag{1.2}$$

Here, index i is the identifying (free) index and since index j, occurs twice, it is the summation index.

We shall adopt this convention throughout the chapters and take the sum whenever a letter appears in a term once in a subscript and once in superscript or if the same two indices are in subscript or are in superscript.

Example 1.3.1. Express the sum $\sum_{i=1}^{3}\sum_{j=1}^{3} a_{ij}u^i v^j$.

Solution: $a_{ij}u^i v^j = \sum_{i=1}^{3}\sum_{j=1}^{3} a_{ij}u^i v^j = \sum_{i=1}^{3} (a_{i1}u^i v^1 + a_{i2}u^i v^2 + a_{i3}u^i v^3)$

$= \sum_{i=1}^{3} a_{i1}u^i v^1 + \sum_{i=1}^{3} a_{i2}u^i v^2 + \sum_{i=1}^{3} a_{i3}u^i v^3$

$= (a_{11}u^1v^1 + a_{21}u^2v^1 + a_{31}u^3v^1) + (a_{12}u^1v^2 + a_{22}u^2v^2 + a_{32}u^3v^2)$
$+ (a_{13}u^1v^3 + a_{23}u^2v^3 + a_{33}u^3v^3)$

1.3.1 Dummy Index

The summation (or dummy) index can be changed at will. Thus, Equation (1.1) can be written in the form $a_{ik}x_k$ if k has the same range of values as j.

We will assume that the summation and identifying indices have ranges of value from 1 to n.

Thus, $a_i x_i$ will represent a linear form

$$a_1x_1 + a_2x_2 + a_3x_3 + \ldots + a_nx_n$$

For example, $\sum_{i=1}^{n}\sum_{k=1}^{n} a_{ik}x^i x^k$ can be written as $a_{ik}x^i x^k$ and here, *i and k* both are dummy indexes.

So, any dummy index can be replaced by any other index with a range of the same numbers.

1.3.2 Free Index

If in an expression an index is not a dummy, i.e., it is not repeated twice, then it is called a *free index*. For example, for $a_{ij}x^j$, the index j is dummy, but index i is free.

1.4 Kronecker Symbols

A particular system of second order denoted by δ^i_j $i, j = 1, 2, \ldots n$, is defined as

$$\begin{aligned} \delta^i_j &= 1 \ \textit{for}\ i = j \\ &= 0 \ \textit{for}\ i \neq j \end{aligned} \tag{1.3}$$

Such a system is called a *Kronecker symbol* or *Kronecker delta.*

For example, δ^i_j, by summation convention is expressed as

$$\delta^i_i = \delta^1_i + \delta^2_i + \cdots . + \delta^n_i = (\delta^1_1) + (\delta^2_2) + \cdots + \delta^n_n = 1 + 1 + \cdots + 1 = n$$

We shall now consider some *properties* of this system.

Property 1.4.1. If $x^1, x^2, \ldots x^n$ are independent variables, then

$$\frac{\partial x^i}{\partial x^j} = 1 \ \textit{for} \ i = j$$
$$= 0 \ \textit{for} \ i \neq j$$

$$\text{Hence,} \ \frac{\partial x^i}{\partial x^j} = \delta^i_j \tag{1.4}$$

Property 1.4.2. From the summation convention, we get

$$\delta^i_i = \delta^1_1 + \delta^2_2 + \cdots .. + \delta^n_n = 1 + 1 + \cdots + 1 = n$$

Similarly, $\delta_{ii} = \delta^{ii} = n$

Property 1.4.3. From the definition of δ^{ij}, taken as an element of unit matrix I, we have

$$I = (\delta^{ij}) = \begin{pmatrix} 1 & 0 & .. & 0 \\ 0 & 1 & .. & 0 \\ .. & .. & .. & .. \\ 0 & 0 & .. & 1 \end{pmatrix}$$

$$\text{Its determinant is } |\delta^{ij}| = \begin{vmatrix} 1 & 0 & .. & 0 \\ 0 & 1 & .. & 0 \\ .. & .. & .. & .. \\ 0 & 0 & .. & 1 \end{vmatrix} = 1$$

Property 1.4.4.

$$\text{Again, } \delta^i_j \, a^j = \delta^i_1 \, a^1 + \delta^i_2 \, a^2 + \cdots + \delta^i_i \, a^i + .. + \delta^i_n \, a^n$$
$$= 0 + 0 + \cdots 1.a^i + 0 + \cdots + .. = a^i \tag{1.5}$$

$$\text{Similarly, } \delta^i_j \, a_i = a_j \tag{1.6}$$

$$\delta^i_j \, a_{ik} = \delta^1_j \, a_{1k} + \delta^2_j \, a_{2k} + \cdots \delta^j_j \, a_{jk}$$

$$+\cdots.+\delta^n_j \, a_{nk} = a_{jk}$$

$$= 0 + 0 \ldots + 1.a_{jk} + 0..+0 = a_{jk}$$

$$\text{Similarly, } \delta^i_j \, a^{jk} = a^{ik} \tag{1.7}$$

Property 1.4.5.

$$\delta^i_j \delta^j_k = \delta^i_1 \delta^1_k + \delta^i_2 \delta^2_k + \cdots \ldots + \delta^i_n \delta^n_k = 1.\delta^1_k + \delta^2_k + \cdots + \delta^n_k = \delta^i_k$$

Also, by definition, $\delta^i_j \delta^j_k = \dfrac{\partial x^i}{\partial x^j} \dfrac{\partial x^j}{\partial x^k} = \dfrac{\partial x^i}{\partial x^k} = \delta^i_l \delta^l_k$

In particular, when $i = k$, we get $\delta^i_j \delta^j_i = \delta^i_i = n$

Remark 1.4.1. If we multiply x^k by δ^i_k, we simply replace index k of x^k with index i and for this reason, δ^i_k is called a *substitution factor.*

Example 1.4.1. Evaluate (a) $\delta^i_j \delta^j_l$ and (b) $\delta^i_j \delta^j_l \delta^l_k$ where the indices take all values from 1 to n.

Solution: (a) We have $\delta^i_j \delta^j_l = \delta^i_1 \delta^1_l + \delta^i_2 \delta^2_l + \cdots + \delta^i_n \delta^n_l$ (1.8a)

Now, $\delta^i_1 \delta^1_l = \delta^1_1 \delta^1_l + \delta^2_1 \delta^1_l + \cdots . + \delta^n_1 \delta^1_l = 1.\delta^1_l = \delta^1_l$

Similarly, other terms of i are $\delta^2_l, \delta^3_l \ldots . \delta^n_l$.

$$\therefore \delta^i_j \delta^j_l = \delta^i_1 \delta^1_l + \delta^i_2 \delta^2_l + \cdots + \delta^i_n \delta^n_l = \delta^1_l + \delta^2_l + \delta^3_l + \cdots . + \delta^n_l = \delta^i_l \tag{1.8b}$$

(b) $\delta^i_j \delta^j_l \delta^l_k = \delta^i_l \delta^l_k = \delta^i_k$ by 1.8b

Example 1.4.2. If x^i and y^i are independent coordinates of a point, it is shown that

$$\frac{\partial x^j}{\partial y^k} \frac{\partial y^k}{\partial x^i} = \delta^j_i$$

Solution: The partial derivative of ϕ in two coordinate systems are different and are connected by the following formula of Differential Calculus:

$$\frac{\partial y}{\partial x}=\frac{\partial \phi}{\partial y^1}\frac{\partial y^1}{\partial x^i}+\frac{\partial \phi}{\partial y^2}\frac{\partial y^2}{\partial x^i}+\cdots.+\frac{\partial \phi}{\partial y^n}\frac{\partial y^n}{\partial x^i}=\frac{\partial \phi}{\partial y^k}\frac{\partial y^k}{\partial x^i}$$

In particular, when $\phi = x^j$, we have $\frac{\partial x^j}{\partial x^i}=\frac{\partial x^j}{\partial y^k}\frac{\partial y^k}{\partial x^i}$ (1.9a)

Since x^j is independent of x^j, $\frac{\partial x^j}{\partial x^i}=0 \ when \ j\neq i$

$$=1 \ for \ i=j \tag{1.9b}$$

Hence, the result follows from (1.9a) and (1.9b).

1.5 Linear Equations

Let us consider n linear equations such that

$$a_{ij}x^j = b_i, \tag{1.10a}$$

where $x^1, x^2, \ldots. x^n$ are n unknown variables.

Let us consider:

For the expansion of det $|a_{ij}|$ in terms of cofactors we have

$$a_{ij}A^{jk} = a\delta_i^k \tag{1.10b}$$

where $a = |a_{ij}|$ and the cofactor of a_{ij} is A^{ij}.

We can derive Cramer's Rule for the solution of the system of n linear equations:

Now, multiplying both sides of (1.10a) by A^{ij}, we get

$$A^{ij}a_{ij}x^j = b_iA^{ij}$$

by (1.10b), we get, $ax^j = b_iA^{ij}$.

From here, we can easily get

$$x^j = \frac{b_i A^{ij}}{a}, \text{ where } a = |a_{ij}| \neq 0.$$

Example 1.5.1. Show that $a_{ij}A^{kj} = a\delta_i^k$, where a is a determinant a_{ij} ie $a = |a_{ij}|$ of order 3 and A^{ij} are cofactors of a_{ij}.
Solution: By expansion of determinants, we have:

$$a_{11}A^{11} + a_{12}A^{12} + a_{13}A^{13} = a$$

$$a_{11}A^{2j} + a_{12}A^{2j} + a_{13}A^{2j} = 0$$

$$a_{11}A^{31} + a_{12}A^{32} + a_{13}A^{33} = 0$$

Which can be written as $a_{1j}A^{1j} = a \quad a_{1j}A^{2j} = 0$ and $a_{1j}A^{3j} = 0$ [we know $a_{ij}A^{ij} = a$].

Similarly, we have $\quad a_{2j}A^{1j} = 0 \quad a_{2j}A^{2j} = a$ and $a_{2j}A^{3j} = 0$

$a_{3j}A^{1j} = 0 \quad a_{3j}A^{2j} = 0$ and $a_{3j}A^{3j} = a$.

Using Kronecker Delta Notation, these can be combined into a single equation:

$$a_{1j}A^{kj} = a\delta_1^k \quad a_{2j}A^{kj} = a\delta_2^k \quad a_{3j}A^{kj} = a\delta_3^k \quad [\delta_j^i = 1, when\ i = j$$
$$= 0,\ when\ j \neq i\ ].$$

All nine of these equations can be combined into $a_{ij}A^{kj} = a\delta_i^k$.

1.6 Results on Matrices and Determinants of Systems

It is known that if the range of the indices of a system of second order are from 1 to n, the number of components is n^2. Systems of second order are organized into three types: a^{ij}, a_{ij}, a_j^i and their matrices,

(a^{ij}), (a_{ij}), (a_j^i):

$$\begin{bmatrix} a^{11} & a^{12}\dots & a^{1n} \\ a^{21} & a^{22}\dots & a^{2n} \\ a^{n1} & a^{n2}\dots & a^{nn} \end{bmatrix}, \begin{bmatrix} a_{11} & a_{12}\dots & a_{1n} \\ a_{21} & a_{22}\dots & a_{2n} \\ a_{n1} & a_{n2}\dots & a_{nn} \end{bmatrix} \text{and} \begin{bmatrix} a_1^1 & a_2^1\dots & a_n^1 \\ a_1^2 & a_2^2\dots & a_n^2 \\ a_1^n & a_2^n\dots & a_n^n \end{bmatrix}$$

each of which is an $n \times n$ matrix.

We shall now establish the following results:

Property 1.6.1. If $a_j^i b_p^j = c_p^i$, then $(a_j^i)(b_p^j) = (c_p^i)$ and $|a_j^i||b_p^j| = |c_p^i|$.

Proof: We shall prove this result by taking the range of the indices from 1 to 2, but the results hold, in general, when they range from 1 to n.

We get $a_j^i b_p^j = a_1^i b_p^1 + a_2^i b_p^2$. Hence, $c_p^i = a_1^i b_p^1 + a_2^i b_p^2$.

$$\therefore \begin{pmatrix} c_1^1 & c_2^1 \\ c_1^2 & c_2^2 \end{pmatrix} = (a_1^i)(b_p^1) + (a_2^i)(b_p^2)$$

$$= \begin{pmatrix} a_1^1 & a_1^1 \\ a_1^2 & a_1^2 \end{pmatrix} \begin{pmatrix} b_1^1 & 0 \\ 0 & b_2^1 \end{pmatrix} + \begin{pmatrix} a_2^1 & a_2^1 \\ a_2^2 & a_2^2 \end{pmatrix} \begin{pmatrix} b_1^2 & 0 \\ 0 & b_2^2 \end{pmatrix}$$

$$= \begin{pmatrix} a_1^1 b_1^1 & a_1^1 b_2^1 \\ a_1^2 b_1^1 & a_1^2 b_2^1 \end{pmatrix} + \begin{pmatrix} a_2^1 b_1^2 & a_2^1 b_2^2 \\ a_2^2 b_1^2 & a_2^2 b_2^2 \end{pmatrix} = \begin{pmatrix} a_1^1 b_1^1 + a_2^1 b_1^2 & a_1^1 b_2^1 + a_2^1 b_2^2 \\ a_1^2 b_1^1 + a_2^2 b_1^2 & a_1^2 b_2^1 + a_2^2 b_2^2 \end{pmatrix}$$

$$= \begin{pmatrix} a_1^1 & a_1^2 \\ a_2^1 & a_2^2 \end{pmatrix} \begin{pmatrix} b_1^1 & b_2^1 \\ b_1^2 & b_2^2 \end{pmatrix} = (a_j^i)(b_p^j)$$

Taking the determinant of both sides, we get $|a_j^i||b_p^j| = |c_p^i|$, as we know $|AB| = |A||B|$.

Property 1.6.2. If $a_{ij}b^{ik} = c_j^k$, then, $(b^{ik})^T(a_{ij}) = (c_j^k)$ and $|a_{ij}||b^{ik}| = |c_j^k|$, where $(b^{ik})^T$ is the transpose of (c_j^k)

Proof: We have $a_{ij}b^{ik} = a_{1j}b^{ik} + a_{2j}b^{ik}$, hence, $c_j^k = a_{1j}b^{1k} + a_{2j}b^{2k}$.

Therefore,

$$\begin{pmatrix} c_1^1 & c_2^1 \\ c_1^2 & c_2^2 \end{pmatrix} = \begin{pmatrix} a_{11}b^{11} + a_{21}b^{21} & a_{12}b^{11} + a_{22}b^{21} \\ a_{11}b^{12} + a_{21}b^{22} & a_{12}b^{12} + a_{22}b^{22} \end{pmatrix}$$

$$= \begin{pmatrix} b^{11} & b^{21} \\ b^{12} & b^{22} \end{pmatrix} \begin{pmatrix} a_{11} & a_{12} \\ a_{21} & a_{22} \end{pmatrix} = (b^{ik})^T(a_{ij})$$

Taking determinants of both sides, we get $|c_j^k| = |b^{ik^T}||a_{ij}| = |b^{ik}||a_{ij}|$ (since $|A^T| = |A|$).

Property 1.6.3. Let the cofactor of the element a^i_j in the determinant $\left|a^i_j\right|$ be denoted by $A^j_{i.}$. Then, by summation convention we have

$$a^i_j A^j_k = a^i_1 A^1_{k.} + a^i_2 A^2_{k.} + \cdots + a^i_n A^n_k = \delta^i_k \left|a^i_j\right| = \delta^i_k a \quad \text{and}$$

$$a^i_j A^k_j = a^1_j A^k_1 + a^2_j A^k_{2.} + \cdots + a^n_j A^k_n = \delta^i_k \left|a^i_j\right| = \delta^k_j a, \text{ where } a = \left|a^i_j\right|.$$

If the cofactor of a_{ij} is represented by A^{kj}, it is expressed by the equation:

$$a_{ij} A^{kj} = a\delta^k_i.$$

If we divide the cofactor A^{kj} of the element of a_{kj} by the value a of the determinant, we form the *normalized cofactor*, represented by:

$$b^{kj} = \frac{1}{a} A^{kj}.$$

The above equation becomes

$$a_{ij} b^{kj} = \delta^k_i$$

Property 1.6.4. Let us consider a system of n linear equations:

$$a^i_j x^j = b^i, i, j = 1, 2, \ldots n$$

for n unknown x^i, where $\left|a^i_j\right| \neq 0$

$A^k_i a^i_j x^j = b^i A^k_i$, where A^k_i is cofactor of a^i_j.

$ax^j = b^i A^k_i \therefore x^j = \dfrac{b^i A^k_i}{a}$, which is called *Cramer's Rule*, for the solution of *n* linear equations.

Property 1.6.5. Considering the transformation $z^i = z^i(y^k)$ and $y^i = y^i(x^k)$, let N function $z^i(y^k)$ be of independent N variables of y^k so that $\left|\dfrac{\partial z^i}{\partial y^k}\right| \neq 0.$

Here, N equation $z^i = z^i(y^k)$ is solvable for the z's in terms terms of y^i's. Similarly, $y^i = y^i(x^k)$ is a solution of y^i in terms of x^i's so that $\left|\frac{\partial y^i}{\partial x^k}\right| \neq 0$.

Now, we have by the chain rule of differentiation that

$$\frac{\partial z^i}{\partial x^k} = \frac{\partial z^i}{\partial y^1}\frac{\partial y^1}{\partial x^k} + \frac{\partial z^i}{\partial y^2}\frac{\partial y^2}{\partial x^k} + \cdots \ldots + \frac{\partial z^i}{\partial y^N}\frac{\partial y^N}{\partial x^k} = \frac{\partial z^i}{\partial y^j}\frac{\partial y^j}{\partial x^k}.$$

Taking the determinant, we get

$$\left|\frac{\partial z^i}{\partial x^k}\right| = \left|\frac{\partial z^i}{\partial y^j}\right|\left|\frac{\partial y^j}{\partial x^k}\right| = \left|\frac{\partial z^i}{\partial y^j}\right|\left|\frac{\partial y^i}{\partial x^j}\right| \tag{1.11}$$

Considering a particular case in which $z^i = x^i$, Equation (1.5) becomes

$$\left|\frac{\partial x^i}{\partial x^k}\right| = \left|\frac{\partial x^i}{\partial y^j}\right|\left|\frac{\partial y^i}{\partial x^j}\right| \quad \text{or} \quad \left|\delta^i_k\right| = \left|\frac{\partial x^i}{\partial y^j}\right|\left|\frac{\partial y^i}{\partial x^j}\right|$$

Or

$$1 = \left|\frac{\partial x^i}{\partial y^j}\right|\left|\frac{\partial y^i}{\partial x^j}\right|$$

$$\Rightarrow \left|\frac{\partial x^i}{\partial y^j}\right| = \frac{1}{\left|\frac{\partial y^i}{\partial x^j}\right|}.$$

This implies that the Jacobian of Direct Transformation is the reciprocal of the Jacobian of Inverse Transformation.

1.7 Differentiation of a Determinant

Consider the determinant $|a^i_j| = a$ and let the element a^i_j be a function of $x_1, x_2 \ldots x_n$, etc. Let A^i_j be the cofactor of a^i_j of det a.

Then, the derivative of a with respect to x_l is given by

$$\frac{\partial a}{\partial x_1}=\begin{vmatrix}\frac{\partial a_1^1}{\partial x_1}\frac{\partial a_2^1}{\partial x_1}\cdots\cdots\frac{\partial a_n^1}{\partial x_1}\\ a_1^2a_2^2\ldots\ldots\ldots a_n^2\\ \ldots\ldots\ldots\ldots\ldots\ldots\\ a_1^na_2^n\ldots\ldots\ldots a_n^n\end{vmatrix}+\begin{vmatrix}a_1^1a_2^1\ldots\ldots\ldots a_n^1\\ \frac{\partial a_1^2}{\partial x_1}\frac{\partial a_2^2}{\partial x_1}\cdots\cdots\frac{\partial a_n^2}{\partial x_1}\\ \ldots\ldots\ldots\ldots\ldots\ldots\\ a_1^na_2^n\ldots\ldots\ldots a_n^n\end{vmatrix}+\ldots\begin{vmatrix}a_1^2a_2^2\ldots\ldots\ldots a_n^2\\ a_1^2a_2^2\ldots\ldots\ldots a_n^2\\ \ldots\ldots\ldots\ldots\ldots\ldots\\ \frac{\partial a_1^n}{\partial x_1}\frac{\partial a_2^n}{\partial x_1}\cdots\cdots\frac{\partial a_n^n}{\partial x_1}\end{vmatrix}.$$

$$=\left(\frac{\partial a_1^1}{\partial x_1}A_1^1+\frac{\partial a_2^1}{\partial x_1}A_1^2+\cdots\frac{\partial a_2^1}{\partial x_1}A_1^n\right)+\left(\frac{\partial a_1^2}{\partial x_1}A_2^1+\frac{\partial a_2^2}{\partial x_1}A_2^2+\cdots\frac{\partial a_n^2}{\partial x_1}A_2^n\right)$$

$$+\cdots+\left(\frac{\partial a_1^n}{\partial x_1}A_n^1+\frac{\partial a_2^n}{\partial x_1}A_n^2+\cdots\frac{\partial a_n^n}{\partial x_1}A_n^n\right)$$

$$=\frac{\partial a_i^1}{\partial x_1}A_1^i+\frac{\partial a_i^2}{\partial x_1}A_2^i+\ldots+.\frac{\partial a_i^n}{\partial x_1}A_n^i=\frac{\partial a_i^j}{\partial x_1}A_j^i$$

Therefore, in general, we can write $\frac{\partial a}{\partial x_p}=\frac{\partial a_i^j}{\partial x_p}A_j^i$.

1.8 Examples

Example 1.8.1. Write the terms contained in $S = a_{ij}x^ix^j$ taking $n = 3$.
Solution: Since the index i (or j) occurs both in subscript and superscript, we first sum on i from 1 to 3, then on j from 1 to 3.

$$\begin{aligned}S &= a_{1j}x^1x^j+a_{2j}x^2x^j+a_{3j}x^3x^j\\ &= (a_{11}x^1x^1+a_{12}x^1x^2+a_{13}x^1x^3)+(a_{21}x^2x^1+a_{22}x^2x^2+a_{23}x^2x^3)\\ &+ (a_{31}x^3x^1+a_{32}x^3x^2+a_{33}x^3x^3)\\ &= a_{11}(x^1)^2+a_{22}(x^2)^2+a_{33}(x^3)^2+(a_{12}+a_{21})x^1x^2+(a_{13}+a_{31})x^1x^3+(a_{32}+a_{23})x^2x^3\end{aligned}$$

Example 1.8.2. Express the sum of $\sum_{i=1}^{3}\sum_{j=1}^{3}\sum_{k=1}^{3}a_{ijk}x^ix^jx^k$.
Solution: Here, the number of terms is $3^3 = 27$.

Since the index *i (or j or k)* occurs both in subscript and superscript, we first sum on *i* from 1 to 3, then on each term of its 3 terms we sum *j* from 1 to 3. This results in 9 terms. Then, on each of the 9 terms we sum *k* from 1 to 3, which results in 27 terms. Like the last example, we sum

$$S = (a_{1jk}x^1x^jx^k + a_{2jk}x^2x^jx^k + a_{3jk}x^3x^jx^k))$$

$$= (a_{11k}x^1x^1x^k + a_{12k}x^1x^2x^k + a_{13k}x^1x^3x^k)$$
$$+\cdots+(a_{31k}x^3x^1x^k + a_{32k}x^3x^2x^k + a_{33k}x^3x^3x^k)$$
$$=[(a_{111}x^1x^1x^1 + a_{122}x^1x^2x^2 + a_{133}x^1x^3x^3)+(a_{121}x^1x^2x^1 + a_{122}x^1x^2x^2 + a_{123}x^1x^2x^3.)]$$
$$+\cdots\ldots\ldots+(\ldots\ldots+a_{32k}x^3x^2x^k + [a_{331}x^3x^3x^1 + a_{332}x^3x^3x^2 + a_{333}x^3x^3x^3)$$
$$=[a_{111}(x^1)^3 + a_{222}(x^2)^3 + a_{333}(x^3)^3 + [a_{123} + a_{132} + a_{231} + a_{213} + a_{321} + a_{312}]x^1x^2x^3$$
$$+[a_{112} + a_{121} + a_{211}](x^1)^2x^2 + [a_{113} + a_{131} + a_{311}](x^1)^2x^3 + [a_{221} + ..](x^2)^2x^1$$
$$+[a_{223} + \cdots.](x^2)^2x^3 + [a_{331} + \cdots.](x^3)^2x^1 + [a_{332} + \cdots.](x^3)^2x^2.$$

Example 1.8.3. If f is a function of n variables x^i, write the differential of f.
Solution: Since $f = f(x^1, x^2, \ldots x^n)$,
from calculus, we have $df = \frac{\partial f}{\partial x^1}dx^1 + \frac{\partial y}{\partial x^2}d\,x^2 + \cdots\ldots + \frac{\partial y}{\partial x^n}d\,x^n = \frac{\partial f}{\partial x^i}d\,x^i$.

Example 1.8.4. (a) If $a_{pq}x^px^q = 0$ for all values of the independent variables $x^1, x^2, \ldots x^n$ and a_{pq}'s are constant, show that $a_{ij} + a_{ji} = 0$.

(b) If $a_{pqr}x^px^qx^r = 0$ for all values of the independent variables $x^1, x^2, \ldots x^n$ and a_{pqr}'s are constant, show that $a_{kij} + a_{kji} + a_{ikj} + a_{jki} + a_{ijk} + a_{jik} = 0$.

Solution: Differentiating: $$a_{pq}\,x^p\,x^q = 0 \tag{1.12a}$$

with respect to x^i

$$a_{pq}x^q\frac{\partial}{\partial x^i}(x^p) + a_{pq}x^p\frac{\partial}{\partial x^i}(x^q) = 0$$
$$\text{or } a_{pq}x^q\delta_i^p + a_{pq}x^p\delta_i^q = 0 \tag{1.12b}$$
$$\text{or } a_{pi}x^p + a_{iq}x^q = 0$$

Differentiating (1.12b), with respect to x^j, we get

$$a_{pi}\frac{\partial}{\partial x^j}(x^p)+a_{iq}\frac{\partial}{\partial x^j}(x^q)=0$$

$$\text{or }\ a_{pi}\delta_j^p+a_{iq}\delta_j^q=0$$

$$\text{or }\ a_{ji}+a_{ij}=0.$$

(b) Differentiating with respect to x^i $\qquad a_{pqr}x^p x^q x^r = 0$

$$a_{pqr}x^q x^r\frac{\partial}{\partial x^i}(x^p)+a_{pqr}x^p x^r\frac{\partial}{\partial x^i}(x^q)+a_{pqr}x^p x^q\frac{\partial}{\partial x^i}(x^r)=0$$

$$or,\ a_{pqr}x^q x^r\delta_i^p+a_{pqr}x^p x^r\delta_i^q+a_{pqr}x^p x^q\delta_i^r=0$$

$$or,\ a_{iqr}x^q x^r+a_{pir}x^p x^r+a_{pqi}x^p x^q=0$$

Differentiating with respect to x^j, we get

$$a_{iqr}\frac{\partial}{\partial x^j}(x^q)x^r+a_{pir}x^p\frac{\partial}{\partial x^j}(x^r)+a_{pqi}\frac{\partial}{\partial x^j}(x^p)x^q$$

$$+\ a_{iqr}(x^q)\frac{\partial}{\partial x^j}(x^r)+a_{pir}\frac{\partial}{\partial x^j}(x^p)(x^r)+a_{pqi}(x^p)\frac{\partial}{\partial x^j}(x^q)=0$$

$$or\ \ a_{iqr}\delta_j^q x^r+a_{pir}x^p\delta_j^r+a_{pqi}\delta_j^p x^q+a_{iqr}(x^q)\delta_j^r+a_{pir}\delta_j^p(x^r)+a_{pqi}(x^p)\delta_j^q=0$$

$$or\ \ a_{ijr}x^r+a_{pij}x^p+a_{jqi}x^q+a_{iqj}(x^q)+a_{jir}(x^r)+a_{pji}(x^p)=0$$

Differentiating in the same way, with respect to x^k we get

$$or\ a_{ijk}+a_{kij}+a_{jki}+a_{ikj}+a_{jik}+a_{kji}=0.$$

Example 1.8.5. If a_j^i is a double system such that $a_k^i a_j^k = \delta_j^i$, show that $|a_j^i| = \pm 1$.

Solution: We have $a_k^i a_j^k = \delta_j^i$, taking determinant $|a_k^i a_j^k| = |\delta_j^i|$,

$$Or, |a_k^i||a_j^k| = |\delta_j^i| \text{ or, } \begin{vmatrix} a_1^1 \, a_2^1 \ldots . a_n^1 \\ a_1^2 \, a_2^2 \ldots \ldots n \\ \ldots\ldots\ldots\ldots \\ a_1^n \, a_2^n \ldots . a_n^n . \end{vmatrix} \begin{vmatrix} a_1^1 \, a_2^1 \ldots . a_n^1 \\ a_1^2 \, a_2^2 \ldots \ldots n \\ \ldots\ldots\ldots\ldots \\ a_1^n \, a_2^n \ldots . a_n^n . \end{vmatrix}$$

$$= \begin{vmatrix} 1 \ 0 \ 0 \ldots 0 \\ 0 \ 1 \ 0 \ldots 0 \\ \ldots\ldots\ldots\ldots \\ 0 \ 0 \ 0 \ldots 1 \end{vmatrix} \quad [since \;\; \det.Of \;\; \delta_j^i \text{ is unit } det = |I|$$

$$\text{Or} \begin{vmatrix} a_1^1 a_2^1 \ldots . a_n^1 \\ a_1^2 a_2^2 \ldots \ldots n \\ \ldots\ldots\ldots\ldots \\ a_1^n a_2^n \ldots . a_n^n \end{vmatrix}^2 = 1, \quad \Rightarrow |a_j^i|^2 = 1 \Rightarrow |a_j^i| = \pm 1$$

Example 1.8.6. If a_j^i is a double system such that $a_k^i a_j^k = \delta_j^i$, show that either $|a_k^i - \delta_k^i| = 0$ or $|a_k^i + \delta_k^i| = 0$.

Solution: From above result $|a_j^i|^2 = 1$

$$\Rightarrow |a_j^i|^2 = |\delta_j^i| \quad \Rightarrow |a_j^i|^2 - |\delta_j^i|^2 = 0$$

$$\Rightarrow (|a_j^i| - |\delta_j^i|)(|a_j^i| + |\delta_j^i|) = 0 \Rightarrow (|a_j^i| - |\delta_j^i|)(|a_j^i| + |\delta_j^i|) = 0$$

$$\Rightarrow \text{either } |a_j^i| - |\delta_j^i| = 0 \text{ or } |a_j^i| + |\delta_j^i| = 0$$

Example 1.8.7. If $\left|a_j^i\right| = \begin{vmatrix} a_1^1 & a_2^1 & a_3^1 \\ a_1^2 & a_2^2 & a_3^2 \\ a_1^3 & a_2^3 & a_3^3 \end{vmatrix}$ and $\left|b_j^i\right| = \begin{vmatrix} b_1^1 & a_2^1 & b_3^1 \\ b_1^2 & b_2^2 & b_3^2 \\ b_1^3 & b_2^3 & b_3^3 \end{vmatrix}$, show that

$$\left|a_j^i\right|\left|b_m^k\right| = \left|c_j^i\right| \text{ where } c_m^i = a_p^i a_m^p \tag{1.13a}$$

Solution: $\left|a_j^i\right|\left|b_m^k\right| = \left|a_j^i\right| e_{kms} b_1^k b_2^m b_3^s$ [we know, $\left|a_m^k\right| = e_{ijk} a_1^i a_2^j a_3^k$] ...(α)
1.13b

$$= (e_{kms}\left|a_j^i\right|) b_1^k b_2^m b_3^s$$

$$= (e_{ijt} a_k^i a_m^j a_s^t) b_1^k b_2^m b_s^s \quad [\left|a_j^i\right| e_{pqr} = e_{ijk} a_p^i a_q^j a_r^k$$

$$= e_{ijt}(a_k^i b_1^k)(a_m^j b_2^m)(a_s^t b_3^s)$$

$$= e_{ijt} c_1^i c_2^j c_3^t = \left|c_j^i\right| \quad \text{[By (1.13a) and by (1.13b)]}$$

The above result can be stated as $\left|a_j^i\right|\left|b_m^k\right| = \left|a_p^i a_m^p\right|$. It is the result of the multiplication of two determinants of the third order.

1.9 Exercises

1. Write out in full the following expression.

   (a) $\delta_j^i a^j$ (b) $\delta_{ij} x^i x^j$ (c) $\delta_{ij}\delta^{jk}$ (d) δ_i^i (e) $a^i = \dfrac{\partial x^i}{\partial y^j} b^j$ (f) $\delta_j^i \delta_l^j \delta_k^l \delta_i^k$

2. Expand the following using the summation convention.

   a) $A_{ij}B^{jk}$ b) $a_j^i x^j$ c) $\dfrac{\partial}{\partial x^k}(\sqrt{g}b^k)$

3. Prove the following.

   a) $\delta_{ij} a_{jk} = a_{ik}$ b) $\delta_{ij} a_{ij} = a_{ik}$ c) $\delta_{ij}\delta_{jk} a_{nm} = a_{im}$ d) $\delta_{ij}\delta^{jk} = \delta_i^k$.

4. Show that $\dfrac{\partial}{\partial x^i}(a_p x^p) = a_i$ for all values of independent variables, $x^1, x^2, \ldots .x^n$, and where x^p's are constants.

5. Calculate

   a) $\dfrac{\partial}{\partial x^k}(a_{ij}x^j)$

   b) $\dfrac{\partial}{\partial x^p}(a_{ijk}x^i x^j x^k)$

6. Using the relation $\frac{\partial x^r}{\partial x^s}\delta^r_s$, show that

$$\frac{\partial}{\partial x^p}(a_{ij}x^i x^j) = (a_{ip} + a_{pi})x^i$$

7. Express each of the following sums using the summation convention:
   (a) $\sum_{i=1}^{n}\sum_{j=1}^{n} u^i v^j$
   (b) $\sum_{i=1}^{n}\sum_{j=1}^{n}\sum_{k=1}^{n} a_{ijk}x^i x^j x^k$
8. Evaluate each of the following (range of indices 1 to n):
   (a) $\delta^i_j A^{kl}_i$ (b) $\delta^i_j \delta^k_l B^{jl}$ (c) $a^i a_j \delta^j_i$ (d) $\delta^i_j \delta^j_l \delta^l_k \delta^k_i$
9. (a) If $x^i = a^i_p y^p$ and $y^i = b^i_q z^q$, show that $x^i = a^i_q b^q_p z^p$.

   (b) If $x^i = a^i_p y^p$ and $z^i = b^i_q z^q$, show that $z^i = a^p_q b^i_p y^q$.
10. If y^i are n independent functions of variables x^i and z^i are n independent functions of y^i and if $u^i = v^j \frac{\partial x^i}{\partial y^j}$ and $v^i = w^j \frac{\partial y^i}{\partial z^j}$ $(i, j = 1, 2, \ldots n)$, then show that $u^i = w^j \frac{\partial}{\partial z^j} x^i$.
11. If $a = |a^i_j|$ and b^j_i is a^{-1} times the cofactor of a^i_j in the determinant of a^i_j, *show that* $b^j_i a^i_l = \delta^j_l$.
12. Prove that $|a^i_j||b^k_l| = |c^i_j|$ where $c^i_l = a^i_p b^p_l$.

2

Tensor Algebra

2.1 Introduction

Tensor Analysis is concerned with the study of abstract objects, called tensors, whose properties are independent of the reference frames used to describe the objects. In a given reference frame, a point is determined by a set of coordinate systems. If the coordinate system is changed, the point is described by the new set of coordinates, but the transformation of coordinates does nothing to the point itself.

This chapter is devoted to a discussion of the transformation of coordinate systems. We discuss the concepts of tensors and their algebra with respect to a certain transformation of coordinates and its law of transformations.

2.2 Scope of Tensor Analysis

In a given reference frame, a vector, A, is determined uniquely by a set of components, A_i. If a new coordinate system is introduced, the same vector, A, is determined by a set of components, B_i, and these new components are related in a definite way to the old ones. It is the law of transformation of components of a vector that is the essence of vector ideas and the same is true of tensors.

Since tensor analysis deals with entities and properties that are independent of the choice of reference frame it forms an ideal tool for the study of natural laws.

The concept of invariance of mathematical objects, under coordinate transformations, permeates the structure of tensor analysis to such an extent that it is important to get at outset a clear notion of the particular brand of invariance we have in mind. We shall suppose that a point is an invariant. In a given reference frame, a point, P, is determined by a set of coordinates, x^i. If the coordinate system is changed, point P is described

Dipankar De. Introduction to Differential Geometry with Tensor Applications, (25–72) © 2022 Scrivener Publishing LLC

by a new set of coordinates, y^i, but the transformation of coordinates does nothing to the point itself. Again, a pair of points (P_1, P_2) determines a vector $\overline{P_1P_2}$. This vector, in a particular reference frame, is uniquely determined by a set of components, A_i. A transformation coordinate does nothing to vector $\overline{P_1P_2}$, but in the new reference frame P_1P_2 is characterized by a different set of components, B_i.

A set of points, such as those forming a curve or surface, is also invariant. The curve may be described in a given coordinate system by an equation which usually changes its form when the coordinates are changed, but the curve itself remains unaltered. In general, an object, whatever it nature, is an invariant provided that it is not altered by a transformation of coordinates.

2.2.1 n-Dimensional Space

An ordered set of n-real numbers $x^1, x^2, \ldots x^n$ is called an n-tuple of real numbers and is denoted by $(x^1, x^2, \ldots x^n)$. The set of all n-tuples of real numbers is said to form an n-dimensional arithmetic continuum and each *n*-tuple is called a point of this continuum. Such a continuum shall be denoted by S_n. Sometimes an S_n is called an *n-dimensional space* because it can be endowed with the structure of an n-dimensional linear space.

The coordinates $(x^1, x^2, \ldots x^n)$ can be assigned to every point in S_n with respect to a chosen coordinate system establishing a 1-1 correspondence between the points of S_n and set of all coordinates like $(x^1, x^2, \ldots x^n)$. If $(x^1, x^2, \ldots x^n)$ are the coordinates of a point P, we shall write x^i as the coordinates of P. The corresponding coordinate system shall be denoted by (x^i).

Definition 2.2.1. A space (or *manifold*) of N dimensions is defined as any set of objects that can be placed in a 1-1 correspondence with the totality of ordered sets of N (real or complex) numbers $x^1, x^2, \ldots x^n$, such that $|x^i - A^i| < k^i \quad (i = 1,2 \ldots N)$, where $A^1 \ldots\ldots A^N$ are constants and $k^1, k^2 \ldots\ldots k^N$ are real numbers.

If the number x^i is real, the N dimensional space is real.

We denote the space of *N*-dimensions with the symbol V_N and we use the term 'points' to denote objects.

Alternately, a set of 'points' *M* is defined to be *manifold* if

i) each point of *M* has an open neighborhood and
ii) has a continuous 1 – 1 map onto an open set of *n*-tuples of real numbers.

There is no implication in these definitions that the concept of distance between the points has any meaning. If a rule is specified for the measurement of the distance between points, the space V_N is called metric.

2.3 Transformation of Coordinates in S_n

Let x^i be the coordinates of a point, P, with respect to a coordinate system and y^i be the coordinates of the same point with respect to another coordinate system and let the two systems be related by the equations:

$$T: y^i = y^i(x^1, x^2, \ldots x^n),\ i = 1,2 \ldots n, \tag{2.1}$$

where values of y^i are single valued continuous functions of x^1, x^2, ... x^n denoted by $y^i(x^1, x^2, \ldots x^n)$ and have continuous partial derivatives up to any desired order and further the determinant

$$\begin{vmatrix} \frac{\partial y^1}{\partial x.^1}, \frac{\partial y^1}{\partial x.^2} \cdots\cdots \frac{\partial y}{\partial x.^n} \\ \frac{\partial y^2}{\partial x.^1}, \frac{\partial y^2}{\partial x.^2} \cdots\cdots \frac{\partial y^2}{\partial x.^n} \\ \cdots\cdots\cdots\cdots\cdots \\ \frac{\partial y^n}{\partial x.^1}, \frac{\partial y^n}{\partial x.^2} \cdots\cdots \frac{\partial y^n}{\partial x.^n} \end{vmatrix} \neq 0 \tag{2.1a}$$

This determinant is called the Jacobian of transformation (2.1) and is denoted by $\left|\frac{\partial y.^i}{\partial x.^j}\right|$ and the Determinant (Jacobian) $J \equiv \left|\frac{\partial y.^i}{\partial x.^j}\right|$ | does not vanish at any point of region R, i.e.,

$$\left|\frac{\partial \phi^i}{\partial x.^j}\right| \neq 0.$$

In virtue of Equation (2.1), functions y^i are independent and Equation (2.1) can be solved for x^i as functions of y^i, giving

$$T^{-1}: x.^i = x^i(y.^1\ y.^2 \ldots y.^n),\ i = 1,2..n, \tag{2.2}$$

where the functions[1] $x^i(y)$ are single valued.

[1]we use the notation $x^i(y)$ and $f(x)$ as $x^i(y^1,y^2,\ldots.y^n)$ and $f(x^1,x^2,\ldots x^n)$ respectively.

It would follow then that not only a single valued inverse[2] of Equation (2.2) exists, but the functions $x^i(y.^i)$ *in (2.2)* are also of class C^1 in some neighborhood of the point under consideration. We shall refer to a class of coordinate transformations with the properties, i.e., when the class of functions are continuous together with their first n partial derivatives, as admissible transformations.

Relations (2.1) and (2.2) are called formulas of transformations of coordinates of S_n. They help to determine the coordinates of any point of S_n with respect to one coordinate system when the coordinates of the same point, with respect to another coordinate system, are known.

Example 2.3.1. We consider a system of equations specifying the relation between the spherical polar coordinates x^i and the rectangular Cartesian coordinates y^i:

$$T:\begin{cases} y^1 = x^1 \sin x^2 \cos x^3 \\ y^2 = x^1 \sin x^2 \sin x^3 \\ y^3 = x^1 \cos x^2 \end{cases}.$$

The Jacobian of this transformation is

$$\begin{vmatrix} \frac{\partial y^1}{\partial x^1} & \frac{\partial y^1}{\partial x^2} & \frac{\partial y^1}{\partial x^3} \\ \frac{\partial y^2}{\partial x^1} & \frac{\partial y^2}{\partial x^2} & \frac{\partial y^2}{\partial x^3} \\ \frac{\partial y^3}{\partial x^1} & \frac{\partial y^3}{\partial x^2} & \frac{\partial y^3}{\partial x^3} \end{vmatrix} = \begin{vmatrix} \sin x^2 \cos x^3 & x^1 \cos x^2 \cos x^3 & -x^1 \sin x^2 \sin x^3 \\ \sin x^2 \sin x^3 & x^1 \cos x^2 \sin x^3 & x^1 \sin x^2 \cos x^3 \\ \cos x^2 & -x^1 \sin x^2 & 0 \end{vmatrix}$$

$$= (x^1)^2 \sin x^2 \neq 0$$

If we suppose that $x^1 > 0$, $0 < x^2 < \pi$, and $0 \leq x^3 < 2\pi$, then $j \neq 0$ and the inverse transformation is given by

[2]we use C^n to denote the class of functions which are continuous and with their first n partial derivatives.

$$x^1 = \sqrt{(y^1)^2 + (y^2)^2 + (y^3)^2}$$

$$T^{-1}: \quad x^2 = \frac{\sqrt{(y^1)^2 + (y^2)^2}}{y^3}.$$

$$x^3 = \tan^{-1}\frac{y^2}{y^1}$$

Example 2.3.2. Show that the equation $x^1 = 4\cos x^2$ in spherical coordinates represents a sphere.
Solution: Let y^i be catrtesian coordinates and x^i be spherical coordinates. Then,

$$y^1 = x^1 \sin x^2 \cos x^3 \; y^2 = x^1 \sin x^2 \sin x^3 \text{ and } y^3 = x^1 \cos x^2$$

$$\therefore (y^1)^2 + (y^2)^2 + (y^3)^2 = (x^1 \sin x^2 \cos x^3)^2 + (x^1 \sin x^2 \sin x^3)^2 + (x^1 \cos x^2)^2$$

$$= (x^1 \sin x^2)^2 \{(\sin x^3)^2 + (\cos x^3)^2\} + (x^1 \cos x^2)^2$$

$$= (x^1 \sin x^2)^2 + (x^1 \cos x^2)^2 = (x^1)^2$$

$$\Rightarrow (x^1)^2 = (y^1)^2 + (y^2)^2 + (y^3)^2$$

$$\therefore x^1 = \sqrt{(y^1)^2 + (y^2)^2 + (y^3)^2}$$

$$\cos x^2 = \frac{y^3}{x^1} = \frac{y^3}{\sqrt{(y^1)^2 + (y^2)^2 + (y^3)^2}}$$

Putting these values in the given equation, we get

$$\sqrt{(y^1)^2 + (y^2)^2 + (y^3)^2} = 4\frac{y^3}{\sqrt{(y^1)^2 + (y^2)^2 + (y^3)^2}}$$

$$(y^1)^2 + (y^2)^2 + (y^3)^2 = 4y^3$$

$$(y^1)^2 + (y^2)^2 + (y^3 - 2)^2 = 4,$$

which is the equation of a sphere in Cartesian coordinates.

2.3.1 Properties of Admissible Transformation of Coordinates

It will be shown that all admissible transformations of coordinates from a group and, hence, every coordinate system in the family can be obtained from a particular transformation by admissible transformations.

There is a relation between the Jacobians of Equations (2.1) and (2.2), the exact nature of which is given below:

Theorem 2.3.1. If J and j' are the Jacobians of transformations (2.1) and (2.2), then $JJ' = 1$.

Proof: We insert the values of x^i from (2.2) in (2.1) and obtain a set of identities, y^i

$$\therefore y^i = y^i[x^1(y^1, y^2, \ldots y^n) \ldots \ldots x^n(y^1, y^2, \ldots y^n)].$$

The differentiation, with respect to y^i, gives

$$\frac{\partial y^i}{\partial y^j} = \delta^i_j = \frac{\partial y^i}{\partial x^\alpha}\frac{\partial x^\alpha}{\partial y^j}, \alpha = 1, 2, \ldots .n,$$

but $\left|\frac{\partial y^i}{\partial x^\alpha}\frac{\partial x^\alpha}{\partial y^j}\right| = \left|\frac{\partial y^i}{\partial x^k}\right| \cdot \left|\frac{\partial x^i}{\partial y^k}\right| = j.j'$. Here, let j and j' are the Jacobian of T and T^{-1}, respectively.

Since $\delta^i_j = 1$, we see that $j.j' = 1$ and we know that $j \neq 0$ in R.

Theorem 2.3.2. The Jacobian of the product transformation is equal to the product of the Jacobians of transformations entering in the product.

The following are admissible transformations T_1 and T_2

$$T_1: y^i = y^i(x^1, x^2, \ldots x^n) \text{ and}$$

$$T_2: z^i = z^i(y^1, y^2, \ldots y^n), \; i = 1, 2, \ldots n$$

Let transformation T_3: $z^i = z^i(y^1, y^2, \ldots y^n) = z^i[y^i(x^1, x^2, \ldots x^n)] = z^i[y^1(x^1, x^2, \ldots x^n)] \ldots y^n(x^1, x^2, \ldots x^n)]$, which is the product of T_1 and T_2.

$$T_3 = T_1 T_2$$

If the Jacobian of T_3 is j_3, we can write $j_3 = \left|\frac{\partial z^i}{\partial y^\alpha}\frac{\partial y^\partial}{\partial x^j}\right| = \left|\frac{\partial z^i}{\partial y^j}\right|\left|\frac{\partial y^i}{\partial x^j}\right| = j_2\, j_1$,

where j_2 and j_1 are the Jacobian of T_2 and T_1, respectively.

Theorem 2.3.3. The set of all admissible transformations of coordinates forms a group.
Proof: It is obvious.

i) The product of two admissible transformations is also a transformation belonging to a set of admissible transformations. Therefore, the closure property of transformation holds.
ii) The product transformation possesses an inverse.
iii) The identity transformation i.e., $x^i = y^i$ exists.
iv) The associative property is also obvious:
$T_3(T_2\, T_1) = (T_3\, T_2)\, T_1$.

Therefore, the admissible transformation forms a group.

2.4 Transformation by Invariance

Let $F(A)$ be a function of a point, A, in a coordinate system (x^i) in n – dimensional manifold V_n and suppose $F(A)$ is a continuous function. In reference frame (x^i), $F(A)$ is assumed to be form $f(x^1, x^2, \ldots x^n)$. Value $F(A)$ depends on A, but not the coordinate system used to represent A, so $F(A)$ is called the scalar function. We introduce a new reference frame (y^i) by means of transformation T:

$$T\colon x^i = x^i(y^1, y^2, \ldots y^n) \tag{2.3}$$

Function $F(A)$ in (y^i) is

$$f\,[x^1(y^1, y^2, \ldots\, y^n) \;\ldots\; \ldots x^n(y^1, y^2, \ldots y^n)] \equiv g(y^1, y^2, \ldots\, y^n) \tag{2.4}$$

The value of $f(x^1, x^2, \ldots x^n)$ at $A(x^1, x^2, \ldots x^n)$ is the same e^3 as $g(y^1, y^2, \ldots\, y^n)$ at $A(y^1, y^2, \ldots\, y^n)$. Then, f is called an *invariant* of V_n with respect

[3]In specific case, F(A) may represent the water level of the sea and f(x) is the function of it in the x-reference frame. g(y) is the representation of F(A) in the y-reference frame.

to transformation T if $f \equiv g$ and we call this substitution transformation, G^0: $f(x(y)) = g(y)$, the *transformation by invariance.*

2.5 Transformation by Covariant Tensor and Contravariant Tensor

Here, we discuss the Law of Transformation of Entities determined by the sets of partial derivatives of Scalar. We consider a continuously differentiable function, $f(x^1, x^2, \ldots x^n)$, representing the scalar $f(A)$ and a transformation of coordinates.

$$T: x^i = x^i(y^1, y^2, \ldots y^n) \tag{2.5}$$

If we form a set of n particle derivatives:

$$\frac{\partial f}{\partial x^1}, \frac{\partial f}{\partial x^2} \cdots \cdot \frac{\partial f}{\partial x^n} \tag{2.6}$$

Now, when we substitute into each function $f\, x^i(x^1, x^2, \ldots x^n)$, the values of x from Equation (2.5), we get

$$g_1(y^1, y^2, \ldots y^n), g_2(y^1, y^2, \ldots y^n), \ldots g_n(y^1, y^2, \ldots y^n). \tag{2.7}$$

It should be noted that the set of functions in (2.6) are not the same as the functions as (2.7). The partial derivatives

$$\frac{\partial f}{\partial y^1}, \frac{\partial f}{\partial y^2} \cdots \cdot \frac{\partial f}{\partial y^n}$$

are computed by the rule for the differentiation of composite functions:

$$G^1: \frac{\partial f}{\partial y^i} = \frac{\partial f}{\partial x^\alpha} \frac{\partial x^\alpha}{\partial y^i} \quad (i, \alpha = 1, 2, \ldots n) \tag{2.8}$$

By transformation T_1 of function $f(x^1, x^2, \ldots x^n)$, where T_1: $x^i = x^i(z^1, z^2, \ldots z^n)$, the set of functions corresponding to (2.6) is determined by law G^1

$$\frac{\partial f}{\partial z^i}=\frac{\partial f}{\partial x^\alpha}\frac{\partial x^\alpha}{\partial z^i}.$$

If we have a set of n functions, $A_1(x), \ldots. A_n(x)$, associated with X-coordinate system, and agree to calculate the corresponding quantities $B_1(y), \ldots. B_n(y)$ in y-Systems by means of *covariant law* G^1,

$$B_i(y)=\frac{\partial x^\alpha}{\partial y^i}A_\alpha(x) \tag{2.9}$$

Set $\{A_i(x)\}$ represents the component of *covariant vector* in X-coordinate system and set $\{B_i(y)\}$ is represented by the same covariant vector in the Y-system.

The displacement vectors in the X-system have its components

$$dx^1, dx^2 \ldots dx^n \tag{2.10}$$

and the same displacement vector when refers to the Y-system, having components

$$dy^1, dy^2 \ldots dy^n, \tag{2.11}$$

where $G^2: dy^i = \dfrac{\partial y^i}{\partial x^\alpha} dx^\alpha \quad (i, \alpha = 1, 2 \ldots n)$.

If we have a set of quantities $A_1(x), \ldots A_n(x)$, then the law G^2, determining the corresponding quantities $B_1(y), \ldots. B_n(y)$, is

$$B_i(y)=\frac{\partial y^i}{\partial x^\alpha}A_\alpha(x) \tag{2.12}$$

Law G^2 is called the *contravariant law* and the sets of quantities transforms in accordance with components of a *contravariant vector.*

We now consider covariants, contravariants, and mixed tensors.

2.6 The Tensor Concept: Contravariant and Covariant Tensors

2.6.1 Covariant Tensors

Definition 2.6.1. Let A_i be a set of n functions of n coordinates, x^i, in a given coordinate system (x^i). Then, A_i are said to form the components of a *covariant vector* if these components transform according to the following rule on change of coordinate systems from (x^i) to another system (y^i):

$$B_i = \frac{\partial x^j}{\partial y^i} A_j \tag{2.13}$$

It is obvious that the gradient of a scalar A, defined $\frac{\partial A}{\partial x^i}$, is an example of a covariant vector.

An example of the gradient of ϕ is denoted by

$$grad\phi \equiv \nabla\phi = \left(i\frac{\partial}{\partial x} + j\frac{\partial}{\partial y} + k\frac{\partial}{\partial z} \right)\phi = \frac{\partial \phi}{\partial x^i}.$$

From Equation (2.13) we can obtain that the expression of A_i in a coordinate system (x^i) in terms of coordinate system (y^i) is as follows:

Multiplying both sides of (2.13) by $\frac{\partial y^i}{\partial x^k}$, we obtain

$$\begin{aligned} \frac{\partial y^i}{\partial x^k} B_i &= \frac{\partial y^i}{\partial x^k}\frac{\partial x^j}{\partial y^i} A_j \\ &= \delta^i_k A_i \quad \left[\text{since} \frac{\partial y^i}{\partial x^k}\frac{\partial x^j}{\partial y^i} = \delta^j_k \right] \\ &= A_k. \end{aligned}$$

$$\text{Thus, } A_k = \frac{\partial y^i}{\partial x^k} B_i \text{ (Covariant Law)} \tag{2.14}$$

Example 2.6.1. If ϕ is a scalar function of coordinates x^i, this shows that $\frac{\partial \phi}{\partial x^i}$ is a covariant vector.

Let us consider the coordinate transformation from x^i *to* y^i.

$$\text{i.e., } y^i = y^i(x^1, x^2, \dots \dots x^n).$$

Therefore, ϕ is a function of coordinates x^i (i.e.,, x^1, x^2 ...x^n) in the first system and a function of coordinates y^i in the second system.

By the Chain Rule:

$$\frac{\partial \phi}{\partial y^i} = \frac{\partial \phi}{\partial x^1}\frac{\partial x^1}{\partial y^i} + \frac{\partial \phi}{\partial x^2}\frac{\partial x^2}{\partial \bar{y}^i} + \cdots \dots + \frac{\partial \phi}{\partial x^n}\frac{\partial x^n}{\partial \bar{y}^i}$$

$$= \frac{\partial \phi}{\partial x^j}\frac{\partial x^j}{\partial y^i}$$

Here, $\frac{\partial \phi}{\partial x^j}$ forms the component of a covariant vector.

The covariant vector is also called a covariant tensor of rank one.

A subscript is always used to indicate covariant components and superscript is always used to indicate contravariant vectors.

2.6.2 Contravariant Vectors

Definition 2.6.2. Let A^i be a set of n functions of n coordinates x^i in a given coordinate system (x^i). Then, A^i are said to form the components of a *contravariant vector* if these components transform according to the following rule on change of a coordinate system from (x^i) to another system (y^i).

$$B^i = \frac{\partial y^i}{\partial x^j} A^j \quad \text{(Contravariant Law)}$$

Therefore, $d\bar{y}^i = \sum_{j=1}^{n} \frac{\partial \bar{y}^i}{\partial x^j} dx^i = \frac{\partial \bar{y}^i}{\partial x^j} dx^j$.

According to the definition, the quantities of dx^i, considered above, form the components of the contravariant vector.

Example 2.6.2. A covariant tensor has components xy, $2y - z^2$, *and* xz in rectangular coordinates. Find its covariant components in spherical coordinates.

Solution: Here, $x^1 = x$, $x^2 = y$, $x^3 = z$ (Cartesian Coordinate System)

$$\bar{x}^1 = r\,\bar{x}^2 = \theta\,, \bar{x} = \theta\,, \bar{x}^3 = \phi \text{ (Spherical Coordinate System)}$$

$$A^1 = xy \text{ , } A^2 = 2y - z^2 \text{ , } A^3 = xz$$

According to the law of transformation, we have

$$\bar{A}_i = \frac{\partial x^j}{\partial \bar{x}^i} A_j, (\text{i} = 1,2,3)$$

and we like to evaluate $\bar{A}^1, \bar{A}^2, \bar{A}^3$, *where* A_1, A_2, *and* A_3, are known.

We know that $x = r \cos \theta \sin \phi$

$y = r \sin \theta \sin \phi$

$z = r \sin \theta.$

$$\text{Now, } \bar{A}_1 = \frac{\partial x^1}{\partial \bar{x}^1} A_1 + \frac{\partial x^2}{\partial \bar{x}^1} A_2 + \frac{\partial x^3}{\partial \bar{x}^1} A_3$$

$$= \frac{\partial x}{\partial r} xy + \frac{\partial y}{\partial r}(2y - z^2) + \frac{\partial x^3}{\partial \bar{x}^1} xz$$

$$= (xy)\frac{\partial}{\partial r}(r\cos\theta\sin\phi) + (2y - z^2)\frac{\partial}{\partial r}(r\sin\theta\sin\phi) + (xz)\frac{\partial}{\partial r}(r\cos\theta)$$

$= (\text{r}\cos\theta\sin\phi.\text{r}\cos\theta\sin\phi)(\cos\theta\sin\phi) + [2r \sin \theta \sin \phi - (r \cos \theta)^2]$ $(r \sin \theta \sin \phi) + (r \cos \theta \sin \phi)r \cos \theta \,.\, \cos \theta$

$== r^2 \cos.^3 \theta sin^3\phi + 2r^2 \sin^2 \theta \sin^2 \phi - r^3 \cos^2 \theta \sin \theta \sin \phi + r^2 cos^3$ $\theta \sin \phi$

$$\bar{A}_2 = \frac{\partial x^1}{\partial \bar{x}^2} A_1 + \frac{\partial x^2}{\partial \bar{x}^2} A_2 + \frac{\partial x^3}{\partial \bar{x}^2} A_3 =$$

$$r^3 \sin\theta\sin\phi\cos\theta\cos^2\phi + 2r^2 \sin\theta\cos\theta\sin^2\phi - r^3\cos^3\theta\sin\phi - r^3\sin^2\theta\cos\theta\cos\phi - .-$$

$$\bar{A}_3 = \frac{\partial x^1}{\partial \bar{x}^3} A_1 + \frac{\partial x^2}{\partial \bar{x}^3} A_2 + .\frac{\partial x^3}{\partial \bar{x}^3} A_3.$$

$$= -r^3 \sin^3 \theta \sin^2 \phi \cos\theta + r \sin\theta\cos\phi(2r\sin\theta\sin\phi - r^2\cos^2\theta) + 0$$

Example 2.6.3. If A^i and B_i are contravariant and covariant vectors, respectively, then the sum of A^iB_i is an invariant.

Proof: We have $\bar{A}^i = \frac{\partial \bar{x}^{\cdot i}}{\partial x^j} A^j$ and, $\bar{B}_i = \frac{\partial x^k}{\partial \bar{x}^i} B_k$,

Hence, $\overline{A}^i\overline{B}_i = \frac{\partial\overline{x}^i}{\partial x^j}A^j\frac{\partial x^k}{\partial\overline{x}^i}B_k$

$$= \frac{\partial\overline{x}^i}{\partial x^j}\frac{\partial x^k}{\partial\overline{x}^i}B_kA^j = \delta_j^k B_k A^j = A^k B_k = A^i B_i, \quad \left[since\ \delta_j^i a^j = a^i\right].$$

This implies the proof.

Example 2.6.4. Prove that there is no distinction between contravariant and covariant vectors if the transformation law is of the form $y^i = a_m^i x^m + b^i$, where a's and b's are constants, such that $a_r^i a_m^i = \delta_m^r$.

Solution:
$$y^i = a_m^i x^m + b^i \tag{2.15a}$$

We get $\frac{\partial y^i}{\partial x^m} = a_m^i$

From (2.15a) $a_r^i y^i = a_r^i a_m^i x^m + a_r^i b^i$

$$= \delta_m^r x^m + a_r^i b^i = x^r + a_r^i b^i$$

$$\therefore x^r = a_r^i y^i - a_r^i b^i$$

$$\therefore \frac{\partial x^r}{\partial y^i} = a_r^i$$

Now, the Contravariant Transformation is $\overline{A}^i = \frac{\partial y^i}{\partial x^m}A^m = a_m^i A^m$

(2.15b)

Now, the Covariant Transformation is $\overline{A}_i = \frac{\partial x^m}{\partial y^i}A^m = a_m^i A_m$ (2.15c)

From (2.15b) and (2.15c), it is shown that there is no distinction between contravariant and covariant tensor by this law of transformation.

Example 2.6.5. If A_i is a covariant vector, is determines whether $\frac{\partial Ai}{\partial x}$ is a tensor.

Solution: Since A_i is a covariant vector, we have

$$\bar{A}_i = \frac{\partial x^k}{\partial y^i} A_k \tag{2.16a}$$

Differentiating both sides with respect to y^j,

$$\frac{\partial \bar{A}_i}{\partial y^j} = \frac{\partial^2 x^k}{\partial y^j \partial y^i} A_k + \left(\frac{\partial x^k}{\partial x^l} \frac{\partial x^l}{\partial y^j} \right) \frac{\partial x^k}{\partial y^j} = \frac{\partial x^k}{\partial y^j} \frac{\partial x^l}{\partial y^j} \frac{\partial x^k}{\partial x^l} + \frac{\partial^2 x^k}{\partial y^j \partial y^i} A_k. \tag{2.16b}$$

From (2.16b), it follows that $\frac{\partial A_i}{\partial x}$ is not a tensor due to the presence of a second term on the right hand side of (2.16b).

Example 2.6.6. If $a_{ij}u^i u^j = 0$ is an invariant for an arbitrary contravariant, vector u^i, show that $a_{ij} + a_{ji} = 0$.

Solution: Since $a_{ij}u^i u^j$ is an invariant, we have

$$\begin{aligned}
\bar{a}_{ij}\bar{u}^i \bar{u}^j &= a_{ij}u^i u^j \\
&= a_{ij} \frac{\partial x^i}{\partial y^p} \bar{u}^p \frac{\partial x^j}{\partial y^q} \bar{u}^q \\
&= a_{ij} \frac{\partial x^i}{\partial y^p} \frac{\partial x^j}{\partial y^q} \bar{u}^p \bar{u}^q \\
&= a_{pq} \frac{\partial x^p}{\partial y^i} \frac{\partial x^q}{\partial y^j} \bar{u}^i \bar{u}^j
\end{aligned}$$

$$\therefore \left(\bar{a}_{ij} - a_{pq} \frac{\partial x^p}{\partial y^i} \frac{\partial x^q}{\partial y^j} \right) \bar{u}^i \bar{u}^j = 0$$

$$\text{Or, } \bar{A}_{ij}\bar{u}^i \bar{u}^j = 0 \tag{2.17a}$$

$$\text{where } \bar{A}_{ij} = \left(\bar{a}_{ij} - a_{pq} \right) \frac{\partial x^p}{\partial y^i} \frac{\partial x^q}{\partial y^j} \tag{2.17b}$$

and $\bar{u}^i$ is an arbitray, following that

$$\bar{A}_{ij} + \bar{A}_{ji} = 0. \tag{2.17c}$$

By (2.17c), we may write (2.17b) as

$$\bar{a}_{ij} - a_{pq}\frac{\partial x^p}{\partial y^i}\frac{\partial x^q}{\partial y^j} + \bar{a}_{ji} - a_{pq}\frac{\partial x^p}{\partial y^j}\frac{\partial x^q}{\partial y^i} = 0$$

$$\bar{a}_{ij} + \bar{a}_{ji} = a_{pq}\frac{\partial x^p}{\partial y^j}\frac{\partial x^q}{\partial y^i} + a_{pq}\frac{\partial x^p}{\partial y^i}\frac{\partial x^q}{\partial y^j}$$

(Replacing the dummy indices (p, q) in the second term of the right hand side by (q, p), respectively)

$$= (a_{pq} + a_{qp})\frac{\partial x^p}{\partial y^i}\frac{\partial x^q}{\partial y^j} \tag{2.17d}$$

From (2.17d), it follows that (a_{pq} + a_{qp}) is a covariant tensor of order 2.

Example 2.6.7. Let S be the set of transformation of covariant vectors and T, T' be two transformations from system (x^i) to system (y^i) and from (y^i) to (z^i), given by

$$T: \bar{A}^i = A^p\frac{\partial y^i}{\partial x^p} \tag{2.18a}$$

$$T': \bar{\bar{A}}^i = \bar{A}^p\frac{\partial z^i}{\partial y^r} \tag{2.18b}$$

Then, product transformation T.T' is

$$\text{T}.T': \bar{\bar{A}}^i = A^p\frac{\partial y^r}{\partial x^p}\frac{\partial z^i}{\partial y^r}, \text{ by (2.18a) and (2.18b)}$$

$$= A^p\frac{\partial z^i}{\partial x^p} \tag{2.18c}$$

It follows that T.$T' \in S$ (2.18d)

Let I be the transformation given by

$$I: A^{i} = A^{i}\frac{\partial x^{p}}{\partial x^{p}}. \tag{2.18e}$$

Then, $IT = TI = T$
⇒ I is the identity transformation and

$$I \in S \tag{2.18f}$$

From (2.18a), we get
$$A^{p} = \bar{A}^{i}\frac{\partial x^{p}}{\partial y^{i}} \tag{2.18g}$$

and its transformation is denoted by T_1.

Since T_1 represents the transformation from (y^i) to (x^i), it is an inverse transformation.

$$T_1 \in S \tag{2.18h}$$

Now, we can show easily that if T'' represents a transformation from (z^i) to (y^i), then

$$(T''.T').\mathrm{T} = T''.(T'.\mathrm{T}) \tag{2.18i}$$

Hence, by (2.18d), (2.18f), (2.18h), and (2.18i), it follows that S forms a group.

2.6.3 Tensor of Higher Order

2.6.3.1 Contravariant Tensors of Order Two

Let A^{ij} be a set of n^2 functions of n coordinates, x^i is in a given system of coordinates (x^i). Then, A^{ij} are said to form the components of a contravariant tensor of order two (or rank 2) if these components transform according to the following rule on change of coordinate system from (x^i) to another system $(\overline{x^i})$.

$$B^{ij} = \frac{\partial y^{i}}{\partial x^{k}}\frac{\partial y^{j}}{\partial x^{p}}A^{kp} \tag{2.19}$$

The formula expressing components A^{ij} in a coordinate system (x^i), in terms of those in another system $(\overline{x}^i)$, is obtained as follows:

Multiplying both sides of (2.12) by $\dfrac{\partial x^r}{\partial y^{i\cdot}}\dfrac{\partial x^s}{\partial y^{j\cdot}}$, we get

$$\begin{aligned}\frac{\partial x^r}{\partial y^{i\cdot}}\frac{\partial x^s}{\partial y^{j\cdot}}B^{ij} &= \frac{\partial x^r}{\partial y^{i\cdot}}\frac{\partial x^s}{\partial y^{j\cdot}}\frac{\partial y^i}{\partial x^k}\frac{\partial y^j}{\partial x^p}A^{kp}\\ &= \frac{\partial x^r}{\partial y^{i\cdot}}\frac{\partial y^i}{\partial x^k}\frac{\partial x^s}{\partial y^{j\cdot}}\frac{\partial y^j}{\partial x^p}A^{kp}\\ &= \delta_k^r\delta_p^s A^{kp}\\ &= A^{rs}.\end{aligned}$$

$$\text{Hence, } A^{rs} = \frac{\partial x^r}{\partial \overline{x}^{i\cdot}}\frac{\partial x^s}{\partial \overline{x}^{j\cdot}}B^{ij} \tag{2.20}$$

The following example shows the existence of a contravariant tensor of second order.

We have $\overline{A}^i = \dfrac{\partial \overline{x}^{\cdot i}}{\partial x^k}A^k$ *and* $\overline{B}^j = \dfrac{\partial \overline{x}^{\cdot j}}{\partial x^p}B^p$, hence $\overline{A}^i\overline{B}^j = \dfrac{\partial \overline{x}^i}{\partial x^k}\dfrac{\partial \overline{x}^j}{\partial x^p}B^pA^k$.

Now, we write $B^pA^k = C^{kp}$ *and* $\overline{A}^i\overline{B}^j = \overline{C}^{ij}$.

The relation can be written as

$$\overline{C}^{ij} = \frac{\partial \overline{x}^i}{\partial x^k}\frac{\partial \overline{x}^j}{\partial x^p}C^{kp}$$

By virtue of (2.20), it follows that C^{ij} or A^iB^j are the components of a contravariant tensor of Order 2.

A contravariant tensor of order r may be similarly defined by considering a system of order r of type $A^{i_1i_2\ldots\ldots i_r}$.

2.6.3.2 Covariant Tensor of Order Two

Let A_{ij} be a set of n^2 functions of n coordinates x^i in a given system of coordinates (x^i). Then, A_{ij} are said to form the components of a covariant tensor of order two (or rank 2) if these components transform according to the following rule on change of coordinates system from (x^i) to another system $(\overline{x}^i)$:

$$B_{ij} = \frac{\partial x^k}{\partial y^i}\frac{\partial x^p}{\partial y^j}A_{kp}. \tag{2.21}$$

The existence of such a tensor may be shown by taking two covariant vectors, A_i and B_j, and proceeding as the above case.

The formula expressing components A_{ij} in a coordinate system (x^i), in terms of those of another system $(\overline{x}^i)$, is as follows:

$$A_{ij} = \frac{\partial \overline{x}^r}{\partial x^i}\frac{\partial \overline{x}^s}{\partial x^j}B_{rs} \tag{2.22}$$

A covariant tensor of order *r* may be similarly defined by considering a system of order *r* of type $A_{i_1 i_2 \ldots\ldots i_r}$.

2.6.3.3 *Mixed Tensors of Order Two*

Definition 2.5.3. Let A^i_j be a set of n^2 functions of *n* coordinates x^i in a given system of coordinates (x^i). Then, A^i_j are said to form the components of a *mixed tensor* of order two (or rank 2) if these components transform according to the following rule on a change of coordinates system from (x^i) to another system (y^i):

$$B^i_j = \frac{\partial y^i}{\partial x^k}\frac{\partial x^p}{\partial y^j}A^k_p \tag{2.23}$$

A mixed tensor of order two sometimes said a mixed tensor of second order with a first order of contravariance and first order of covariance.

Similarly, A^i_j in a system (x^i), in terms of those in another system (y^i), *is* as follows:

$$A^i_j = \frac{\partial x^i}{\partial y^r}\frac{\partial y^s}{\partial x^j}A^r_s$$

Example 2.6.8. Show that the Kronecker delta is a mixed tensor of order two.

Solution: If δ^i_j transforms to δ_j^{-i} in coordinate system (y^i) by the law for mixed tensors of order two, then $\delta_j^{-i} = \frac{\partial y^j}{\partial x^m}\frac{\partial x^l}{\partial y^i}\delta^m_l = \frac{\partial y^j}{\partial x^m}\frac{\partial x^m}{\partial y^i} = \frac{\partial y^j}{\partial y^i} = \delta^j_i$, hence δ^j_i is a mixed tensor of order two, having the same components in every coordinate system.

Definition 2.6.4. The sets of n^{r+s} quantities, in the x-coordinate system by the expressions $A^{j_1 j_2 \ldots\ldots j_s}_{i_1 i_2 \ldots i_r}(x)$ is a *mixed tensor, covariant rank r, and contravariant rank s*, provided the corresponding quantities $B^{j_1 j_2 \ldots\ldots j_s}_{i_1 i_2 \ldots i_r}(y)$ in the y-coordinate system are given by the law:

$$B^{j_1 j_2 \ldots\ldots j_s}_{i_1 i_2 \ldots i_r} = \frac{\partial x^{\alpha_1}}{\partial y^{i_1}} \frac{\partial x^{\alpha_2}}{\partial y^{i_2}} \cdots \frac{\partial x^{\alpha_r}}{\partial y^{i_r}} \frac{\partial y^{j_1}}{\partial x^{\beta_1}} \frac{\partial y^{j_1}}{\partial x^{\beta_1}} \frac{\partial y^{j_2}}{\partial x^{\beta_2}} \cdots \frac{\partial y^{j_s}}{\partial x^{\beta_s}} A^{\beta_1 \beta_2 \ldots\ldots \beta_s}_{\alpha_1 \alpha_{2.} \ldots\ldots \alpha_r} \tag{2.24}$$

From the above structure of formulas, we can deduce an important theorem:

If all components of a tensor vanish in one coordinate system, then they necessarily vanish in all other admissible coordinate systems.

Definition 2.6.5. A tensor whose components are all zero in a coordinate system is called a zero tensor, so we can define the zero tensor as a tensor whose components are all zero in every coordinate system.

2.7 Algebra of Tensors

The addition and subtraction of two tensors of the same type is a tensor of the same type.

Theorem 2.7.1. If the components of tensor are all zero in one coordinate system, then they are also zero in every other coordinate system.
Proof: Let the components $A^{i_1 i_2 \ldots\ldots i_p}_{j_1 j_2 \ldots\ldots j_q}$ of a tensor of type (p, q) be all zero in coordinate system (x^i) and its components in another coordinate system (y^i) by $\bar{A}^{i_1 i_2 \ldots\ldots i_p}_{j_1 j_2 \ldots\ldots j_q}$. Then, by (2.24), we get

$$\bar{A}^{i_1 i_2 \ldots\ldots i_p}_{j_1 j_2 \ldots\ldots j_q} = \frac{\partial y^{i_1}}{\partial x^{r_1}} \frac{\partial y^{i_2}}{\partial x^{r_2}} \cdots \frac{\partial y^{i_p}}{\partial x^{r_p}} \frac{\partial x^{t_1}}{\partial y^{j_1}} \frac{\partial x^{t_2}}{\partial y^{j_2}} \cdots \frac{\partial x^{t_q}}{\partial y^{j_q}} A^{r_1 r_2 \ldots\ldots r_p}_{t_1 t_2 \ldots\ldots t_q}, \tag{2.25}$$

but $A^{r_1 r_2 \ldots\ldots r_p}_{t_1 t_2 \ldots\ldots t_q} = 0$ *and* it follows from (2.25) that $\bar{A}^{i_1 i_2 \ldots\ldots i_p}_{j_1 j_2 \ldots\ldots j_q} = 0$, hence proved.

A tensor whose components are all zero in every coordinate system is called a *zero tensor*.

Theorem 2.7.2. If $A^{i_1 i_2 \ldots\ldots i_p}_{j_1 j_2 \ldots\ldots j_q}$ and $B^{i_1 i_2 \ldots\ldots i_p}_{j_1 j_2 \ldots\ldots j_q}$ are components of two tensors of type (p, q), then $A^{i_1 i_2 \ldots\ldots i_p}_{j_1 j_2 \ldots\ldots j_q} + B^{i_1 i_2 \ldots\ldots i_p}_{j_1 j_2 \ldots\ldots j_q}$ are the components of another tensor of type (p, q).

Proof: Let $A^{i_1 i_2 \ldots\ldots i_p}_{j_1 j_2 \ldots\ldots j_q}$ and $B^{i_1 i_2 \ldots\ldots i_p}_{j_1 j_2 \ldots\ldots j_q}$ be two tensors of type (p, q) in coordinate system (x^i) and $\overline{A}^{i_1 i_2 \ldots\ldots i_p}_{j_1 j_2 \ldots\ldots j_q}$ and $\overline{B}^{i_1 i_2 \ldots\ldots i_p}_{j_1 j_2 \ldots\ldots j_q}$ be their components in another coordinate system (y^i).

By (2.25) we can write

$$\overline{A}^{i_1 i_2 \ldots\ldots i_p}_{j_1 j_2 \ldots\ldots j_q} + \overline{B}^{i_1 i_2 \ldots\ldots i_p}_{j_1 j_2 \ldots\ldots j_q} = \frac{\partial y^{i_1}}{\partial x^{r_1}}\frac{\partial y^{i_2}}{\partial x^{r_2}}\cdots\frac{\partial y^{i_p}}{\partial x^{r_p}}\frac{\partial x^{t_1}}{\partial x_{j_1}}\frac{\partial x^{t_2}}{\partial y^{j_2}}\cdots\frac{\partial x^{\cdot t_q}}{\partial y^{j_q}}\left(A^{i_1 i_2 \ldots\ldots i_p}_{j_1 j_2 \ldots\ldots j_q} + B^{i_1 i_2 \ldots\ldots i_p}_{j_1 j_2 \ldots\ldots j_q}\right).$$

In (x^i) system, $A^{i_1 i_2 \ldots\ldots i_p}_{j_1 j_2 \ldots\ldots j_q} + B^{i_1 i_2 \ldots\ldots i_p}_{j_1 j_2 \ldots\ldots j_q}$ is the tensor of (p, q), hence proved.

Similarly, it can be proven that $A^{i_1 i_2 \ldots\ldots i_p}_{j_1 j_2 \ldots\ldots j_q} - B^{i_1 i_2 \ldots\ldots i_p}_{j_1 j_2 \ldots\ldots j_q}$ are another tensor of type (p, q).

Theorem 2.7.3. If $A^{i_1 i_2 \ldots\ldots i_p}_{j_1 j_2 \ldots\ldots j_q}$ are the components of a tensor of type (p, q) and ϕ is a scalar, then $\phi A^{i_1 i_2 \ldots\ldots i_p}_{j_1 j_2 \ldots\ldots j_q}$ are the components of another tensor of type (p, q).

Proof: Let $A^{i_1 i_2 \ldots\ldots i_p}_{j_1 j_2 \ldots\ldots j_q}$ be the components of a tensor of type (p, q) in (x^i) and $\overline{A}^{i_1 i_2 \ldots\ldots i_p}_{j_1 j_2 \ldots\ldots j_q}$ be their components in another coordinate system (y^i). Let ϕ and $\overline{\phi}$ be a scalar in the respective system of coordinates.

Then, we have

$$\overline{\phi A}^{i_1 i_2 \ldots\ldots i_p}_{j_1 j_2 \ldots\ldots j_q} = \overline{\phi} = \frac{\partial y^{i_1}}{\partial x^{r_1}}\frac{\partial y^{i_2}}{\partial x^{r_2}}\cdots\frac{\partial y^{i_p}}{\partial x^{r_p}}\frac{\partial x^{t_1}}{\partial y^{j_1}}\frac{\partial x^{t_2}}{\partial y^{j_2}}\cdots\frac{\partial x^{\cdot t_q}}{\partial y^{j_q}}A^{r_1 r_2 \ldots\ldots r_p}_{t_1 t_2 \ldots\ldots t_q}$$

$$= \phi\frac{\partial y^{i_1}}{\partial x^{r_1}}\frac{\partial y^{i_2}}{\partial x^{r_2}}\cdots\frac{\partial y^{i_p}}{\partial x^{r_p}}\frac{\partial x^{t_1}}{\partial y^{j_1}}\frac{\partial x^{t_2}}{\partial y^{j_2}}\cdots\frac{\partial x^{\cdot t_q}}{\partial y^{j_q}}A^{r_1 r_2 \ldots\ldots r_p}_{t_1 t_2 \ldots\ldots t_q}, \text{ since } \phi = \overline{\phi}$$

$$= \frac{\partial y^{i_1}}{\partial x^{r_1}}\frac{\partial y^{i_2}}{\partial x^{r_2}}\cdots\frac{\partial y^{i_p}}{\partial x^{r_p}}\frac{\partial x^{t_1}}{\partial y^{j_1}}\frac{\partial x^{t_2}}{\partial y^{j_2}}\cdots\frac{\partial x^{t_q}}{\partial y^{j_q}}\phi A^{r_1 r_2 \ldots\ldots r_p}_{t_1 t_2 \ldots\ldots t_q}$$

It follows that $\phi A^{i_1 i_2 \ldots\ldots i_p}_{j_1 j_2 \ldots\ldots j_q}$ are the components of tensor type (p, q).

2.7.1 Equality of Two Tensors of Same Type

Two tensors of the same type are said to be equal in the same coordinate system if their components in this system are equal to each.

If $A^{i_1 i_2 \ldots\ldots i_p}_{j_1 j_2 \ldots\ldots j_q}$ and $B^{i_1 i_2 \ldots\ldots i_p}_{j_1 j_2 \ldots\ldots j_q}$ are components of two equal tensors in the same coordinate system, then we must have $A^{i_1 i_2 \ldots\ldots i_p}_{j_1 j_2 \ldots\ldots j_q} = B^{i_1 i_2 \ldots\ldots i_p}_{j_1 j_2 \ldots\ldots j_q}$.

If two tensors are equal in a certain coordinate system, then they are equal in every coordinate system.

Remark 2.7.1.

(i) If $a_{ij} A^i B^j = 0$ for two distinct arbitrary vectors, A^i and B^j, then $a_{ij} = 0$.

(ii) The two tensors are said to be equal in a certain coordinate system if they are of the same type and their corresponding components in the coordinate system are equal.

(iii) The order of the indices of a tensor are important. Consider tensors A^{ij} and A^{ji}. $A^{ij} - A^{ji}$ is a tensor, but not generally a zero tensor.

Cases i) $A^{ij} = A^{ji}$ and ii) $A^{ij} = -A^{ji}$ are discussed in the next section.

2.8 Symmetric and Skew-Symmetric Tensors

2.8.1 Symmetric Tensors

Definition 2.8.1. A tensor is said to be *symmetric* with respect to two contravariant (or two covariant) indices if its components remain unchanged on an interchange of the two indices.

Thus, tensor A^{ijk} is symmetric if $A^{ijk} = A^{jik} = A^{ikj} = A^{kji}$ for every pair of indices.

The definition of symmetry of tensors obviously would not be satisfactory if the symmetry of its components are not persevered under the transformation of coordinates.

So, it can be easily shown that if a tensor is symmetric with respect to two covariant or contravariant indices in any coordinate system, then it remains unchanged in any other coordinate system.

That is, the symmetric property remains unchanged under coordinate transformation.

2.8.2 Skew-Symmetric Tensors

Definition 2.8.2. A tensor is said to be *skew-symmetric* (or anti-symmetric) with respect to a pair of contravariant (or covariant) indices if the components change sign on an interchange of the pair of indices.

Thus, tensor A^{ij} is skew-symmetric if $A^{ij} = -A^{ji}$ for every i and j.

In general, the tensor is $A^{ijk}_{lm} = -A^{jik}_{lm}$.

It should be noted that skew-symmetry cannot be defined for a tensor with respect to two indices of which one is contravariant and the other is covariant.

This also happens in the case of symmetric tensors, but in an exceptional case it is provided.

We have $\delta^i_j = \delta^j_i$.

Theorem 2.8.1. The components of a tensor of type (0,2) can be expressed as the sum of a symmetric tensor and a skew-symmetric tensor of the same type.

Proof: Let a_{ij} be the components of a tensor of type (0,2).

We can express a_{ij} as follows:

$$\begin{aligned} a_{ij} &= \frac{1}{2}(a_{ij} + a_{ji}) + \frac{1}{2}(a_{ij} - a_{ji}) \\ &= A_{ij} + B_{ij}, \end{aligned}$$

where $A_{ij} = \frac{1}{2}(a_{ij} + a_{ji})$ and $B_{ij} = \frac{1}{2}(a_{ij} - a_{ji})$.

Since a_{ij} is a tensor of type (0,2), a_{ji} is also a tensor of type (0,2). Hence, both A_{ij} and B_{ij} are tensor of type (0,2). Since $A_{ij} = A_{ji}$ *and* $B_{ij} = -B_{ji}$, tensor A_{ji} is symmetric and B_{ij} is skew-symmetric, hence proved.

This is also true for a (2,0) type tensor, i.e., contravariant tensor of the second order.

Example 2.8.1. Show that:

(i) A symmetric tensor of the second order has only $\frac{1}{2}n(n+1)$ different components.
(ii) A skew symmetric tensor of the second order has only $\frac{1}{2}n(n-1)$ different non-zero components.

Solution:

(i) Let A^{ij} be a symmetric tensor of order two so that $A^{ij} = A^{ji}$. If each of the indices *i* and *j* take values 1 to *n*, then A^{ij} will have n^2 components. Out of these n^2 components, n components $A^{11}, A^{22}, \ldots \ldots A^{nn}$, *and* $A^{22} \ldots\ldots A^{nn}$ are independent.

Thus, the remaining components are $(n^2 - n)$, which can be taken in pairs, since, $A^{12} = A^{21}$, $A^{31} = A^{13}$, etc.

Hence, the total number of independent components are $n+\frac{1}{2}(n^2-n)=\frac{1}{2}n(n+1)$.

(ii) Let A^{ij} be a skew-symmetric tensor of order two so that $A^{ij} = -A^{ji}$. As above, A^{ij} have n^2 components. Out of these, *n* components $A^{11}, A^{22}, \ldots \ldots A^{nn}$ are all zero. $[A^{11} = -A^{11}]$

Omitting these, there are $(n^2 - n)$ components which are independent and can be taken pair wise (ignoring the sign).

Hence, the total number of independent non-zero components is $\frac{1}{2}(n^2-n)=\frac{1}{2}n(n-1)$.

Example 2.8.2. If T_i is the component of a covariant vector, show that $\left(\frac{\partial T_i}{\partial x^j}-\frac{\partial T_j}{\partial x^i}\right)$ are components of a skew-symmetric covariant tensor of order two.

Solution: Since T_i is a covariant vector, by the covariant law of transformation,

$$\overline{T_i}=\frac{\partial x^j}{\partial y^i}T_j.$$

Partially differentiating with respect to y^i, we get

$$\begin{aligned}
&=\frac{\partial \overline{T_i}}{\partial y^k}=\frac{\partial}{\partial y^k}\left(\frac{\partial x^j}{\partial y^i}T_j\right)\\
&=\frac{\partial^2 x^j}{\partial y^k \partial y^i}T_j+\frac{\partial x^j}{\partial y^i}\frac{\partial T_j}{\partial y^k}+\\
&=\frac{\partial^2 x^j}{\partial y^k \partial y^i}T_j+\frac{\partial x^j}{\partial y^i}\frac{\partial x^l}{\partial y^k}\frac{\partial T_j}{\partial x^l}
\end{aligned} \tag{2.26a}$$

Similarly, $\frac{\partial \overline{T_k}}{\partial y^i}=\frac{\partial^2 x^j}{\partial y^k \partial y^i}T_j+\frac{\partial x^j}{\partial y^k}\frac{\partial x^l}{\partial y^i}\frac{\partial T_j}{\partial x^l}$ (interchanging *i and k*).

Now interchanging *j and l* in the 2nd term,

$$\frac{\partial \overline{T_k}}{\partial y^i}=\frac{\partial^2 x^j}{\partial y^k \partial y^i}T_j+\frac{\partial x^j}{\partial y^k}\frac{\partial x^l}{\partial y^i}\frac{\partial T_l}{\partial x^j} \tag{2.26b}$$

Subtracting (2.26b) from (2.26a),

$$\begin{aligned}
\frac{\partial \overline{T_l}}{\partial y^k}-\frac{\partial \overline{T_k}}{\partial y^i} &=\left(\frac{\partial^2 x^j}{\partial y^k \partial y^i}T_j+\frac{\partial x^j}{\partial y^i}\frac{\partial x^l}{\partial y^k}\frac{\partial T_j}{\partial x^l}\right)\\
&\quad-\left(\frac{\partial^2 x^j}{\partial y^k \partial y^i}T_j+\frac{\partial x^j}{\partial y^k}\frac{\partial x^l}{\partial y^i}\frac{\partial T_l}{\partial x^j}\right)\\
&=\frac{\partial x^j}{\partial y^i}\frac{\partial x^l}{\partial y^k}\frac{\partial T_j}{\partial x^l}-\frac{\partial x^j}{\partial y^k}\frac{\partial x^l}{\partial y^i}\frac{\partial T_l}{\partial x^j}\\
&=\frac{\partial x^j}{\partial y^i}\frac{\partial x^l}{\partial y^k}\left(\frac{\partial T_j}{\partial x^l}-\frac{\partial T_l}{\partial x^j}\right),
\end{aligned}$$

which obeys the Law of Covariant Tensor of rank 2.

Therefore, $\frac{\partial T_i}{\partial y^k}-\frac{\partial T_k}{\partial y^i}$ is a covariant tensor of order 2.

Now, we have to show $\frac{\partial T_i}{\partial y^k}-\frac{\partial T_k}{\partial y^i}$ is a skew-symmetric tensor.

Let $C_{ik}=\frac{\partial T_k}{\partial y^i}-\frac{\partial T_k}{\partial y^i}$

$$\therefore\ C_{ki}=\frac{\partial T_k}{\partial y^i}-\frac{\partial T_k}{\partial y^i}=-\left(\frac{\partial T_k}{\partial y^i}-\frac{\partial T_k}{\partial y^i}\right)=-C_{ik}.$$

Hence, $\frac{\partial T_k}{\partial y^i}-\frac{\partial T_k}{\partial y^i}$ is a component of a skew-symmetric tensor of 2nd order.

Example 2.8.3. If a tensor T_{ijk} is symmetric in the first two indices from the left and skew-symmetric in the second and third indices from the left, show that $T_{ijk}=0$.

Solution: Here, $T_{ijk}=T_{jik}$. Since it is symmetric, with respect to is two indices from the left and *i and j*

$=-T_{jki}$, it is skew-symmetric with respect to *k*& *i*

$=-T_{kji}$, it is symmetric with respect to *j* &*k*.

$=-(-T_{kij})$, it is skew-symmetric with respect to *j*& *i*

$=T_{kij}$

$=T_{ikj}$, it is symmetric with respect to *k* & *i*

$=-T_{ijk}$, it is skew-symmetric with respect to *k* & *j*.

$\therefore\ T_{ijk}=-T_{ijk}$

or $2T_{ijk}=0$

$\Rightarrow T_{ijk}=0$

Example 2.8.4.

(a) If a_{ij} are constants, calculate $\frac{\partial}{\partial x^k}(a_{ij}x^ix^j)$.

(b) If a_{ij} *is symmetric* or skew symmetric, calculate it.

(c) Find $\frac{\partial}{\partial x^k\partial x^l}(a_{ij}x^ix^j)$ if a_{ij} *is symmetric.*

Solution: $\frac{\partial}{\partial x^k}(a_{ij}x^ix^j)=a_{ij}\frac{\partial}{\partial x^k}(x^ix^j)=a_{ij}x^i\frac{\partial}{\partial x^k}(x^j)+a_{ij}x^j\frac{\partial}{\partial x^k}(x^i)$

$=a_{ij}x^i\delta_k^j+a_{ij}x^j\delta_k^i=a_{ik}x^i+a_{kj}x^i=a_{jk}x^j+a_{kj}x^j$. (since *j* is a dummy index)

(b) If a_{jk} *is symmetric*, then $a_{jk}=a_{kj}$; $\frac{\partial}{\partial x^k}(a_{ij}x^ix^j)=2a_{jk}x^j$

If a_{jk} *is skew symmetric*, then $a_{jk} = -a_{kj}$; $\frac{\partial}{\partial x^k}(a_{ij}x^i x^j) = 0$

(c) From (b), if a_{ij} is symmetric, $\frac{\partial}{\partial x^k}(a_{ij}x^i x^j) = 2a_{jk}x^j$.

Now, $\frac{\partial}{\partial x^k \partial x^l}(a_{ij}x^i x^j) = \frac{\partial}{\partial x^l}(2a_{jk}x^j) = \frac{\partial}{\partial x^l}(2a_{lk}x^l)$ and i is a dummy index.

$$=2a_{lk}$$

Example 2.8.5. If T_{ijkl} is a tensor which satisfies the relations

$T_{ijkl} + T_{ijlk} = 0$; ....(i) $T_{ijkl} + T_{jikl} = 0$;.....(ii) and $T_{ijkl} + T_{iklj} + T_{iljk} = 0$, ... (iii)

then show that $T_{ijkl} = T_{klij}$.

Solution: Putting $i = l, l = k, k = j$ *and* $j = l$ in (iii), we get

$$T_{lijk} + T_{ljki} + T_{lkij} = 0 \quad \text{(iv)}$$

Now, putting, $i = k, j = l$, *and* $k = j$ in (iii), we have

$$T_{kijl} + T_{kjli} + T_{klij} = 0 \quad \text{(v)}$$

Lastly, putting $i = j$ *and* $j = l$ in (iii), we get

$$T_{jikl} + T_{jkli} + T_{jlik} = 0, \quad \text{(vi)}$$

Now, adding (iii), (iv), (v), and (vi), we have by (i) and (ii),

$$2T_{iklj} + 2T_{ljki} = 0$$
$$\text{or } T_{iklj} = -T_{ljki}$$
$$=T_{ljik} \quad \text{(vii)}$$

Now, putting $k = j, l = k, j = l$ in (vii), we get

$$T_{ijkl} = T_{klij}.$$

Example 2.8.6. If a_{ij} ($\neq 0$) are the components of a covariant tensor of order 2, such that $ba_{ij} + ca_{ji} = 0$, where b and c are non-zero scalars, show that either $b = c$ and a_{ij} is skew-symmetric or that $b = -c$ and a_{ij} are symmetric.
Solution: Given $ba_{ij} + ca_{ji} = 0$, we have $ba_{ij} = -ca_{ji}$ ....(i)
Multiplying both sides of (i) by b, we get

$$b^2 a_{ij} = -bca_{ji} = -c(ba_{ji})$$

$$= -c(-a_{ij})$$

$$= c^2 a_{ij}$$

$$\text{Or} \quad (b^2 - c^2)a_{ij} = 0$$

$$\Rightarrow b^2 - c^2 = 0, \text{ because } a_{ij} \neq 0$$
$$\text{or } b = \pm c.$$

Case I: When $b = c$ from (i),

$$ba_{ij} = -ca_{ji} = ca_{ij}$$
$$\Rightarrow a_{ji} = -a_{ij},$$

which implies that a_{ji} is skew symmetric tensor.

Case II: When $b = -c$ from (i),

$$ba_{ij} = -ca_{ji} = -ca_{ij}$$

$$\Rightarrow a_{ji} = a_{ij},$$

which implies that a_{ji} is a symmetric tensor.

2.9 Outer Multiplication and Contraction

2.9.1 Outer Multiplication

If A^{ij} is a contravariant tensor of order two and B_{kl} is a covariant tensor of order two, then their product is a mixed tensor C^{ij}_{kl} of order four such that

$$\bar{C}_{kl}^{ij} = \bar{A}^{ij}\bar{B}_{kl} = \left(\frac{\partial \bar{x}^i}{\partial x^p}\frac{\partial \bar{x}^j}{\partial x^q}A^{pq}\right)\left(\frac{\partial x^r}{\partial \bar{x}^k}\frac{\partial x^s}{\partial \bar{x}^l}B_{rs}\right)$$
$$= \frac{\partial \bar{x}^i}{\partial x^p}\frac{\partial \bar{x}^j}{\partial x^q}\frac{\partial x^r}{\partial \bar{x}^k}\frac{\partial x^s}{\partial \bar{x}^l}A^{pq}B_{rs}$$
$$= \frac{\partial \bar{x}^i}{\partial x^p}\frac{\partial \bar{x}^j}{\partial x^q}\frac{\partial x^r}{\partial \bar{x}^k}\frac{\partial x^s}{\partial \bar{x}^l}C_{rs}^{pq},$$

but this is the law of transformation of a mixed tensor of order four. Therefore, $\bar{C}_{kl}^{ij}$ is a mixed tensor of order four. Such products are called outer products of two tensors.

Theorem 2.9.1. If $A_{j_1 j_2 \ldots\ldots j_q}^{i_1 i_2 \ldots\ldots i_p}$ and $B_{l_1 l_2 \ldots\ldots l_s}^{k_1 k_2 \ldots\ldots k_r}$ are components of two tensors of type (p, q) and (r, s), respectively (r and s not being zero), then the quantities $A_{j_1 j_2 \ldots\ldots j_q}^{i_1 i_2 \ldots\ldots i_p}$ *and* $B_{l_1 l_2 \ldots\ldots l_s}^{k_1 k_2 \ldots\ldots k_r}$ are the components of a tensor of type $(p + r, q + s)$.

The proof can be completed by using (2.25) and proceeding exactly in the same manner as in the case of the above result.

Example 2.9.1. Show that the product of two tensors P_j^i and Q_t^{rs} are tensors of order five (tensor of type (3,2)).
Solution: By law of covariant and contravariant tensor transformation, we get

$$\bar{P}_j^l = \frac{\partial y^i}{\partial x^\beta}\frac{\partial x^\alpha}{\partial y^j}P_\alpha^\beta \text{ and } \bar{Q}_t^{rs} = \frac{\partial y^r}{\partial x^\gamma}\frac{\partial y^s}{\partial x^\delta}\frac{\partial x^\rho}{\partial y^t}Q_\rho^{\gamma\delta}$$

$$\bar{P}_j^l\bar{Q}_t^{rs} = \frac{\partial y^i}{\partial x^\beta}\frac{\partial x^\alpha}{\partial y^j}\frac{\partial y^r}{\partial x^\gamma}\frac{\partial y^s}{\partial x^\delta}\frac{\partial x^\rho}{\partial y^t}P_\alpha^\beta Q_\rho^{\gamma\delta}.$$

This is the law of transformation of tensor order five, so the product of the given tensor is of order five.

Example 2.9.2. If P^i and Q^j are two contravariant vectors, then prove that their outer product P^iQ^j is a tensor of order two, but that the converse is not true.

Solution: We have

$$\bar{P}^i = P^k \frac{\partial y^i}{\partial x^k}$$

$$\bar{Q}^j = Q^l \frac{\partial y^i}{\partial x^l}.$$

Taking on the outer product of P^i, Q^j,

$$\bar{P}^i \bar{Q}^j \frac{\partial y^i}{\partial x^k} \frac{\partial y^i}{\partial x^l} P^k Q^l. \tag{i}$$

Let $\bar{A}^{ij} = \bar{P}^i \bar{Q}^j$ and $A^{kl} = P^k Q^l$,

Hence (i) can be written as $\bar{A}^{ij} = \frac{\partial y^i}{\partial x^k} \frac{\partial y^i}{\partial x^l} A^{kl}$.

$\Rightarrow A^{kl}$ is a contravariant tensor of order 2.

Conversely, let us consider Euclidean space E_2 and A^{ij}, defined as

$$A^{ij} = 1 \; if \, i = j$$

$$= 0, \; if \, i \neq j.$$

If possible, let the outer product of C^i and D^j be $A^{ij} = C^i D^j$,

where $$C^1 D^1 = A^{11} = 1, \Rightarrow C^1 \neq 0 \; (let)$$

$$C^1 D^2 = A^{12} = 0, \Rightarrow D^2 = 0$$

$$C^2 D^2 = A^{22} = 1,$$

$$\text{since } D^2 = 0 \Rightarrow C^2 D^2 = 0, \; but \; C^2 D^2 = 1,$$

which contradicts our assumption. Therefore, the converse is not always true.

2.9.2 Contraction of a Tensor

Consider a mixed tensor $\bar{A}_l^{ijk} = \frac{\partial \bar{x}^i}{\partial x^p} \frac{\partial \bar{x}^j}{\partial x^q} \frac{\partial \bar{x}^k}{\partial x^r} \frac{\partial x^s}{\partial \bar{x}^l}$.

In this, the covariant index l is a covariant index i, so that

$$\bar{A}_l^{ijk} = \frac{\partial \bar{x}^i}{\partial x^p}\frac{\partial \bar{x}^j}{\partial x^q}\frac{\partial \bar{x}^k}{\partial x^r}\frac{\partial x^s}{\partial \bar{x}^i}A_s^{pqr} = \frac{\partial \bar{x}^j}{\partial x^q}\frac{\partial \bar{x}^k}{\partial x^r}\frac{\partial x^s}{.\partial x^p}A_s^{pqr}$$

$$= \frac{\partial \bar{x}^j}{\partial x^q}\frac{\partial \bar{x}^k}{\partial x^r}\delta_p^s A_s^{pqr} = \frac{\partial \bar{x}^j}{\partial x^q}\frac{\partial \bar{x}^k}{\partial x^r}A_p^{pqr}.$$

This shows that $\bar{A}_l^{ijk}$ is a contravariant tensor of order two.

This process of getting a tensor of lower order (reduced by 2) by putting a covariant index equal to a contravariant index and performing the summation indicated is known as *contraction.*

The type of resulting tensor is reduced by 2 and type (3,1) is reduced to (3-1,1-1)=(2,0), i.e., a contravariant of order 2.

Theorem 2.9.2. If $A_{j_1 j_2 \ldots j_q}^{i_1 i_2 \ldots i_p}$ are the components of a tensor of type (p,q), $p \neq 0$, *and* $q \neq 0$, then the quantities obtained by replacing any upper index i_p and lower index j_q by the same index i_p and performing summation over i_p are the components of a tensor of type $(p-1, q-1)$.

The proof is similar to that of the above result of the theorem.

If, in a mixed tensor of contravariant of order s and covariant of order r, we equate a covariant and a contravariant index and sum with respect to that index, then the resulting set of n^{r+s-2} sums is a mixed tensor covariant of order $r-1$ and contravariant of order $s-1$.

2.9.3 Inner Product of Two Tensors

Given the tensors A_k^{ij} and B_{qr}^p, if we first form their outer product $A_k^{ij} B_{qr}^p$ and contract this by putting $p = k$, then the result is A_k^{ij} and B_{qr}^k, which is also a tensor called the inner product of the given tensors.

Hence, the inner product of two tensors is obtained by first taking their outer product and then by contracting it.

Example 2.9.3. Show that any inner product of tensors C_j^i and D_t^{rs} is a tensor of rank three.

Solution: Using the transformation laws of tensors for C_j^i and D_t^{rs},

$$\bar{C}_j^i = \frac{\partial y^i}{\partial \mathrm{x}^k}\frac{\partial x^p}{\partial y^j}C_p^k \quad \text{(i) and } \bar{D}_t^{rs} = \frac{\partial y^r}{\partial \mathrm{x}^l}\frac{\partial y^s}{\partial \mathrm{x}^m}\frac{\partial x^n}{\partial y^t}D_n^{lm}. \quad \text{(ii)}$$

The inner product of C_r^i and D_t^{rs} is

$$\bar{C}_r^i \bar{D}_t^{rs} = \frac{\partial y^i}{\partial \mathrm{x}^{\mathrm{k}}} \frac{\partial x^p}{\partial y^r} \frac{\partial y^r}{\partial \mathrm{x}^{\mathrm{l}}} \frac{\partial y^s}{\partial \mathrm{x}^{\mathrm{m}}} \frac{\partial x^n}{\partial y^t} \mathrm{D}_n^{lm} \mathrm{C}_p^k$$

$$= \frac{\partial y^i}{\partial \mathrm{x}^{\mathrm{k}}} \frac{\partial y^s}{\partial \mathrm{x}^{\mathrm{m}}} \frac{\partial x^n}{\partial y^t} \delta_l^p \, \mathrm{D}_n^{lm} \mathrm{C}_p^k$$

$$= \frac{\partial y^i}{\partial \mathrm{x}^{\mathrm{k}}} \frac{\partial y^s}{\partial \mathrm{x}^{\mathrm{m}}} \frac{\partial x^n}{\partial y^t} \mathrm{D}_n^{lm} \mathrm{C}_l^k.$$

Hence, the inner product of tensors C_r^i and D_t^{rs} is a tensor of rank 3.

Similarly, putting $i = t$ in the product of (i) and (ii) we get that $\bar{C}_r^i \bar{D}_i^{rs}$ is found to be a tensor of rank 3.

Similarly, in other cases in the same process, we get the tensor of rank 3 by product of these two tensors.

Example 2.9.4. If P^i and Q^j are the the components of two contravariant tensors of rank one respectively, then the n^2 quantites P^iQ^j are the components of a contravariant tensor of order two.

Solution: P^i and Q^j are the components of a contravariant tensor of rank one, respectively, by law of transformation

$$\bar{P}^i = \frac{\partial y^i}{\partial x^\alpha} P^\alpha \text{ and } \bar{Q}^j = \frac{\partial y^j}{\partial x^\beta} Q^\beta.$$

Multiplying these, we get

$$\bar{P}^i \bar{Q}^j = \frac{\partial y^i}{\partial x^\alpha} P^\alpha \frac{\partial y^j}{\partial x^\beta} Q^\beta = \frac{\partial y^i}{\partial x^\alpha} \frac{\partial y^j}{\partial x^\beta} P^\alpha Q^\beta = \frac{\partial y^i}{\partial x^\alpha} \frac{\partial y^j}{\partial x^\beta} R^{\alpha\beta},$$

writing $P^\alpha Q^\beta = R^{\alpha\beta}$ and $\bar{P}^i \bar{Q}^j = \bar{R}^{\alpha\beta}$

$$\therefore \bar{R}^{\alpha\beta} = \frac{\partial y^i}{\partial x^\alpha} \frac{\partial y^j}{\partial x^\beta} R^{\alpha\beta}$$

Hence, in virtue of (2.19) it follows that $P^\alpha Q^\beta$ *or* $R^{\alpha\beta}$ are the components of a contravariant tensor of second order.

2.10 Quotient Law of Tensors

We have seen that the product of two tensors is a tensor. Let us assume that the product of two quantities is a tensor and if one of them is a tensor, is the other quantity a tensor?

A simple test is provided by the Quotient Law, which states that if the inner product of a set of functions with an arbitrary tensor is a tensor, then these sets of functions are the components of a tensor.

Quotient Law plays an important role in Tensor Calculus and its application. The name quotient law is, in a certain sense, appropriate because the application of this law produces a tensor from two tensors, just as the operation of the division of two numbers produces a number, namely their quotient.

Quotient Law for Tensors:

i) For Tensors of type (0,1), if, relative to every system of coordinates there is a set of functions $B_i (i = 1,2, \dots n)$, such that $A^i B_i$ is an invariant for any tensor A^i of type (1,0), then B_i will form the components of a tensor of type (0,1).

ii) For tensors of type (1,0), if, relative to every system of coordinates there is a set of functions $A^i (i = 1,2, \dots n)$, such that $A^i B_i$ is an invariant for any tensor B_i of type (0,1), then A^i will form the components of a tensor of type (1,0).

iii) For tensors of type (0,2), if, relative to every system of coordinates there is a set of functions a_{ik} $(i, k = 1,2, \dots n)$, such that $a_{ik} A^i B^k$ is an invariant for any contravariant tensors A^i and B^k, then a_{ik} will form the components of a tensor of type (0,2).

iv) For tensors of type (0,2), if relative to every system of coordinates there is a set of functions a_{ik} $(i, k = 1,2, \dots n)$, such that $a^{ik} A_i B_k$ is an invariant for any covariant tensors A_i and B_k, then a^{ik} will form the components of a tensor of type (2,0).

v) Quotient Law in General Form: Let $A^{i_1 i_2 \dots\dots i_p}_{j_1 j_2 \dots\dots j_q}$ be n^{p+q} quantities in a certain reference frame. If $A^1_{i_1}, A^2_{i_2} \dots . A^r_{i_r} B^{j_1}_1 B^{j_2}_2 \dots\dots B^{j_s}_s$ are the components of arbitrary covariant and contravariant vectors, respectively, such that

$$A^{i_1 i_2 \dots\dots i_r i_{r+1} \dots i_p}_{j_1 j_2 \dots\dots j_s j_{s+1} \, j_q} A^1_{i_1} A^2_{i_2} \dots . A^r_{i_r} B^{j_1}_1 B^{j_2}_2 \dots\dots B^{j_s}_s \quad (r \le p, s \le q)$$

are components of a tensor of type $(p - r, q - s)$ and $A^{i_1 i_2 \cdots i_p}_{j_1 j_2 \cdots j_q}$ will form the components of a tensor of type (p, q).

Example 2.10.1. Show that the expression $P(i, j, k)$ is a covariant tensor of rank three if $P(i, j, k)Q^k$ is a covariant tensor of rank two and Q^k is contravariant vector.

Solution: Let $P(i, j, k)$ be in the coordinate system (x^i) and it transformed into coordinate system (y^i).

Given that $P(i, j, k)Q^k$ is a covariant tensor of rank two, then

$$\bar{P}(\mathrm{i},\mathrm{j},\mathrm{k})\bar{Q}^k = \frac{\partial x^\alpha}{\partial y^i}\frac{\partial x^\beta}{\partial y^j}P(\alpha,\beta,\gamma)Q^\gamma$$

Since Q^k is a contravariant vector,

$$\text{then } \bar{Q}^k = \frac{\partial y^k}{\partial x^\gamma}Q^\gamma$$

$$\text{or } Q^\gamma - \frac{\partial_x{}^\gamma}{\partial y^k}\bar{Q}^k.$$

Substitute it in the above expression and

$$\bar{P}(\mathrm{i},\mathrm{j},\mathrm{k})\bar{Q}^k = \frac{\partial x^\alpha}{\partial y^i}\frac{\partial x^\beta}{\partial y^j}P(\alpha,\beta,\gamma)\frac{\partial x^\gamma}{\partial y^k}\bar{Q}^k = \frac{\partial x^\alpha}{\partial y^i}\frac{\partial x^\beta}{\partial y^j}\frac{\partial x^\gamma}{\partial y^k}P(\alpha,\beta,\gamma)\bar{Q}^k.$$

$$\bar{P}(\mathrm{i},\mathrm{j},\mathrm{k}) = \frac{\partial x^\alpha}{\partial y^i}\frac{\partial x^\beta}{\partial y^j}\frac{\partial x^\gamma}{\partial y^k}P(\alpha,\beta,\gamma) \text{ since } \bar{Q}^k \text{ is arbitrary,}$$

hence $P(i, j, k)$ is a covariant tensor of rank three.

Example 2.10.2. Show that the expression $P(i, j, k)$ is a covariant tensor if its inner product with an arbitrary tensor Q^{jl}_k is a tensor.

Solution: Let $$P(i,j,k)Q^{jl}_k = T^l_i \qquad \text{(i)}$$

If $P(i, j, k)$ is a (0,3) tensor, we consider by contraction and inner product, the new tensor to be a (1,1) type.

In the coordinate system (y^i), (i) is transformed to become:

$$\bar{P}(p,q,r)\bar{Q}_r^{qs} = \bar{T}_p^s \tag{ii}$$

By Law of Tensor Transformation, we get from (i) and (ii): (expressing $\bar{Q}_r^{qs}$ and $\bar{T}_p^s$ with the Law of Tensor Transformation)

$$\bar{P}(p,q,r)\frac{\partial y^q}{\partial x^j}\frac{\partial y^s}{\partial x^l}\frac{\partial x^k}{\partial y^r}Q_k^{jl} = \frac{\partial y^s}{\partial x^i}\frac{\partial x^l}{\partial y^p}T_i^l = \tag{iii}$$

Multiplying (i) by $\frac{\partial y^s}{\partial x^l}\frac{\partial x^i}{\partial y^p}$ and subtracting from (iii), we get

$$\left\{\bar{P}(p,q,r)\frac{\partial y^q}{\partial x^j}\frac{\partial y^s}{\partial x^l}\frac{\partial x^k}{\partial y^r} - P(i,j,k)\frac{\partial y^s}{\partial x^l}\frac{\partial x^i}{\partial y^p}\right\}Q_k^{jl} = 0.$$

Since Q_k^{jl} is an arbitrary tensor, the expression within the bracket must be zero,

implying that $\bar{P}(p,q,r)\frac{\partial y^q}{\partial x^j}\frac{\partial y^s}{\partial x^l}\frac{\partial x^k}{\partial y^r} = P(i,j,k)\frac{\partial y^s}{\partial x^l}\frac{\partial x^i}{\partial y^p}$

$$\bar{P}(p,q,r) = \frac{\partial x^i}{\partial y^p}\frac{\partial x^j}{\partial y^q}\frac{\partial y^r}{\partial x^k}P(i,j,k).$$

This is the Law of Tensor of Order 3 Transformation, hence *P*(*i*, *j*, *k*) is a tensor of order 3 (*i*, *j* as covariant indices and *k* as a contravariant index.).

2.11 Reciprocal Tensor of a Tensor

Let a_{ij} be a symmetric tensor of type (0,2), satisfying the condition $|a_{ij}| \neq 0$.

We denote by b^{ij} the cofactor of a_{ij} in $|a_{ij}|$, divided by $|a_{ij}|$,

i.e., $b^{ij} = \frac{cofactor\ of\ a_{ij}\ in\,|a_{ij}|}{|a_{ij}|}$ (we know $a_{ij}A_{ij} = |a_{ij}| = a$).

It is known from the Theory of Determinants that

$$a_{ij}b^{ik} = 1,\ when\ k = j$$

$$= 0,\ when\ k \neq j,$$

hence $a_{ij}b^{ik} = \delta_j^k$.

Definition 2.11.1. If b^{ij} is the reciprocal tensor of a_{ij}, then a_{ij} is the *reciprocal* (or *conjugate*) *tensor* of b^{ij}. Tensors a_{ij} and b^{ij} of type (0,2) and (2,0), respectively, are called mutually reciprocal tensors or mutually conjugate tensors if $a_{ij}b^{ki} = \delta_i^k$.

If b^{ij} is the reciprocal tensor of tensor a_{ij}, then a_{ij} is the reciprocal tensor of tensor b^{ij}.

Since $a_{ij}b^{ki} = \delta_i^k$, it follows that

$$\therefore\ |a_{ij}||b^{ki}| = |\delta_j^k| \neq 0.$$

We define another tensor G_{ij} so that

$$G_{ij} = \frac{cofactor\ of\ b^{ij}\ in\,|b^{ij}|}{|b^{ij}|}.$$

We know from the property of determinants that $G_{ij}b^{ik} = \delta_j^k$.
Multiplying by a_{pk}, we get $a_{pk}G_{ij}b^{ik} = \delta_j^k a_{pk}$

$$\text{or}\quad G_{ij}\delta_p^i = a_{pj}$$

$$G_{pj} = a_{pj} \tag{2.27}$$

Since G_{ij} is the reciprocal tensor of b^{ij} and from (2.27) we have to write a_{ij} as the reciprocal tensor of b^{ij}, if b^{ij} is the reciprocal tensor of a_{ij}, then a_{ij} is the reciprocal tensor of b^{ij}.

If the relation $a_{ij}b^{ik} = \delta_j^k$ is satisfied, then a_{ij} and b^{ij} are reciprocal tensors to each other.

Example 2.11.1. If a_{ij} and a^{ij} are reciprocal symmetric tensors of order 2, show that

$$a^{ij}\frac{\partial a_{ij}}{\partial x} + a_{ij}\frac{\partial a^{ij}}{\partial x} = 0.$$

Hence, show that $\dfrac{\partial(\log a)}{\partial x} = -a_{ij}\dfrac{\partial a^{ij}}{\partial x}$.

Solution: We have $a^{ij}a_{ij} = n$.

Differentiating with respect to x^k, we get

$$a^{ij}\frac{\partial a_{ij}}{\partial x}+a_{ij}\frac{\partial a^{ij}}{\partial x}=0.$$

From above we get $a^{ij}\dfrac{\partial a_{ij}}{\partial x}=-a_{ij}\dfrac{\partial a^{ij}}{\partial x}$. (i)

If $a = |a_{ij}|$ and by definition of reciprocal of a, we get

$$\left(\frac{1}{a}\right)\frac{\partial a}{\partial x^k}=\frac{cofactor\ of\ a_{ij}\ of\ |a_{ij}|}{a}\left(\frac{\partial a_{ij}}{\partial x^k}\right)$$

$$=a^{ij}\frac{\partial a_{ij}}{\partial x^k},\text{ by definition of } a^{ij}$$

$$\text{or }\frac{\partial(\log a)}{\partial x}=-a_{ij}\frac{\partial a^{ij}}{\partial x}.$$

2.12 Relative Tensor, Cartesian Tensor, Affine Tensor, and Isotropic Tensors

2.12.1 Relative Tensors

Definition 2.12.1. A set of n^{p+q} components of $C^{i_1 i_2 \ldots i_p}_{j_1 j_2 \ldots\ldots j_q}$, in a coordinate system (x^i), transform according to the following formula when referring to another coordinate system (y^i):

$$\bar{C}^{i_1 i_2 \ldots i_p}_{j_1 j_2 \ldots\ldots j_q}=J^{\omega}\frac{\partial y^{i_1}}{\partial x^{k_1}}\frac{\partial y^{i_2}}{\partial x^{k_2}}\cdots\frac{\partial y^{i_p}}{\partial x^{k_p}}\cdot\frac{\partial x^{l_1}}{\partial y^{j_1}}\frac{\partial x^{l_2}}{\partial y^{j_2}}\cdots\frac{\partial x^{l_q}}{\partial y^{j_2}}C^{k_1 k_2 \ldots k_p}_{l_1 l_2 \ldots\ldots l_q} \tag{2.28}$$

is called a *relative tensor* of order p+q and weight ω, where J is the Jacobian of the transformation: $J=\left|\dfrac{\partial x}{\partial y}\right|$.

If $\omega = 1$, the relative tensor is called a *tensor density.* If $\omega = 0$, then a tensor is said to be *absolute.* In the case of a relative on two sides, the equations must be of the same weight.

A relative tensor of order zero is called a relative scalar. A relative scalar of weight one is called a scalar density and a relative scalar of weight zero is called an absolute scalar.

Some operations of relative tensors:

(a) Relative tensors of the same type and weight may be added and the sum is a relative tensor of the same type and weight.
(b) Relative tensors may be multiplied by the weight of the product, being the sum of the weights of tensors entering in the product.
(c) The operation of contraction on a relative tensor yields a relative tensor of the same weight as the original tensor.

We distinguish the mixed tensor from relative tensors and the term *absolute tensor* is frequently used to designate the mixed tensor.

Example 2.12.1. Prove that the scalar product of a relative covariant vector of weight ω_1 and a contavariant vector of weight ω_2 is a relative scalar vector of weight $\omega_1 + \omega_2$.
Solution: Let A^i be the components of relative contravariant vector of weight ω_1.

Then, $\bar{A}^i = A^r \dfrac{\partial y^i}{\partial x^r}\left|\dfrac{\partial x}{\partial y}\right|^{\omega_1} = A^r \dfrac{\partial y^i}{\partial x^r} j^{\omega_1}$.

When B_i are the components of a relative covariant vector of weight ω_2, then

$$\bar{B}_i = B_s \frac{\partial x^s}{\partial y^i}\left|\frac{\partial x}{\partial y}\right|^{\omega_2} = B_s \frac{\partial x^s}{\partial y^i} j^{\omega_2}.$$

Now,

$$\bar{A}^i\bar{B}_i = A^r B_s \frac{\partial y^i}{\partial x^r}\frac{\partial x^s}{\partial y^i} j^{\omega_1} j^{\omega_2} = A^r B_s \frac{\partial x^s}{\partial x^r} j^{\omega_1+\omega_2} = A^r B_s \delta_r^s j^{\omega_1+\omega_2} = A^r B^r j^{\omega_1+\omega_2}.$$

It follows that the product is a relative vector of weight $\omega_1 + \omega_2$.

Example 2.12.2. If A^{ij} and A_{ij} are components of relative tensors of weight w, show that

$$|\bar{A}^{ij}| = |A^{ij}|\left|\frac{\partial x}{\partial y}\right|^{w-2}$$

$$|\bar{A}_{ij}| = |A_{ij}|\left|\frac{\partial x}{\partial y}\right|^{w+2}.$$

Solution: By the definition of a relative tensor, we have

$$\bar{A}^{ij}=A^{pq}\frac{\partial y^i}{\partial x^p}\frac{\partial y^j}{\partial x}\left|\frac{\partial x}{\partial y}\right|^w \tag{i}$$

$$\bar{A}_{ij}=A_{pq}\frac{\partial x^p}{\partial y^i}\frac{\partial x^q}{\partial y^j}\left|\frac{\partial x}{\partial y}\right|^w. \tag{ii}$$

From (i) and (ii), $\left|\frac{\partial y^i}{\partial x^p}\right|=\left|\frac{\partial y^j}{\partial xq}\right|=\left|\frac{\partial y}{\partial x}\right|$, and taking determinants on both sides, we get

$$|\bar{A}^{ij}|=|A^{pq}|\left|\frac{\partial y}{\partial x}\right|\left|\frac{\partial y}{\partial x}\right|\left|\frac{\partial x}{\partial y}\right|^w$$

$$|\bar{A}_{ij}|=|A_{pq}|\left|\frac{\partial x}{\partial y}\right|\left|\frac{\partial x}{\partial y}\right|\left|\frac{\partial x}{\partial y}\right|^w,$$

but $$\left|\frac{\partial y}{\partial x}\right|=\left|\frac{\partial x}{\partial y}\right|^{-1} \text{ and } |A_{pq}|=|\bar{A}_{ij}|.$$

Hence, $$|\bar{A}^{ij}|=|A^{ij}|\left|\frac{\partial x}{\partial y}\right|^{w-2} \text{ and}$$

$$|\bar{A}_{ij}|=|A_{ij}|\left|\frac{\partial x}{\partial y}\right|^{w+2}.$$

Example 2.12.3. If a_{ij} is a covariant tensor of order 2 and $|a_{ij}| = a$, then show that $\sqrt{a}$ is a relative tensor of order 0 and weight 1.
Solution: By Transformation Law, we get

$$\bar{a}_{ij}=\frac{\partial x^k}{\partial y^i}\frac{\partial x^p}{\partial y^j}a_{kp}.$$

Taking determinants on both sides of it, we get

$$|\bar{a}_{ij}| = \left|\frac{\partial x^k}{\partial y^i}\right|\left|\frac{\partial x^p}{\partial y^j}\right||a_{kp}|$$

or

$$\bar{a} = a\left|\frac{\partial x}{\partial y}\right|^2$$

$$\therefore \sqrt{\bar{a}} = \sqrt{a}\left|\frac{\partial x}{\partial y}\right|, \qquad \text{(taking the square root on both sides)}$$

$\Rightarrow \sqrt{a}$ is a relative tensor of order 0 and weight 1.

2.12.2 Cartesian Tensors

Definition 2.12.2. A tensor of Euclidian Space E^n, obtained by orthogonal transformation of coordinate axes, is called a *Cartesian Tensor.* Thus, a Cartesian Tensor of rank p in a three-dimensional Euclidean space is a set of 3^p components which transform by the rule

$$\bar{P}^{i_1 i_2 \ldots i_p}_{j_1 j_2 \ldots\ldots j_p} = \frac{\partial y^{i_1}}{\partial x^{k_1}}\frac{\partial y^{i_2}}{\partial x^{k_2}}\cdots\frac{\partial y^{i_p}}{\partial x^{k_p}}\cdot\frac{\partial x^{l_1}}{\partial y^{j_1}}\frac{\partial x^{l_2}}{\partial y^{j_2}}\cdots\cdot\frac{\partial x^{l_p}}{\partial y^{j_p}}P^{k_1 k_2 \ldots k_p}_{l_1 l_2 \ldots\ldots l_p} \qquad (2.29)$$

Under orthogonal transformations,
$T : y^i = a^i_j x^j$ and (a^i_j) is orthogonal so that $a^i_j \neq 0$.

2.12.3 Affine Tensor

Definition 2.12.3. Tensors corresponding to admissible coordinate changes by transformation

$$T : y^i = a^i_j x^j, \qquad (2.30)$$

where $|a^i_j| \neq 0$. A transformation that takes rectangular coordinates (x^i) to an oblique axes system (y^i) is called an *affine tensor.*

Thus, affine tensors are defined in the class of all such oblique coordinate systems.

The Jacobian matrices of T and T^{-1} *are* $J = \left[\dfrac{\partial y^i}{\partial x^j}\right]_{rr} = \left[a_j^i\right]_{rr}$ *and* $J = \left[\dfrac{\partial x^i}{\partial y^j}\right]_{rr} = \left[b_j^i\right]_{rr}$.

The laws for affine tensors are:

$$\begin{aligned} \text{Contravariant tensor:} \quad & \bar{A}^i = a_l^i A^l, \quad \bar{A}^{ij} = a_l^i a_m^j A^{lm} \\ \text{Covariant tensor:} \quad & \bar{A}_i = b_i^l A_l, \quad \bar{A}_{ij} = b_i^l b_j^m A_{lm} \\ \text{Mixed tensor:} \quad & \bar{A}_j^i = a_l^i a_j^m A_m^l, \quad \bar{A}_{jk}^i = a_l^i a_j^m a_k^n A_m^l \end{aligned} \tag{2.31}$$

2.12.4 Isotropic Tensor

Definition 2.12.4. A Cartesian Tensor whose components remain unchanged under rotation of axes is called an *isotropic tensor.*

Let $\omega = (\omega_1, \omega_2, \omega_3)$ be a vector and $P = (b_{ij})$ an arbitrary orthogonal transformation.

Let $\bar{\omega} = (\bar{\omega}_1, \bar{\omega}_2, \bar{\omega}_3)$ be the transformed vector under P.

Then, $\bar{\omega} = P\,\omega$.

Since ω is an isotopic tensor, $\bar{\omega} = \omega$.

If we substitute this in the above expression, we get $P\omega = \omega$.

We get
$$P\omega - \omega = 0 \tag{2.32}$$

$$\Rightarrow (P - 1)\,\omega = 0, \qquad \text{where 0 is a null vector.}$$

It is clear from Equation (2.32) that it has only one solution: $\omega = 0$. Thus, there is no isotropic tensor of rank one except the null vector.

2.12.5 Pseudo-Tensor

Definition 2.12.5. Pseudo-tensors are usually discussed in terms of mechanics.

In V_4 we consider two vectors: $\begin{cases} a\,b\,c\,d \\ p\,q\,r\,s \end{cases}$. From these we can form 6 determinants as:

$$\begin{vmatrix} a & b \\ p & q \end{vmatrix} = aq - bp$$

$$\text{or } c_{ij} = a_i b_j - a_j b_i.$$

It is shown in V_4 that there are six independent components (determinants) of tensors c_{ij}. If we transform it to any coordinate system, consider the law of transformation of this set of 6 components to the corresponding set of components in the new system of coordinates. These six components form a *pseudo-tensor.*

2.13 Examples

Example 2.13.1. A Covariant Vector has components $2x$, $y^2 - z$, *and* z^2 in rectangular coordinates. Determine its covariant components in cylindrical coordinates.

Solution: Let the rectangular coordinates be (x^i) as $x^1 = x$, $x^2 = y$, *and* $x^3 = z$.

In (x^i) the covariant vector components are $A_1 = 2x = 2x^1$, $A_2 = y^2 - z = (x^2)^2 - x^3$, *and* $A_3 = z^2 = (x^3)^2$.

We know the cylindrical coordinates related with rectangular coordinates as

$x^1 = y^1 cos y^2$, $x^2 = y^1 sin y^2$, *and* $x^3 = y^3$, where $y^1 = r$, $y^2 = \theta$, and $y^3 = z$ (let) $[x^1 = r cos\theta$, $x^2 = r sin\theta$, *and* $x^3 = z]$

Now apply the Law of Covariant Transformation: $\bar{A}_i = \dfrac{\partial x^j}{\partial y^i} A_j, i = 1,2,3$

$$\text{For } i = 1, \bar{A}_i = \frac{\partial x^j}{\partial y^1} A_j = \frac{\partial x^1}{\partial y^1} A_1 + \frac{\partial x^2}{\partial y^1} + A_2 + \frac{\partial x^3}{\partial y^1} A_3$$

$$= 2x^1 cosy^2 + \{(x^2)^2 - x^3\} siny^2 + 0 = 2rcos^2\theta + sin\theta\{r^2 sin^2\theta - z\}$$

$$\text{For } i = 2, \bar{A}_2 = \frac{\partial x^j}{\partial y^2} A_j = \frac{\partial x^1}{\partial y^2} A_1 + \frac{\partial x^2}{\partial y^2} A_2 + \frac{\partial x^3}{\partial y^2} A_3$$

$$= -y^1 siny^2 (2rcos\theta) + y^1 cosy^2 \{(rsin\theta)2 - z\} + 0$$

$$= -rsin\,\theta(2rcos\theta) + rcos\theta(r^2 sin^2\theta - z)$$

$$\text{For } i = 3, \bar{A}_3 = \frac{\partial x^j}{\partial y^3} A_j = \frac{\partial x^1}{\partial y^3} A_1 + \frac{\partial x^2}{\partial y^3} A_2 + \frac{\partial x^3}{\partial y^3} A_3 = 0 + 0 + 1.z^2 = z^2.$$

Example 2.13.2. If a_{ij} is a skew-symmetric tensor, prove that $(\delta_j^i\delta_l^k + \delta_l^i\delta_j^k)a_{ik} = 0$.

Solution: Here, a_{ij} is a skew symmetric tensor, so $a_{ij} = -a_{ji}$.

$$\begin{aligned}\text{Now, } (\delta_j^i\delta_l^k + \delta_l^i\delta_j^k)a_{ik} &= \delta_j^i\delta_l^k a_{ik} + \delta_l^i\delta_j^k a_{ik}\\ &= \delta_j^i a_{il} + \delta_l^i a_{ij}\\ &= a_{jl} + a_{lj} = a_{jl} + (-a_{jl})\\ &= 0 \text{ (since } a_{ij} \text{ is a skew-symmetric tensor)}\end{aligned}$$

Example 2.13.3. If B_{ij} are components of a Covariant Tensor of second order and C^i and D^j are components of two contravariant vectors, show that $B_{ij}C^iD^j$ is an invariant.

Solution: Since B_{ij} are components of a Covariant Tensor of second order and C^i *and* D^j are components of two contravariant vectors:

We have $\bar{B}_{ij} = \dfrac{\partial x^k}{\partial y^i}\dfrac{\partial x^p}{\partial y^j}B_{kp}$ and $\bar{C}^i = \dfrac{\partial y^i}{\partial x^r}C^r, \bar{D}^j = \dfrac{\partial y^j}{\partial y^s}D^s$.

Now,

$$\begin{aligned}\bar{B}_{ij}\bar{C}^i\bar{D}^j &= \frac{\partial x^k}{\partial y^i}\frac{\partial x^p}{\partial y^j}B_{kp}\frac{\partial y^i}{\partial x^r}C^r\frac{\partial y^j}{\partial y^s}D^s\\ &= \frac{\partial x^k}{\partial y^i}\frac{\partial x^p}{\partial y^j}\frac{\partial y^i}{\partial x^r}\frac{\partial y^j}{\partial y^s}B_{kp}C^rD^s\\ &= \frac{\partial x^k}{\partial y^i}\frac{\partial y^i}{\partial x^r}\frac{\partial x^p}{\partial y^j}\frac{\partial y^j}{\partial y^s}B_{kp}C^rD^s\\ &= \delta_r^k\delta_s^p\, B_{kp}C^rD^s\\ &= B_{rp}C^rD^p = B_{ij}C^iD^j.\end{aligned}$$

It follows that $B_{ij}C^iD^j$ is an invariant.

Example 2.13.4. Show that the determinant of a tensor of type (1,1) is an invariant.

Solution: Let A_j^i be a tensor of type (1,1).

Then, $$\bar{A}^i_j = \frac{\partial y^i}{\partial \mathrm{x}^k}\frac{\partial x^p}{\partial y^j}A^k_p.$$

taking matrix on both sides, we get

Hence, $$\left(\bar{A}^i_j\right) = \left(\frac{\partial y^i}{\partial \mathrm{x}^k}\right)\left(A^k_p\right)\left(\frac{\partial x^p}{\partial y^j}\right). \qquad \text{(i)}$$

Taking determinants on both sides of i,

$$\left|\bar{A}^i_j\right| = \left|\frac{\partial y^i}{\partial \mathrm{x}^k}\right|\left|\frac{\partial x^p}{\partial y^j}\right|\left|A^k_p\right|$$

$$= \left|A^k_p\right|, \text{ since } .j\,.\,J' = 1, \text{ where } j = \left|\frac{\partial y^i}{\partial \mathrm{x}^k}\right| \; and\ J'$$

$$= \left|A^i_j\right|$$

$\Rightarrow \left|A^i_j\right|$ is an invariant.

Example 2.13.5. If $A^i_{jk}B^{jk} = C^i$, where C^i is a contravariant vector and B^{jk} is an arbitrary symmetric tensor, show that $A^i_{jk} + A^i_{jk}$ is a tensor.

Solution: Let $$\bar{A}^i_{jk}\bar{B}^{jk} = \bar{C}^i \qquad \text{(i)}$$

Now, $$\bar{B}^{jk} = \frac{\partial y^j}{\partial x^{p\cdot}}\frac{\partial y^k}{\partial x^{q\cdot\cdot}}B^{pq}$$

and $\bar{C}^i = \frac{\partial y^i}{\partial x^{s\cdot\cdot}}C^s$

and (i) can be written as

$$\bar{A}^i_{jk}\frac{\partial y^j}{\partial x^{p\cdot}}\frac{\partial y^k}{\partial x^{q\cdot\cdot}}B^{pq} = \frac{\partial y^i}{\partial x^{s\cdot\cdot}}C^s = \frac{\partial y^i}{\partial x^{s\cdot\cdot}}A^s_{pq}B^{pq} \;\; (\text{since } A^i_{jk}B^{jk} = C^i \; is\ given)$$

or $$\left(\bar{A}^i_{jk}\frac{\partial y^j}{\partial x^{p\cdot}}\frac{\partial y^k}{\partial x^{q\cdot\cdot}} - \frac{\partial y^i}{\partial x^{s\cdot\cdot}}A^s_{pq}\right)B^{pq} = 0 \qquad \text{(ii)}$$

Since B^{jk} is an arbitrary symmetric tensor, let one of the components of the tensor be non-zero, say $B^{mn} \neq 0$ and the rest be zero. Here, $B^{mn} = B^{nm}$ *also.*

From (ii) we get

$$\left(\bar{A}^i_{jk}\frac{\partial y^j}{\partial x^{m\cdot}}\frac{\partial y^k}{\partial x^{n\cdot\cdot}} - \frac{\partial y^i}{\partial x^{s\cdot\cdot}}A^s_{mn}\right)B^{mn} + \left(\bar{A}^i_{jk}\frac{\partial y^j}{\partial x^{n\cdot}}\frac{\partial y^k}{\partial x^{m\cdot\cdot}} - \frac{\partial y^i}{\partial x^{s\cdot\cdot}}A^s_{nm}\right)B^{nm} = 0$$

or $$\left(\bar{A}^i_{jk}\frac{\partial y^j}{\partial x^{m\cdot}}\frac{\partial y^k}{\partial x^{n\cdot\cdot}} + \bar{A}^i_{jk}\frac{\partial y^j}{\partial x^{n\cdot\cdot}}\frac{\partial y^k}{\partial x^{m\cdot\cdot}} - \frac{\partial y^i}{\partial x^{s\cdot\cdot}}A^s_{mn} - \frac{\partial y^i}{\partial x^{s\cdot\cdot}}A^s_{mn}\right)B^{mn} = 0 \tag{iii}$$

Since $B^{mn} \neq 0$, from (iii) we get

$$\bar{A}^i_{jk}\frac{\partial y^j}{\partial x^{m\cdot}}\frac{\partial y^k}{\partial x^{n\cdot\cdot}} + \bar{A}^i_{jk}\frac{\partial y^j}{\partial x^{n\cdot}}\frac{\partial y^k}{\partial x^{m\cdot\cdot}} = \frac{\partial y^i}{\partial x^{s\cdot\cdot}}A^s_{mn} + \frac{\partial y^i}{\partial x^{s\cdot\cdot}}A^s_{nm}.$$

Replacing the indices *j and k by k* and *j*, respectively, from the 2nd term of left hand side, we get

$$\bar{A}^i_{jk}\frac{\partial y^j}{\partial x^{m\cdot}}\frac{\partial y^k}{\partial x^{n\cdot\cdot}} + \bar{A}^i_{kj}\frac{\partial y^j}{\partial x^{m\cdot}}\frac{\partial y^k}{\partial x^{n\cdot\cdot}} = \frac{\partial y^i}{\partial x^{s\cdot\cdot}}A^s_{mn} + \frac{\partial y^i}{\partial x^{s\cdot\cdot}}A^s_{nm}$$

$$(\bar{A}^i_{jk} + \bar{A}^i_{kj})\frac{\partial y^j}{\partial x^{m\cdot}}\frac{\partial y^k}{\partial x^{n\cdot\cdot}} = \frac{\partial y^i}{\partial x^{s\cdot\cdot}}(A^s_{mn} + A^s_{nm}) \tag{iv}$$

Multiplying both sides of (iv) by $\frac{\partial x^m}{\partial y^{p\cdot}}\frac{\partial x^n}{\partial y^{q\cdot\cdot}}$, we get

$$(\bar{A}^i_{jk} + \bar{A}^i_{kj})\frac{\partial y^j}{\partial x^{m\cdot}}\frac{\partial y^k}{\partial x^{n\cdot\cdot}}\frac{\partial x^m}{\partial y^{p\cdot}}\frac{\partial x^n}{\partial y^{q\cdot\cdot}} = \frac{\partial x^m}{\partial y^{p\cdot}}\frac{\partial x^n}{\partial y^{q\cdot}}\frac{\partial y^i}{\partial x^{s\cdot\cdot}}(A^s_{mn} + A^s_{nm})$$

or $$(\bar{A}^i_{jk} + \bar{A}^i_{kj})\delta^j_p\delta^k_q = \frac{\partial x^m}{\partial y^{p\cdot}}\frac{\partial x^n}{\partial y^{q\cdot\cdot}}\frac{\partial y^i}{\partial x^{s\cdot\cdot}}(A^s_{mn} + A^s_{nm})$$

or $$\bar{A}^i_{pq} + \bar{A}^i_{qp} = \frac{\partial x^m}{\partial y^{p\cdot}}\frac{\partial x^n}{\partial y^{q\cdot\cdot}}\frac{\partial y^i}{\partial x^{s\cdot\cdot}}(A^s_{mn} + A^s_{nm}) \tag{v}$$

From (v), it follows that $A^i_{jk} + A^i_{kj}$ is a tensor of type (1,2).

Example 2.13.6. If $a_{ij}u^i u^j$ is an invariant for an arbitrary covariant vector u^i, show that $a_{ij} + a_{ji}$ is a tensor.
Solution: Since $a^{ij}u^i u^j$ is an invariant, we have

$$\bar{a}_{ij}\bar{u}^i\bar{u}^j = a_{ij}u^i u^j \text{, (by the property of invariant)}$$

$$= a_{ij}\frac{\partial x^i}{\partial y^k}\bar{u}^k\frac{\partial x^j}{\partial y^l}\bar{u}^l$$ (replacing dummy indices *i, j, k, l by k, l, i, j*, respectively)

$$= a_{kl}\frac{\partial x^k}{\partial y^i}\bar{u}^i\frac{\partial x^l}{\partial y^j}\bar{u}^j$$

$$= a_{kl}\frac{\partial x^k}{\partial y^i}\frac{\partial x^l}{\partial y^j}\bar{u}^i\bar{u}^j$$

$$\left(\bar{a}_{ij} - a_{kl}\frac{\partial x^k}{\partial y^i}\frac{\partial x^l}{\partial y^j}\right)u^i u^j = 0$$

$$\text{or } \bar{B}_{ij}\bar{u}^i\,\bar{u}^j = 0 \tag{i}$$

$$\text{where } \bar{B}_{ij} = \left(\bar{a}_{ij} - a_{kl}\frac{\partial x^k}{\partial y^i}\frac{\partial x^l}{\partial y^j}\right). \tag{ii}$$

Since $\bar{u}^i{}'s$ are arbitrary, then we can write $\bar{B}_{ij} + \bar{B}_{ji} = 0$ (iii)
In virtue of (ii), (iii) can be expressed as

$$\bar{a}_{ij} - a_{kl}\frac{\partial x^k}{\partial y^i}\frac{\partial x^l}{\partial y^j} + \bar{a}_{ji} - a_{kl}\frac{\partial x^k}{\partial y^j}\frac{\partial xl}{\partial y^i} = 0$$

or $\bar{a}_{ij} + \bar{a}_{ji} = a_{kl}\frac{\partial x^k}{\partial y^i}\frac{\partial x^l}{\partial y^j} + a_{kl}\frac{\partial x^k}{\partial y^j}\frac{\partial x^l}{\partial y^i}$ (changing indices *k and l* in the 2nd term)

$$= a_{kl}\frac{\partial x^k}{\partial y^i}\frac{\partial x^l}{\partial y^j} + a_{lk}\frac{\partial x^k}{\partial y^i}\frac{\partial x^l}{\partial y^j}$$

$$= (a_{kl} + a_{lk})\frac{\partial x^k}{\partial y^i}\frac{\partial x^l}{\partial y^j}$$

It follows that $a_{ij} + a_{ji}$ is a covariant tensor of order 2.

Example 2.13.7. If the relations $A_{ijk}\, u^i u^j u^k = 0$ hold for any arbitrary contravariant vector u^i, show that $A_{ijk} + A_{jki} + A_{kij} + A_{jik} + A_{ikj} + A_{kji} = 0$ where A_{ijk} are constants for all *i, j, and k.*

Solution: Let $P = A_{ijk}\, u^i u^j u^k = 0.$

It may also be written as $P = A_{pqr}\, u^p u^q u^r = 0.$

Since u^r is an arbitrary contravariant vector, we get:

$$\frac{\partial}{\partial u^i}P = A_{iqr}u^q u^r + A_{pir}u^p u^r + A_{pqi}u^q u^p = 0$$

$$\frac{\partial^2}{\partial u^j\,\partial u^i}P = A_{ijr}u^r + A_{iqj}u^q + A_{jir}u^r + A_{pij}u^p + A_{jqi}u^q + A_{pji}u^p = 0.$$

Again, differentiating with respect to u^k, we get

$$\frac{\partial^3}{\partial u^k \partial u^j \partial u^i}P = A_{ijk} + A_{ikj} + A_{jik} + A_{kij} + A_{jki} + A_{kji} = 0$$

Hence $\Rightarrow A_{ijk} + A_{ikj} + A_{jik} + A_{kij} + A_{jki} + A_{kji} = 0$

2.14 Exercises

1. Discuss the transformation in which the coordinates y^i are rectangular Cartesian:

$$y^1 = x^1 cos x^2$$

$$(a)\ y^2 = x^1 sin x^2$$

$$y^3 = x^3$$

$$(b)\ y^1 = \frac{1}{\sqrt{6}}x^1 + \frac{2}{\sqrt{6}}x^2 + \frac{1}{\sqrt{6}}x^3 \quad , \quad y^2 = \frac{1}{\sqrt{2}}x^1 - \frac{1}{\sqrt{3}}x^2 + \frac{1}{\sqrt{3}}x^3 \quad , \quad y^3 = \frac{1}{\sqrt{2}}x^1 - \frac{1}{\sqrt{2}}x^3$$

2. If f is a scalar function of coordinates x^j, then prove that:
   i) $\frac{\partial f}{\partial x^j}$ is a covariant vector
   ii) dx^j is a contavariant vector
3. Show that if A_i is a covariant vector, then $\frac{\partial A_i}{\partial x^j}$ is not a tensor.
4. If the relation $x^i_j A_i = 0$ holds for an arbitrary covariant vector A_i, show that $x^i_j = 0$.
5. If C^i_{jk} is an an arbitrary mixed tensor and $\mathrm{B}(i, j, k)\mathrm{C}^i_{jk}$ an invariant, prove that B(i, j, k) ias a tensor of type B^{jk}_i.
6. If $A^i_{jk}B^{jk} = C^i$, where C^i is a contravariant vector and B^{jk} is an arbitrary symmetric tensor, show that $A^i_{jk} + A^i_{kj}$ is a tensor. Hence, deduce that if A^i_{jk} is symmetric in j and k, then A^i_{jk} is a tensor.
7. If $A^i_{jk}B^{jk} = C^i$, where C^i is a contravariant vector and B^{jk} is an arbitrary skew- symmetric tensor, show that $A^i_{jk} + A^i_{kj}$ is a tensor. Hence, deduce that if A^i_{jk} is skew-symmetric in *j and k*, then A^i_{jk} is a tensor.
8. If a_{ij} is a tensor show that A^{ij}, the cofactor of a_{ij} in $|a_{ij}|$ divided by $|a_{ij}| \neq 0$, is a tensor.
9. If the relation $a^{ij}v_i v_j = 0$ holds for an arbitrary covariant vector v_i, show that $a^{ij} + a^{ji} = 0$.
10. If a_{ij} is a skew-symmetric tensor, prove that $(\delta^i_j \delta^k_l + \delta^i_l \delta^k_j) a_{ik} = 0$.
11. If $v_i \neq 0$ are the components of tensor of type (0,2) and if the equation

$$fv_{ij} + gv_{ij} = 0 \text{ holds,}$$

then prove that either $f = g$ and v_{ij} is skew-symmetric or $f = -g$ and v_{ij} is symmetric.

12. If a covariant vector has components $\left(\frac{x^1}{x^2}, \frac{x^2}{x^1}\right)$ in rectangular Cartesian coordinates (x^1, x^2), find its components in polar coordinates (r, θ).
13. If B^{ij} are components of a Contravariant tensor of second order and C_i, D_j are components of two covariant vectors, show that $B^{ij}C_i D_j$ is an invariant.
14. If $T_{ijkl}A^iB^jA^kB^l = 0$ holds for arbitrary contravariant vectors A^i, B^j, prove that $T_{ijkl} + T_{kjil} + T_{ilkj} + T_{klij} = 0$.
15. If A^{ij}_{kl} is skew-symmetric with respect to k *and* l and if B^{ij} is defined by the equation $B^{ij} = A^{ij}_{kl}C^{kl}$ as a tensor for arbitrary skew-symmetric tensor C^{kl}, prove that A^{ij}_{kl} is a tensor.
16. If a^{ij} is a contravariant tensor such that $|a^{ij}| \neq 0$, show that $|a^{ij}|$ is a relative invariant of weight -2.
17. If A_{ij} is a skew-symmetric tensor and B^i is a contravariant vector, then show that $A_{ij}\, B^i\, B^j = 0$.
18. If a_{ij} is a tensor such that $|a_{ij}| \neq 0$ and b^{ij} is the cofactor of a_{ij} in $|a_{ij}|$, examine whether b^{ij} is a relative tensor.
19. Prove that $A_{ij}B^iC^j$ is an invariant if B^i *and* C^j are vectors and A_{ij} is a tensor of order 2.

3

Riemannian Metric

3.1 Introduction

In an n-dimensional Euclidean space, E_n, with a rectangular Cartesian coordinate system, we can express the square of distance, ds, between the two points x^i and $x^i + dx^i$. We have seen that the expression of distance (dx^i) between two points is different for different coordinate systems. On 10th June 1854, Riemann (1826-1866) submitted his thesis on the hypothesis of the underlying foundations of geometry for print (published posthumously in 1867), so that the mathematical world could recognize the role played by the metric (distance) concepts of geometry. This idea of Metric Geometry (which was later named after himself) generalized all ideas by assuming that the distance between two neighboring points is independent of the coordinate system.

In the preceding chapter we discussed algebraic operations on tensors in S_n. Each of these operations on tensors produce a tensor, but the partial differentiation of a covariant vector does not give a tensor. That is, the operation of partial differentiation does not always produce a tensor. A question therefore arises for whether a new type of differentiation can be defined in S_n, which when applied to a tensor produces another tensor. The answer to this question is not affirmative unless additional features can be built into the structure of S_n.

A space which admits an object called an affine connection possesses a sufficient structure to permit the operation of tensor calculus within it. Riemannian space is necessarily endowed with an affine connection. Therefore, for the development of Tensor Calculus, we can consider a Riemannian space. In Riemannian spaces a new type of differentiation can be defined simply and in this space, Tensor Calculus has important applications for physics and engineering, especially in the theory of relativity.

We suppose in the remainder of this chapter that our tensors are defined in metric manifolds and that the element of arc (ds) is given by the quadratic form $ds^2 = g_{ij}(\mathrm{x})\mathrm{d}x^i\mathrm{d}x^j$, where $g_{ij}'s$ are functions belonging to C^1.

Dipankar De. Introduction to Differential Geometry with Tensor Applications, (73–96) © 2022 Scrivener Publishing LLC

We also assume that the symmetric tensor $g_{ij}(x)$ is such that $|g_{ij}| \neq 0$ at any point of the region under discussion, but do not assume that our manifold is necessarily Euclidean.

In this chapter we mainly discuss metric tensors or Riemannian metrics and n-dimensional spaces characterized by this metric, called Riemannian spaces. We also discuss curvilinear coordinates.

3.2 The Metric Tensor

We introduced the idea of n-dimensional space, En, by extending our familiar concepts of ordinary Euclidean Geometry. We used the generalized formula of Pythagoras, $|x| = \sqrt{(x^i x^i)}$, where x^i are the components of vector x, referred to as a set of orthogonal Cartesian axes. Let us consider a displacement (dx^i) between the two pair of points $A(x^i)$ and $B(x^i + dx^i)$, where the coordinates are orthogonal Cartesian, therefore applying Pythagoras's formula for the square of the distance AB.

In the expression

$$ds^2 = dx^i dx^i, \tag{3.1}$$

ds is the element of arc (AB) in E_n.

Change the coordinate to a new coordinate (y^i) by transformation:

$$x^i = x^i(y^1, \ldots y^n). \tag{3.2}$$

(3.1) can be written as

$$\therefore ds^2 = \frac{\partial x^i}{\partial y^\alpha}\frac{\partial x^i}{\partial y^\beta} dy^\alpha dy^\beta \left(\text{since } dx^i = \frac{\partial x^i}{\partial y^\alpha} dy^\alpha \right). \tag{3.3}$$

Thus, we can write the square of the element arc ds in y – reference frame as a quadratic form

$$ds^2 = g_{\alpha\beta}\, dy^\alpha dy^\beta, \tag{3.4}$$

where

$$g_{\alpha\beta}(y) = \frac{\partial x^i}{\partial y^\alpha}\frac{\partial x^i}{\partial y^\beta} \tag{3.5}$$

This is a function of (y^i) and they are obviously symmetric with respect to indices α and β.

Since the arc distance (ds) is an invariant and product of contravariant vector dy^α *and* dy^β (each has order one) and by quotient law can be a contravariant of order two, the set of functions $g_{\alpha\beta}(y)$ is a symmetric tensor. This is called a *metric tensor* or *first fundamental tensor.*

All essential metric properties of a Euclidean space are completely determined by this tensor.

The space, S_n, endowed with such a structure is called a *Riemannian space*[1] and is denoted by V_n.

Next, we suppose that the tensors are defined in metric manifolds and the element of arc (ds) is given by the quadratic form $ds^2 = g_{ij}(x)\, dx^i dx^j$, where the g_{ij} are functions belonging to C^1.

We also assume that $g_{ij}(x)$ is such that $|g_{ij}| \neq 0$ at any point of the considered region.

3.3 Conjugate Tensor

Let g be the determinant $|g_{ij}|$ and G^{ij} be the cofactor of g_{ij} in g.

Define the function of g^{ij} by the relation

$$g^{ij} = \frac{G^{ij}}{g}.$$

Since g_{ij} and G_{ij} are symmetric in subscripts, the functions g^{ij} will be symmetric in superscripts.

$$\therefore g_{ij}g^{ij} = g_{ij}\frac{G^{ij}}{g} = \frac{g}{g} = 1$$

Here,
$$g_{ij}g^{lj} = g_{ij}\frac{G^{lj}}{g} = 1 \; when\; l = i$$

$$= 0 \; and\; when\; l \neq i.$$

[1] German mathematician Bernhard Riemann (1826-1866), along with Weierstrass, laid the foundations of complex analysis. Riemann introduced the concept of integration and made basic contributions to number theory and mathematical analysis. He developed Riemannian Geometry which formed the mathematical base for Einstein's Relativity Theory.

$$g^{lj} g_{ij} = \delta_i^l \tag{3.6}$$

If u^j is an arbitrary contravariant tensor, then the inner product with the tensor g_{ij} will be an arbitrary covariant tensor due to contraction,

$$\text{i.e., } g_{ij} u^j = v_i \tag{3.7}$$

$$\therefore g^{lj} v_l = g^{lj} g_{lj} u^j = u^j,$$

which is a contravariant tensor of order one.

Therefore, by quotient law, g^{ij} are the components of a contravariant tensor of order two.

Hence, g^{ij} is a symmetric contravariant tensor which is called the *conjugate tensor or second fundamental tensor.*

In view of (3.6), the relation between g^{ij} *and* g_{ij} is reciprocal. As such, first and second fundamental tensors are also called *reciprocal tensors.*

Example 3.1.1. Find the conjugate metric tensor in a Riemannian space V_3, in which the distance (ds) is given by

$$ds^2 = 5(dx^1)^2 + 3(dx^2)^2 + 4(dx^3)^2 - 3dx^1dx^2 - 3dx^2dx^1 + 2dx^2dx^3 + 2dx^3dx^2.$$

Solution: Here, $g_{11} = 5, g_{22} = 3, g_{33} = 4, g_{12} = g_{21} = -3, g_{32} = g_{23} = 2, g_{13} = g_{31} = 0,$

$$\text{therefore } g = \begin{vmatrix} 5 & -3 & 0 \\ -3 & 3 & 2 \\ 0 & 2 & 4 \end{vmatrix} = 4$$

$$g^{11} = \frac{\textit{cofactor of } g_{11} \textit{ in } g}{g} = \frac{8}{4} = 2\text{, similarly, } g^{33} = \frac{6}{4} = \frac{3}{2},$$

$$g^{12} = 3, g^{23} = -\frac{5}{2} \quad g^{13} = -\frac{3}{2}$$

$$g^{22} = \frac{\textit{cofactor } g_{22} \textit{ in } g}{g} = \frac{20}{4} = 5$$

$$g^{ij} = \begin{vmatrix} 2 & 3 & -\frac{3}{2} \\ 3 & 5 & -\frac{5}{2} \\ -\frac{3}{2} & -\frac{5}{2} & \frac{3}{2} \end{vmatrix}.$$

3.4 Associated Tensors

We have the fundamental tenors g_{ij} and g^{ij}, which allow us a number of new combinations.

Let u^j and v_j be a contravariant and a covariant tensor. We now define two tensors u_i and v^i as follows:

$$u_i = g_{ij}\, u^j \tag{3.8}$$

$$v^i = g^{ij} v_j \tag{3.9}$$

The inner product of tensor u^j with fundamental tensor g_{ij} is another covariant tensor u_i, which is called the associated tensor of u^j.

Similarly, we have $v^i = g^{ij} v_j$.

Hence, v^i is the associated tensor of v_j.

Thus, the indices of any tensor can be lowered or raised by forming its inner product with either of the fundamental tensors g_{ij} or g^{ij}.

Thus, $$g^{ij} u_j = g^{ij}\, g_{jl}\, u^l = \delta^i_l\, u^l = u^i \tag{3.10}$$

Now, from (3.10), it follows that the associate to u_i is u^i.

Thus, if u_i is the associate to u^i, then u^i is the associate to u_i. Hence, u^i and u_i are mutually associative and are associate tensors.

Definition 3.4.1 A tensor obtained by the process of inner multiplication of any tensor $A^{i_1 \ldots i_r}_{j_1 \ldots j_s}$ with either of the fundamental tensors g_{ij} or g^{ij} is called a tensor *associated* with the given tensor.

In S_n, the scalar product of two vectors of opposite variances could only be formed as A_iB^i, but in V_n it is possible to extend as

$$A_iB^i = g_{ij}A^jB^i = A^jB_j = g^{ij}A_iB_j.$$

The above formula enables us to introduce the notion of length or magnitude of a vector in V_n.

The procedure of raising and lowering indices is clearly reversible. The position occupied by the raised (or lowered) index is indicated by a dot. In general, such systems as $g^{\alpha i}A_{j\alpha} = A^i_{j.}$ and $g^{i\alpha}A_{\alpha j} = A^{i}_{.j}$ are different. They are identical when $A_{ij} = A_{ji}$ because they are symmetric tensors.

Theorem 3.4.1. The Metric tensor g_{ij} is a covariant symmetric tensor of rank two.
Proof: The metric is given by

$$ds^2 = g_{ij}(x)\, dx^i dx^j \tag{3.11a}$$

Let x^i be the coordinates in X-coordinate system and y^i be the coordinates in Y-coordinate system.

The metric $ds^2 = g_{ij}\, dx^i dx^j$ transforms to $ds^2 = \overline{g_{ij}}\, dy^i dy^j$ and since distance is invariant,

$$ds^2 = g_{ij}\, dx^i dx^j = \overline{g_{ij}}\, dy^i dy^j. \tag{3.11b}$$

Step 1. We have to show that dx^i is a covariant vector.

$$\text{If } y^i = y^i(x^1, x^2, \ldots x^n)$$

$$\therefore dy^i = dy^i(x^1, x^2, \ldots x^n) = \frac{\partial y^i}{\partial x^1}dx^1 + \frac{\partial y^i}{\partial x^2}dx^2 + \cdots + \frac{\partial y^i}{\partial x^n}dx^n$$

$$= \frac{\partial y^i}{\partial x^k}dx^k, \text{ which is the Law of Contravariant Vectors.}$$

Step 2. We have to show that g_{ij} is a covariant tensor of rank two.

$$dy^i = \frac{\partial y^i}{\partial x^k}dx^k \quad \text{and} \quad dy^j = \frac{\partial y^j}{\partial x^l}dx^l$$

From (3.11b),
$$g_{ij}\, dx^i dx^j = \overline{g_{ij}} \frac{\partial y^i}{\partial x^k} dx^k \frac{\partial y^j}{\partial x^l} dx^l$$

$$= \overline{g_{ij}} \frac{\partial y^i}{\partial x^k} \frac{\partial y^j}{\partial x^l} dx^k dx^l$$

$$g_{kl}\, dx^k dx^l = \overline{g_{ij}} \frac{\partial y^i}{\partial x^k} \frac{\partial y^j}{\partial x^l} dx^k dx^l$$

or
$$\left(g_{kl} - \overline{g_{ij}} \frac{\partial y^i}{\partial x^k} \frac{\partial y^j}{\partial x^l} \right) dx^k dx^l = 0$$

or
$$g_{kl} - \overline{g_{ij}} \frac{\partial y^i}{\partial x^k} \frac{\partial y^j}{\partial x^l} = 0, \text{ since } dx^k \text{ and } dx^l \text{ are arbitrary,}$$

or
$$g_{kl} - \overline{g_{ij}} \frac{\partial y^i}{\partial x^k} \frac{\partial y^j}{\partial x^l}.$$

Therefore, $\overline{g_{ij}} = g_{kl} \frac{\partial x^k}{\partial y^i} \frac{\partial x^l}{\partial y^j}$ and g_{ij} is a covariant tensor of rank two.

Step 3. We have to show that g_{ij} is a symmetric tensor.

Let $g_{ij} = A_{ij} + B_{ij}$.

Let $A_{ij} = \frac{1}{2}(a_{ij} + a_{ji})$ *and* $h_{ij} = \frac{1}{2}(a_{ij} - a_{ji})$.

Here, $A_{ij} = A_{ji}$ as a symmetric tensor and $B_{ij} = -B_{ji}$ as a skew-symmetric tensor.

Since B_{ij} is a skew-symmetric tensor, implying that $B_{ij} = 0$, therefore, g_{ij} is a symmetric tensor.

Hence, metric tensor g_{ij} is a covariant symmetry tensor of rank two.

Theorem 3.4.2. Prove that $g_{ij}\, dx^i dx^j$ is invariant.

Proof: Let x^i be the coordinates in X-coordinate system and y^i be the coordinates in Y-coordinate system of a point.

Since g_{ij} is a covariant tensor of rank two,

then
$$\overline{g_{ij}} = g_{kl} \frac{\partial x^k}{\partial y^i} \frac{\partial x^l}{\partial y^j}$$

or
$$\overline{g_{ij}} - g_{kl} \frac{\partial x^k}{\partial y^i} \frac{\partial x^l}{\partial y^j} = 0,$$

or $\left(\overline{g_{ij}} - g_{kl}\frac{\partial x^k}{\partial y^i}\frac{\partial x^l}{\partial y^j}\right)\mathrm{d}x^i\mathrm{d}x^j = 0$ (since dx^i , $\mathrm{d}x^j$ are arbitrary).

$$g_{ij}\,\mathrm{d}x^i\mathrm{d}x^j = g_{kl}\frac{\partial x^k}{\partial y^i}\frac{\partial x^l}{\partial y^j}\mathrm{d}x^i\mathrm{d}x^j = g_{kl}\frac{\partial x^k}{\partial y^i}\mathrm{d}x^i\frac{\partial x^l}{\partial y^j}\mathrm{d}x^j = g_{kl}\mathrm{d}x^k\mathrm{d}x^l$$

This implies that $g_{ij}\,\mathrm{d}x^i\mathrm{d}x^j$ is invariant.

Example 3.4.1. Find the components of the first and second fundamental tensors in spherical coordinates.
Solution: Let (x^1, x^2, x^3) *and* $(\overline{x}^1, \overline{x}^2, \overline{x}^3)$ be the rectangular and spherical coordinates of a point, respectively,

$$\text{i.e., } x^1 = x,\ x^2 = y,\ \textit{and } x^3 = z,$$

where $x = r\,sin\theta cos\phi$, $y = rsin\theta sin\phi$, *and* $z = rcos\theta$.

Let g_{pq} and g_{ij} be the metric tensors in Cartesian and spherical coordinates, respectively.

$$\text{Then, } (ds)^2 = (dx^1)^2 + (dx^2)^2 + (dx^3)^2$$

$$= g_{pq}dx^p dx^q$$

$$g_{11} = 1 = g_{22} = g_{33};\ g_{12} = g_{13} = g_{32} = 0$$

On transformation,

$$\overline{g_{ij}} = \frac{\partial x^p}{\partial \overline{x}^l}\frac{\partial x^q}{\partial \overline{x}^j} = \frac{\partial x^p}{\partial \overline{x}^i}\frac{\partial x^q}{\partial \overline{x}^j}g_{pq} = \frac{\partial x^1}{\partial \overline{x}^i}\frac{\partial x^1}{\partial \overline{x}^j}g_{11} + \frac{\partial x^2}{\partial \overline{x}^i}\frac{\partial x^2}{\partial \overline{x}^j}g_{22} + \frac{\partial x^3}{\partial \overline{x}^i}\frac{\partial x^3}{\partial \overline{x}^j}g_{33}.$$

Putting i = j = 1,

$$= \bar{g}_{11} = \frac{\partial x^p}{\partial \bar{x}^i}\frac{\partial x^q}{\partial \bar{x}^j} g_{pq}$$

$$= \frac{\partial x^1}{\partial \bar{x}^1}\frac{\partial x^1}{\partial \bar{x}^1} g_{11} + \frac{\partial x^2}{\partial \bar{x}^1}\frac{\partial x^2}{\partial \bar{x}^1} g_{22} + \frac{\partial x^3}{\partial \bar{x}^1}\frac{\partial x^3}{\partial \bar{x}^1} g_{33}$$

$$= \left(\frac{\partial x^1}{\partial r}\right)^2 + \left(\frac{\partial x^2}{\partial r}\right)^2 + \left(\frac{\partial x^3}{\partial r}\right)^2$$

$$= (sin\theta cos\phi)^2 + (sin\theta sin\phi)^2 + (cos\theta)^2$$

$$= sin^2\theta + cos^2\theta = 1.$$

Putting i = j = 2,

$$\bar{g}_{22} = \frac{\partial x^p}{\partial \bar{x}^i}\frac{\partial x^q}{\partial \bar{x}^j} g_{pq}$$

$$= \frac{\partial x^1}{\partial \bar{x}^2}\frac{\partial x^1}{\partial \bar{x}^2} g_{11} + \frac{\partial x^2}{\partial \bar{x}^2}\frac{\partial x^2}{\partial \bar{x}^2} g_{22} + \frac{\partial x^3}{\partial \bar{x}^2}\frac{\partial x^3}{\partial \bar{x}^2} g_{33}$$

$$= \left(\frac{\partial x^1}{\partial \theta}\right)^2 + \left(\frac{\partial x^2}{\partial \theta}\right)^2 + \left(\frac{\partial x^3}{\partial \theta}\right)^2$$

$$= (rcos\theta cos\phi)^2 + (rcos\theta sin\phi)^2 + (-rsin\theta)^2$$

$$= r^2cos^2\theta + r^2sin^2\theta = r^2$$

$$\bar{g}_{33} = \frac{\partial x^p}{\partial \bar{x}^i}\frac{\partial x^q}{\partial \bar{x}^j} g_{pq}$$

$$= \frac{\partial x^1}{\partial \bar{x}^3}\frac{\partial x^1}{\partial \bar{x}^3} g_{11} + \frac{\partial x^2}{\partial \bar{x}^3}\frac{\partial x^2}{\partial \bar{x}^3} g_{22} + \frac{\partial x^3}{\partial x^3}\frac{\partial x^3}{\partial x^3} g_{33}$$

$$= \left(\frac{\partial x^1}{\partial \phi}\right)^2 + \left(\frac{\partial x^2}{\partial \phi}\right)^2 + \left(\frac{\partial x^3}{\partial \phi}\right)^2$$

$$= (rsin\theta . - sin\phi)^2 + (r\ sin\theta cos\phi)^2 + 0$$

$$= r^2 sin^2\theta.$$

Hence, the first fundamental tensor, written in matrix form, is

$$\begin{bmatrix} 1 & 0 & 0 \\ 0 & r^2 & 0 \\ 0 & 0 & r^2 sin^2\theta \end{bmatrix}$$

$$g = \left| \begin{bmatrix} 1 & 0 & 0 \\ 0 & r^2 & 0 \\ 0 & 0 & r^2 sin^2\theta \end{bmatrix} \right| = r^4 sin^2\theta.$$

The cofactors in g are

$$G_{11} = r^4 sin^2\theta, G_{22} = r^2 sin^2\theta, G_{33} = r^2, G_{ij} = 0 \; for \; i \neq j \; and \; also \; g^{ij} = \frac{G_{ij}}{g}.$$

Hence, the second fundamental tensor in matrix form is $\begin{bmatrix} 1 & 0 & 0 \\ 0 & \frac{1}{r^2} & 0 \\ 0 & 0 & \frac{1}{r^2 sin^2\theta} \end{bmatrix}$.

Example 3.4.2. Find the components of the metric tensor and the conjugate tensor in cylindrical coordinates.
Solution: Here, let (x^1, x^2, x^3) *and* $(\overline{x}^1, \overline{x}^2, \overline{x}^3)$ be the rectangular and cylindrical coordinates of a point, respectively,

$$\text{i.e., } x^1 = x,\; x^2 = y,\; x^3 = z, \; and \; \overline{x}^1 = \;,, \overline{x}^2 = \phi, \overline{x}^3 = z,$$

where $x = \rho \cos \phi$, $y = \rho \sin\phi$, *and* $z = z$.

Let g_{pq} *and* $\overline{g}_{ij}$ be the metric tensors in Cartesian and spherical coordinates, respectively.

$$g_{11} = g_{22} = g_{33} = 1, \; g_{ij} = 0 \; for \; i \neq j$$

$$g_{11} = 1, g_{22} = {}^2, \quad g_{33} = 1, g_{ij} = 0 \; for \; i \neq j.$$

Metric tensor of first fundamental is $\begin{bmatrix} 1 & 0 & 0 \\ 0 & p^2 & 0 \\ 0 & 0 & 1 \end{bmatrix}$ $and \; g = \rho^2.$

Also, cofactors of g are given by $G_{11} = \rho^2$, $G_{22} = 1$, $G_{33} = \rho^2$, $G_{12} = G_{13} = \cdots. = G_{32} = 0$.

The components of the conjugate tensor are given by $g^{ij} = \frac{G_{ij}}{g}$.

Hence, the second fundamental metric tensor is $\begin{bmatrix} 1 & 0 & 0 \\ 0 & \frac{1}{\rho^2} & 0 \\ 0 & 0 & 1 \end{bmatrix}$.

Alternatively,

$$y^1 = x^1 \cos x^2$$

$$y^2 = x^1 \sin x^2$$

$$y^3 = x^3$$

$$(dy^1) = (\cos x^2 \, dx^1 - x^1 sinx^2 dx^2)$$

$$(dy^1) = (\sin x^2 \, dx^1 + x^1 cosx^2 dx^2)$$

$$dy^3 = dx^3$$

$$\therefore ds^2 = (dy^1)^2 + (dy^2)^2 + (dy^3)^2 = (\cos x^2 \, dx^1 - x^1 sinx^2 dx^2)^2$$

$$+ (\sin x^2 \, dx^1 + x^1 cosx^2 dx^2)^2 + (dx^3)^2$$

$$= (dx^1)^2 + (x^1)^2 (dx^2)^2 + (dx^3)^2.$$

$$g_{11} = 1 \; g_{22} = (x^1)^2, g_{33} = 1, g_{ij} = 0 \; for \; i \neq j$$

$$g = |g_{ij}| = \begin{vmatrix} 1 & 0 & 0 \\ 0 & (x^1)^2 & 0 \\ 0 & 0 & 1 \end{vmatrix} = (x^1)^2$$

$$g^{11} = \frac{cofactor\ of\ g_{11} in\ g}{g} = \frac{(x^1)^2}{(x^1)^2} = 1$$

$$g^{22} = \frac{1}{(x^1)^2},$$

$$g^{33} = \frac{(x^1)^2}{(x^1)^2} = 1 \quad g^{ij} = 0\ for\ i \neq j$$

The second fundamental metric tensor is $\begin{pmatrix} 1 & 0 & 0 \\ 0 & \frac{1}{(x^1)^2} & 0 \\ 0 & 0 & 1 \end{pmatrix}$ (*here,* $\rho = x^1$).

3.5 Length of a Vector

3.5.1 Length of Vector

The square of the magnitude of a covariant vector A_i is defined by $g^{ij}A_iA_j$.

The square of the magnitude of a contravariant vector A^j is defined by $g_{ij}A^iA^j$.

Since

$$A_iB^i = g_{ij}A^jB^i = A^jB_j = g^{ij}A_iB_j, \tag{3.12}$$

it follows that the associate vectors A_i and A^i are of the same length.

In details, by the inner product rule, we have from (3.12)

$$A_iA^i = g_{ij}A^jA^i = A^jA_j = g^{ij}A_iA_j.$$

If the vector with A_i and A^i as its covariant and contravariant components is denoted by vector A and $|A|$ denotes its magnitude, then the length of the vector is $\mathbf{A} = (\mathbf{A.A})^{\frac{1}{2}} = (g_{ij}A^jA^i)^{\frac{1}{2}}$.

Hence, the magnitude or length of the vector is $A = (g_{ij}A^jA^i)^{\frac{1}{2}} = (g^{ij}A_iA_j)^{\frac{1}{2}} = (A_iA^i)^{\frac{1}{2}}$.

For example, the length of a vector $(A^1, 0,0)$ in 3-dimensions is $\sqrt{g_{11}A^1A^1} = \sqrt{g_{11}}A^1$. Similarly, the length of a vector $(0, A^2, 0)$ is $\sqrt{g_{22}}A^2$

and the length of a vector 0, (0, 0, A^3) is $\sqrt{g_{33}}A^3$. Hence, the physical components of vector A^i are $(\sqrt{g_{11}}A^1, \sqrt{g_{22}}A^2, \sqrt{g_{33}}A^3)$.

3.5.2 Unit Vector

A vector A_i is said to be a *unit vector* or a covariant vector of unit length if

$$g^{ij}A_iA_j = 1 \tag{3.13}$$

while A vector A^i is said to be a unit vector or a covariant vector of unit length if

$$g_{ij}A^jA^i = 1 \tag{3.14}$$

Example 3.5.1. In V_4, with line element

$$ds^2 = -(dx^1)^2 - (dx^2)^2 - (dx^3)^2 + c^2(dx^4)2,$$

show that the vector $\left(\sqrt{2},0,0\frac{\sqrt{3}}{c}\right)$ is a unit vector.

Solution: Here, $g_{11} = -1$, $g_{22} = -1$, $g_{33} = -1$, $g_{44} = c^2$, *and* $g_{ij} = 0$ for other g_{ij}.

Let A^i denote the components of the vector $\left(\sqrt{2},0,0\frac{\sqrt{3}}{c}\right)$.

$$\text{Then, } A^1 = \sqrt{2},\ A^2 = 0,\ A^3 = 0,\ and\ A^4 = \frac{\sqrt{3}}{c},$$

$$\text{so } g_{ij}A^iA^j = g_{11}(A^1)^2 + g_{22}(A^2)^2 + g_{33}(A^3)^2 + g_{44}(A^4)^2$$

$$= -1.(\sqrt{2})^2 + 0 + 0 + c^2\left(\frac{\sqrt{3}}{c}\right)^2$$

$$= -2 + c^2 \cdot \frac{3}{c^2} = -2 + 3 = 1.$$

Therefore, the length of A is 1 and A is a unit vector.

The V_4 with the above metric is called *Minkowski Space-Time* of the Special Theory of Relativity.

3.5.3 Null Vector

A vector A_i is said to be a *null vector* or a covariant vector of unit length if

$$g^{ij}A_iA_j = 0, \tag{3.15}$$

while A vector A^i is said to be a unit vector or a covariant vector of unit length if

$$g_{ij}A^jA^i = 0. \tag{3.16}$$

Since the components of a zero vector are all zero, it follows that zero vectors are a null vector, but conversely is not necessarily true.

Example 3.5.2. Show that in V_4 with the line element given by

$$ds^2 = -(dx^1)^2 - (dx^2)^2 - (dx^3)^2 + c^2(dx^4)^2,$$

the vector $\left(-1,-1,-1,\frac{\sqrt{3}}{c}\right)$ is a null vector in the space.

Solution: Here, $g_{11} = -1$, $g_{22} = -1$, $g_{33} = -1$, $g_{44} = c^2$, *and* $g_{ij} = 0$ for other g_{ij}.

Let A^i denote the components of the vector $\left(-1,-1,-1,\frac{\sqrt{3}}{c}\right)$ and $A^1 = -1, A^2 = -1, A^3 = -1, A^4 = \frac{\sqrt{3}}{c}$.

Then,

$$g_{ij}A^iA^j = g_{11}(A^1)^2 + g_{22}(A^2)^2 + g_{33}(A^3)^2 + g_{44}(A^4)^2 = -1(-1)^2$$
$$-1(-1)^2 - 1(-1)^2 + c^2\left(\frac{\sqrt{3}}{c}\right) = -1-1-1+c^2\frac{3}{c2} = -3+3 = 0.$$

This shows that it is not necessary that the components of a null vector are all zero components.

3.6 Angle Between Two Vectors

The angle θ between two non-null covariant vectors, A^i and B^i is defined by

$$\cos\theta = \frac{g_{ij}A^iB^j}{\sqrt{g_{ij}A^iA^j}\sqrt{g_{ij}B^iB^j}}. \tag{3.17}$$

The angle θ between two non-null covariant vectors, A_i and B_i, is given by

$$\cos\theta = \frac{g^{ij} A_i B_j}{\sqrt{g^{ij} A_i A_j}} \frac{1}{\sqrt{g^{ij} B_i B_j}}. \tag{3.18}$$

It is to be noted that if two vectors are such that one of them is a null vector or both of them are null vectors, then the angle between them is not defined.

If A and B are two non-null vectors and A^i, A_i; B^i, B_i, respectively, are contravariant and covariant components, then the angle θ between A and B is given by

$$\cos\theta = \frac{A^i B_i}{\sqrt{A^j A_j}\sqrt{B^k B_k}}. \tag{3.19}$$

3.6.1 Orthogonality of Two Vectors

Two vectors A^i and B^i are said to be orthogonal if

$$g_{ij} A^i B^j = 0, \tag{3.20}$$

while two vectors, A_i and B_i, are said to be orthogonal if

$$g^{ij} A_i B_j = 0. \tag{3.21}$$

It follows from (3.20) and (3.21) that the angle between two non-null orthogonal vectors, A^i and B^i, is $\frac{\pi}{2}$. It also follows from (3.20) and (3.21) that a null vector, A^i or A_i, is self-orthogonal.

Example 3.6.1. If a_{ij} is a symmetric tensor of type (0,2) and A^i and B^i are unit vectors orthogonal to a vector C^i satisfying the conditions

$$a_{ij} A^i - \omega g_{ij} A^i + \sigma g_{ij} C^i = 0$$

and $a_{ij}B^i - \omega' g_{ij}B^i + \sigma' g_{ij}C^i = 0$, where $\omega \neq \omega'$, show that A^i and B^i are orthogonal and $a_{ij}A^iB^j = 0$.
Solution: According to the given conditions:

$$g_{ij}A^iA^j = 1 \qquad \text{(i)} \qquad g_{ij}B^iB^j = 1 \qquad \text{(ii)}$$

$$g_{ij}A^iC^j = 0 \qquad \text{(iii)} \qquad g_{ij}B^iC^j = 0 \qquad \text{(iv)}$$

$$a_{ij}A^i - \omega g_{ij}A^i + \sigma g_{ij}C^i = 0, \text{ multiplying by } B^j, \text{ we get}$$

$$a_{ij}A^iB^j - \omega g_{ij}A^iB^j + \sigma g_{ij}C^iB^j = 0$$

$$\text{or } a_{ij}A^iB^j - \omega g_{ij}A^iB^j = 0 \qquad \text{(v)}$$

Similarly, multiplying the relations

$$a_{ij}B^i - \omega' g_{ij}B^i + \sigma' g_{ij}C^i = 0 \text{ by } A^j,$$

we get

$$a_{ij}B^iA^j - \omega' g_{ij}B^iA^j + \sigma' g_{ij}C^iA^j = 0,$$

$$\text{or } a_{ij}B^iA^j - \omega' g_{ij}B^iA^j = 0,$$

$$\text{or } a_{ij}A^iB^j - \omega' g_{ij}A^iB^j = 0 \text{ (since } a_{ij} \text{ and } g_{ij} \text{ are symmetric).} \qquad \text{(vi)}$$

Subtracting (vi) from (v), we get $(\omega' - \omega)g_{ij}A^iB^j = 0$ or $g_{ij}A^iB^j = 0$ (since $\omega \neq \omega'$).

Hence, A^i and B^i are orthogonal. From (vi), using this result, we get $a_{ij}A^iB^j = 0$.

3.7 Hypersurface

Let $u^1, u^2, \ldots. u^m$ be parameters and n equations

$$x^i = x^i(u^1, u^2, \ldots.u^m); \quad i = 1,2 \ldots n, n > m \qquad (3.22)$$

define m-dimensional subspace V_m of V_n. If we eliminate the m parameters $u^1u^2, \ldots, u^m$ from these n equations, we will get $(n - m)$ equations in $x^{i'}s$, which represent the m-dimensional curve in V_n.

Similarly, the n equations $x^i = x^i(u^1, u^2)$ represent a two-dimensional subspace of V_n. If we eliminate the parameters u^1 and u^2, we get $n - 2$

equations in $x^{i'}s$, which represent a two-dimensional curve in V_n. This two-dimensional curve defines a subspace denoted by V_2 of V_n.

Definition 3.7.1 $x^i = x^i(u^1 u^2, \ldots, u^{n-1})$ represent the $(n-1)$ dimensional subspace V_{n-1} of V_n. If we eliminate the parameters $u^1u^2, \ldots, u^{n-1}$, we get only one equation in $x^{i'}s$ which represent. the $n - 1$ dimensional curve in V_n. Thus, this particular curve is called a *hypersurface* in V_n.

Let φ be a scalar function of coordinates x^i. Then,

$$\varphi(x^i) = \varphi(x^i, x^i, \ldots. x^i.) = constant \tag{3.23}$$

determines a family of hypersurfaces of V_n.

A parametric hypersurface is a hypersurface on which one particular coordinate x^i is constant, while the others vary. Let us call it the x^i-hypersurface with equation $x^i = c =$ constant.

Definition 3.7.2 If, in a V_n, there are n families of hypersurfaces such that, at every point, each hypersurface is orthogonal to the $n - 1$ hypersurface of the other families which pass through that point, they are said to form an *n –polyorthogonal system of hypersurfaces.*

Definition 3.7.3 A family of curves, one of which passes through each point of V_n, is called a *Congruence of Curves.*

Definition 3.7.4 An *orthogonal ennuple* in a Riemannian V_n consists of an n mutually orthogonal congruence of curves.

3.8 Angle Between Two Coordinate Hypersurfaces

$$\text{Let } \theta(x^i) = \text{constant} \tag{i}$$

$$\text{and } \varphi(x^i) = \text{constant} \tag{ii}$$

represent families of hypersurfaces.

Differentiating (i) with respect to x^i: $\frac{\partial\theta}{\partial x^i}dx^i = 0$ implies that $\frac{\partial\theta}{\partial x^i}$ is orthogonal to dx^i. Since dx^i is tangential to hypersurface (i), we conclude that $\frac{\partial\theta}{\partial x^i}$ is normal to $\theta(x^i) =$ constant.

Similarly, $\frac{\partial\varphi}{\partial x^i}$ is normal to $\varphi(x^i) =$ constant.

If ω is the angle between the hypersurfaces, then ω is the angle between their respective normals also.

$$\text{Hence, } cos\omega = \frac{g^{ij}\dfrac{\partial\theta\,\partial\varphi}{\partial x^i\,\partial x^j}}{\sqrt{g^{ij}\dfrac{\partial\theta\,\partial\theta}{\partial x^i\,\partial x^j}}\sqrt{g^{ij}\dfrac{\partial\varphi\,\partial\varphi}{\partial x^i\,\partial x^j}}}.$$

If we take

$$\theta = x^r = C_1 \tag{iii}$$

$$\varphi = x^s = C_2 \tag{iv}$$

$$\therefore cos\omega = \frac{g^{ij}\dfrac{\partial\theta\,\partial\varphi}{\partial x^i\,\partial x^j}}{\sqrt{g^{ij}\dfrac{\partial\theta\,\partial\theta}{\partial x^i\,\partial x^j}}\sqrt{g^{ij}\dfrac{\partial\varphi\,\partial\varphi}{\partial x^i\,\partial x^j}}} = \frac{g^{ij}\dfrac{\partial x^r\,\partial x^s}{\partial x^i\,\partial x^j}}{\sqrt{g^{ij}\dfrac{\partial x^r\,\partial x^r}{\partial x^i\,\partial x^j}}\sqrt{g^{ij}\dfrac{\partial x^s\,\partial x^s}{\partial x^i\,\partial x^j}}} \tag{v}$$

$$= \frac{g^{ij}\delta_i^r\delta_j^s}{\sqrt{g^{ij}\delta_i^r\delta_j^r}\sqrt{g^{ij}\delta_i^s\delta_j^s}}\, cos\omega = \frac{g^{rs}}{\sqrt{g^{rr}}\sqrt{g^{ss}}}$$

If ω_{ij} is the angle between the hypersurfaces of parameters x^i and x^j, then by (v),

$$cos\omega_{ij} = \frac{g^{ij}}{\sqrt{g^{ii}}\sqrt{g^{jj}}}. \tag{vi}$$

If the hypersurfaces are orthogonal, then $\omega_{ij} = \dfrac{\pi}{2}$.
From (vi), we get $cos\omega_{ij} = \cos\dfrac{\pi}{2} = 0,$

$$\therefore \frac{g^{ij}}{\sqrt{g^{ii}}\sqrt{g^{jj}}} = 0 \Rightarrow g^{ij} = 0.$$

If the hypersurfaces of parameters of x^i and x^j are orthogonal, then $g^{ij} = 0$.

Example 3.8.1. Show that the angle between the vectors (1,0,0,0) and $\left(\sqrt{2},0,0\frac{\sqrt{3}}{c}\right)$, with c being a constant, in a space with line element given by $ds^2 = -(dx^1)^2 - (dx^2)^2 - (dx^3)^2 + c^2(dx^4)^2$, is not real.
Solution: $ds^2 = g_{ij}dx^i dx^j$,
where $g_{11} = -1$, $g_{22} = -1$, $g_{33} = -1$, $g_{44} = c^2$

$$A^1 = 1,\ A^2 = 0,\ A^3 = 0,\ A^4 = 0$$

$$B^1 = \sqrt{2}, B^2 = 0, B^3 = 0, B^4 = \frac{\sqrt{3}}{c}.$$

We know $cos\theta = \frac{g_{ij}A^i B^j}{\sqrt{g_{ij}A^i A^j}\sqrt{g_{ij}B^i B^j}}$.

Now, $\quad g_{ij}A^i A^j = (-1).\ 1^2 + (-1).\ 0 + (-1).\ 0 + c^2.0 = -1$

$$g_{ij}B^i B^j = (-1).(\sqrt{2})^2 + (-1).0 + (-1).0 + c^2.\left(\frac{\sqrt{3}}{c}\right)^2 = -2+3=1$$

$$g_{ij}A^i B^j = (-1).1.\sqrt{2} + (-1).0.0.0 + (-1).0 + c^2.0.\left(\frac{\sqrt{3}}{c}\right) = -\sqrt{2}$$

$cos\theta = \frac{-\sqrt{2}}{\sqrt{-1}\sqrt{-1}}$, which not a real number.
Hence, the angle between the two vectors is not real.

Example 3.8.2. In E_3 (x^i) are orthogonal Cartesian coordinates and consider a transformation

$$x^1 = y^1 sin y^2 cos y^3$$

$$x^2 = y^1 sin y^2 sin y^3$$

$$x^3 = y^1 cosy^2,$$

where (y^i) are spherical polar coordinates ($y^1 = r$, $y^2 = \theta$, $y^1 = \varphi$). What are the metric coefficients of $g_{ij}(y)$?
Solution: Now

$$x^1 = dy^1(siny^2cosy^3) + y^1(cosy^2cosy^3dy^2 - siny^2siny^3dy^3)$$

$$dx^2 = dy^1(siny^2siny^3) + y^1(cosy^2siny^3dy^2 + siny^2cosy^3dy^3)$$

$$dx^3 = dy^1cosy^2 - y^1siny^2dy^2.$$

Since x^i is an orthogonal Cartesian coordinate system,

$$\begin{aligned}(ds)^2 &= (dx^1)^2 + (dx^2)^2 + (dx^3)^2 \\ &= \{dy^1(siny^2cosy^3) + y^1(cosy^2cosy^3dy^2 - siny^2siny^3dy^3)\}^2 \\ &+\{dy^1(siny^2siny^3) + y^1(cosy^2siny^3dy^2 + siny^2cosy^3dy^3)\}^2 \\ &+ \{dy^1cosy^2 - y^1siny^2dy^2\}^2\end{aligned}$$

$$\begin{aligned}&= (dy^1)^2Sin^2y^2 + (y^1)^2\{cos^2y^2(dy^2)^2 + sin^2y^2(dy^3)^2 + cos^2y^2(dy^1)^2 \\ &+ (y^1)^2Sin^2y^2(dy^2)^2 - 2y^1cosy^2siny^2dy^1dy^2 \\ &= (dy^1)^2 + (y^1)^2(dy^2)^2 + (y^1)^2sin^2y^2(dy^3)^2 - 2y^1cosy^2siny^2dy^1dy^2\end{aligned}$$

Here, $g_{11} = 1$, $g_{22} = (y^1)^2 = r^2$, $g_{33} = (y^1)^2sin^2y^2 = r^2sin^2\theta$, $g_{12} = -2y^1cosy^2siny^2 = -2rcos\theta sin\theta$ and $g_{13} = g_{21} = g_{23} = g_{31} = g_{32} = 0$.

Example 3.8.3. Let g^{ij} and g_{ij} be the fundamental metric tensors and reciprocal tensors, respectively. Show that $g^{ij}\frac{\partial g_{ij}}{\partial x^k} + g_{ij}\frac{\partial g^{ij}}{\partial x^k} = 0$ and $\frac{\partial \log g}{\partial x^k} = g^{ij}\frac{\partial g_{ij}}{\partial x^k} = -g_{ij}\frac{\partial g^{ij}}{\partial x^k}$, where $g = |g_{ij}|$.
Solution:
We know $g_{ij}g^{jk} = \delta_i^k$, so $g_{ij}g^{ij} = n$.

Partially differentiating

$$g^{ij}\frac{\partial g_{ij}}{\partial x^k} + g_{ij}\frac{\partial g^{ij}}{\partial x^k} = 0, \tag{i}$$

we get a cofactor of $g_{12} = (-)^{1+2}\begin{vmatrix} g_{21} & g_{23} \\ g_{31} & g_{33} \end{vmatrix} = G^{12}$

$$\therefore\ g_{11}G^{11}g_{12}G^{12} + g_{13}G^{13} = g$$

$$g_{ij}G^{ij} = g \tag{ii}$$

$$\text{or } \frac{\partial g}{\partial g_{ij}} = G^{ij} \tag{iii}$$

from (ii), multiplying g^{ik}

$$g^{ik}g_{ij}G^{ij} = gg^{ik},$$

$$\text{or } \delta_j^k G^{ij} = gg^{ik},$$

$$\text{or } G^{ik} = gg^{ik} \tag{iv}$$

Now, $\dfrac{\partial g}{\partial x^k} = \dfrac{\partial g}{\partial g_{ij}}\dfrac{\partial g_{ij}}{\partial x^k} = G^{ij}\dfrac{\partial g_{ij}}{\partial x^k}$ and by (iii),

$$\frac{\partial g}{\partial x^k} = gg^{ij}\frac{\partial g_{ij}}{\partial x^k} \quad \text{by (iv)}$$

$$g^{ij}\frac{\partial g_{ij}}{\partial x^k} = \frac{1}{g}\frac{\partial g}{\partial x^k} = \frac{\partial(\log g)}{\partial x^k}$$

$$\frac{\partial(\log g)}{\partial x^k} = -g_{ij}\frac{\partial g^{ij}}{\partial x^k}$$

$$\therefore \frac{\partial \log g}{\partial x^k} = g^{ij}\frac{\partial g_{ij}}{\partial x^k} = -g_{ij}\frac{\partial g^{ij}}{\partial x^k}.$$

Example 3.8.4. If $g_{ij} = 0\ for\ i \neq j$, prove that $g^{ii} = \dfrac{1}{g_{ii}}$ (no summation on i).

Solution: Here, $|g_{ij}| = \begin{vmatrix} g_{11} & 0 & 0 \ldots\ldots 0 \\ 0 & g_{22} & 0 \ldots\ldots 0 \\ 0 & 0 & 0 \ldots\ldots g_{nn} \end{vmatrix} = g_{11}g_{22}\cdots\cdots g_{nn}$

$$g^{ii} = \frac{\textit{cofactor of } g_{ii} \textit{in } g}{|g_{ij}|} = \frac{g_{11}g_{22}\cdots g_{i-1i-1}g_{i+1i+1}\cdots\cdots g_{nn}}{g_{11}g_{22}\cdots g_{ii}\cdots g_{nn}} = \frac{1}{g_{ii}}.$$

Example 3.8.5. If u^i and v^i are two orthogonal vectors, show that $(g_{lj}g_{ki} - g_{lk}g_{ji})u^lv^iu^jv^k = 1$

Solution: By condition, we get

$$g_{ij}u^iu^j = 1, \qquad \text{(i)}$$

$$g_{ij}v^iv^j = 1, \qquad \text{(ii)}$$

$$\text{and orthogonality } g_{ij}u^iv^j = 0 \qquad \text{(iii)}$$

Now, we have $(g_{lj}g_{ki} - g_{lk}g_{ji})u^lv^iu^jv^k$

$$= g_{lj}g_{ki}u^lv^iu^jv^k - g_{lk}g_{ji}u^lv^iu^jv^k$$

$$= g_{lj}u^lu^jg_{ki}v^iv^k - g_{lk}u^lv^kg_{ji}u^jv^i$$

$$= 1.1 - 0.0 = 1.$$

Example 3.8.6. If a^i and b^i are two non-null vectors such that $g_{ij}u^iu^j = g_{ij}v^iv^j$, where $u^i = a^i + b^i$ and $v^i = a^i - b^i$, show that a^i and b^i are orthogonal.

Solution: We have $g_{ij}u^iu^j = g_{ij}(a^i + b^i)(a^j + b^j)$

$$= g_{ij}a^ia^i + g_{ij}b^ib^i + 2g_{ij}a^ib^j \; (g_{ij} \textit{ is symmetric}) \qquad \text{(i)}$$

Similarly,

$$g_{ij}v^iv^j = g_{ij}a^ia^i + g_{ij}b^ib^i - 2g_{ij}a^ib^j \qquad \text{(ii)}$$

Since $g_{ij}u^iu^j = g_{ij}v^iv^j$ are given,

From (i) and (ii) we get $4g_{ij}a^ib^j = 0$

Or $g_{ij}a^ib^j = 0$, imply that a^i and b^i are orthogonal.

3.9 Exercises

1. If the metric is given by

$$ds^2 = 5(dx^1)^2 + 3(dx^2)^2 + 4(dx^3)^2 - 6dx^1dx^2 + 4dx^2dx^3,$$

evaluate g and g^{ij}.

2. Let E_3 be covered by orthogonal Cartesian coordinates x^i and let

$$x^1 = y^1 cos y^2,$$

$$x^2 = y^1 sin y^2, \text{ and}$$

$x^3 = y^3$ represent a transformation to cylindrical coordinates y^i. Find the expression for ds^2 in cylindrical coordinates.

3. Determine the metric tensor and the conjugate metric tensor in cylindrical coordinates.

4. If the metric is given by

(i) $ds^2 = (dx^1)^2 - 2(dx^2)^2 + 3(dx^3)^2 - 8dx^2dx^3$

(ii) $ds^2 = \dfrac{dr^2}{1-\dfrac{r^2}{a^2}} + r^2(d\theta^2 + sin^2\theta d\varphi^2),$

evaluate g and g^{ij}.

5. In E_3 is covered by orthogonal Cartesian coordinates x^i, let

$$x^1 = y^1y^2 cos y^3$$

$$x^2 = y^1y^2 sin y^3, \text{ and}$$

$$x^3 = \sqrt{(y^1)^2 - (y^2)^2}$$

represent a transformation to parabolic coordinates y^i. Find the expression for ds^2.

6. Prove that g^{ij} is a symmetrical contravariant tensor of type (2,0).

7. Show that in V_4 with line element $ds^2 = -(dx^1)^2 - (dx^2)^2 - (dx^3)^2 + c^2(dx^4)^2$

each of the following vectors is a unit vector:

$$\text{(i)} \left(1,1,0,\frac{\sqrt{3}}{c}\right)$$

$$\text{(ii)} \ (1,0,0,0)$$

8. Find the conjugate metric tensor in a Riemannian metric space, V_2, in which the distance (ds) is given by $ds^2 = (dx^1)^2 + 2\cos\alpha(dx^1)(dx^2) + (dx^2)^2$.
9. Show that $ds^2 = (du)^2 + [u^2 + (a^2)](du)^2$, where $y^1 = u\cos v$ and $y^2 = u\sin v$, $y^3 = av$.
10. If $u^i = a^i + b^i$, where a^i and b^i are two orthogonal unit vectors, show that the square of the length of vector u^i is 2.

4
Tensor Calculus

4.1 Introduction

We mentioned that in S_n partial differentiation of a tensor does not, in general, produce a tensor. With the introduction of a metric in S_n, the environment has changed and the question arises as to whether in V_n a new operation of differentiation can be introduced so that when applied to a tensor it produces another tensor. The answer is affirmative, but in achieving its affirmation the fundamental tensors are again essential.

With the motivation to build up expressions involving the derivatives of a tensor which again produce components of a tensor, in 1869 E.B. Christoffel introduced certain combinations of partial derivatives of the fundamental tensor g_{ij}, which proved useful in the development of the Calculus of Tensors. A new operation of differentiation may be introduced with the help of two functions formed in terms of the partial derivatives of the components of the fundamental tensor. They are *Christoffel symbols of the first and second kind* denoted respectively by $[i, j, k]$ and $\begin{Bmatrix} l \\ i\ j \end{Bmatrix}$. We introduce in this chapter the combinations of partial derivatives of the fundamental tensor $g_{ij}(x)$, which will prove useful in the development of the calculus of tensors.

4.2 Christoffel Symbols

Let us construct a set of functions denoted by symbol

$$[ij,k] = \frac{1}{2}\left(\frac{\partial g_{ik}}{\partial x^j} + \frac{\partial g_{jk}}{\partial x^i} - \frac{\partial g_{ij}}{\partial x^k}\right) \quad (i,j,k = 1,2,3) \tag{4.1}$$

and call them *Christoffel (3-index) symbols of the first kind.*

Dipankar De. Introduction to Differential Geometry with Tensor Applications, (97–142) © 2022 Scrivener Publishing LLC

$$\text{The set of functions} \begin{Bmatrix} k \\ i\ j \end{Bmatrix} = g^{k\alpha}[ij,\alpha], \tag{4.2}$$

where $g^{k\alpha}$ is the contravariant tensor constructed with $g_{ij}'s$ are called *Christoffel symbols of the second kind.*

No summation is indicated in the Christoffel symbols of the first kind, but summation is to be made over l in the Christoffel symbols of the second kind.

4.2.1 Properties of Christoffel Symbols

In this section the following properties of the symbols will be proved.

Property 4.2.1. The Christoffel symbols $[ij,k] = \frac{1}{2}\left(\frac{\partial g_{ik}}{\partial x^j} + \frac{\partial g_{jk}}{\partial x^i} - \frac{\partial g_{ij}}{\partial x^k}\right)$ and $\begin{Bmatrix} k \\ i\ j \end{Bmatrix} = g^{k\alpha}[ij,\alpha]$ are symmetric with respect to indices i and j.

Proof:

(i) Interchanging the indices i and j in (4.1),

$$\begin{aligned}
[ji,k] &= \frac{1}{2}\left(\frac{\partial g_{jk}}{\partial x^i} + \frac{\partial g_{ik}}{\partial x^j} - \frac{\partial g_{ji}}{\partial x^k}\right) \\
&= \frac{1}{2}\left(\frac{\partial g_{ik}}{\partial x^j} + \frac{\partial g_{jk}}{\partial x^i} + -\frac{\partial g_{ij}}{\partial x^k}\right) \quad \text{(since } g_{ij} \textit{ is } \text{a symmetric tensor)} \\
&= [ij,k]
\end{aligned} \tag{4.3}$$

$$\begin{aligned}
\text{(ii)} \begin{Bmatrix} k \\ j\ i \end{Bmatrix} &= g^{k\alpha}[ji,\alpha] = g^{k\alpha}[ij,\alpha] \text{ by } (i) \\
&= \begin{Bmatrix} k \\ i\ j \end{Bmatrix}
\end{aligned} \tag{4.4}$$

Property 4.2.2.

$$\text{i) } [ij,h] = g_{kh}\begin{Bmatrix} k \\ i\ j \end{Bmatrix} \tag{4.5}$$

$$\text{ii) } \frac{\partial g_{ij}}{\partial x^k} = g_{lj}\begin{Bmatrix} l \\ i\ k \end{Bmatrix} + g_{li}\begin{Bmatrix} l \\ j\ k \end{Bmatrix}; \tag{4.6}$$

$$\text{iii) } \frac{\partial g^{im}}{\partial x^l} = -g^{ij}\begin{Bmatrix} m \\ l\ j \end{Bmatrix} - g^{km}\begin{Bmatrix} i \\ k\ l \end{Bmatrix} \tag{4.7}$$

$$\text{iv) } \frac{\partial g_{ij}}{\partial x^k} = [ik,j] + [jk,i] \tag{4.8}$$

Proof:

(i) By definition, $\begin{Bmatrix} k \\ i\ j \end{Bmatrix} = g^{k\alpha}[ij,\alpha]$.

Multiplying both sides by $g_{kh,}$ we get

$$g_{kh}\begin{Bmatrix} k \\ i\ j \end{Bmatrix} = g_{kh}g^{k\alpha}[ij,\alpha] = \delta_h^{\alpha}[ij,\alpha] = [ij,h]$$

and (ii) by (iv), we have $\dfrac{\partial g_{ij}}{\partial x^k} = [ik,j] + [jk,i]$

$$= g_{lj}\begin{Bmatrix} l \\ i\ k \end{Bmatrix} + g_{li}\begin{Bmatrix} l \\ j\ k \end{Bmatrix} \text{(using 2}^{\text{nd}}\text{ Christoffel symbol).}$$

(iii) Since $g_{lj}g^{ij} = \delta_l^i$, differentiating with respect to x^k, we get

$$\frac{\partial g_{lj}}{\partial x^k}g^{ij} + g_{lj}\frac{\partial g^{ij}}{\partial x^k} = 0.$$

Multiplying by g^{lm}, we get

$$g^{lm}g^{ij}\frac{\partial g_{lj}}{\partial x^k}+g^{lm}g_{lj}\frac{\partial g^{ij}}{\partial x^k}=0$$

$$g^{lm}g_{lj}\frac{\partial g^{ij}}{\partial x^k}=-g^{lm}g^{ij}\frac{\partial g_{lj}}{\partial x^k}$$

$$\delta_j^m\frac{\partial g^{ij}}{\partial x^k}=-g^{lm}g^{ij}\frac{\partial g_{lj}}{\partial x^k}=-g^{lm}g^{ij}\{[lk,j]+[jk,l]\}$$

$$=-g^{lm}\{g^{ij}[lk,j]\}-g^{ij}\{g^{lm}[jk,l]\}$$

$$o,\frac{\partial g^{im}}{\partial x^k}=-g^{lm}\begin{Bmatrix} i \\ l\ k\end{Bmatrix}-g^{ij}\begin{Bmatrix} m \\ j\ k\end{Bmatrix}.$$

Interchanging m and j, we get $\frac{\partial g^{ij}}{\partial x^k}=-g^{lj}\begin{Bmatrix} i \\ l\ k\end{Bmatrix}-g^{im}\begin{Bmatrix} j \\ m\ k\end{Bmatrix}$

$$\frac{\partial g^{ij}}{\partial x^k}=-g^{jl}\begin{Bmatrix} i \\ l\ k\end{Bmatrix}-g^{im}\begin{Bmatrix} j \\ m\ k\end{Bmatrix}\quad [\text{ since } g^{lj}=g^{jl}].$$

(iv) $[ik,j]=\frac{1}{2}\left(\frac{\partial g_{ij}}{\partial x^k}+\frac{\partial g_{kj}}{\partial x^i}-\frac{\partial g_{ik}}{\partial x^j}\right)$

and $[jk,i]=\frac{1}{2}\left(\frac{\partial g_{ji}}{\partial x^k}+\frac{\partial g_{ki}}{\partial x^j}-\frac{\partial g_{jk}}{\partial x^i}\right).$

Adding these two relations,

$$[ik,j]+[jk,i]=\frac{1}{2}\left(\frac{\partial g_{ij}}{\partial x^k}+\frac{\partial g_{kj}}{\partial x^i}-\frac{\partial g_{ik}}{\partial x^j}\right)+\frac{1}{2}\left(\frac{\partial g_{ji}}{\partial x^k}+\frac{\partial g_{ki}}{\partial x^j}-\frac{\partial g_{jk}}{\partial x^i}\right)$$

$$=\frac{1}{2}.2\frac{\partial g_{ij}}{\partial x^k}=\frac{\partial g_{ij}}{\partial x^k}.$$

Property 4.2.3. If $g=|g_{ij}|\neq 0$, then $\begin{Bmatrix} i \\ i\ j\end{Bmatrix}=\frac{\partial}{\partial x^j}(log\sqrt{g}).$

Proof: By the definition of reciprocal tensor

$$\text{we know, } g^{ij} = \frac{cofactor\ of\ g_{ij} in |g_{ij}|}{|g_{ik}|} = \frac{G_{ij}}{g}, \tag{i}$$

where G^{ij} is a cofactor of g_{ij} and $g = |g_{ij}| = \begin{vmatrix} g_{11}\ g_{12} \cdots \cdots g_{1n} \\ g_{21}\ g_{22} \cdots \cdots g_{2n} \\ \cdots \cdots \cdots \cdots \cdots \cdots \cdots \\ g_{n1}\ g_{n2} \cdots \cdots g_{nn} \end{vmatrix}$.

We know $g_{ij}g^{ij} = 1$. (ii)

From (i) and (ii), it is implied that $g = g_{ij}G_{ij}$. (iii)

Since the derivative of the determinant is obtained by differentiating each row of it separately, keeping the other rows the same, and summing the resulting all determinants, thus

$$\frac{\partial g}{\partial x^k} = \begin{vmatrix} \frac{\partial g_{11}}{\partial x^k}\frac{\partial g_{12}}{\partial x^k} \cdots \cdots \frac{\partial g_{1n}}{\partial x^k} \\ g_{21}\ g_{22} \cdots \cdots g_{2n} \\ \cdots \cdots \cdots \cdots \cdots \cdots \cdots \\ g_{n1}\ g_{n2} \cdots \cdots g_{nn} \end{vmatrix} + \cdots \ldots + \begin{vmatrix} g_{11}\ g_{12} \cdots \cdots g_{1n} \\ g_{21}\ g_{22} \cdots \cdots g_{2n} \\ \cdots \cdots \cdots \cdots \cdots \cdots \cdots \\ \frac{\partial g_{n1}}{\partial x^k}\frac{\partial g_{n2}}{\partial x^k} \cdots \cdots \frac{\partial g_{nn}}{\partial x^k} \end{vmatrix}$$

$$= \left(\frac{\partial g_{11}}{\partial x^k} G_{11} + \frac{\partial g_{12}}{\partial x^k} G_{12} + \cdots .. + \frac{\partial g_{1n}}{\partial x^k} G_{1n} \right) + \cdots .$$

$$+ \left(\frac{\partial g_{n1}}{\partial x^k} G_{n1} + \frac{\partial g_{n2}}{\partial x^k} G_{n2} + \cdots .. + \frac{\partial g_{nn}}{\partial x^k} G_{nn} \right)$$

$$= \sum_{i,j=1}^{n} \frac{\partial g_{ij}}{\partial x^k} G_{ij}$$

$$= \sum_{i,j=1}^{n} \frac{\partial g_{ij}}{\partial x^k} g g^{ij} \text{ (from (i))}$$

$$= g g^{ij} \frac{\partial g_{ij}}{\partial x^k}$$

$$= g g^{ij} \{[ik, j] + [jk, i]\} \quad \text{by (4.8)}$$

$$= g \begin{Bmatrix} i \\ i\ k \end{Bmatrix} + g \begin{Bmatrix} j \\ j\ k \end{Bmatrix} = 2g \begin{Bmatrix} i \\ i\ k \end{Bmatrix}$$

or $\frac{1}{2g}\frac{\partial g}{\partial x^k} = \left\{ \begin{matrix} i \\ i\ k \end{matrix} \right\}$

$$\left\{ \begin{matrix} i \\ i\ j \end{matrix} \right\} = \frac{1}{2g}\frac{\partial g}{\partial x^j} = \frac{\partial}{\partial x^j}(log\sqrt{g}). \tag{4.9}$$

Example 4.2.1. If $(ds)^2 = (dr)^2 + r^2(d\theta)^2 + r^2 sin^2\theta(d\phi)^2$, find the values of

$$\text{(i) } [22,1] \text{ and } [13,3] \text{ and } \left\{ \begin{matrix} 1 \\ 2\ 2 \end{matrix} \right\} \text{ and } \left\{ \begin{matrix} 3 \\ 1\ 3 \end{matrix} \right\}$$

Solution: Here $x^1 = r$, $x^2 = \theta$, $x^3 = \phi$,

$$g_{11} = 1,\ g_{22} = r^2,\ g_{33} = r^2 sin^2\theta \text{ and } g_{ij} = 0,\ when\ i \neq j$$

$$\text{Det}(g_{ij}) = \begin{vmatrix} 1 & 0 & 0 \\ 0 & r^2 & 0 \\ 0 & 0 & r^2 sin^2\theta \end{vmatrix} = r^4 sin^2\theta.$$

$$\text{Here, } g \neq 0,\ \therefore g^{11} = \frac{cofactor\ of\ g_{ij} in\ |g_{11}|}{|g_{ik}|} = \frac{r^4 sin^2\theta}{r^4 sin^2\theta} = 1,$$

$$g^{22} = \frac{cofactor\ of\ g_{ij} in\ |g_{22}|}{|g_{ik}|} = \frac{r^4 sin^2\theta}{r^4 sin^2\theta} = \frac{1}{r^2}$$

$$g^{33} = \frac{cofactor\ of\ g_{ij} in\ |g_{33}|}{|g_{ik}|} = \frac{r^2}{r^4 sin^2\theta} = \frac{1}{r^2 sin^2\theta}$$

$$g^{ij} = \frac{cofactor\ of\ g_{ij} in\ |g_{ij}|}{|g_{ik}|} = 0,\ since\ all\ cofactors\ of\ g_{ij} = 0, i \neq j.$$

(i) We know, $[ij,k]=\frac{1}{2}\left(\frac{\partial g_{ik}}{\partial x^j}+\frac{\partial g_{jk}}{\partial x^i}-\frac{\partial g_{ij}}{\partial x^k}\right)$ and $\begin{Bmatrix} k \\ i\ j \end{Bmatrix}=g^{k\alpha}[ij,\alpha]$

$$\therefore [22,1]=\frac{1}{2}\left(\frac{\partial g_{21}}{\partial x^2}+\frac{\partial g_{21}}{\partial x^2}-\frac{\partial g_{22}}{\partial x^1}\right)=-\frac{1}{2}\frac{\partial g_{22}}{\partial x^1}=-\frac{1}{2}\frac{\partial r^2}{\partial r}=-\frac{1}{2}2r=-r$$

$$[13,3]=\frac{1}{2}\left(\frac{\partial g_{13}}{\partial x^3}+\frac{\partial g_{33}}{\partial x^1}-\frac{\partial g_{13}}{\partial x^3}\right)=\frac{1}{2}\frac{\partial g_{33}}{\partial x^1}=\frac{1}{2}\frac{\partial r^2 sin^2\theta}{\partial r}=\frac{1}{2}2rsin^2\theta=-rsin^2\theta.$$

(ii) $\begin{Bmatrix} k \\ i\ j \end{Bmatrix}=g^{k\alpha}[ij,\alpha]$

$$\therefore \begin{Bmatrix} 1 \\ 2\ 2 \end{Bmatrix}=g^{1\alpha}[22,\alpha]=g^{11}[22,1]+g^{12}[22,2]+g^{13}[22,3]$$

$$=1.(-\mathrm{r})+0+0=-\mathrm{r}$$

and $\begin{Bmatrix} 3 \\ 1\ 3 \end{Bmatrix}=g^{3\alpha}[1\,3,\alpha]=g^{31}[1\,3,1]+g^{32}[1\,3,2]+g^{33}[1\,3,3]=0+0+\frac{1}{r^2 sin^2\theta}rsin^2\theta=\frac{1}{r}.$

Example 4.2.2. Show that if $g_{ij}=0,\ i\neq j$, *then*

(a) $\begin{Bmatrix} k \\ i\ j \end{Bmatrix}=0$, *whenever i, j, k, are distinct*

(b) $\begin{Bmatrix} i \\ i\ i \end{Bmatrix}=\frac{1}{2}\frac{\partial log g_{ii}}{\partial x^i}$

(c) $\begin{Bmatrix} i \\ i\ j \end{Bmatrix}=\frac{1}{2}\frac{\partial log g_{ii}}{\partial x^j}$

(d) $\begin{Bmatrix} i \\ j\ j \end{Bmatrix}=\frac{1}{2}\frac{\partial g_{jj}}{\partial x^i}$

Solution: We know $[ij,k]=\frac{1}{2}\left(\frac{\partial g_{ik}}{\partial x^j}+\frac{\partial g_{jk}}{\partial x^i}-\frac{\partial g_{ij}}{\partial x^k}\right)$ (4.10)

(a) When $i=j=k$, $[ii,i]=\frac{1}{2}\frac{\partial g_{ii}}{\partial x^i}$.

(b) When $i=j\neq k$, (4.10) becomes $[ii,k]=\frac{1}{2}\left(\frac{\partial g_{ik}}{\partial x^i}+\frac{\partial g_{ik}}{\partial x^i}-\frac{\partial g_{ii}}{\partial x^k}\right)$.

Since $g_{ik}=0$, $i\neq k$, it becomes $[ii,k]=\frac{1}{2}\left(\frac{\partial g_{ik}}{\partial x^i}+\frac{\partial g_{ik}}{\partial x^i}-\frac{\partial g_{ii}}{\partial x^k}\right)=-\frac{1}{2}\frac{\partial g_{ii}}{\partial x^k}$

$$\text{and } [jj,i]=-\frac{1}{2}\frac{\partial g_{jj}}{\partial x^i}.$$

(c) When $i=k\neq j$, $[ij,k]=\frac{1}{2}\left(\frac{\partial g_{ii}}{\partial x^j}+\frac{\partial g_{ji}}{\partial x^i}-\frac{\partial g_{ij}}{\partial x^i}\right)$

$$=\frac{1}{2}\frac{\partial g_{ii}}{\partial x^j}, \quad \text{as } g_{ij}=0.$$

(d) When $i\neq k\neq j$, $[ii,k]=\frac{1}{2}\left(\frac{\partial g_{ii}}{\partial x^j}+\frac{\partial g_{ji}}{\partial x^i}-\frac{\partial g_{ij}}{\partial x^k}\right)$

$$=0\text{, as } g_{ij}=0\text{, } g_{jk}=0,$$

(i) as I, j, and k are distinct, i.e., $i\neq k\neq j$, $\left\{\begin{matrix}k\\ i\ j\end{matrix}\right\}=g^{kl}[ij,l]$

$$=0 \text{ as } g^{kl}=0\text{, } k\neq l$$

(ii) $\left\{\begin{matrix}i\\ i\ i\end{matrix}\right\}=g^{ii}[ii,i]=g^{ii}\frac{1}{2}\frac{\partial g_{ii}}{\partial x^i}=\frac{1}{2g_{ii}}\frac{\partial g_{ii}}{\partial x^i}$ as $g_{ii}=\frac{1}{g^{ii}}$

$$=\frac{1}{2}\frac{\partial \log g_{ii}}{\partial x^i}$$

(iii) $i=k\neq j$, $\left\{\begin{matrix}i\\ i\ j\end{matrix}\right\}=g^{ii}[ij,i]=\frac{1}{g^{ii}}[ij,i]=\frac{1}{g^{ii}}\frac{1}{2}\frac{\partial g_{ii}}{\partial x^j}$

$$=\frac{1}{2}\frac{\partial \log g_{ii}}{\partial x^j}$$

(iv) $j=k\neq i$ $\left\{\begin{matrix}i\\ j\ j\end{matrix}\right\}=g^{ii}[jj,i]=\frac{1}{g^{ii}}\left(-\frac{1}{2}\frac{\partial g_{jj}}{\partial x^i}\right)$

$$\begin{Bmatrix} i \\ j\ j \end{Bmatrix} = -\frac{1}{2g^{ii}}\frac{\partial g_{jj}}{\partial x^i}$$

Example 4.2.3. If $|g_{ij}| \neq 0$, show that

$$g_{\alpha\beta}\frac{\partial}{\partial x^j}\begin{Bmatrix} \beta \\ i\ k \end{Bmatrix} = \frac{\partial}{\partial x^j}[i\,k,\alpha] - \begin{Bmatrix} \beta \\ i\ k \end{Bmatrix}([\beta\ j,\alpha]+[\alpha\ j,\beta]).$$

Solution: Christoffel's symbol of 2nd kind

$$\begin{Bmatrix} \beta \\ i\ k \end{Bmatrix} = g^{\beta\alpha}[ik,\alpha]$$

Multiplying both sides by $g_{\alpha\beta}$, we get

$$g_{\alpha\beta}\begin{Bmatrix} \beta \\ i\ k \end{Bmatrix} = g_{\alpha\beta}g^{\beta\alpha}[ik,\alpha]$$

$$g_{\alpha\beta}\begin{Bmatrix} \beta \\ i\ k \end{Bmatrix} = [ik,\alpha] \text{ as } g_{\alpha\beta}g^{\beta\alpha} = 1.$$

Differentiating both sides with respect to x^j, we get

$$g_{\alpha\beta}\frac{\partial}{\partial x^j}\begin{Bmatrix} \beta \\ i\ k \end{Bmatrix} + \begin{Bmatrix} \beta \\ i\ k \end{Bmatrix}\frac{\partial g_{\alpha\beta}}{\partial x^j} = \frac{\partial}{\partial x^j}[ik,\alpha].$$

$$\text{Since } \frac{\partial g_{\alpha\beta}}{\partial x^j} = [\alpha j,\beta]+[\beta j,\alpha], \qquad \text{[By (4.8)]}$$

$$\therefore g_{\alpha\beta}\frac{\partial}{\partial x^j}\begin{Bmatrix}\beta\\ i\ k\end{Bmatrix}+\begin{Bmatrix}\beta\\ i\ k\end{Bmatrix}([\alpha j,\beta]+[\beta j,\alpha])=\frac{\partial}{\partial x^j}[ik,\alpha]$$

$$\therefore g_{\alpha\beta}\frac{\partial}{\partial x^j}\begin{Bmatrix}\beta\\ i\ k\end{Bmatrix}=\frac{\partial}{\partial x^j}[ik,\alpha)-\begin{Bmatrix}\beta\\ i\ k\end{Bmatrix}([\alpha j,\beta]+[\beta j,\alpha]).$$

Example 4.2.4. Show that the only non-vanishing Christoffel symbols of the second for V_2 with line element

$(ds)^2 = (dx^1)^2 + sin^2x^1(dx^2)^2$ are $\begin{Bmatrix}1\\ 2\ 2\end{Bmatrix}=-sinx^1cosx^1$,

$\begin{Bmatrix}2\\ 1\ 2\end{Bmatrix}=\begin{Bmatrix}2\\ 2\ 1\end{Bmatrix}=cotx^1.$

Solution: $g_{11} = 1$, $g_{22} = sin^2x^1$ and $g_{12} = g_{21} = 0$ and $|g|=\begin{vmatrix}1 & 0\\ 0 & sin^2x^1\end{vmatrix}=sin^2x^1$

$$g^{11}=\frac{cofactor\ of\ g_{11}\ in\ |g|}{|g|}=\frac{sin^2x^1}{sin^2x^1}=1.$$

Similarly, $g^{22}=\dfrac{1}{sin^2x^1}$ and $g^{12} = g^{21} = 0$.

$[ij,k]=\dfrac{1}{2}\left(\dfrac{\partial g_{ik}}{\partial x^j}+\dfrac{\partial g_{jk}}{\partial x^i}-\dfrac{\partial g_{ij}}{\partial x^k}\right)$, we get $[22,1]=\dfrac{1}{2}\left(\dfrac{\partial g_{21}}{\partial x^2}+\dfrac{\partial g_{21}}{\partial x^2}-\dfrac{\partial g_{22}}{\partial x^1}\right)$

$$=-\frac{1}{2}\frac{\partial g_{22}}{\partial x^1}=-\frac{1}{2}\frac{\partial sin^2x^1}{\partial x^1}=-\frac{1}{2}2sinx^1cosx^1=-sinx^1cosx^1$$

$$\begin{Bmatrix}1\\ 2\ 2\end{Bmatrix}=g^{1\alpha}[22,\alpha]=g^{11}[22,1]+0+0=1.(-sinx^1cosx^1)=-sinx^1cosx^1$$

$$[12,2]=\frac{1}{2}\left(\frac{\partial g_{12}}{\partial x^2}+\frac{\partial g_{22}}{\partial x^1}-\frac{\partial g_{12}}{\partial x^2}\right)=\frac{1}{2}\frac{\partial g_{22}}{\partial x^1}=\frac{1}{2}\frac{\partial sin^2x^1}{\partial x^1}=sinx^1cosx^1$$

$$\begin{Bmatrix} 2 \\ 1\ 2 \end{Bmatrix}=\begin{Bmatrix} 2 \\ 2\ 1 \end{Bmatrix}=g^{2\alpha}[21,\alpha]=g^{22}[21,2]=g^{22}[12,2]=\frac{1}{sin^2x^1}sinx^1cosx^1=cotx^1$$

Example 4.2.5. If a^{ij} are components of a symmetric tensor, show that

$$a^{jk}[ij,k]=\frac{1}{2}a^{jk}\frac{\partial g_{jk}}{\partial x^i},$$

where g_{jk} have their usual meaning.

Solution: We know $\frac{\partial g_{jk}}{\partial x^i}=[ji,k]+[ki,j]$.

Multiplying both sides by a^{jk}, we get

$$a^{jk}\frac{\partial g_{jk}}{\partial x^i}=a^{jk}[ji,k]+a^{jk}[ki,j]$$

$$a^{jk}[ji,k]+a^{kj}[ki,j]=a^{jk}\frac{\partial g_{jk}}{\partial x^i} \text{ as } a^{jk}=a^{kj},$$

$$a^{jk}[ji,k]+a^{jk}[ji,k]=a^{jk}\frac{\partial g_{jk}}{\partial x^i},$$

$$\text{or } 2a^{jk}[ji,k]=a^{jk}\frac{\partial g_{jk}}{\partial x^i},$$

$$\text{or } 2a^{jk}[ij,k]=a^{jk}\frac{\partial g_{jk}}{\partial x^i},$$

$$a^{jk}[ij,k] = \frac{1}{2}a^{jk}\frac{\partial g_{jk}}{\partial x^i}$$

Example 4.2.6. Prove that $\frac{\partial}{\partial x^j}\left(\sqrt{g}g^{ij}\right)+\sqrt{g}\begin{Bmatrix} i \\ j\ k \end{Bmatrix}g^{jk}=0.$

Solution:
$$\frac{\partial}{\partial x^j}\left(\sqrt{g}g^{ij}\right)=\frac{\partial\sqrt{g}}{\partial x^j}g^{ij}+\sqrt{g}\frac{\partial g^{ij}}{\partial x^j}$$

$$=\frac{\partial\sqrt{g}}{\partial x^j}g^{ij}+\sqrt{g}\left[-g^{jp}\begin{Bmatrix} i \\ p\ j \end{Bmatrix}-g^{it}\begin{Bmatrix} j \\ t\ j \end{Bmatrix}\right]$$

$$=\frac{\partial\sqrt{g}}{\partial x^j}g^{ij}-\sqrt{g}\begin{Bmatrix} i \\ p\ j \end{Bmatrix}g^{jp}-\sqrt{g}g^{it}\begin{Bmatrix} j \\ t\ j \end{Bmatrix}$$

$$=\frac{\partial\sqrt{g}}{\partial x^j}g^{ij}-\sqrt{g}\begin{Bmatrix} i \\ p\ j \end{Bmatrix}g^{jp}-\sqrt{g}g^{it}\frac{1}{2g}\frac{\partial g}{\partial x^t}$$

$$=\frac{1}{2\sqrt{g}}\frac{\partial g}{\partial x^i}g^{ij}-\frac{1}{2\sqrt{g}}\frac{\partial g}{\partial x^t}g^{it}-\sqrt{g}\begin{Bmatrix} i \\ p\ j \end{Bmatrix}g^{jp}$$

$$=-\sqrt{g}\begin{Bmatrix} i \\ p\ j \end{Bmatrix}g^{jp}$$

$$\therefore \frac{\partial}{\partial x^j}\left(\sqrt{g}g^{ij}\right)+\sqrt{g}\begin{Bmatrix} i \\ p\ k \end{Bmatrix}g^{jk}=0$$

Example 4.2.7. Calculate the non-vanishing Christoffel symbols corresponding to the metric

$$(ds)^2 = -(dx^1)^2 - (dx^2)^2 - (dx^3)^2 + f(x^1, x^2, x^3)(dx^4)^2.$$

Solution: Comparing the expression $ds^2 = g_{ij}dx^idx^j$ (i)
with $(ds)^2 = -(dx^1)^2 - (dx^2)^2 - (dx^3)^2 + f(x^1, x^2, x^3)(dx^4)^2$,
we get $g_{11} = -1, g_{22} = -1, g_{33} = -1, \textit{and } g_{44} = f(x^1, x^2, x^3) = f, \textit{(let)}$

$$g_{ij} = 0 \textit{ for } i \neq j$$

$$\text{and } g = \begin{vmatrix} -1 & 0 & 0 & 0 \\ 0 & -1 & 0 & 0 \\ 0 & 0 & -1 & 0 \\ 0 & 0 & 0 & f \end{vmatrix} = -f.$$

We know $g^{ij} = \dfrac{cofactor\ of\ g_{ij}\ in\ g}{|g|}$

$$\therefore g^{11} = \frac{cofactor\ of\ g_{11}\ in\ g}{|g|} = \frac{f}{-f} = -1,\ g^{22} = -1,\ g^{33} = -1,\ g^{44} = \frac{1}{f}\ and\ g^{ij}$$
$$= 0\ for\ i \neq j.$$

Since g^{11}, g^{22}, *and* g^{33} are constants, we find that all non-vanishing Christoffel symbols of the 1st and 2nd kinds are

$$[14,4] = \frac{1}{2}\frac{\partial g_{44}}{\partial x^1} = \frac{1}{2}\frac{\partial f}{\partial x^1},\ \begin{Bmatrix} 4 \\ 1\ 4 \end{Bmatrix} = \begin{Bmatrix} 4 \\ 4\ 1 \end{Bmatrix} = g^{4\alpha}[14,\alpha] = g^{44}[14,4]$$
$$= \frac{1}{f}\frac{1}{2}\frac{\partial f}{\partial x^1} = \frac{\partial f}{\partial x^1} log\sqrt{f}$$

$$[24,4] = \frac{1}{2}\frac{\partial g_{44}}{\partial x^2} = \frac{1}{2}\frac{\partial f}{\partial x^2} \qquad \begin{Bmatrix} 4 \\ 2\ 4 \end{Bmatrix} = \begin{Bmatrix} 4 \\ 4\ 2 \end{Bmatrix} = g^{44}[24,4] = \frac{\partial}{\partial x^2} log\sqrt{f}$$

$$[34,4] = \frac{1}{2}\frac{\partial g_{44}}{\partial x^3} = \frac{1}{2}\frac{\partial f}{\partial x^3} \qquad \begin{Bmatrix} 4 \\ 3\ 4 \end{Bmatrix} = \begin{Bmatrix} 4 \\ 4\ 3 \end{Bmatrix} = g^{44}[34,4] = \frac{\partial}{\partial x^3} log\sqrt{f}$$

$$[44,1] = -\frac{1}{2}\frac{\partial g_{44}}{\partial x^1} = -\frac{1}{2}\frac{\partial f}{\partial x^1} \qquad \begin{Bmatrix} 1 \\ 4\ 4 \end{Bmatrix} = g^{11}[44,1] = -1.-\frac{1}{2}\frac{\partial f}{\partial x^1} = \frac{1}{2}\frac{\partial f}{\partial x^1}$$

$$[44,2] = -\frac{1}{2}\frac{\partial g_{44}}{\partial x^2} = -\frac{1}{2}\frac{\partial f}{\partial x^3} \qquad \begin{Bmatrix} 2 \\ 4\ 4 \end{Bmatrix} = g^{22}[44,2] = \frac{1}{2}\frac{\partial f}{\partial x^2}$$

$$[44,3] = -\frac{1}{2}\frac{\partial g_{44}}{\partial x^3} = -\frac{1}{2}\frac{\partial f}{\partial x^3} \qquad \begin{Bmatrix} 3 \\ 4\ 4 \end{Bmatrix} = g^{33}[44,3] = \frac{1}{2}\frac{\partial f}{\partial x^3}$$

4.3 Transformation of Christoffel Symbols

The fundamental tensors g_{ij} and g^{ij} are the functions of coordinates x^i and $[ij, k]$ and $\begin{Bmatrix} i \\ j\ k \end{Bmatrix}$ are also the functions of coordinates x^i. Let $\bar{g}_{ij}, \bar{g}^{ij}, \overline{[ij,k]}$, and $\overline{\begin{Bmatrix} l \\ j\ k \end{Bmatrix}}$ occur in another coordinate system, y^i.

4.3.1 Law of Transformation of Christoffel Symbols of 1st Kind

Let $[ij, k]$ be the functions of coordinates x^i and $\overline{[ij,k]}$ be in another coordinate system, y^i. Then,

$$\overline{[ij,k]} = \frac{1}{2}\left(\frac{\partial \bar{g}_{ik}}{\partial y^j} + \frac{\partial \bar{g}_{jk}}{\partial y^i} - \frac{\partial \bar{g}_{ij}}{\partial y^k}\right) \tag{4.11}$$

Since $\bar{g}_{ij}$ is a covariant tensor of order 2, then

$$\bar{g}_{ij} = \frac{\partial x^p}{\partial y^i}\frac{\partial x^q}{\partial y^j} g_{pq} \tag{4.12}$$

Differentiating it with respect to y^k, we get

$$\begin{aligned}
\frac{\partial \bar{g}_{ij}}{\partial y^k} &= \frac{\partial}{\partial y^k}\left(\frac{\partial x^p}{\partial y^i}\frac{\partial x^q}{\partial y^j} g_{pq}\right) \\
&= \frac{\partial}{\partial y^k}\left(\frac{\partial x^p}{\partial y^i}\frac{\partial x^q}{\partial y^j}\right) g_{pq} + \frac{\partial x^p}{\partial y^i}\frac{\partial x^q}{\partial y^j}\frac{\partial g_{pq}}{\partial y^k} \\
&= \frac{\partial^2 x^p}{\partial y^k y^i}\frac{\partial x^q}{\partial y^j} g_{pq} + \frac{\partial x^p}{\partial y^i}\frac{\partial^2 x^q}{\partial y^j y^k} g_{pq} + \frac{\partial x^p}{\partial y^i}\frac{\partial x^q}{\partial y^j}\frac{\partial g_{pq}}{\partial x^r}\frac{\partial x^r}{\partial y^k}
\end{aligned} \tag{4.13}$$

Interchanging *i*, *k* and *p*, *r* in the last term of (iii),

$$\frac{\partial \bar{g}_{kj}}{\partial y^i} = \frac{\partial^2 x^p}{\partial y^i y^k}\frac{\partial x^q}{\partial y^j} g_{pq} + \frac{\partial x^p}{\partial y^k}\frac{\partial^2 x^q}{\partial y^j y^i} g_{pq} + \frac{\partial x^r}{\partial y^i}\frac{\partial x^q}{\partial y^j}\frac{\partial g_{rq}}{\partial x^p}\frac{\partial x^p}{\partial y^k} \quad (4.14)$$

Interchanging *j*, *k* and *q*, *r* in the last term of (iii),

$$\frac{\partial \bar{g}_{ik}}{\partial y^j} = \frac{\partial^2 x^p}{\partial y^j y^i}\frac{\partial x^q}{\partial y^k} g_{pq} + \frac{\partial x^p}{\partial y^i}\frac{\partial^2 x^q}{\partial y^k y^j} g_{pq} + \frac{\partial x^p}{\partial y^i}\frac{\partial x^r}{\partial y^j}\frac{\partial g_{pr}}{\partial x^q}\frac{\partial x^q}{\partial y^k} \quad (4.15)$$

Substituting the values of (4.13), (4.14), and (4.15) in (4.11), we get

$$\overline{[ij,k]} = \frac{1}{2}\left[2\frac{\partial^2 x^p}{\partial y^j y^i}\frac{\partial x^q}{\partial y^k} g_{pq} + \frac{\partial x^r}{\partial y^i}\frac{\partial x^p}{\partial y^j}\frac{\partial x^q}{\partial y^k}\left(\frac{\partial g_{pr}}{\partial x^q} + \frac{\partial g_{qr}}{\partial x^p} - \frac{\partial g_{pq}}{\partial x^r}\right)\right]$$

$$= \frac{\partial^2 x^p}{\partial y^j y^i}\frac{\partial x^q}{\partial y^k} g_{pq} + \frac{\partial x^p}{\partial y^i}\frac{\partial x^q}{\partial y^j}\frac{\partial x^r}{\partial y^k}\frac{1}{2}\left(\frac{\partial g_{pr}}{\partial x^q} + \frac{\partial g_{qr}}{\partial x^p} - \frac{\partial g_{pq}}{\partial x^r}\right)$$

$$= \frac{\partial^2 x^p}{\partial y^i y^j}\frac{\partial x^q}{\partial y^k} g_{pq} + \frac{\partial x^p}{\partial y^i}\frac{\partial x^q}{\partial y^j}\frac{\partial x^r}{\partial y^k}[pq,r] \quad (4.16)$$

This is a law of transformation of Christoffel's symbol of the 1st kind, but it is not the transformation of any tensor due to the presence of the 1st term of (4.16), so Christoffel's symbol of the 1st kind is not a tensor.

The 1st term will vanish identically if the coordinate transformation is affine, that is if $y^i = c^i_j x^j$ and $c^i_j{}'s$ are constants.

4.3.2 Law of Transformation of Christoffel Symbols of 2nd Kind

Let $g^{kl}[ij,k] = \begin{Bmatrix} k \\ i \ j \end{Bmatrix}$ be a function of coordinate x^i and $\bar{g}^{kl}\overline{[ij,l]} = \overline{\begin{Bmatrix} k \\ i \ j \end{Bmatrix}}$ be a function of coordinate y^i. Then, from (4.16) we have

$$\overline{[ij,l]} = \frac{\partial^2 x^p}{\partial y^i y^j}\frac{\partial x^q}{\partial y^l} g_{pq} + \frac{\partial x^p}{\partial y^i}\frac{\partial x^q}{\partial y^j}\frac{\partial x^r}{\partial y^l}[pq,r].$$

Since, g^{kl} is a contravariant tensor of order 2,

$$\therefore \overline{g}^{kl} = \frac{\partial y^k}{\partial x^s}\frac{\partial y^l}{\partial x^t} g^{st} \tag{4.17}$$

$$\text{Now, } \overline{g}^{kl}\overline{[ij,l]} = \frac{\partial y^k}{\partial x^s}\frac{\partial y^l}{\partial x^t} g^{st} \frac{\partial^2 x^p}{\partial y^i y^j}\frac{\partial x^q}{\partial y^l} g_{pq} + \frac{\partial y^k}{\partial x^s}\frac{\partial y^l}{\partial x^t} g^{st} \frac{\partial x^p}{\partial y^i}\frac{\partial x^q}{\partial y^j}\frac{\partial x^r}{\partial y^l}[pq,r]$$

$$= \frac{\partial y^k}{\partial x^s}\left(\frac{\partial y^l}{\partial x^t}\frac{\partial y^q}{\partial y^l}\right) g^{st} \frac{\partial^2 x^p}{\partial y^i y^j} g_{pq} + \frac{\partial y^k}{\partial x^s}\left(\frac{\partial y^l}{\partial x^t}\frac{\partial x^r}{\partial y^l}\right)\frac{\partial x^p}{\partial y^i}\frac{\partial x^q}{\partial y^j} g^{st}[pq,r]$$

$$= \frac{\partial y^k}{\partial x^s}\delta_t^q g^{st}\frac{\partial^2 x^p}{\partial y^i y^j} g_{pq} + \frac{\partial y^k}{\partial x^s}\delta_t^r \frac{\partial x^p}{\partial y^i}\frac{\partial x^q}{\partial y^j} g^{st}[pq,r]$$

$$= \frac{\partial y^k}{\partial x^s} g^{sq}\frac{\partial^2 x^p}{\partial y^i y^j} g_{pq} + \frac{\partial y^k}{\partial x^s}\frac{\partial x^p}{\partial y^i}\frac{\partial x^q}{\partial y^j} g^{sr}[pq,r]$$

as $\delta_t^q g^{st} = g^{sq}$ and $\delta_t^r g^{st} = g^{sr}$

$$\overline{g}^{kl}\overline{[ij,l]} = \frac{\partial y^k}{\partial x^s}\frac{\partial^2 x^p}{\partial y^i y^j} g^{sq} g_{pq} + \frac{\partial y^k}{\partial x^s}\frac{\partial x^p}{\partial y^i}\frac{\partial x^q}{\partial y^j} g^{sr}[pq,r],$$

as $g^{sq}g_{pq} = \delta_p^s$ and $g^{sr}[pq,r] = \begin{Bmatrix} s \\ p\ q \end{Bmatrix}$

$$= \frac{\partial y^k}{\partial x^s}\frac{\partial^2 x^p}{\partial y^i y^j}\delta_p^s + \frac{\partial y^k}{\partial x^s}\frac{\partial x^p}{\partial y^i}\frac{\partial x^q}{\partial y^j}\begin{Bmatrix} s \\ p\ q \end{Bmatrix} \tag{4.18}$$

$$\therefore \overline{\begin{Bmatrix} k \\ i\ j \end{Bmatrix}} = \frac{\partial y^k}{\partial x^s}\frac{\partial^2 x^s}{\partial y^i y^j} + \frac{\partial y^k}{\partial x^s}\frac{\partial x^p}{\partial y^i}\frac{\partial x^q}{\partial y^j}\begin{Bmatrix} s \\ p\ q \end{Bmatrix}$$

This is the Law of Transformation of Christoffel's Symbol of the 2nd Kind, but it is not the Law of Transformation of any tensor, so Christoffel's symbol of the 2nd kind is not a tensor unless the coordinate transformation is affine.

Multiplying (4.18) by $\frac{\partial x^m}{\partial y^k}$, summing with respect to the common value $k = y$, we obtain

$$\overline{\begin{Bmatrix} \gamma \\ l\ j \end{Bmatrix}} \frac{\partial x^m}{\partial y^\gamma} = \frac{\partial x^m}{\partial y^\gamma}\frac{\partial y^\gamma}{\partial x^s}\frac{\partial^2 x^s}{\partial y^i y^j} + \frac{\partial x^m}{\partial y^\gamma}\frac{\partial y^\gamma}{\partial x^s}\frac{\partial x^p}{\partial y^i}\frac{\partial x^q}{\partial y^j}\begin{Bmatrix} s \\ p\ q \end{Bmatrix}.$$

Since $\frac{\partial x^m}{\partial x^s} = \delta_s^m$ and $\frac{\partial x^m}{\partial x^p} = \delta_p^m$, the above expression is

$$\frac{\partial^2 x^m}{\partial y^i y^j} = \overline{\begin{Bmatrix} \gamma \\ l\ j \end{Bmatrix}} \frac{\partial x^m}{\partial y^\gamma} - \frac{\partial x^p}{\partial y^i} - \frac{\partial x^q}{\partial y^j}\begin{Bmatrix} m \\ p\ q \end{Bmatrix} \tag{4.19}$$

Obviously *y and x* can be interchanged and from (4.19) we get

$$\frac{\partial^2 y^m}{\partial x^i x^j} = \overline{\begin{Bmatrix} \gamma \\ l\ j \end{Bmatrix}} \frac{\partial y^m}{\partial x^\gamma} - \frac{\partial y^p}{\partial x^i}\frac{\partial y^q}{\partial x^j}\begin{Bmatrix} m \\ p\ q \end{Bmatrix}. \tag{4.20}$$

These formulas (4.19) and (4.20) were deduced in different ways by Christoffel in a memoir concerned with the study of quadratic differential forms. We will make use of these formulas to define the operation of *tensorial differentiation.*

4.4 Covariant Differentiation of Tensor

We know the set of partial derivatives, $\frac{\partial f}{\partial x^i}$, of a scalar function $f(x^1, x^1, \ldots x^1)$ represent a covariant vector since $\frac{\partial f}{\partial y^i} = \frac{\partial f}{\partial x^k}\frac{\partial x^k}{\partial y^i}$, but if we form the set of partial derivatives $\frac{\partial}{\partial y^j}\left(\frac{\partial f}{\partial y^i}\right)$ of the covariant vector $\frac{\partial f}{\partial y^i}$, we get

$$\begin{aligned} \frac{\partial^2 f}{\partial y^i \partial y^j} &= \frac{\partial}{\partial y^j}\left(\frac{\partial f}{\partial x^k}\frac{\partial x^k}{\partial y^i}\right) \\ &= \frac{\partial^2 f}{\partial x^k \partial x^l}\frac{\partial x^l}{\partial y^j}\frac{\partial x^k}{\partial y^i} + \frac{\partial f}{\partial x^k}\frac{\partial^2 x^k}{\partial y^i \partial y^j}, \end{aligned}$$

which shows that the set of $\dfrac{\partial^2 f}{\partial y^i \partial y^j}$ does not transform according to a tensorial derivative because of the presence of $\dfrac{\partial f}{\partial x^k}\dfrac{\partial^2 x^k}{\partial y^i \partial y^j}$. That is, the set of partial derivatives of a covariant vector, in general, are not a tensor.

4.4.1 Covariant Derivative of Covariant Tensor

If we have a covariant vector $A_k(x)$, then

$$B_i(y) = \frac{\partial x^k}{\partial y^i} A_k.$$

Differentiating partially, $$\frac{\partial B_i}{\partial y^j} = \frac{\partial x^k}{\partial y^i}\frac{\partial x^l}{\partial y^j}\frac{\partial A_k}{\partial x^l} + \frac{\partial^2 x^k}{\partial y^i \partial y^j} A_k, \tag{4.21}$$

which shows that the derivative of a vector does not form a tensor unless the coordinate transformation $x^i = x^i(y)$ is affine.

If we insert $\dfrac{\partial^2 x^k}{\partial y^i \partial y^j}$ from (4.19) into (4.21), we get

$$\frac{\partial B_i}{\partial y^j} = \frac{\partial x^k}{\partial y^i}\frac{\partial x^l}{\partial y^j}\frac{\partial A_k}{\partial x^l} + \left[\overline{\begin{Bmatrix} \gamma \\ l\ j \end{Bmatrix}}\frac{\partial x^k}{\partial y^\gamma} - \frac{\partial x^p}{\partial y^i}\frac{\partial x^q}{\partial y^j}\begin{Bmatrix} k \\ p\ q \end{Bmatrix}\right] A_k.$$

Since, $\dfrac{\partial x^k}{\partial y^\gamma} A_k = B_\gamma$, we have on rearranging

$$\therefore \frac{\partial B_i}{\partial y^j} = \frac{\partial x^k}{\partial y^i}\frac{\partial x^l}{\partial y^j}\frac{\partial A_k}{\partial x^l} + \overline{\begin{Bmatrix} \gamma \\ l\ j \end{Bmatrix}} B_\gamma - \frac{\partial x^p}{\partial y^i}\frac{\partial x^q}{\partial y^j}\begin{Bmatrix} k \\ p\ q \end{Bmatrix} A_k$$

$$\text{or } \frac{\partial B_i}{\partial y^j} - \overline{\begin{Bmatrix} \gamma \\ l\ j \end{Bmatrix}} B_\gamma = \left(\frac{\partial A_\alpha}{\partial x^\beta} - \begin{Bmatrix} \gamma \\ \alpha\ \beta \end{Bmatrix} A_\gamma\right)\frac{\partial x^\alpha}{\partial y^i}\frac{\partial x^\beta}{\partial y^j} \tag{4.22}$$

from which it is clear that the set of n^2 functions $\dfrac{\partial A_i}{\partial x^j} - \begin{Bmatrix} \alpha \\ i\ j \end{Bmatrix} A_\alpha$ obeys the Law of Transformation of a Covariant Tensor of Rank 2.

Definition 4.4.1. The set of n^2 functions $\dfrac{\partial A_i}{\partial x^j} - \begin{Bmatrix} \alpha \\ i\ j \end{Bmatrix} A_\alpha$ defines covariant x^j as a derivative of a covariant tensor A_i (with respect to g_{ij}).

We denote the covariant x^j as a derivative of A_i by the symbol $A_{i,j}$.

$$\therefore A_{i,j} \equiv \frac{\partial A_i}{\partial x^j} - \begin{Bmatrix} \alpha \\ i\ j \end{Bmatrix} A_\alpha \tag{4.23}$$

4.4.2 Covariant Derivative of Contravariant Tensor

If we start with a contravariant vector A^α and differentiate the relation

$$B^i(y) = \frac{\partial y^i}{\partial x^\alpha} A^\alpha(x), \text{ we obtain}$$

$$\frac{\partial B^i}{\partial y^j} = \frac{\partial y^i}{\partial x^\alpha}\frac{\partial x^\beta}{\partial y^j}\frac{\partial A^\alpha}{\partial x^\beta} + \frac{\partial^2 y^i}{\partial x^\alpha \partial x^\beta}\frac{\partial x^\beta}{\partial y^j} A^\alpha$$

and using (4.20), we get

$$\frac{\partial B^i}{\partial y^j} + \begin{Bmatrix} i \\ \gamma\ j \end{Bmatrix} B^\gamma = \left(\frac{\partial A^\alpha}{\partial x^\beta} + \begin{Bmatrix} \alpha \\ \gamma\ \beta \end{Bmatrix} A^\gamma \right) \frac{\partial y^i}{\partial x^\alpha}\frac{\partial x^\beta}{\partial y^j}. \tag{4.24}$$

Thus, the set of n^2 functions $\dfrac{\partial A^i}{\partial x^j} + \begin{Bmatrix} i \\ \alpha\ j \end{Bmatrix} A^\alpha$ forms a mixed tensor of rank 2.

Definition 4.4.2. The set of n^2 functions $\dfrac{\partial A^i}{\partial x^j} + \begin{Bmatrix} i \\ \alpha\ j \end{Bmatrix} A^\alpha$ represents the covariant x^j derivative (with respect to g_{ij}) of the contravariant tensor A^i.

Thus,

$$A^i_{,j} \equiv \frac{\partial A^i}{\partial x^j} + \left\{ \begin{matrix} i \\ \alpha \; j \end{matrix} \right\} A^\alpha \tag{4.25}$$

It should be noted that the covariant derivative of a tensor of type (1,0) is a tensor of type (1,1).

The equation of (4.24) can be written as

$$\bar{A}^i_{,j} = A^i_{,j} \frac{\partial y^i}{\partial x^\alpha} \frac{\partial x^\beta}{\partial y^j} \tag{4.26}$$

From the above relation, it follows that the covariant derivative of tensor of type (1,0) is a tensor of type (1,1).

4.4.3 Covariant Derivative of Tensors of Type (0,2)

Let A_{ij} be the components of a tensor of type (0,2).

$$\text{Then, } \bar{A}_{ij} = \frac{\partial x^p}{\partial y^i} \frac{\partial x^q}{\partial y^j} A_{pq}. \tag{4.27}$$

Differentiating (4.27) with respect to y^k, we get

$$\begin{aligned} \frac{\partial \bar{A}_{ij}}{\partial y^k} &= \frac{\partial x^p}{\partial y^i} \frac{\partial x^q}{\partial y^j} \frac{\partial A_{pq}}{\partial y^k} + \frac{\partial}{\partial y^k}\left(\frac{\partial x^p}{\partial y^i} \frac{\partial x^q}{\partial y^j} \right) A_{pq} \\ &= \frac{\partial x^p}{\partial y^i} \frac{\partial x^q}{\partial y^j} \frac{\partial A_{pq}}{\partial y^k} + \frac{\partial^2 x^p}{\partial y^k \partial y^i} \frac{\partial x^q}{\partial y^j} A_{pq} + \frac{\partial x^p}{\partial y^i} \frac{\partial^2 x^q}{\partial y^k \partial y^j} A_{pq}. \end{aligned} \tag{4.28}$$

We know $\frac{\partial^2 y^m}{\partial x^i x^j} = \overline{\left\{ \begin{matrix} \gamma \\ l \; J \end{matrix} \right\}} \frac{\partial y^m}{\partial x^\gamma} - \frac{\partial y^p}{\partial x^i} \frac{\partial y^q}{\partial x^j} \left\{ \begin{matrix} m \\ p \; q \end{matrix} \right\}$ from (4.20) and using this in the 2nd term of (4.28), we get

$$\frac{\partial^2 x^p}{\partial y^k \partial y^i}\frac{\partial x^q}{\partial y^j}A_{pq} = \frac{\partial x^q}{\partial y^j}A_{lq}\left[\overline{\begin{Bmatrix} h \\ l\ k \end{Bmatrix}}\frac{\partial y^l}{\partial x^h} - \frac{\partial y^p}{\partial x^i}\frac{\partial y^r}{\partial x^j}\begin{Bmatrix} l \\ p\ r \end{Bmatrix}\right]$$

$$= A_{lq}\overline{\begin{Bmatrix} h \\ l\ k \end{Bmatrix}}\frac{\partial y^l}{\partial x^h}\frac{\partial x^q}{\partial y^j} - \frac{\partial x^q}{\partial y^j}\frac{\partial y^p}{\partial x^i}\frac{\partial y^r}{\partial x^j}\begin{Bmatrix} l \\ p\ r \end{Bmatrix}A_{lq}$$

$$A_{pq}\frac{\partial^2 x^p}{\partial y^k \partial y^i}\frac{\partial x^q}{\partial y^j} = A_{lq}\overline{\begin{Bmatrix} h \\ l\ k \end{Bmatrix}}\frac{\partial x^l}{\partial y^h}\frac{\partial x^q}{\partial y^j} - \frac{\partial x^q}{\partial y^j}\frac{\partial x^p}{\partial y^i}\frac{\partial x^r}{\partial y^j}\begin{Bmatrix} l \\ p\ r \end{Bmatrix}A_{lq} \tag{4.29}$$

$$= \overline{\begin{Bmatrix} h \\ l\ k \end{Bmatrix}}\bar{A}_{hj} - \frac{\partial x^q}{\partial y^j}\frac{\partial x^p}{\partial y^i}\frac{\partial x^r}{\partial y^j}\begin{Bmatrix} l \\ p\ r \end{Bmatrix}A_{lq}$$

$$\left(\text{since } \bar{A}_{hj} = \frac{\partial x^l}{\partial y^h}\frac{\partial x^q}{\partial y^j}A_{lq}, by\ (4.27)\right)$$

and

$$A_{pq}\frac{\partial^2 x^q}{\partial y^k \partial y^i}\frac{\partial x^p}{\partial y^j} = A_{pl}\frac{\partial^2 x^l}{\partial y^k \partial y^i}\frac{\partial x^p}{\partial y^j} = A_{pl}\frac{\partial x^p}{\partial y^j}\left[\overline{\begin{Bmatrix} h \\ J\ k \end{Bmatrix}}\frac{\partial x^l}{\partial y^h} - \frac{\partial x^q}{\partial y^j}\frac{\partial x^r}{\partial y^k}\begin{Bmatrix} l \\ q\ r \end{Bmatrix}\right]$$

$$A_{pq}\frac{\partial^2 x^q}{\partial y^k \partial y^i}\frac{\partial x^p}{\partial y^j} = \overline{\begin{Bmatrix} h \\ J\ k \end{Bmatrix}}\bar{A}_{ih} - \frac{\partial x^p}{\partial y^i}\left[\frac{\partial x^q}{\partial y^j}\frac{\partial x^r}{\partial y^k}\begin{Bmatrix} l \\ q\ r \end{Bmatrix}A_{pl}\right] \tag{4.30}$$

Substituting the value of (4.29) and (4.28) in equation (4.28), we get

$$\frac{\partial \bar{A}_{ij}}{\partial y^k} = \frac{\partial x^p}{\partial y^i}\frac{\partial x^q}{\partial y^j}\frac{\partial A_{pq}}{\partial y^k} + \overline{\begin{Bmatrix} h \\ i\ k \end{Bmatrix}}\bar{A}_{hj} - \frac{\partial x^q}{\partial y^j}\frac{\partial x^p}{\partial y^i}\frac{\partial x^r}{\partial y^j}\begin{Bmatrix} l \\ p\ r \end{Bmatrix}A_{lq} + \overline{\begin{Bmatrix} h \\ j\ k \end{Bmatrix}}\bar{A}_{ih} - \frac{\partial x^p}{\partial y^i}\left[\frac{\partial x^q}{\partial y^j}\frac{\partial x^r}{\partial y^k}\begin{Bmatrix} l \\ q\ r \end{Bmatrix}A_{pl}\right]$$

$$= \frac{\partial x^q}{\partial y^j}\frac{\partial x^p}{\partial y^i}\frac{\partial x^r}{\partial y^j}\left[\frac{\partial A_{pq}}{\partial x^r} - \begin{Bmatrix} l \\ p\ r \end{Bmatrix}A_{lq}\right] + \overline{\begin{Bmatrix} h \\ l\ k \end{Bmatrix}}\bar{A}_{hj} + \overline{\begin{Bmatrix} h \\ j\ k \end{Bmatrix}}\bar{A}_{ih} - \frac{\partial x^p}{\partial y^i}\left[\frac{\partial x^q}{\partial y^j}\frac{\partial x^r}{\partial y^k}\begin{Bmatrix} l \\ q\ r \end{Bmatrix}A_{pl}\right]$$

$$\frac{\partial \bar{A}_{ij}}{\partial y^k} - \overline{\begin{Bmatrix} h \\ l\ k \end{Bmatrix}}\bar{A}_{hj} - \overline{\begin{Bmatrix} h \\ j\ k \end{Bmatrix}}\bar{A}_{ih} = \left[\frac{\partial A_{pq}}{\partial x^r} - \begin{Bmatrix} l \\ p\ r \end{Bmatrix}A_{lq} - \begin{Bmatrix} l \\ q\ r \end{Bmatrix}A_{pl}\right]\frac{\partial x^q}{\partial y^j}\frac{\partial x^p}{\partial y^i}\frac{\partial x^r}{\partial y^k}.$$

We denote

$$A_{pq,r} = \frac{\partial A_{pq}}{\partial x^r} - \begin{Bmatrix} l \\ p\ r \end{Bmatrix}A_{lq} - \begin{Bmatrix} l \\ q\ r \end{Bmatrix}A_{pl} \quad \therefore \bar{A}_{ij,k} = A_{pq,r}\frac{\partial x^q}{\partial y^i}\frac{\partial x^p}{\partial y^j}\frac{\partial x^r}{\partial y^k}. \tag{4.31}$$

Definition 4.4.3. The set of functions $\frac{\partial A_{ij}}{\partial x^k} - \begin{Bmatrix} l \\ i\ k \end{Bmatrix} A_{lj} - \begin{Bmatrix} l \\ j\ k \end{Bmatrix} A_{il}$ represents the covariant x^k derivative (with respect to g_{ij}) of the covariant tensor A_{ij}. Thus,

$$A_{ij,k} = \frac{\partial A_{ij}}{\partial x^k} - \begin{Bmatrix} l \\ i\ k \end{Bmatrix} A_{lj} - \begin{Bmatrix} l \\ j\ k \end{Bmatrix} A_{il} \tag{4.32}$$

It is to be noted that the covariant derivative of a tensor of type (0,2) is a tensor of type (0,3).

4.4.4 Covariant Derivative of Tensors of Type (2,0)

Let A^{ij} be the components of a tensor of type (2,0).

Then,

$$\bar{A}^{\alpha\beta} = \frac{\partial y^\alpha}{\partial x^i}\frac{\partial y^\beta}{\partial x^j} A^{ij} \tag{4.33}$$

Differentiating (4.33) with respect to y^γ, we get

$$\frac{\partial}{\partial y^\gamma}\bar{A}^{\alpha\beta} = \frac{\partial A^{ij}}{\partial x^k}\frac{\partial x^k}{\partial y^\gamma}\frac{\partial y^\alpha}{\partial x^i}\frac{\partial y^\beta}{\partial x^j} + A^{ij}\frac{\partial^2 y^\alpha}{\partial x^k \partial x^i}\frac{\partial x^k}{\partial y^\gamma}\frac{\partial y^\beta}{\partial x^j} + A^{ij}\frac{\partial y^\alpha}{\partial x^i}\frac{\partial^2 y^\beta}{\partial x^k \partial x^j}\frac{\partial x^k}{\partial y^\gamma}$$

$$= \frac{\partial A^{ij}}{\partial x^k}\frac{\partial y^\alpha}{\partial x^i}\frac{\partial y^\beta}{\partial x^j}\frac{\partial x^k}{\partial y^\gamma} + A^{ij}\frac{\partial x^k}{\partial y^\gamma}\frac{\partial y^\beta}{\partial x^j}\left[\begin{Bmatrix} r \\ k\ i \end{Bmatrix}\frac{\partial y^\alpha}{\partial x^r} - \frac{\partial y^\gamma}{\partial x^i}\frac{\partial y^\gamma}{\partial x^k}\overline{\begin{Bmatrix} \alpha \\ \gamma\ \sigma \end{Bmatrix}}\right]$$

$$+ A^{ij}\frac{\partial y^\alpha}{\partial x^i}\frac{\partial x^k}{\partial y^\gamma}\left[\begin{Bmatrix} r \\ k\ j \end{Bmatrix}\frac{\partial y^\beta}{\partial x^r} - \frac{\partial y^\gamma}{\partial x^j}\frac{\partial y^\sigma}{\partial x^k}\overline{\begin{Bmatrix} \beta \\ \gamma\ \sigma \end{Bmatrix}}\right]$$

$$= \frac{\partial A^{ij}}{\partial x^k}\frac{\partial y^\alpha}{\partial x^i}\frac{\partial y^\beta}{\partial x^j}\frac{\partial x^k}{\partial y^\gamma} + A^{ij}\frac{\partial y^\alpha}{\partial x^r}\frac{\partial x^k}{\partial y^\gamma}\frac{\partial y^\beta}{\partial x^j}\begin{Bmatrix} r \\ k\ i \end{Bmatrix}$$

$$- A^{ij}\frac{\partial y^\gamma}{\partial x^i}\frac{\partial y^\sigma}{\partial x^k}\frac{\partial x^k}{\partial y^\gamma}\frac{\partial y^\beta}{\partial x^j}\overline{\begin{Bmatrix} \alpha \\ \gamma\ \sigma \end{Bmatrix}} + A^{ij}\frac{\partial y^\alpha}{\partial x^i}\frac{\partial y^\beta}{\partial x^r}\frac{\partial x^k}{\partial y^\gamma}\begin{Bmatrix} r \\ k\ j \end{Bmatrix}$$

$$- A^{ij}\frac{\partial y^\alpha}{\partial x^i}\frac{\partial y^\gamma}{\partial x^j}\frac{\partial y^\sigma}{\partial x^k}\frac{\partial x^k}{\partial y^\gamma}\overline{\begin{Bmatrix} \beta \\ \gamma\ \sigma \end{Bmatrix}}$$

$$= \frac{\partial A^{ij}}{\partial x^k}\frac{\partial y^\alpha}{\partial x^i}\frac{\partial y^\beta}{\partial x^j}\frac{\partial x^k}{\partial y^\gamma} + A^{rj}\frac{\partial y^\alpha}{\partial x^i}\frac{\partial x^k}{\partial y^\gamma}\frac{\partial y^\beta}{\partial x^j}\begin{Bmatrix} i \\ k \;\; r \end{Bmatrix}$$

$$-A^{ij}\frac{\partial y^\gamma}{\partial x^i}\delta^\sigma_\gamma\frac{\partial y^\beta}{\partial x^j}\overline{\begin{Bmatrix} \alpha \\ \gamma \;\; \sigma \end{Bmatrix}} + A^{ir}\frac{\partial y^\alpha}{\partial x^i}\frac{\partial y^\beta}{\partial x^j}\frac{\partial x^k}{\partial y^\gamma}\begin{Bmatrix} j \\ k \;\; r \end{Bmatrix}$$

$$-A^{ij}\frac{\partial y^\alpha}{\partial x^i}\frac{\partial y^\gamma}{\partial x^j}\delta^\sigma_\gamma\overline{\begin{Bmatrix} \beta \\ \gamma \;\; \sigma \end{Bmatrix}}$$

(interchanging dummy index r and i in the 2nd term and index r and j in the 4th term)

$$= \left(\frac{\partial A^{ij}}{\partial x^k} + A^{rj}\begin{Bmatrix} i \\ k \;\; r \end{Bmatrix} + A^{ir}\begin{Bmatrix} j \\ k \;\; r \end{Bmatrix}\right)\frac{\partial y^\alpha}{\partial x^i}\frac{\partial y^\beta}{\partial x^j}\frac{\partial x^k}{\partial y^\gamma}$$

$$-A^{ij}\frac{\partial y^\sigma}{\partial x^i}\frac{\partial y^\beta}{\partial x^j}\overline{\begin{Bmatrix} \alpha \\ \gamma \;\; \sigma \end{Bmatrix}} - A^{ij}\frac{\partial y^\alpha}{\partial x^i}\frac{\partial y^\sigma}{\partial x^j}\overline{\begin{Bmatrix} \beta \\ \gamma \;\; \sigma \end{Bmatrix}}$$

or, $$\frac{\partial \bar{A}^{\alpha\beta}}{\partial y^\gamma} + \bar{A}^{\sigma\beta}\overline{\begin{Bmatrix} \alpha \\ \sigma\gamma \end{Bmatrix}} + \bar{A}^{\alpha\sigma}\overline{\begin{Bmatrix} \beta \\ \gamma \;\; \sigma \end{Bmatrix}}$$

$$= \left(\frac{\partial A^{ij}}{\partial x^k} + A^{rj}\begin{Bmatrix} i \\ k \;\; r \end{Bmatrix} + A^{ir}\begin{Bmatrix} j \\ k \;\; r \end{Bmatrix}\right)\frac{\partial y^\alpha}{\partial x^i}\frac{\partial y^\beta}{\partial x^j}\frac{\partial x^k}{\partial y^\gamma}. \tag{4.34}$$

It follows that $\frac{\partial A^{ij}}{\partial x^k} + A^{rj}\begin{Bmatrix} i \\ k \;\; r \end{Bmatrix} + A^{ir}\begin{Bmatrix} j \\ k \;\; r \end{Bmatrix}$ are components of tensors of type (2,1).

Definition 4.4.4. The set of functions $\frac{\partial A^{ij}}{\partial x^k} + A^{rj}\begin{Bmatrix} i \\ k \;\; r \end{Bmatrix} + A^{ir}\begin{Bmatrix} j \\ k \;\; r \end{Bmatrix}$ represents the covariant x^k derivative (with respect to g_{ij}) of the contravariant tensor A^{ij}. Thus,

$$A^{ij}_{,k} = \frac{\partial A^{ij}}{\partial x^k} + A^{rj}\begin{Bmatrix} i \\ k \;\; r \end{Bmatrix} + A^{ir}\begin{Bmatrix} j \\ k \;\; r \end{Bmatrix} \tag{4.35}$$

It is to be noted that the covariant derivative of a tensor of type (2,0) is a tensor of type (2,1).

(4.34) can be written as $\bar{A}^{ij}_{,k} = A^{ij}_{,k} \dfrac{\partial y^{\alpha}}{\partial x^{i}} \dfrac{\partial y^{\beta}}{\partial x^{j}} \dfrac{\partial x^{k}}{\partial y^{\gamma}}$. (4.36)

Thus, the covariant derivative of tensor type (p, q) is a tensor of $(p, q + 1)$.

4.4.5 Covariant Derivative of Mixed Tensor of Type (s, r)

Relation (4.34) can be extended in an obvious way to mixed tensors. Thus, we define the covariant x^k derivative (with respect to g_{ij}) of the mixed tensor $A^{j_1 j_1 \cdots j_s}_{i_1 i_2 \ldots i_r}$ by the formula

$$A^{j_1 j_1 \cdots j_s}_{i_1 i_2 \ldots i_r, k} = \frac{\partial A^{j_1 j_1 \cdots j_s}_{i_1 i_2 \ldots i_r}}{\partial x^k} + A^{l j_2 \cdots j_s}_{i_1 i_2 \ldots i_r} \begin{Bmatrix} j_1 \\ k \ l \end{Bmatrix} + A^{j_1 l j_3 \cdots j_s}_{i_1 i_2 \ldots i_r} \begin{Bmatrix} j_2 \\ k \ l \end{Bmatrix}$$

$$+ \cdots \ldots + A^{j_1 j_2 \cdots j_{s-1} l}_{i_1 i_2 \ldots i_r} \begin{Bmatrix} j_s \\ k \ l \end{Bmatrix} - A^{j_1 j_3 \cdots j_s}_{l i_1 \ldots i_r} \begin{Bmatrix} l \\ i_1 \ k \end{Bmatrix}$$

$$A^{j_1 j_2 \cdots j_s}_{i_1 l i_3 \ldots i_r} \begin{Bmatrix} l \\ i_2 \ k \end{Bmatrix} - \cdots \ldots - A^{j_1 j_2 \cdots j_s}_{i_1 i_2 \ldots i_{r-1} l} \begin{Bmatrix} l \\ i_r \ k \end{Bmatrix}.$$

It can be easily shown that $A^{j_1 j_1 \cdots j_s}_{i_1 i_2 \ldots i_r, k}$ are the components of a tensor of type (s,r+1).

If A is a tensor of rank zero, we define its covariant derivative to be the ordinary derivative. Thus, $A_{,l} = \dfrac{\partial A}{\partial x^l}$. We also note that if gij's are constants, the Christoffel symbols vanish identically, hence the covariant derivative reduces to ordinary derivatives.

It follows, for Type of (1,1) tensor: $A^i_{j,k} = \dfrac{\partial A^i_j}{\partial x^k} + A^l_j \begin{Bmatrix} i \\ k \ l \end{Bmatrix} - A^i_l \begin{Bmatrix} l \\ j \ k \end{Bmatrix}$

(4.38)

4.4.6 Covariant Derivatives of Fundamental Tensors and the Kronecker Delta

The following theorem shows the nature of the behavior of the tensors g_{ij}, g^{ij}, and δ^i_j under covariant differentiation.

Ricci's Theorem 4.4.1. Fundamental tensors and Kronecker Deltas behave in covariant differentiation as though they were constants, i.e., $g_{ij,k} = 0$, $g^{ij}_{,k} = 0$, and $\delta^i_{j,k} = 0$.

Proof:

(i) We have $g_{ij,k} = \frac{\partial g_{ij}}{\partial x^k} - \begin{Bmatrix} l \\ i\ k \end{Bmatrix} g_{lj} - \begin{Bmatrix} l \\ j\ k \end{Bmatrix} g_{il}$ (using 4.32)

$$= \frac{\partial g_{ij}}{\partial x^k} - [ik,j] - [jk,i]$$

$$= \frac{\partial g_{ij}}{\partial x^k} - \frac{1}{2}\left\{\left(\frac{\partial g_{ij}}{\partial x^k} + \frac{\partial g_{kj}}{\partial x^i} - \frac{\partial g_{ik}}{\partial x^j}\right) + \left(\frac{\partial g_{ij}}{\partial x^k} + \frac{\partial g_{ki}}{\partial x^j} - \frac{\partial g_{jk}}{\partial x^i}\right)\right\}$$

$$= \frac{\partial g_{ij}}{\partial x^k} - \frac{1}{2}.2\frac{\partial g_{ij}}{\partial x^k} = 0$$

(ii) Now, we use the formula for covariant derivatives of contravariant tensors of order 2

$$g^{ij}_{,k} = \frac{\partial g^{ij}}{\partial x^k} + g^{rj}\begin{Bmatrix} i \\ k\ r \end{Bmatrix} + g^{ir}\begin{Bmatrix} j \\ k\ r \end{Bmatrix} \text{ (by 4.35).}$$

We know, $g_{ij}g^{jk} = \delta_i^k$

$$\therefore \frac{\partial (g_{ij}g^{jk})}{\partial x^m} = 0$$

$$\text{or } \frac{\partial g_{ij}}{\partial x^m} g^{jk} + \frac{\partial g^{jk}}{\partial x^m} g_{ij} = 0.$$

Multiplying both sides by g^{il},

$$\frac{\partial g_{ij}}{\partial x^m} g^{jk} g^{il} + \frac{\partial g^{jk}}{\partial x^m} g_{ij} g^{il} = 0,$$

$$\text{or } \delta_j^l \frac{\partial g^{jk}}{\partial x^m} = -g^{jk} g^{il} \frac{\partial g_{ij}}{\partial x^m},$$

$$\text{or } \frac{\partial g^{lk}}{\partial x^m} = -g^{jk} g^{il} \frac{\partial g_{ij}}{\partial x^m} = -g^{jk} g^{il}\{[im,j] + [jm,i]\}$$

$$= -g^{il}\begin{Bmatrix} k \\ i\ m \end{Bmatrix} - g^{jk}\begin{Bmatrix} l \\ j\ m \end{Bmatrix},$$

or, $\frac{\partial g^{ij}}{\partial x^k} = -g^{rj}\begin{Bmatrix} i \\ r\ k \end{Bmatrix} - g^{ri}\begin{Bmatrix} j \\ r\ k \end{Bmatrix}$ (changing *l to j*, *k* to *i*, *m to k*, and change the dummy index on the right hand side *r for i* and 2nd term *r for j*)

$$\therefore \frac{\partial g^{ij}}{\partial x^k} + g^{rj}\begin{Bmatrix} i \\ r\ k \end{Bmatrix} + g^{ri}\begin{Bmatrix} j \\ r\ k \end{Bmatrix} = 0$$

Hence, $g^{ij}_{,k} = 0$.

(iii) $\delta^i_{k,j} = \frac{\partial \delta^k_j}{\partial^j_x} - \begin{Bmatrix} l \\ j\ k \end{Bmatrix}\delta^i_l + \begin{Bmatrix} i \\ j\ l \end{Bmatrix}\delta^l_k = 0 - \begin{Bmatrix} i \\ j\ k \end{Bmatrix} + \begin{Bmatrix} i \\ j\ k \end{Bmatrix} = 0$ using (4.38).

It is known that $g_{ij}A^i = A_j$ where A^i is a contravariant vector. Therefore, $(g_{ij}A^i)_{,k}$ must give the same result as $A_{j,k}$. To decide whether this equality is ensured, it is necessary to know the behavior of a product of two tensors under covariant differentiation. We now consider the sum, difference, and product of the tensors.

4.4.7 Formulas for Covariant Differentiation

It is easy to deduce from the structure of Formula (4.37) that the rules for covariant differentiation of sums and products of tensors are identical with ordinary differentiation. If we consider that

$A^{j_1 j_2 \cdots j_s}_{i_1 i_2 \ldots\ldots i_r}(x)$ and $B^{j_1 j_2 \cdots j_s}_{i_1 i_2 \ldots\ldots i_r}(x)$ are two tensors, then the formula

$$\left(A^{j_1 j_2 \cdots j_s}_{i_1 i_2 \ldots\ldots i_r} + B^{j_1 j_2 \cdots j_s}_{i_1 i_2 \ldots\ldots i_r}\right)_{,l} = A^{j_1 j_2 \cdots j_s}_{i_1 i_2 \ldots\ldots i_r,l} + B^{j_1 j_2 \cdots j_s}_{i_1 i_2 \ldots\ldots i_r,l}$$

follows directly from (4.37).

Now, the derivative of the outer and inner products are given by the rules:

$$\left(A^{j_1 j_2 \cdots j_s}_{i_1 i_2 \ldots\ldots i_r} B^{j_{s-1} \cdots j_v}_{i_{r-1} \ldots\ldots i_w}\right)_{,l} = A^{j_1 j_2 \cdots j_s}_{i_1 i_2 \ldots\ldots i_r} B^{j_{s+1} \cdots j_v}_{i_{r+1} \ldots\ldots i_w,l} + A^{j_1 j_2 \cdots j_s}_{i_1 i_2 \ldots\ldots i_r,l} B^{j_{s+1} \cdots j_v}_{i_{r+1} \ldots\ldots i_w,}$$

$$\left(A^{j_1j_2\cdots j_{s-1}\alpha}_{i_1i_2\ldots\ldots i_r} B^{j_{s-1}\cdots j_v}_{i_{r-1}\ldots\ldots i_{w-1}\alpha}\right)_{,l} = A^{j_1j_2\cdots j_{s-1}\alpha}_{i_1i_2\ldots\ldots i_r,l} B^{j_{s+1}j_{s+2}\cdots j_v}_{i_{r+1}i_{r+2}\ldots\ldots i_{w-1}\alpha} + A^{j_1j_2\cdots j_{s-1}\alpha}_{i_1i_2\ldots\ldots i_r} B^{j_{s+1}j_{s+2}\cdots j_v}_{i_{r+1}i_{r+2}\ldots\ldots i_{w-1}\alpha,l}$$

Now that we consider $A^{j_1j_2}B_{i_1i_2} = C^{j_1j_2}_{i_1i_2}$, we have

$$\begin{aligned}
C^{j_1j_2}_{i_1i_2,l} &= \frac{\partial C^{j_1j_2}_{i_1i_2}}{\partial x^l} - \begin{Bmatrix}\alpha\\ i_1\ l\end{Bmatrix} C^{j_1j_2}_{\alpha i_2} - \begin{Bmatrix}\alpha\\ i_2\ l\end{Bmatrix} C^{j_1j_2}_{i_1\alpha} + \begin{Bmatrix}j_1\\ \alpha\ l\end{Bmatrix} C^{\alpha j_2}_{i_1i_2} + \begin{Bmatrix}j_2\\ \alpha\ l\end{Bmatrix} C^{j_1\alpha}_{i_1i_2}\\
&= A^{j_1j_2}\left(\frac{\partial B_{i_1i_2}}{\partial x^l} - \begin{Bmatrix}\alpha\\ i_1\ l\end{Bmatrix} B_{\alpha i_2} - \begin{Bmatrix}\alpha\\ i_2\ l\end{Bmatrix} B_{i_1\alpha}\right)\\
&\quad + B_{i_1i_2}\left(\frac{\partial A^{j_1j_2}}{\partial x^l} - \begin{Bmatrix}j_1\\ \alpha\ l\end{Bmatrix} A^{\alpha j_2} - \begin{Bmatrix}j_2\\ \alpha\ l\end{Bmatrix} A^{j_1\alpha}\right)\\
&= A^{j_1j_2}B_{i_1i_2,l} + B_{i_1i_2}A^{j_1j_2}_{,l}
\end{aligned}$$

$$\therefore \left(A^{j_1j_2}B_{i_1i_2}\right)_{,l} = A^{j_1j_2}B_{i_1i_2,l} + B_{i_1i_2}A^{j_1j_2}_{,l} \tag{4.39}$$

4.4.8 Covariant Differentiation of Relative Tensors

The covariant x^l derivative of relative tensors are defined as follows. If $f(x)$ is a relative tensor scalar of weight W such that $g(y) = f(x)\left|\frac{\partial x^i}{\partial y^j}\right|^W$, then

$$f_{,l} \equiv \frac{\partial f}{\partial x^l} - Wf\begin{Bmatrix}\alpha\\ l\ \alpha\end{Bmatrix} \tag{4.40}$$

This set of functions represents a relative vector of weight W.

$$\text{If } A^{j_1j_2\cdots j_s}_{i_1i_2\ldots\ldots i_r,l} \equiv \frac{\partial A^{j_1j_2\cdots j_s}_{i_1i_2\ldots\ldots i_r}}{\partial\, xl} - WA^{j_1j_2\cdots j_s}_{i_1i_2\ldots\ldots i_r}\begin{Bmatrix}\alpha\\ i\ l\end{Bmatrix} - \begin{Bmatrix}\alpha\\ i_1\ l\end{Bmatrix} A^{j_1j_2\cdots j_s}_{\alpha i_2\ldots\ldots i_r} - \cdots - \begin{Bmatrix}\alpha\\ i_r\ l\end{Bmatrix} A^{j_1j_2\cdots j_s}_{i_1i_2\ldots\ldots\alpha}$$
$$+\begin{Bmatrix}j_1\\ \alpha\ l\end{Bmatrix} A^{\alpha j_2\cdots j_s}_{i_1i_2\ldots\ldots i_r} + \cdots\cdots + \begin{Bmatrix}j_s\\ \alpha\ l\end{Bmatrix} A^{j_1j_1j_2\cdots\alpha}_{i_1i_2\ldots\ldots i_r}.$$

Example 4.4.1. Show that $(g_{ij}A^i)_{,k} = A_{j,k}$.
Solution: We have $(g_{ij}A^i)_{,k} = g_{ij,k}A^i + g_{ij}A^i_{,k}$ (using Equation 4.39)

$$= g_{ij}A^i_{,k}, \qquad \text{since } g_{ij,k} = 0$$

$$= g_{ij}\left(\frac{\partial A^i}{\partial x^k} + A^t\begin{Bmatrix} i \\ t\ k \end{Bmatrix}\right) \tag{i}$$

$$\text{Now, } \frac{\partial A_j}{\partial x^k} = \frac{\partial (g_{ij}A^i)}{\partial x^k} = \frac{\partial g_{ij}}{\partial x^k}A^i + \frac{\partial A^i}{\partial x^k}g_{ij}$$

$$= [ik,j]A^i + [jk,i]A^i + \frac{\partial A^i}{\partial x^k}g_{ij}$$

$$\therefore \frac{\partial A^i}{\partial x^k}g_{ij} = \frac{\partial A_j}{\partial x^k} - [ik,j]A^i - [jk,i]A^i \tag{ii}$$

By (i) and (ii) we get

$$(g_{ij}A^i)_{,k} = \frac{\partial A_j}{\partial x^k} - [ik,j]A^i - [jk,i]A^i + g_{ij}A^t\begin{Bmatrix} i \\ t\ k \end{Bmatrix}$$

$$= \frac{\partial A_j}{\partial x^k} - [ik,j]A^i - [jk,i]A^i + [tk,j]A^t$$

$$= \frac{\partial A_j}{\partial x^k} - [jk,i]A^i$$

$$= \frac{\partial A_j}{\partial x^k} - g^{li}A_l[jk,i]$$

$$= \frac{\partial A_j}{\partial x^k} - A_l\begin{Bmatrix} l \\ j\ k \end{Bmatrix} = A_{j,k}, \text{ hence proved.}$$

Example 4.4.2. Proved that if A^{ij} is a symmetric tensor, then

$$A^j_{i,j} = \frac{1}{\sqrt{g}}\frac{\partial\left(A^j_i\sqrt{g}\right)}{\partial x^j} - \frac{1}{2}A^{jk}\frac{\partial g_{jk}}{\partial x^i}.$$

Solution: Since A^{ij} is a symmetric tensor, we get $A^{ij} = A^{ji}$.

We know that $A^j_{i,k} = \frac{\partial A^j_i}{\partial x^k} + \begin{Bmatrix} j \\ l\ j \end{Bmatrix} A^l_i - \begin{Bmatrix} l \\ i\ k \end{Bmatrix} A^j_l$.

Putting $k = j$, we get $A^j_{i,j} = \frac{\partial A^j_i}{\partial x^j} + \begin{Bmatrix} j \\ l\ j \end{Bmatrix} A^l_i - \begin{Bmatrix} l \\ i\ j \end{Bmatrix} A^j_l$

$$= \frac{\partial A^j_i}{\partial x^j} + A^l_i \frac{\partial\left(\log\sqrt{g}\right)}{\partial x^l} - A^j_l g^{hl}[ij,h]$$

$$= \frac{\partial A^j_i}{\partial x^j} + A^j_i \frac{1}{2g}\frac{\partial g}{\partial x^j} - A^{jh}[ij,h] \qquad \left[\frac{\partial\left(\log\sqrt{g}\right)}{\partial x^j} = \frac{1}{2}\frac{\partial(\log g)}{\partial x^j} = \frac{1}{2g}\frac{\partial g}{\partial x^j}\right]$$

$$= \frac{1}{\sqrt{g}}\left(\sqrt{g}\frac{\partial A^j_i}{\partial x^j} + A^j_i \frac{1}{2\sqrt{g}}\frac{\partial g}{\partial x^j}\right) - A^{jh}[ij,h$$

$$= \frac{1}{\sqrt{g}}\frac{\partial}{\partial x^j}\left(\sqrt{g}A^j_i\right) - A^{jh}[ij,h]$$

$$A^j_{i,j} = \frac{1}{\sqrt{g}}\frac{\partial}{\partial x^j}\left(\sqrt{g}A^j_i\right) - A^{jh}[ij,h],$$

$$\text{but } A^{jk}[ij,k] = \frac{1}{2}A^{jk}\left(\frac{\partial g_{ik}}{\partial x^j} + \frac{\partial g_{jk}}{\partial x^i} - \frac{\partial g_{ij}}{\partial x^k}\right)$$

$$= \frac{1}{2}\left(A^{jk}\frac{\partial g_{ik}}{\partial x^j} + A^{jk}\frac{\partial g_{jk}}{\partial x^i} - A^{jk}\frac{\partial g_{ij}}{\partial x^k}\right)$$

$$= \frac{1}{2}A^{jk}\frac{\partial g_{jk}}{\partial x^i}$$

$$\therefore A^j_{i,j} = \frac{1}{\sqrt{g}}\frac{\partial}{\partial x^j}\left(\sqrt{g}A^j_i\right) - \frac{1}{2}A^{jk}\frac{\partial g_{jk}}{\partial x^i}.$$

Example 4.4.3. Prove that $\begin{Bmatrix} k \\ i\ j \end{Bmatrix}_a - \begin{Bmatrix} k \\ i\ j \end{Bmatrix}_b$ are components of a tensor of rank 3 where $\begin{Bmatrix} k \\ i\ j \end{Bmatrix}_a$ *and* $\begin{Bmatrix} k \\ i\ j \end{Bmatrix}_b$ are the Christoffel symbols formed from symmetric tensors a_{ij} and b_{ij}.

Solution: We know from (4.19)

$$\frac{\partial^2 x^m}{\partial y^i y^j} = \overline{\begin{Bmatrix} k \\ i\ J \end{Bmatrix}} \frac{\partial x^m}{\partial y^k} - \frac{\partial x^p}{\partial y^i}\frac{\partial x^q}{\partial y^j}\begin{Bmatrix} m \\ p\ q \end{Bmatrix}$$

$$\overline{\begin{Bmatrix} k \\ i\ J \end{Bmatrix}} \frac{\partial x^m}{\partial y^k} - \frac{\partial^2 x^m}{\partial y^i y^j} + \frac{\partial x^p}{\partial y^i}\frac{\partial x^q}{\partial y^j}\begin{Bmatrix} m \\ p\ q \end{Bmatrix}$$

$$\text{or } \overline{\begin{Bmatrix} k \\ i\ J \end{Bmatrix}} = \left[\frac{\partial^2 x^m}{\partial y^i y^j} + \frac{\partial x^p}{\partial y^i}\frac{\partial x^q}{\partial y^j}\begin{Bmatrix} m \\ p\ q \end{Bmatrix}\right]\frac{\partial y^k}{\partial x^{m.}}$$

Using this equation, we can write

$$\overline{\begin{Bmatrix} k \\ i\ J \end{Bmatrix}}_a = \left[\frac{\partial^2 x^m}{\partial y^i y^j} + \frac{\partial x^p}{\partial y^i}\frac{\partial x^q}{\partial y^j}\begin{Bmatrix} m \\ p\ q \end{Bmatrix}_a\right]\frac{\partial y^k}{\partial x^m}$$

$$\overline{\begin{Bmatrix} k \\ i\ J \end{Bmatrix}}_b = \left[\frac{\partial^2 x^m}{\partial y^i y^j} + \frac{\partial x^p}{\partial y^i}\frac{\partial x^q}{\partial y^j}\begin{Bmatrix} m \\ p\ q \end{Bmatrix}_h\right]\frac{\partial y^k}{\partial x^m}.$$

Subtracting, we get

$$\overline{\begin{Bmatrix} k \\ i\ J \end{Bmatrix}_a - \begin{Bmatrix} k \\ i\ J \end{Bmatrix}_b} = \left(\begin{Bmatrix} m \\ p\ q \end{Bmatrix}_a - \begin{Bmatrix} m \\ p\ q \end{Bmatrix}_b\right)\frac{\partial x^p}{\partial y^i}\frac{\partial x^q}{\partial y^j}\frac{\partial y^k}{\partial x^m}$$

Putting $\begin{Bmatrix} m \\ p\ q \end{Bmatrix}_a - \begin{Bmatrix} m \\ p\ q \end{Bmatrix}_b = A^m_{pq}$, we can write the equation

$$\bar{A}^k_{ij} = A^m_{pq} \frac{\partial x^p}{\partial y^i} \frac{\partial x^q}{\partial y^j} \frac{\partial y^k}{\partial x^m}.$$

$\begin{Bmatrix} k \\ i\ j \end{Bmatrix}_a - \begin{Bmatrix} k \\ i\ j \end{Bmatrix}_b$ are the components of a tensor of order 3.

Example 4.4.4. If A^{ij} is a skew-symmetric tensor, show that

$$A^{ij}_{,i} = \frac{1}{\sqrt{g}} \frac{\partial}{\partial x^i}\left(\sqrt{g} A^{ij}\right).$$

Solution: We know

$$A^{ij}_{,k} = \frac{\partial A^{ij}}{\partial x^k} + A^{rj}\begin{Bmatrix} i \\ k\ r \end{Bmatrix} + A^{ir}\begin{Bmatrix} j \\ k\ r \end{Bmatrix}.$$

Putting $k = i$, we get

$$A^{ij}_{,i} = \frac{\partial A^{ij}}{\partial x^i} + A^{rj}\begin{Bmatrix} i \\ i\ r \end{Bmatrix} + A^{ir}\begin{Bmatrix} j \\ i\ r \end{Bmatrix}$$

$$= \frac{\partial A^{ij}}{\partial x^i} + A^{rj}\begin{Bmatrix} i \\ i\ r \end{Bmatrix} \quad \left(\text{If } A^{ij} \text{ is skew symmetric tensor, then } A^{jk}\begin{Bmatrix} i \\ j\ k \end{Bmatrix} = 0\right)$$

$$= \frac{\partial A^{ij}}{\partial x^i} + A^{rj}\frac{\partial}{\partial x^r}\left(log\sqrt{g}\right)$$

$$= \frac{\partial A^{ij}}{\partial x^i} + A^{rj}\frac{1}{\sqrt{g}}\frac{\partial\sqrt{g})}{\partial x^r}$$

$$= \frac{1}{\sqrt{g})}\left(\sqrt{g}\frac{\partial A^{ij}}{\partial x^i} + A^{ij}\frac{\partial\sqrt{g}}{\partial x^i}\right) \quad \text{(replacing dummy index } r \text{ } with\ i\text{)}$$

$$= \frac{1}{\sqrt{g}}\frac{\partial}{\partial x^i}\left(A^{ij}\sqrt{g}\right)$$

Example 4.4.5. If A^{ijk} is a skew-symmetric tensor, show that $\frac{1}{\sqrt{g}}\frac{1}{\partial x^i}\left(\sqrt{g}A^{ijk}\right)$ is a tensor.

Solution: We have

$$A^{ijk}_{,l} = \frac{\partial A^{ijk}}{\partial x^l} + A^{rjk}\begin{Bmatrix} i \\ r\ l \end{Bmatrix} + A^{irk}\begin{Bmatrix} j \\ r\ l \end{Bmatrix} + A^{ijr}\begin{Bmatrix} k \\ r\ l \end{Bmatrix},$$

hence $A^{ijl}_{,l} = \frac{\partial A^{ijl}}{\partial x^l} + A^{rjl}\begin{Bmatrix} i \\ r\ l \end{Bmatrix} + A^{irl}\begin{Bmatrix} j \\ r\ l \end{Bmatrix} + A^{ijr}\begin{Bmatrix} l \\ r\ l \end{Bmatrix}.$

We know $A^{irl}\begin{Bmatrix} j \\ r\ l \end{Bmatrix} = 0$

$$\therefore A^{ijl}_{,l} = \frac{\partial A^{ijl}}{\partial x^l} + A^{ijr}\begin{Bmatrix} l \\ r\ l \end{Bmatrix}, \quad \left[\text{Also } \begin{Bmatrix} l \\ r\ l \end{Bmatrix} = \frac{\partial}{\partial x^r}\left(\log\sqrt{g}\right)\right]$$

$$= \frac{\partial A^{ijl}}{\partial x^l} + A^{ijr}\frac{\partial}{\partial x^r}\left(\log\sqrt{g}\right)$$

$$= \frac{\partial A^{ijl}}{\partial x^l} + \frac{1}{\sqrt{g}}A^{ijr}\frac{\partial}{\partial x^r}\sqrt{g}$$

$$= \frac{1}{\sqrt{g}}\frac{\partial}{\partial x^r}\left(A^{ijr}\sqrt{g}\right) = \frac{1}{\sqrt{g}}\frac{\partial}{\partial x^k}\left(A^{ijk}\sqrt{g}\right).$$

Since $A^{ijl}_{,l}$ is a tensor, the right hand side $\frac{1}{\sqrt{g}}\frac{\partial}{\partial x^k}\left(A^{ijk}\sqrt{g}\right)$ is also a tensor.

Example 4.4.6. If A_{ij} is a skew-symmetric tensor of rank two, show that

$$A_{ij,k} + A_{jk,i} + A_{ki,j} = \frac{\partial A_{ij}}{\partial x^k} + \frac{\partial A_{jk}}{\partial x^i} + \frac{\partial A_{ki}}{\partial x^j}.$$

Solution: We know that

$$A_{ij,k} = \frac{\partial A_{ij}}{\partial x^k} - \begin{Bmatrix} l \\ i\ k \end{Bmatrix} A_{lj} - \begin{Bmatrix} l \\ j\ k \end{Bmatrix} A_{il}$$

$$A_{jk,i} = \frac{\partial A_{jk}}{\partial x^i} - \begin{Bmatrix} l \\ j\ i \end{Bmatrix} A_{lk} - \begin{Bmatrix} l \\ j\ k \end{Bmatrix} A_{jl}$$

$$A_{ki,j} = \frac{\partial A_{ki}}{\partial x^j} - \begin{Bmatrix} l \\ k\ j \end{Bmatrix} A_{li} - \begin{Bmatrix} l \\ i\ j \end{Bmatrix} A_{kl}$$

$$A_{ij,k} + A_{jk,i} + A_{ki,j} = \frac{\partial A_{ij}}{\partial x^k} - \begin{Bmatrix} l \\ i\ k \end{Bmatrix} A_{lj} - \begin{Bmatrix} l \\ j\ k \end{Bmatrix} A_{il} + \frac{\partial A_{jk}}{\partial x^i} - \begin{Bmatrix} l \\ j\ i \end{Bmatrix} A_{lk}$$

$$+ \frac{\partial A_{ki}}{\partial x^j} - \begin{Bmatrix} l \\ k\ j \end{Bmatrix} A_{li} - \begin{Bmatrix} l \\ i\ j \end{Bmatrix} A_{kl}$$

$$= \frac{\partial A_{ij}}{\partial x^k} + \frac{\partial A_{jk}}{\partial x^i} + \frac{\partial A_{ki}}{\partial x^j} - \left(\begin{Bmatrix} l \\ i\ k \end{Bmatrix} A_{lj} + \begin{Bmatrix} l \\ k\ i \end{Bmatrix} A_{jl} + \right)$$

$$- \left(\begin{Bmatrix} l \\ k\ j \end{Bmatrix} A_{li} + \begin{Bmatrix} l \\ j\ k \end{Bmatrix} A_{il} \right) - \left(\begin{Bmatrix} l \\ j\ i \end{Bmatrix} A_{lk} + \begin{Bmatrix} l \\ i\ j \end{Bmatrix} A_{kl} \right).$$

Since $\begin{Bmatrix} l \\ i\ k \end{Bmatrix} = \begin{Bmatrix} l \\ k\ i \end{Bmatrix}$, etc.

$$= \frac{\partial A_{ij}}{\partial x^k} + \frac{\partial A_{jk}}{\partial x^i} + \frac{\partial A_{ki}}{\partial x^j} - \begin{Bmatrix} l \\ i\ k \end{Bmatrix} (A_{lj} + A_{jl}) - \begin{Bmatrix} l \\ k\ j \end{Bmatrix} (A_{li} + A_{il}) - \begin{Bmatrix} l \\ i\ j \end{Bmatrix} (A_{kl} + A_{lk}),$$

Since A_{ij} is a skew-symmetric tensor of rank two, then $A_{lj} + A_{jl} = 0$, $A_{li} + A_{il} = 0$, $A_{kl} + A_{lk} = 0$

$$\therefore A_{ij,k} + A_{jk,i} + A_{ki,j} = \frac{\partial A_{ij}}{\partial x^k} + \frac{\partial A_{jk}}{\partial x^i} + \frac{\partial A_{ki}}{\partial x^j}.$$

4.5 Gradient, Divergence, and Curl

The tensorial nature of partial derivatives is a very useful feature. We can apply it to extend the scope of classical operations of vector analysis. The equations in theoretical physics are expressed with the help of a small number of operators called gradient, divergence, curl, and Lapacian.

4.5.1 Gradient

The partial derivative of an invariant ϕ (a scalar function of a coordinate) is a covariant vector which is called the gradient of ϕ and is denoted by $grad\phi$ or $\nabla\phi$.

$$\nabla\phi = grad\phi = \frac{\partial\phi}{\partial x^i}$$

Indeed, the nabla operator applied to a scalar field, which is a tensor field of type (0,0), produces a tensor field of type (0,1).

4.5.2 Divergence

Let A^i be a contravariant vector. Then, its covariant derivative, given by

$$A^i_{,j} = \frac{\partial A^i}{\partial x^j} + \begin{Bmatrix} i \\ \alpha \;\; j \end{Bmatrix} A^\alpha,$$

is a tensor of type (1,1). If we contract the indices i and j, we get the tensor $A^i_{,i}$, which is the tensor of type (0,0), the invariant.

The divergence of contravariant vector A^i is defined by

$$divA^i = \frac{\partial A^i}{\partial x^i} + \begin{Bmatrix} i \\ \alpha \;\; i \end{Bmatrix} A^\alpha, \text{ which is sometimes written as } A^i_{,i}.$$

Now, we derive an expression for the divergence of a contravariant vector A^i

$$divA^i = A^i_{,i} = \frac{1}{\sqrt{g}}\frac{\partial}{\partial x^k}\left(\sqrt{g}A^k\right), \text{ where } g = |g_{ij}|. \tag{4.41}$$

Proof:

$$\begin{aligned} A^i_{,i} &= \frac{\partial A^i}{\partial x^i} + \begin{Bmatrix} i \\ \alpha \;\; i \end{Bmatrix} A^\alpha \\ &= \frac{\partial A^i}{\partial x^i} + \frac{\partial}{\partial x^\alpha}\left(log\sqrt{g}\right)A^\alpha \\ &= \frac{\partial A^k}{\partial x^k} + \frac{\partial}{\partial x^k}\left(log\sqrt{g}\right)A^k \\ &= \frac{1}{\sqrt{g}}\frac{\partial}{\partial x^k}\left(\sqrt{g}A^k\right) \end{aligned}$$

Divergence of a Covariant Vector: Let us consider an arbitrary covariant vector A_i. Then, $g^{jk}A_{j,k}$ is an invariant. It is the divergence of A_i and is denoted by $divA_i$, therefore, $div\ A_i = g^{jk}A_{j,k.}$

The divergence of contravariant vector A^i is said to be the divergence of A_i.

$$div\ A_i = g^{ik}A_{j,k} = (g^{ik}A_j)_{,k}\ , as\ g^{jk}_{,k} = 0$$
$$= A^k_{,k} = divA^k = divA^i.$$

If A_i and A^i are the covariant and contravariant components of same vector, then $div\ A_i = divA^i$.

Theorem 4.5.1. Let A be an arbitrary vector and ϕ and φ be a scalar function of coordinates x^i. Then,

(i) $div(\phi A) = grad(\phi)A + \phi \mathrm{div} A$

(ii) $\nabla(\phi\varphi) = \phi\nabla\varphi + \varphi\nabla\phi$

(iii) $\nabla^2(\phi\varphi) = \phi\nabla^2\varphi + \varphi\nabla^2\phi + 2\nabla\phi.\nabla\varphi$

(iv) $div(\varphi\nabla\phi) = \varphi\nabla^2\phi + \nabla\phi.\nabla\varphi$

Proof:

(i) We know $divA^i = A^i_{,i} = \dfrac{1}{\sqrt{g}}\dfrac{\partial}{\partial x^i}\left(\sqrt{g}A^i\right)$ (4.42a)

$$\therefore div(\phi A^i) = (\phi A^i)_{,i} = \frac{1}{\sqrt{g}}\frac{\partial}{\partial x^i}\left(\sqrt{g}\phi A^i\right)$$
$$= \frac{1}{\sqrt{g}}\left[\frac{\partial}{\partial x^i}\left(\sqrt{g}A^i\right)\phi + \sqrt{g}A^i\frac{\partial\phi}{\partial x^i}\right]$$
$$= A^i\frac{\partial\phi}{\partial x^i} + \phi\frac{1}{\sqrt{g}}\frac{\partial}{\partial x^i}\left(\sqrt{g}A^i\right)$$
$$= grad(\phi)A + \phi divA\left[\text{since } divA = \frac{1}{\sqrt{g}}\frac{\partial}{\partial x^k}\left(\sqrt{g}A^k\right)\right]$$

$$div(\phi A) = grad(\phi)A + \phi divA \tag{4.42b}$$

(ii) By definition of gradient,

$$\nabla(\phi\varphi) = \frac{\partial(\phi\varphi)}{\partial x^i} = \varphi\frac{\partial\phi}{\partial x^i} + \phi\frac{\partial\varphi}{\partial x^i}$$

$$\therefore \nabla(\phi\varphi) = \phi\nabla\varphi + \varphi\nabla\phi \tag{4.42c}$$

(iii) Taking the divergence on both sides in Equation (4.42c),

$$div\nabla(\phi\varphi) = div[\phi\nabla\varphi + \varphi\nabla\phi]$$

$$\text{or } \nabla^2(\phi\varphi) = div(\phi\nabla\varphi) + div(\varphi\nabla\phi)$$

$$= \nabla(\phi)\nabla\varphi + \phi div\nabla\varphi + \nabla(\varphi)\nabla\phi + \varphi div\nabla\phi \qquad \text{(by 4.42b)}$$

$$= \varphi div\nabla\phi + \phi div\nabla\varphi + 2\nabla(\varphi)\nabla\phi$$

(iv) Replacing A by $\nabla\varphi$ in (4.42b), we get

$$div(\phi\nabla\varphi) = grad(\phi)\nabla\varphi + \phi \text{div}(\nabla\varphi)$$

$$= \nabla\phi.\nabla\varphi + \varphi\nabla^2\phi.$$

4.5.2.1 Divergence of a Mixed Tensor (1,1)

Let A^i_j be a arbitrary tensor of type (1,1). Then, the *div* of A^i_j

$$divA^i_j = A^i_{j,i}.$$

We know $A^i_{j,k} = \frac{\partial A^i_j}{\partial x^k} + A^l_j \begin{Bmatrix} i \\ k\ l \end{Bmatrix} - A^i_l \begin{Bmatrix} l \\ j\ k \end{Bmatrix}$ (using 4.38).

Putting $k = i$, we get,

$$A^i_{j,i} = \frac{\partial A^i_j}{\partial x^i} + A^l_j \begin{Bmatrix} i \\ i\ l \end{Bmatrix} - A^i_l \begin{Bmatrix} l \\ j\ i \end{Bmatrix}$$

$$= \frac{1}{\sqrt{g}}\left(\sqrt{g}\frac{\partial A^i_j}{\partial x^i} + A^i_j \frac{\partial \sqrt{g}}{\partial x^i}\right) - A^i_l \begin{Bmatrix} l \\ j\ i \end{Bmatrix} \left[\begin{Bmatrix} i \\ i\ l \end{Bmatrix} = \frac{\partial}{\partial x^l}\left(\log\sqrt{g}\right)\right]$$

$$= \frac{\partial A^i_j}{\partial x^i} + A^i_j \frac{\partial\left(\log\sqrt{g}\right)}{\partial x^i} - A^i_l \begin{Bmatrix} l \\ j\ i \end{Bmatrix},$$ changing dummy index i by l from the

2nd term,

$$= \frac{1}{\sqrt{g}} \frac{\partial\left(\sqrt{g}A^i_j\right)}{\partial x^i} - A^i_l \begin{Bmatrix} l \\ j\ i \end{Bmatrix}.$$

Example 4.5.1. Prove that $divA^{ij} = \nabla_j = A^{ij}_{,j} = \frac{1}{\sqrt{g}}\frac{\partial}{\partial x^j}\left(\sqrt{g}A^{ij}\right) + \begin{Bmatrix} j \\ a\ k \end{Bmatrix} A^{i\alpha}.$

Find the expression for A^{ij} when A^{ij} is skew-symmetric.
Solution: We know

$$A^{ij}_{,k} = \frac{\partial A^{ij}}{\partial x^k} + A^{rj}\begin{Bmatrix} i \\ k\ r \end{Bmatrix} + A^{ir}\begin{Bmatrix} j \\ k\ r \end{Bmatrix}.$$

Putting $j = k$, we get

$$A^{ij}_{,j} = \frac{\partial A^{ij}}{\partial x^j} + A^{rj}\begin{Bmatrix} i \\ k\ r \end{Bmatrix} + A^{ir}\begin{Bmatrix} j \\ j\ r \end{Bmatrix}$$

$$= \frac{\partial A^{ij}}{\partial x^j} + A^{rj}\begin{Bmatrix} i \\ k\ r \end{Bmatrix} + A^{ir}\frac{\partial}{\partial x^r}\left(log\sqrt{g}\right) \quad \left(\text{using}\begin{Bmatrix} l \\ r\ l \end{Bmatrix} = \frac{\partial}{\partial x^r}\left(log\sqrt{g}\right)\right)$$

$$= \frac{\partial A^{ij}}{\partial x^j} + \frac{A^{ir}}{\sqrt{g}}\frac{\partial}{\partial x^r}\left(\sqrt{g}\right) + A^{rj}\begin{Bmatrix} i \\ k\ r \end{Bmatrix}$$

$$= \frac{\partial A^{ij}}{\partial x^j} + \frac{A^{ij}}{\sqrt{g}}\frac{\partial}{\partial x^j}\left(\sqrt{g}\right) + A^{jk}\begin{Bmatrix} i \\ k\ j \end{Bmatrix} \quad \text{replacing dummy index } r \text{ by } k.$$

$$= \frac{1}{\sqrt{g}}\left(A^{ij}\sqrt{g}\right) + A^{jk}\begin{Bmatrix} i \\ k\ j \end{Bmatrix} \quad \text{(i)}$$

2nd part: If A^{ij} is skew-symmetric, then $A^{ij} = -A^{ji}$.

Interchanging indices *j and k*, we get

$$A^{jk}\begin{Bmatrix} i \\ k\ j \end{Bmatrix} = A^{kj}\begin{Bmatrix} i \\ j\ k \end{Bmatrix} = -A^{jk}\begin{Bmatrix} i \\ j\ k \end{Bmatrix} = -A^{kj}\begin{Bmatrix} i \\ k\ j \end{Bmatrix} \Rightarrow A^{jk}\begin{Bmatrix} i \\ k\ j \end{Bmatrix} = 0.$$

Putting this value in (i), we get $divA^{ij} = A^{ij}_{,j} = \frac{1}{\sqrt{g}}\frac{\partial}{\partial x^j}\left(A^{ij}\sqrt{g}\right)$.

Example 4.5.2. If the components A_1, A_2, *and* A_3 of a vector in cylindrical coordinates, x^1, x^2, *and* x^3 are x^1, $x^3\, sinx^2$, *and* $e^{x^2} cosx^3$, prove that

$$divA_i = 2 + \frac{x^3}{(x^1)^2}cosx^2 - e^{x^2}sinx^3.$$

Solution: $(ds)^2 = (dx^1)^2 + (x^1)^2 (dx^2)^2 + (dx^3)^2$

Then, $g_{11} = 1$, $g_{22} = (x^1)^2$, $g_{33} = 1$, $g_{ij} = 0$ *when* $i \neq j$

$$\therefore\ detg = g = |g_{ij}| = \begin{vmatrix} 1 & 0 & 0 \\ 0 & x^1 & 0 \\ 0 & 0 & 1 \end{vmatrix} = \left(x^1\right)^2$$

We get $g^{11} = 1$, $g^{22} = \frac{1}{(x^1)^2}$, $g^{33} = 1$, *and* $g^{ij} = 0$, *when* $i \neq j$.

$$A_1 = x^1, A_2 = x^3 sinx^2, A_3 = e^{x^2} cosx^3$$

$$A^1 = g^{1i} A_1 = g^{11} A_1 = x^1$$

$$A^2 = g^{2i} A_i = g^{22} A_2 = \frac{1}{(x^1)^2} x^3 sinx^2$$

$$A^3 = g^{3i} A_i = g^{33} A_3 = e^{x^2} cosx^3$$

$$\therefore divA^i = divA_i = \frac{1}{\sqrt{g}} \frac{\partial}{\partial x^k}\left(\sqrt{g} A^k\right)$$

$$= \frac{1}{x^1}\left[\frac{\partial}{\partial x^1}\left(\sqrt{g} A^1\right) + \frac{\partial}{\partial x^2}\left(\sqrt{g} A^2\right) + \frac{\partial}{\partial x^3}\left(\sqrt{g} A^3\right)\right]$$

$$= \frac{1}{x^1}\left[\frac{\partial}{\partial x^1}(x^1 x^1) + \frac{\partial}{\partial x^2}\left(x^1 \frac{1}{(x^1)^2} x^3 sinx^2\right) + \frac{\partial}{\partial x^3}(x^1 e^{x^2} cosx^3)\right]$$

$$= \frac{1}{x^1}\left[2x^1 + \frac{x^3}{x^1} \frac{\partial}{\partial x^2}(sinx^2) + \frac{\partial}{\partial x^3}(x^1 e^{x^2} cosx^3)\right]$$

$$= \frac{1}{x^1}\left[2x^1 + \frac{x^3}{x^1} cosx^2 - x^1 e^{x^2} sinx^3)\right]$$

$$= 2 + \frac{x^3}{(x^1)^2} cosx^2 - e^{x^2} sinx^3$$

Example 4.5.3. Prove that in a V_2 with line element $(ds)^2 = (dx^1)^2 + (x^1)^2 (dx^2)^2$, the divergence of the covariant vector with components $x^1 cos2x^2$, $-(x^1)^2 sin2x^2$ is zero.

Solution: $g_{11} = 1$, $g_{22} = (x^1)^2$, $g_{ij} = 0$ *when* $i \neq j$

$$detg = g = |g_{ij}| = (x^1)^2$$

$$g^{11} = 1, g^{22} = \frac{1}{(x^1)^2}$$

$$A_1 = x^1 cos2x^2,\ A_2 = -(x^1)^2 sin2x^2.$$

$$A^1 = g^{1i} A_1 = g^{11} A_1 = x^1 cos2x^2$$

$$A^2 = g^{2i}A_1 = g^{22}A_2 = \frac{1}{(x^1)^2}(-(x^1)^2 sin2x^2) = -sin2x^2$$

$$\therefore divA^i = divA_i = \frac{1}{\sqrt{g}}\frac{\partial}{\partial x^k}\left(\sqrt{g}A^k\right)$$

$$= \frac{1}{x^1}\left[\frac{\partial}{\partial x^1}\left(\sqrt{g}A^1\right) + \frac{\partial}{\partial x^2}\left(\sqrt{g}A^2\right)\right]$$

$$= \frac{1}{x^1}\left(\frac{\partial}{\partial x^1}(x^1x^1cos2x^2\right) + \frac{1}{x^1}\frac{\partial}{\partial x^2}(x^1.-sin2x^2)$$

$$= \frac{1}{x^1}2x^1.cos2x^2 + \frac{1}{x^1}(-x^1).2.cos2x^2$$

$$= 2.cos2x^2 - .2.cos2x^2 = 0$$

If the divergence of a covariant vector or contravariant vector vanishes identically, then the vector i said to be *conservative.*

4.5.3 Laplacian of an Invariant

Let ϕ be an invariant and $A_i = \frac{\partial \phi}{\partial x^i}$. Then, Ai is a covariant vector. Hence, $g^{ij}A_j = A^i$ is a contravariant vector. The divergence of this contravariant vector is called the *Laplacian* of ϕ.

The Laplacian of an invariant ϕ is denoted by $\nabla^2\phi$.

$$\text{Thus, } \nabla^2\phi = \text{div (grad}\phi) \tag{4.43}$$

An Expression for $\nabla^2\phi$:

$$\nabla^2\phi = \frac{1}{\sqrt{g}}\frac{\partial}{\partial x^k}\left(\sqrt{g}g^{kj}\frac{\partial \phi}{\partial x^j}\right)$$

Proof: We know $A^i_{,i} = \frac{1}{\sqrt{g}} = \frac{\partial}{\partial x^k}\left(\sqrt{g}A^k\right)$. (i)

$$\text{Let } A^i = g^{ij}A_j = g^{ij}\frac{\partial \phi}{\partial x^j}.$$

(i) takes the form $A^i_{,i} = \frac{1}{\sqrt{g}}\frac{\partial}{\partial x^k}\left(\sqrt{g}g^{kj}\frac{\partial \phi}{\partial x^j}\right)$.

Now, by definition, $\nabla^2\phi = A^i_{,i}$

$$\therefore \nabla^2\phi = \frac{1}{\sqrt{g}}\frac{\partial}{\partial x^k}\left(\sqrt{g}g^{kj}\frac{\partial \phi}{\partial x^j}\right) \tag{4.44}$$

4.5.4 Curl of a Covariant Vector

A skew-symmetric tensor of type (0,2) formed from a covariant vector, assuming particular importance in E_3, where it can be identified with a vector. We now introduce such a skew-symmetric tensor in, which is called the *curl* of the covariant vector.

Let us consider a covariant vector A_i. Then, $A_{i,j}$ is a tensor of type (0,2). Hence, $A_{j,i}$ is also a tensor of type (0,2). Consequently, $A_{i,j} - A_{j,i}$ is a tensor of type (0,2), which is evidently skew- symmetric. This skew-symmetric tensor is called the curl of vector A_i and is denoted by $curl A_i$.

An Expression for the Curl of Covariant Vector A_i

$$curl\ A_i = A_{i,j} - A_{j,i} = \frac{\partial A_i}{\partial x^j} - \frac{\partial A_j}{\partial x^i} \tag{4.45}$$

Proof: We have $A_{i,j} = \frac{\partial A_i}{\partial x^j} - \begin{Bmatrix} r \\ i\ j \end{Bmatrix}$

$$\text{and } A_{j,i} = \frac{\partial A_j}{\partial x^i} - \begin{Bmatrix} r \\ j\ i \end{Bmatrix}$$

$$\therefore curl\ A_i = A_{i,j} - A_{j,i} = \frac{\partial A_i}{\partial x^j} - \begin{Bmatrix} r \\ i\ j \end{Bmatrix} - \frac{\partial A_j}{\partial x^i} + \begin{Bmatrix} r \\ j\ i \end{Bmatrix}$$

$$= \frac{\partial A_i}{\partial x^j} - \frac{\partial A_j}{\partial x^i} \quad \left(\because \begin{Bmatrix} r \\ i\ j \end{Bmatrix} = \begin{Bmatrix} r \\ j\ i \end{Bmatrix}\right)$$

Example 4.5.4. Show that *curl grad*$\phi = 0$ for invariant ϕ.

Solution: Let $A_i = \frac{\partial \phi}{\partial x^i} = grad\phi$.

Then, $\frac{\partial A_i}{\partial x^j} = \frac{\partial\left(\frac{\partial \phi}{\partial x^i}\right)}{\partial x^j} = \frac{\partial^2 \phi}{\partial x^j \partial x^i}$ and $\frac{\partial A_j}{\partial x^i} = \frac{\partial\left(\frac{\partial \phi}{\partial x^j}\right)}{\partial x^i} = \frac{\partial^2 \phi}{\partial x^i \partial x^j}$.

Now, $curlA_i = \frac{\partial A_i}{\partial x^j} - \frac{\partial A_j}{\partial x^i} = \frac{\partial^2 \phi}{\partial x^j \partial x^i} - \frac{\partial^2 \phi}{\partial x^i \partial x^j} = 0$

or $curlA_i = curlgrad\phi = 0$.

Theorem 4.5.2. A necessary and sufficient condition is that the curl of the vector field vanishes if the vector field is gradient.

Proof: Let A_i be a covariant vector. Let the curl of the vector A_i vanish so that

$$curlA_i = A_{i,j} - A_{j,i} = \frac{\partial A_i}{\partial x^j} - \frac{\partial A_j}{\partial x^i} = 0.$$

We have to show that $A_i = \nabla \phi$ where ϕ is a scalar.

$$curlA_i = 0 \Rightarrow \frac{\partial A_i}{\partial x^j} - \frac{\partial A_j}{\partial x^i} = 0$$

$$\Rightarrow \frac{\partial A_i}{\partial x^j} dx^j = \frac{\partial A_j}{\partial x^i} dx^i$$

$$\Rightarrow dA_i = \frac{\partial}{\partial x^i}(A_j dx^i)$$

$$\Rightarrow A_i = \int \frac{\partial}{\partial x^i}(A_j\, dx^i) = \frac{\partial}{\partial x^i} \int (A_j dx^i),$$

but if $\int(A_j dx^i)$ is a scalar quantity, let $\int(A_j dx^i) = \phi$. Then,

$$A_i = \frac{\partial \phi}{\partial x^i} = grad\phi.$$

$\Rightarrow$ the vector A_i is a gradient.

Conversely, let vector A_i be a gradient such that $A_i = grad\phi$, i.e., we have to show that $curlA_i = curlgrad\phi = 0$.

Conversely,

$$A_i = \nabla \phi \Rightarrow A_i = \frac{\partial \phi}{\partial x^i}$$

$$\frac{\partial A_i}{\partial x^j} = \frac{\partial^2 \phi}{\partial x^j \partial x^i} = \frac{\partial^2 \phi}{\partial x^i \partial x^j}$$

$$\frac{\partial A_i}{\partial x^j} - \frac{\partial A_j}{\partial x^i} = \frac{\partial^2 \phi}{\partial x^j \partial x^i} - \frac{\partial^2 \phi}{\partial x^i \partial x^j} = 0$$

$$\text{or } curl\ A_i = A_{i,j} - A_{j,i} = \frac{\partial A_i}{\partial x^j} - \frac{\partial A_j}{\partial x^i} = 0$$

If the vector field is gradient, the curl of the vector field vanishes.

Example 4.5.5. If A_{ij} are the components of the curl of a covariant vector, prove that

$$A_{ij,k} + A_{ki,j} + A_{jk,i} = 0.$$

Solution: Let A_{ij} be the components of the curl of a covariant vector B_i. We have

$$A_{ij} = curlB_i = B_{i,j} - B_{j,i}.$$

Interchanging indices i and j, $A_{ji} = B_{j,i} - B_{i,j} = (B_{i,j} - B_{j,i}) = -A_{ij}$.
$\Rightarrow A_{ij}$ is skew-symmetric.

$$A_{ij} = curlB_i = B_{i,j} - B_{j,i} = \frac{\partial B_i}{\partial x^j} - \frac{\partial B_j}{\partial x^i}$$

$$A_{ij,k} = \frac{\partial A_{ij}}{\partial x^k} = \frac{\partial^2 B_i}{\partial x^k \partial x^j} - \frac{\partial^2 B_j}{\partial x^k \partial x^i} \tag{i}$$

$$\text{Now, } A_{ij,k} = \frac{\partial A_{ij}}{\partial x^k} - \begin{Bmatrix} l \\ i\ k \end{Bmatrix} A_{lj} - \begin{Bmatrix} l \\ j\ k \end{Bmatrix} A_{il} \tag{ii}$$

$$= \frac{\partial^2 B_i}{\partial x^k \partial x^j} - \frac{\partial^2 B_j}{\partial x^k \partial x^i} - \begin{Bmatrix} l \\ i\ k \end{Bmatrix} A_{lj} - \begin{Bmatrix} l \\ j\ k \end{Bmatrix} A_{il} \text{ (by (i))} \tag{iii}$$

Putting $i = j$, $j = k$, $k = i$ in (iii), we get

$$A_{jk,i} = \frac{\partial^2 B_j}{\partial x^i \partial x^k} - \frac{\partial^2 B_k}{\partial x^i \partial x^j} - \begin{Bmatrix} l \\ j\ i \end{Bmatrix} A_{lk} - \begin{Bmatrix} l \\ k\ i \end{Bmatrix} A_{jl} \tag{iv}$$

Putting $i = j$, $j = k$, $k = i$ in (iv), we get

$$A_{ki,j} = \frac{\partial^2 B_k}{\partial x^j \partial x^i} - \frac{\partial^2 B_i}{\partial x^j \partial x^i} - \begin{Bmatrix} l \\ k\ j \end{Bmatrix} A_{li} - \begin{Bmatrix} l \\ i\ j \end{Bmatrix} A_{kl} \tag{v}$$

Adding (ii), (iv), and (v), we get

$$\begin{aligned} & A_{ij,k} + A_{ki,j} + A_{jk,i} \\ &= \frac{\partial^2 B_i}{\partial x^k \partial x^j} - \frac{\partial^2 B_j}{\partial x^k \partial x^i} - \begin{Bmatrix} l \\ i\ k \end{Bmatrix} A_{lj} - \begin{Bmatrix} l \\ j\ k \end{Bmatrix} A_{il} \\ &+ \frac{\partial^2 B_j}{\partial x^i \partial x^k} - \frac{\partial^2 B_k}{\partial x^i \partial x^j} - \begin{Bmatrix} l \\ j\ i \end{Bmatrix} A_{lk} - \begin{Bmatrix} l \\ k\ i \end{Bmatrix} A_{jl} \\ &+ \frac{\partial^2 B_k}{\partial x^j \partial x^i} - \frac{\partial^2 B_i}{\partial x^j \partial x^i} - \begin{Bmatrix} l \\ k\ j \end{Bmatrix} A_{li} - \begin{Bmatrix} l \\ i\ j \end{Bmatrix} A_{kl} = 0. \end{aligned}$$

The result is equivalently written as

$$\nabla_k A_{ij} + \nabla_i A_{jk} + \nabla_j A_{ki} = 0.$$

4.6 Exercises

1. Prove that all Christoffel symbols vanish at a point, if and only if g_{ij}'s are all constant at the point.
2. If A^{ij} is a skew-symmetric tensor, show that $A^{ik}\left\{ \begin{matrix} i \\ j\ k \end{matrix} \right\} = 0.$
3. If A^{pq} is a symmetric tensor, show that $A^{qm}[i,q,m] = \frac{1}{2} A^{qm} \frac{\partial g_{qm}}{\partial x^i}.$
4. Show that the number of independent components of the Christoffel symbols in a V_n is, at most, $\frac{1}{2}n^2(n+1).$
5. Calculate the non-zero Christoffel symbols corresponding to metrics

$$\text{(i) } (ds)^2 = (dx^1)^2 + (x^1)^2(dx^2)^2 + (dx^3)^2$$

$$\text{(i) } (ds)^2 = (dx^1)^2 + f(x^1, x^2)(dx^2)^2 + (dx^3)^2$$

$$\text{(ii) } (ds)^2 = a^2[(sinhy^1)^2 + (siny^2)^2]\{(dy^1)^2 + (dy^2)^2] + (dy^3)^2$$

6. Calculate the Christoffel symbols of the 2nd kind in (i) rectangular, (ii) cylindrical, and (iii) spherical coordinates.
7. Show that $A^{ij}_{,i} = \frac{1}{\sqrt{g}} \frac{\partial\left(A_i^j \sqrt{g}\right)}{\partial x^i}$, where Ai^j is a skew-symmetric tensor.
8. (i) If A_{ij} is a symmetric tensor, show that $A_{ij,k}$ is symmetric in i and j.
(ii) If A_{ij} is a symmetric tensor such that $A_{ij,k} = A_{ik,j}$, show that $A_{ij,k}$ is a symmetric tensor.
9. If A^i is a contravariant vector such that $A^i_{,k} = a_k A^i$ where a_k is a covariant vector, show that a_k is gradient.
10. Prove that $A^r_{ijk,l} = \frac{\partial A^r_{ijk}}{\partial x} - \left\{ \begin{matrix} \alpha \\ i\ l \end{matrix} \right\} A^r_{\alpha jk} - \left\{ \begin{matrix} \alpha \\ j\ l \end{matrix} \right\} A^r_{i\alpha k} - \left\{ \begin{matrix} \alpha \\ k\ l \end{matrix} \right\} A^r_{ij\alpha} - \left\{ \begin{matrix} r \\ \alpha\ l \end{matrix} \right\} A^\alpha_{ijk}$

is a tensor.

11. If the contravariant components A^i of a vector in spherical coordinates r, θ, *and* ϕ are r, $2\cos\theta$, *and* $-\phi$, find $curl A_i$.
12. If ϕ is scalar, then $g^{ij}\phi_{,ij}$ is a scalar and is equal to $\frac{1}{\sqrt{g}}\frac{\partial}{\partial x^k}\left(\sqrt{g}\,g^{kj}\frac{\partial\phi}{\partial x^j}\right)$.

5
Riemannian Geometry

5.1 Introduction

The covariant derivative of a tensor is, in general, a tensor. If the resulting tensor is again subjected to covariant differentiation, then again we get a tensor and so on. In the case of an invariant, the operation of covariant differentiation is commutative, but a tensor of order greater than or equal to one of the operation is not commutative. This is due to a peculiarity in which the operation is undertaken, namely a Riemannian space.

Characteristic peculiarity of such a space in a certain tensor is called the curvature tensor whose components can be expressed with the help of components of fundamental tensors. The operation of covariant differentiation is not, in general, commutative in a Riemannian space.

We discuss in this chapter Riemannian-Christoffel tensors, Ricci's tensor and its properties, and discuss Riemannian and Euclidean spaces.

5.2 Riemannian-Christoffel Tensor

A sufficient condition for the equality of mixed partial derivatives $\frac{\partial^2 u}{\partial x \partial y}$ and $\frac{\partial^2 u}{\partial y \partial x}$, of a function u(x,y), is that u(x,y) be of class C^2, but this is not sufficient to ensure the equality of mixed covariant derivatives. Indeed, it is shown that if the order of covariant differentiation is to be immaterial, our tensors must be defined over a particular metric manifold X, for which a certain tensor of rank four, made up entirely of the g_{ij}'s, vanishes. This tensor is known as the Riemann-Christoffel tensor. It plays a basic role in many investigations of Differential Geometry and other branches of engineering sciences.

The covariant derivative of a tensor is a tensor; hence it can be differentiated again to obtain a new tensor which is called the 2nd *covariant derivative* of the given tensor.

Dipankar De. Introduction to Differential Geometry with Tensor Applications, (143–176) © 2022 Scrivener Publishing LLC

Consider the covariant x^j derivative of A_i (with respect to g_{ij})

$$A_{i,j} = \frac{\partial A_i}{\partial x^j} - \begin{Bmatrix} \alpha \\ i\ j \end{Bmatrix} A_\alpha \tag{5.1}$$

If $a_{ij} = A_{i,j,}$ *then* $a_{ij,k} = A_{i,jk}$.

Now, differentiate (5.1) covariantly with respect to x^k, their results should be a tensor.

$$\begin{aligned} A_{i,jk} &= \frac{\partial A_{i,j}}{\partial x^k} - \begin{Bmatrix} \alpha \\ i\ k \end{Bmatrix} A_{\alpha,j} - \begin{Bmatrix} \alpha \\ j\ k \end{Bmatrix} A_{i,\alpha} \\ &= \frac{\partial}{\partial x^k}\left(\frac{\partial A_i}{\partial x^j} - \begin{Bmatrix} \alpha \\ i\ j \end{Bmatrix} A_\alpha \right) - \begin{Bmatrix} \alpha \\ i\ k \end{Bmatrix}\left(\frac{\partial A_\alpha}{\partial x^j} - \begin{Bmatrix} \beta \\ \alpha\ j \end{Bmatrix} A_\beta \right) \\ &- \begin{Bmatrix} \alpha \\ j\ k \end{Bmatrix}\left(\frac{\partial A_i}{\partial x^\alpha} - \begin{Bmatrix} \gamma \\ i\ \alpha \end{Bmatrix} A_\gamma \right) \end{aligned} \tag{5.2}$$

and

$$\begin{aligned} A_{i,kj} &= \frac{\partial}{\partial x^j}\left(\frac{\partial A_i}{\partial x^k} - \begin{Bmatrix} \alpha \\ i\ k \end{Bmatrix} A_\alpha \right) - \begin{Bmatrix} \alpha \\ i\ j \end{Bmatrix}\left(\frac{\partial A_i}{\partial x^k} - \begin{Bmatrix} \beta \\ \alpha\ k \end{Bmatrix} A_\beta \right) \\ &- \begin{Bmatrix} \alpha \\ k\ j \end{Bmatrix}\left(\frac{\partial A_i}{\partial x^\alpha} - \begin{Bmatrix} \gamma \\ i\ \alpha \end{Bmatrix} A_\gamma \right) \end{aligned} \tag{5.3}$$

Now (5.2) and (5.3) can be differentiated as indicated:

$$\begin{aligned} A_{i,jk} &= \frac{\partial^2 A_i}{\partial x^k \partial x^j} - \frac{\partial \begin{Bmatrix} \alpha \\ i\ j \end{Bmatrix}}{\partial x^k} A_\alpha - \begin{Bmatrix} \alpha \\ i\ j \end{Bmatrix} \frac{\partial A_\alpha}{\partial x^k} \\ &- \begin{Bmatrix} \alpha \\ i\ k \end{Bmatrix} \frac{\partial A_\alpha}{\partial x^j} + \begin{Bmatrix} \alpha \\ i\ k \end{Bmatrix}\begin{Bmatrix} \beta \\ \alpha\ j \end{Bmatrix} A_\beta \\ &- \begin{Bmatrix} \alpha \\ j\ k \end{Bmatrix} \frac{\partial A_i}{\partial x^\alpha} + \begin{Bmatrix} \alpha \\ j\ k \end{Bmatrix}\begin{Bmatrix} \gamma \\ i\ \alpha \end{Bmatrix} A_\gamma \end{aligned} \tag{5.4}$$

$$A_{i,kj} = \frac{\partial^2 A_i}{\partial x^j \partial x^k} - \frac{\partial \begin{Bmatrix} \alpha \\ i\ k \end{Bmatrix}}{\partial x^j} A_\alpha - \begin{Bmatrix} \alpha \\ i\ k \end{Bmatrix} \frac{\partial A_\alpha}{\partial x^j}$$

$$- \begin{Bmatrix} \alpha \\ i\ j \end{Bmatrix} \frac{\partial A_\alpha}{\partial x^k} + \begin{Bmatrix} \alpha \\ i\ j \end{Bmatrix} \begin{Bmatrix} \beta \\ \alpha\ k \end{Bmatrix} A_\beta$$

$$- \begin{Bmatrix} \alpha \\ k\ j \end{Bmatrix} \frac{\partial A_i}{\partial x^\alpha} + \begin{Bmatrix} \alpha \\ k\ j \end{Bmatrix} \begin{Bmatrix} \gamma \\ i\ \alpha \end{Bmatrix} A_\gamma \tag{5.5}$$

If we subtract (5.5) from (5.4), we get

$$A_{i,jk} - A_{i,kj} = \begin{Bmatrix} \alpha \\ i\ k \end{Bmatrix} \begin{Bmatrix} \beta \\ \alpha\ j \end{Bmatrix} A_\beta - \frac{\partial \begin{Bmatrix} \alpha \\ i\ j \end{Bmatrix}}{\partial x^k} A_\alpha - \begin{Bmatrix} \alpha \\ i\ j \end{Bmatrix} \begin{Bmatrix} \beta \\ \alpha\ k \end{Bmatrix} A_\beta + \frac{\partial \begin{Bmatrix} \alpha \\ i\ k \end{Bmatrix}}{\partial x^j} A_\alpha$$

and the interchanging of α *and* β in the first terms of each preceding line gives

$$A_{i,jk} - A_{i,kj} = \left[\frac{\partial \begin{Bmatrix} \alpha \\ i\ k \end{Bmatrix}}{\partial x^j} - \frac{\partial \begin{Bmatrix} \alpha \\ i\ j \end{Bmatrix}}{\partial x^k} + \begin{Bmatrix} \beta \\ i\ k \end{Bmatrix} \begin{Bmatrix} \alpha \\ \beta\ j \end{Bmatrix} - \begin{Bmatrix} \beta \\ i\ j \end{Bmatrix} \begin{Bmatrix} \alpha \\ \beta\ k \end{Bmatrix} \right] A_\alpha . \tag{5.6}$$

Since A_α is a covariant tensor of rank 1 and on right hand side the difference of two $A_{i,jk} - A_{i,kj}$ is a covariant tensor of rank 3, then by Quotient Law we conclude that within in the bracket of (5.6), there is a mixed tensor of rank 4, i.e.,

$$\frac{\partial \begin{Bmatrix} \alpha \\ i\ k \end{Bmatrix}}{\partial x^j} - \frac{\partial \begin{Bmatrix} \alpha \\ i\ j \end{Bmatrix}}{\partial x^k} + \begin{Bmatrix} \beta \\ i\ k \end{Bmatrix} \begin{Bmatrix} \alpha \\ \beta\ j \end{Bmatrix} - \begin{Bmatrix} \beta \\ i\ j \end{Bmatrix} \begin{Bmatrix} \alpha \\ \beta\ k \end{Bmatrix} = R^\alpha_{ijk} .$$

If the left hand side of (5.6) is to vanish, that is, if the order of covariant derivative is to be immaterial, then $R^{\alpha}_{ijk} = 0$, but since A_{α} is arbitrary, in general $R^{\alpha}_{ijk} \neq 0$, so that the order of covariant differentiation is not immaterial.

Since Christoffel symbols are functions of g_{ij}'s, it follows that tensor R^{α}_{ijk} is formed exclusively from g_{ij} and its derivatives up to second order.

This tensor R^{α}_{ijk} was first introduced by G.B. Riemann and E.D. Christoffel and called the *Riemann-Christoffel Tensor.* It is also named the *Riemann-Christoffel Curvature Tensor* of type (1,3) or curvature tensor.

It is clear from (5.6) that a necessary and sufficient condition for the validity of the inversion of the order of covariant differentiation is so that the vector $\mathrm{R}^{\alpha}_{ijk}$ vanishes identically.

$$R^{i}_{jkl} \equiv \begin{vmatrix} \dfrac{\partial}{\partial x^k} & \dfrac{\partial}{\partial x^l} \\ \begin{Bmatrix} i \\ j\ k \end{Bmatrix} & \begin{Bmatrix} i \\ j\ l \end{Bmatrix} \end{vmatrix} + \begin{vmatrix} \begin{Bmatrix} \alpha \\ j\ k \end{Bmatrix} & \begin{Bmatrix} \alpha \\ j\ l \end{Bmatrix} \\ \begin{Bmatrix} \alpha \\ j\ k \end{Bmatrix} & \begin{Bmatrix} \alpha \\ j\ l \end{Bmatrix} \end{vmatrix} \tag{5.7}$$

is called the mixed Riemann-Christoffel tensor or the Riemann-Christoffel tensor of the second kind.

The associated tensor

$$R_{ijkl} \equiv g_{i\alpha} R^{\alpha}_{jkl} \tag{5.8}$$

is known as the covariant Riemannian-Christoffel tensor, or the Riemannian-Christoffel tensor of the 1st kind.

$$R_{ijkl} \equiv \begin{vmatrix} \dfrac{\partial}{\partial x^k} & \dfrac{\partial}{\partial x^l} \\ [jk,i] & [jl,i] \end{vmatrix} + \begin{vmatrix} \begin{Bmatrix} \alpha \\ j\ k \end{Bmatrix} & \begin{Bmatrix} \alpha \\ j\ l \end{Bmatrix} \\ [ik,\alpha] & [il,\alpha] \end{vmatrix} \tag{5.9}$$

A formula which is a special case of (5.6) was established by Ricci and is given here.

$$A_{i_1 \ldots i_m, jk\cdot} - A_{i_1 \ldots i_m, kj\cdot} = \textstyle\sum_{\alpha=1}^{m} A_{i_1 \ldots i_{\alpha-1} h i_{\alpha+1}\cdot} \ldots . i_{m}, . R^{h}_{i_\alpha jk}$$

In the special case of a tensor of rank 2:

$$A_{ij,kl} - A_{ij,lk} = A_{i\alpha} R^{\alpha}_{jkl} + A_{\alpha j} R^{\alpha}_{ikl}.$$

Next we consider a contravariant vector A^i

Then,
$$A^i_{,j} = \frac{\partial A^i}{\partial x^j} + \begin{Bmatrix} i \\ \alpha \; j \end{Bmatrix} A^{\alpha} \tag{5.10}$$

$$(a^i_j = A^i_{,j}, \; then \; a^i_{j,k} = A^i_{,jk}. from \; (5.1) \; we \; get \; the \; following)$$

Now,

$$(A^i_{,j})_{,k} = a^i_{j,k} = \frac{\partial a^i_j}{\partial x^k} + a^{\alpha}_j \begin{Bmatrix} i \\ k \; a \end{Bmatrix} - a^i_l \begin{Bmatrix} \alpha \\ j \; k \end{Bmatrix}$$

$$= \frac{\partial}{\partial x^k}\left(\frac{\partial A^i}{\partial x^j} + \begin{Bmatrix} i \\ \alpha \; j \end{Bmatrix} A^{\alpha}\right) + A^{\alpha}_{,j} \begin{Bmatrix} i \\ k \; \alpha \end{Bmatrix} - A^i_{,\alpha} \begin{Bmatrix} \alpha \\ j \; k \end{Bmatrix}$$

$$= \frac{\partial}{\partial x^k}\left(\frac{\partial A^i}{\partial x^j} + \begin{Bmatrix} i \\ \alpha \; j \end{Bmatrix} A^{\alpha}\right) + \left(\frac{\partial A^{\alpha}}{\partial x^j} + \begin{Bmatrix} \alpha \\ \alpha \; j \end{Bmatrix} A^{\alpha}\right) \begin{Bmatrix} i \\ k \; \alpha \end{Bmatrix}$$
$$- \left(\frac{\partial A^i}{\partial x^{\alpha}} + \begin{Bmatrix} i \\ \alpha \; \alpha \end{Bmatrix} A^{\alpha}\right) \begin{Bmatrix} \alpha \\ j \; k \end{Bmatrix}$$

$$A^i_{,jk} = \frac{\partial^2 \; (A^i)}{\partial x^k \; \partial x^j} + \frac{\partial \begin{Bmatrix} i \\ \alpha \; j \end{Bmatrix}}{\partial x^k} A^{\alpha} + \begin{Bmatrix} i \\ \alpha \; j \end{Bmatrix} \frac{\partial A^{\alpha}}{\partial x^k} + \frac{\partial A^{\alpha}}{\partial x^j} \begin{Bmatrix} i \\ k \; \alpha \end{Bmatrix}$$
$$+ \begin{Bmatrix} i \\ k \; \alpha \end{Bmatrix} \begin{Bmatrix} \alpha \\ m \; j \end{Bmatrix} A^m - \frac{\partial A^i}{\partial x^{\alpha}} \begin{Bmatrix} \alpha \\ j \; k \end{Bmatrix} - \begin{Bmatrix} \alpha \\ j \; k \end{Bmatrix} \begin{Bmatrix} i \\ \alpha \; m \end{Bmatrix} A^m \tag{5.11}$$

Interchanging *j and k*, we get from (5.11)

$$A^i_{,kj} = \frac{\partial^2 (A^i)}{\partial x^j \partial x^k} + \frac{\partial \begin{Bmatrix} i \\ \alpha\, k \end{Bmatrix}}{\partial x^j} A^\alpha + \begin{Bmatrix} i \\ \alpha\ k \end{Bmatrix} \frac{\partial A^\alpha}{\partial x^k}$$
$$+ \frac{\partial A^\alpha}{\partial x^k} \begin{Bmatrix} i \\ j\ \alpha \end{Bmatrix} + \begin{Bmatrix} i \\ j\ \alpha \end{Bmatrix} \begin{Bmatrix} \alpha \\ m\ k \end{Bmatrix} A^m$$
$$- \frac{\partial A^i}{\partial x^\alpha} \begin{Bmatrix} \alpha \\ k\ j \end{Bmatrix} - \begin{Bmatrix} \alpha \\ k\ j \end{Bmatrix} \begin{Bmatrix} i \\ \alpha\ m \end{Bmatrix} A^m \tag{5.12}$$

Using Young's Theorem for calculus of several variables, we get $\frac{\partial^2 (A^i)}{\partial x^k \partial x^j} = \frac{\partial^2 (A^i)}{\partial x^j \partial x^k}$

$$\therefore A^i_{,jk} - A^i_{,kj} = \left(\frac{\partial \begin{Bmatrix} i \\ \alpha\ j \end{Bmatrix}}{\partial x^k} - \frac{\partial \begin{Bmatrix} i \\ \alpha\ k \end{Bmatrix}}{\partial x^j} \right) A^\alpha$$
$$+ \left(\begin{Bmatrix} m \\ \alpha\ j \end{Bmatrix} \begin{Bmatrix} i \\ m\ k \end{Bmatrix} - \begin{Bmatrix} m \\ k\ \alpha \end{Bmatrix} \begin{Bmatrix} i \\ j\ m \end{Bmatrix} \right) A^\alpha$$

(interchanging α and m in the last two terms of (5.12))

$$= \left[\left(\frac{\partial \begin{Bmatrix} i \\ \alpha\ j \end{Bmatrix}}{\partial x^k} - \frac{\partial \begin{Bmatrix} i \\ \alpha\ k \end{Bmatrix}}{\partial x^j} \right) + \left(\begin{Bmatrix} m \\ \alpha\ j \end{Bmatrix} \begin{Bmatrix} i \\ m\ k \end{Bmatrix} - \begin{Bmatrix} m \\ k\ \alpha \end{Bmatrix} \begin{Bmatrix} i \\ j\ m \end{Bmatrix} \right) \right] A^\alpha$$

$$= R^i_{akj} A^\alpha$$

$$= -R^i_{ajk} A^\alpha$$

$$\text{Now, } g_{hl}R^l_{ijk} = g_{hl}\left[\left(\frac{\partial\begin{Bmatrix} l \\ i\ k\end{Bmatrix}}{\partial x^j} - \frac{\partial\begin{Bmatrix} l \\ i\ j\end{Bmatrix}}{\partial x^k}\right) + \left(\begin{Bmatrix} m \\ i\ k\end{Bmatrix}\begin{Bmatrix} l \\ m\ j\end{Bmatrix} - \begin{Bmatrix} m \\ j\ i\end{Bmatrix}\begin{Bmatrix} l \\ k\ m\end{Bmatrix}\right)\right]$$

$$= \frac{\partial}{\partial x^j}\left[g_{hl}\begin{Bmatrix} l \\ i\ k\end{Bmatrix}\right] - \begin{Bmatrix} l \\ i\ k\end{Bmatrix}\frac{\partial g_{hl}}{\partial x^j} - \frac{\partial}{\partial x^k}\left[g_{hl}\begin{Bmatrix} l \\ i\ j\end{Bmatrix}\right]$$

$$+ \begin{Bmatrix} l \\ i\ j\end{Bmatrix}\frac{\partial g_{hl}}{\partial x^k} + g_{hl}\begin{Bmatrix} m \\ i\ k\end{Bmatrix}\begin{Bmatrix} l \\ m\ j\end{Bmatrix} - g_{hl}\begin{Bmatrix} m \\ j\ i\end{Bmatrix}\begin{Bmatrix} l \\ k\ m\end{Bmatrix}$$

$$= \frac{\partial}{\partial x^j}\left[g_{hl}\begin{Bmatrix} l \\ i\ k\end{Bmatrix}\right] - \frac{\partial}{\partial x^k}\left[g_{hl}\begin{Bmatrix} l \\ i\ j\end{Bmatrix}\right] - \begin{Bmatrix} l \\ i\ k\end{Bmatrix}([\mathrm{hj},l] + [l\,\mathrm{j},\mathrm{h}])$$

$$+ \begin{Bmatrix} l \\ i\ j\end{Bmatrix}([\mathrm{hk},l] + [l\,\mathrm{k},\mathrm{h}]) + [\mathrm{mj},\mathrm{h}]\begin{Bmatrix} m \\ i\ k\end{Bmatrix} - [km,h]\begin{Bmatrix} m \\ j\ i\end{Bmatrix}$$

$$= \frac{\partial}{\partial x^j}[ik,h] - \frac{\partial}{\partial x^k}[ij,h] - \begin{Bmatrix} l \\ i\ k\end{Bmatrix}([\mathrm{hj},l] + [l\,\mathrm{j},\mathrm{h}])$$

$$+ \begin{Bmatrix} l \\ i\ k\end{Bmatrix}([\mathrm{hk},l] + [l\,\mathrm{k},\mathrm{h}]) + [l\,\mathrm{j},\mathrm{h}]\begin{Bmatrix} l \\ i\ k\end{Bmatrix} - [kl,h]\begin{Bmatrix} l \\ j\ i\end{Bmatrix}$$

$$= \frac{\partial}{\partial x^j}[ik,h] - \frac{\partial}{\partial x^k}[ij,h] + \begin{Bmatrix} l \\ i\ j\end{Bmatrix}([\mathrm{hk},l] - \begin{Bmatrix} l \\ i\ k\end{Bmatrix}([\mathrm{hj},l] \tag{5.13}$$

The tensor of type (0,4) is defined as

$$R_{hijk} = g_{hl}R^l_{ijk}. \tag{5.14}$$

Equation (5.13) is called Riemann-Christoffel Curvature Tensor of type (0,4).

If we multiply this equation by $g^{h\beta}$ and sum, we get

$$g^{h\beta}R_{hijk} = g^{h\beta}g_{hl}R^l_{ijk} = \delta^\beta_l R^l_{ijk} = R^\beta_{ijk} \tag{5.15}$$

The Riemann-Christoffel tensor of the second kind is obtained by raising the first covariant index in the tensor R_{hijk}.

5.3 Properties of Riemann-Christoffel Tensors

If we determine the properties of the set of functions defining the Riemann-Christoffel tensor of the first kind, we expand the determinants in (5.9) and insert Christoffel's symbols into it. We get

$$
\begin{aligned}
R_{ijkl} &= \begin{vmatrix} \dfrac{\partial}{\partial x^k} & \dfrac{\partial}{\partial x^l} \\ [jk,i] & [jl,i] \end{vmatrix} + \begin{vmatrix} \begin{Bmatrix} \alpha \\ j\ k \end{Bmatrix} & \begin{Bmatrix} \alpha \\ j\ l \end{Bmatrix} \\ [ik,\alpha] & [il,\alpha] \end{vmatrix} \\
&= \frac{\partial}{\partial x^k}[jl,i] - \frac{\partial}{\partial x^l}[jk,i] + \begin{Bmatrix} \alpha \\ j\ k \end{Bmatrix}[il,\alpha] - \begin{Bmatrix} \alpha \\ j\ l \end{Bmatrix}[ik,\alpha] \\
&= \frac{\partial}{\partial x^k}\frac{1}{2}\left(\frac{\partial g_{li}}{\partial x^j} + \frac{\partial g_{ji}}{\partial x^l} - \frac{\partial g_{jl}}{\partial x^i}\right) - \frac{\partial}{\partial x^l}\frac{1}{2}\left(\frac{\partial g_{ki}}{\partial x^j} + \frac{\partial g_{ji}}{\partial x^k} - \frac{\partial g_{jk}}{\partial x^i}\right) \\
&\quad + g_{pa}\begin{Bmatrix} \alpha \\ j\ k \end{Bmatrix}\begin{Bmatrix} p \\ i\ l \end{Bmatrix} - g_{pa}\begin{Bmatrix} p \\ i\ k \end{Bmatrix}\begin{Bmatrix} \alpha \\ j\ l \end{Bmatrix} \\
&= \frac{1}{2}\left(\frac{\partial^2 g_{li}}{\partial x^k \partial x^j} + \frac{\partial^2 g_{ji}}{\partial x^k \partial x^l} - \frac{\partial^2 g_{jl}}{\partial x^k \partial x^i} - \frac{\partial^2 g_{kt}}{\partial x^l \partial x^j} - \frac{\partial^2 g_{ji}}{\partial x^l \partial x^k} + \frac{\partial^2 g_{jk}}{\partial x^l \partial x^i}\right) \\
&\quad + g_{pa}\begin{Bmatrix} \alpha \\ j\ k \end{Bmatrix}\begin{Bmatrix} p \\ i\ l \end{Bmatrix} - g_{pa}\begin{Bmatrix} p \\ i\ k \end{Bmatrix}\begin{Bmatrix} \alpha \\ j\ l \end{Bmatrix} \\
&= \frac{1}{2}\left(\frac{\partial^2 g_{li}}{\partial x^k \partial x^j} - \frac{\partial^2 g_{jl}}{\partial x^k \partial x^i} - \frac{\partial^2 g_{ki}}{\partial x^l \partial x^j} + \frac{\partial^2 g_{jk}}{\partial x^l \partial x^i}\right) + g_{pa}\begin{Bmatrix} \alpha \\ j\ k \end{Bmatrix}\begin{Bmatrix} p \\ i\ l \end{Bmatrix} \\
&\quad - g_{pa}\begin{Bmatrix} p \\ i\ k \end{Bmatrix}\begin{Bmatrix} \alpha \\ j\ l \end{Bmatrix}
\end{aligned}
\tag{5.16}
$$

We get the following equations from (5.16):

(a) $R_{jikl} = -R_{ijkl}$

(b) $R_{ijlk} = -R_{ijkl}$

(c) $R_{klij} = R_{ijkl}$

(d) $R_{ijkl} + R_{iklj} + R_{iljk} = 0$

(e) $R^i_{jkl} + R^i_{klj} + R^i_{ljk} = 0$

The Riemann-Christoffel tensor of (a) and (b) are skew-symmetric with respect to the first two and last two indices, respectively, and (c) is symmetric with respect to a group of two indices. It follows from these identities that distinct, nonvanishing components of R_{jikl} are of three types:

(i) Symbols with two distinct identities, i.e., R_{ijij}
(ii) Symbols with three distinct identities, i.e., R_{ijik}
(iii) Symbols with four distinct identities, i.e., R_{ijkl}.

This is to easily verify that the total number of n distinct non-vanishing components of

$$R_{ijkl} = \binom{n}{4} = \frac{n(n-1)(n-2)(n-3)}{12}.$$

Property 5.3.1. The number of distinct non-vanishing components of the covariant curvature tensor does not exceed $\frac{1}{12}n^2(n^2-1)$.

Proof: We know that Riemann-Christoffel curvature tensor satisfies 4 of the following properties:

$$\left.\begin{aligned} &\text{i) } R_{ijkl} = -R_{ijkl} \\ &\text{ii) } R_{ijkl} = -R_{ijlk} \end{aligned}\right\}(\text{Skew}-\text{Symmetry Property})$$

$$\text{iii) } R_{ijkl} = R_{klij} \quad (\text{Block Symmetry Property})$$

$$\text{iv) } R_{ijkl} + R_{iklj} + R_{iljk} = 0, \quad (\textit{Cyclic Property})$$

Due to the above properties, all n^4 components of R_{ijkl} are not all non-vanishing independent components. The following 4 cases may arise:

Case 1. All 4 indices are the same, i.e., R_{iiii}
Case 2. There are only 2 distinct indices, i.e., R_{ijij}
Case 3. Three indices of R_{ijkl} are distinct, i.e., R_{ijik}
Case 4. When all indices i,j,k,l all are different

Case 1. We consider R_{iiii}. it is self-skew symmetric, i.e., $R_{iiii} = -R_{iiii}$ [by i) and ii)].

$$\Rightarrow 2R_{iiii} = 0$$

$$\Rightarrow R_{iiii} = 0$$

$\Rightarrow R_{iiii}$ *has* no non-vanishing component.

Case 2. The two indices, i and j, can be selected for arrangement. For I, $\begin{pmatrix} n \\ 1 \end{pmatrix} = n$ *ways* and for j, $\begin{pmatrix} n-1 \\ 1 \end{pmatrix} = (n-1)$ *ways*, so the total number of selections = $n(n - 1)$.

Since R_{ijkl} is skew-symmetric in the first two and last two indices, i.e.

$$R_{ijij} = -R_{jiij} = -(-R_{jiji}) = R_{jiji}.$$

⇒If two indices are interchanged, these two components are the same, i.e., the number of components would be half.

For symmetry properties, no change of components means no reduction.

Moreover, for cyclic

$$R_{ijij} + R_{iijj} + R_{ijji} = R_{ijij} + 0 - R_{ijij} = 0 \text{ (by (ii) and (iii))},$$

so, due to the cyclic property, there is no deduction.

The total number of components $= \frac{1}{2}n(n-1)$.

Case 3. We consider that there are 3 distinct indices *i, j, and k*.

As in Case 2, the number of selections of components

$$= \begin{pmatrix} n \\ 1 \end{pmatrix}\begin{pmatrix} n-1 \\ 1 \end{pmatrix}\begin{pmatrix} n-2 \\ 1 \end{pmatrix} = n(n-1)(n-2).$$

Due to the symmetric property, $R_{ijik} = R_{ikij}$ the number of selections is reduced to $\frac{1}{2}n(n-1)(n-2)$.

Due to skew-symmetric property there is no reduction. Since $R_{iijk} = 0$, $R_{ijik} = 0$ also occur for the cyclic property, there is no reduction (since Bianchi identity is self-satisfied).

Case 4. When *i, j, k, and l* have four distinct indices, *i* can be chosen by $\begin{pmatrix} n \\ 1 \end{pmatrix}\begin{pmatrix} n-1 \\ 1 \end{pmatrix}\begin{pmatrix} n-2 \\ 1 \end{pmatrix}\begin{pmatrix} n-3 \\ 1 \end{pmatrix} = n(n-1)(n-2)(n-3)$ ways.

By skew symmetric property, we have

$$R_{ijkl} = -R_{jikl} \text{ and } R_{ijkl} = -R_{ijlk}.$$

The number of independent non vanishing components reduces to $\frac{1}{4}n(n-1)(n-2)(n-3)$.

Also for the symmetry property, i.e. ($R_{ijkl} = R_{klij}$), the number of non-vanishing components reduces to

$$\frac{1}{8}n(n-1)(n-2)(n-3)$$

and for cyclic property, we get $R_{ijkl} + R_{iklj} + R_{iljk} = 0$ or $R_{ijkl} = -(R_{iklj} + R_{iljk})$.

Therefore, one among every three can be expressed in terms of the remaining two terms. Hence, it would be reduced by $\frac{2}{3}$ th number of terms, so the total number of components

$$=\frac{2}{3}\cdot\frac{1}{8}n(n-1)(n-2)(n-3)$$
$$=\frac{1}{12}n(n-1)(n-2)(n-3).$$

Combining all cases, the total number of Components of Curvature Tensor of order 4

$$=0+\frac{1}{2}n(n-1)+\frac{1}{2}n(n-1)(n-2)+\frac{1}{12}n(n-1)(n-2)(n-3)$$
$$=\frac{1}{2}n(n-1)\left\{1+(n-2)+\frac{1}{6}(n-2)(n-3)\right\}$$
$$=\frac{1}{2}n(n-1)\frac{1}{6}(6n-6+n^2-5n+6)$$
$$=\frac{1}{12}n(n-1)(n^2+n)=\frac{1}{12}n^2(n^2-1).$$

Theorem 5.3.1

(α) $R^i_{jkl} = -R^i_{jlk}$

(β) $R^i_{jkl} + R^i_{klj} + R^i_{ljk} = 0$

(γ) $R^i_{ikl} = 0$

(δ) $R_{ijkl} = -R_{jikl}$

Proof: (α) We know

$$R^i_{jkl} = \begin{vmatrix} \dfrac{\partial}{\partial x^k} & \dfrac{\partial}{\partial x^l} \\ \begin{Bmatrix} i \\ j\ k \end{Bmatrix} & \begin{Bmatrix} i \\ j\ l \end{Bmatrix} \end{vmatrix} + \begin{vmatrix} \begin{Bmatrix} i \\ j\ k \end{Bmatrix} & \begin{Bmatrix} i \\ j\ l \end{Bmatrix} \\ \begin{Bmatrix} \alpha \\ j\ k \end{Bmatrix} & \begin{Bmatrix} \alpha \\ j\ l \end{Bmatrix} \end{vmatrix}$$

$$= \frac{\partial}{\partial x^k}\begin{Bmatrix} i \\ j\ l \end{Bmatrix} - \frac{\partial}{\partial x^l}\begin{Bmatrix} i \\ j\ k \end{Bmatrix} + \begin{Bmatrix} i \\ j\ k \end{Bmatrix}\begin{Bmatrix} \alpha \\ j\ l \end{Bmatrix} - \begin{Bmatrix} i \\ j\ l \end{Bmatrix}\begin{Bmatrix} \alpha \\ j\ k \end{Bmatrix}$$

Interchanging k and l we get from above equation

$$R^i_{jlk} = \frac{\partial}{\partial x^l}\begin{Bmatrix} i \\ j\ k \end{Bmatrix} - \frac{\partial}{\partial x^k}\begin{Bmatrix} i \\ j\ l \end{Bmatrix} + \begin{Bmatrix} i \\ j\ l \end{Bmatrix}\begin{Bmatrix} a \\ jk \end{Bmatrix} - \begin{Bmatrix} i \\ j\ k \end{Bmatrix}\begin{Bmatrix} \alpha \\ j\ l \end{Bmatrix}$$
$$= -R^i_{jkl}.$$

(β) We know

$$R^i_{jkl} = \frac{\partial}{\partial x^k}\begin{Bmatrix} i \\ j\ l \end{Bmatrix} - \frac{\partial}{\partial x^l}\begin{Bmatrix} i \\ j\ k \end{Bmatrix} + \begin{Bmatrix} i \\ j\ k \end{Bmatrix}\begin{Bmatrix} \alpha \\ j\ l \end{Bmatrix} - \begin{Bmatrix} i \\ j\ l \end{Bmatrix}\begin{Bmatrix} \alpha \\ j\ k \end{Bmatrix}. \qquad \text{(i)}$$

Similarly,

$$R^i_{klj} = \frac{\partial}{\partial x^l}\begin{Bmatrix} i \\ k\ j \end{Bmatrix} - \frac{\partial}{\partial x^j}\begin{Bmatrix} i \\ k\ l \end{Bmatrix} + \begin{Bmatrix} i \\ k\ l \end{Bmatrix}\begin{Bmatrix} \alpha \\ j\ l \end{Bmatrix} - \begin{Bmatrix} i \\ k\ j \end{Bmatrix}\begin{Bmatrix} \alpha \\ k\ l \end{Bmatrix} \cdots\cdot \qquad \text{(ii)}$$

and

$$R^i_{ljk} = \frac{\partial}{\partial x^j}\begin{Bmatrix} i \\ l\ k \end{Bmatrix} - \frac{\partial}{\partial x^k}\begin{Bmatrix} i \\ l\ j \end{Bmatrix} + \begin{Bmatrix} i \\ l\ j \end{Bmatrix}\begin{Bmatrix} \alpha \\ j\ l \end{Bmatrix} - \begin{Bmatrix} i \\ l\ k \end{Bmatrix}\begin{Bmatrix} \alpha \\ l\ j \end{Bmatrix} \qquad \text{(iii)}$$

Adding (i), (ii), and (iii), we get $R^i_{jkl} + R^i_{klj} + R^i_{ljk} = 0$

γ) We know $R^i_{jkl} = \frac{\partial}{\partial x^k}\begin{Bmatrix} i \\ j\ l \end{Bmatrix} - \frac{\partial}{\partial x^l}\begin{Bmatrix} i \\ j\ k \end{Bmatrix} + \begin{Bmatrix} i \\ j\ k \end{Bmatrix}\begin{Bmatrix} \alpha \\ j\ l \end{Bmatrix} - \begin{Bmatrix} i \\ j\ l \end{Bmatrix}\begin{Bmatrix} \alpha \\ j\ k \end{Bmatrix}$.

Putting $j = i$, we get from above

$$R^i_{ikl} = \frac{\partial}{\partial x^k}\begin{Bmatrix} i \\ i\ l \end{Bmatrix} - \frac{\partial}{\partial x^k}\begin{Bmatrix} i \\ i\ l \end{Bmatrix} + \begin{Bmatrix} i \\ i\ l \end{Bmatrix}\begin{Bmatrix} \alpha \\ i\ k \end{Bmatrix} - \begin{Bmatrix} i \\ i\ l \end{Bmatrix}\begin{Bmatrix} \alpha \\ i\ k \end{Bmatrix}$$

[in 2nd term interchange dummy indices k and l]

$= 0$

δ) We know,

$$R_{ijkl} = \frac{1}{2}\left(\frac{\partial^2 g_{li}}{\partial x^k \partial x^j} - \frac{\partial^2 g_{jl}}{\partial x^k \partial x^i} - \frac{\partial^2 g_{ki}}{\partial x^l \partial x^j} + \frac{\partial^2 g_{jk}}{\partial x^l \partial x^i} \right.$$

$$+ g_{p\alpha}\begin{Bmatrix} \alpha \\ j\ k \end{Bmatrix}\begin{Bmatrix} p \\ i\ l \end{Bmatrix} - g_{p\alpha}\begin{Bmatrix} p \\ i\ k \end{Bmatrix}\begin{Bmatrix} \alpha \\ j\ l \end{Bmatrix} \qquad \text{(from 5.16)}$$

Interchanging i and j, we get

$$R_{jikl} = \frac{1}{2}\left(\frac{\partial^2 g_{lj}}{\partial x^k \partial x^i} - \frac{\partial^2 g_{il}}{\partial x^k \partial x^j} - \frac{\partial^2 g_{kj}}{\partial x^l \partial x^i} + \frac{\partial^2 g_{ik}}{\partial x^l \partial x^j} \right)$$

$$+ g_{p\alpha}\begin{Bmatrix} \alpha \\ i\ k \end{Bmatrix}\begin{Bmatrix} p \\ j\ l \end{Bmatrix} - g_{p\alpha}\begin{Bmatrix} p \\ j\ k \end{Bmatrix}\begin{Bmatrix} \alpha \\ i\ l \end{Bmatrix}$$

(interchanging dummy index p and α)

$$= \frac{1}{2}\left(\frac{\partial^2 g_{lj}}{\partial x^k \partial x^i} - \frac{\partial^2 g_{il}}{\partial x^k \partial x^j} - \frac{\partial^2 g_{kj}}{\partial x^l \partial x^i} + \frac{\partial^2 g_{ik}}{\partial x^l \partial x^j} \right)$$

$$+ g_{\alpha p}\begin{Bmatrix} p \\ i\ k \end{Bmatrix}\begin{Bmatrix} \alpha \\ j\ l \end{Bmatrix} - g_{\alpha p}\begin{Bmatrix} \alpha \\ j\ k \end{Bmatrix}\begin{Bmatrix} p \\ i\ l \end{Bmatrix} = -R_{ijkl}$$

Similarly, we can easily prove by using the above equations.

Theorem 5.3.2

(i) $R_{klij} = R_{ijkl}$
(ii) $R_{ijkl} + R_{iklj} + R_{iljk} = 0$

Proof: (i) We know,

$$R_{ijkl} = \frac{1}{2}\left(\frac{\partial^2 g_{li}}{\partial x^k \partial x^j} - \frac{\partial^2 g_{jl}}{\partial x^k \partial x^i} - \frac{\partial^2 g_{ki}}{\partial x^l \partial x^j} + \frac{\partial^2 g_{jk}}{\partial x^l \partial x^i}\right)$$
$$+ g_{p\alpha}\begin{Bmatrix} \alpha \\ j\ k \end{Bmatrix}\begin{Bmatrix} p \\ i\ l \end{Bmatrix} - g_{p\alpha}\begin{Bmatrix} p \\ i\ k \end{Bmatrix}\begin{Bmatrix} \alpha \\ j\ l \end{Bmatrix} \cdots \qquad \text{(i)}$$

Interchange *i and k* in (i) and we get

$$R_{kjil} = \frac{1}{2}\left(\frac{\partial^2 g_{lk}}{\partial x^i \partial x^j} - \frac{\partial^2 g_{jl}}{\partial x^i \partial x^k} - \frac{\partial^2 g_{ik}}{\partial x^l \partial x^j} + \frac{\partial^2 g_{ji}}{\partial x^l \partial x^k}\right)$$
$$+ g_{p\alpha}\begin{Bmatrix} \alpha \\ j\ i \end{Bmatrix}\begin{Bmatrix} p \\ k\ l \end{Bmatrix} - g_{p\alpha}\begin{Bmatrix} p \\ k\ i \end{Bmatrix}\begin{Bmatrix} \alpha \\ j\ l \end{Bmatrix}$$

Now, interchange j and *l* and we get

$$R_{klij} = \frac{1}{2}\left(\frac{\partial^2 g_{jk}}{\partial x^i \partial x^l} - \frac{\partial^2 g_{lj}}{\partial x^i \partial x^k} - \frac{\partial^2 g_{ik}}{\partial x^j \partial x^l} + \frac{\partial^2 g_{ji}}{\partial x^j \partial x^k}\right)$$
$$+ g_{p\alpha}\begin{Bmatrix} \alpha \\ l\ i \end{Bmatrix}\begin{Bmatrix} p \\ k\ j \end{Bmatrix} - g_{p\alpha}\begin{Bmatrix} p \\ k\ i \end{Bmatrix}\begin{Bmatrix} \alpha \\ l\ j \end{Bmatrix}$$

Since the symmetric property of g_{ij} and 2nd Christoffel's symbols,

$$R_{klij} = \frac{1}{2}\left(\frac{\partial^2 g_{li}}{\partial x^k \partial x^j} - \frac{\partial^2 g_{jl}}{\partial x^k \partial x^i} - \frac{\partial^2 g_{ki}}{\partial x^l \partial x^j} + \frac{\partial^2 g_{jk}}{\partial x^l \partial x^i}\right)$$
$$+ g_{p\alpha}\begin{Bmatrix} \alpha \\ j\ k \end{Bmatrix}\begin{Bmatrix} p \\ i\ l \end{Bmatrix} - g_{p\alpha}\begin{Bmatrix} p \\ i\ k \end{Bmatrix}\begin{Bmatrix} \alpha \\ j\ l \end{Bmatrix}$$

$= R_{ijkl}$, which is known as the block symmetric property.

(ii) From (i),

$$R_{ijkl} = \frac{1}{2}\left(\frac{\partial^2 g_{li}}{\partial x^k \partial x^j} - \frac{\partial^2 g_{jl}}{\partial x^k \partial x^i} - \frac{\partial^2 g_{ki}}{\partial x^l \partial x^j} + \frac{\partial^2 g_{jk}}{\partial x^l \partial x^i}\right)$$
$$+ g_{p\alpha}\begin{Bmatrix} \alpha \\ j\ k \end{Bmatrix}\begin{Bmatrix} p \\ i\ l \end{Bmatrix} - g_{p\alpha}\begin{Bmatrix} p \\ i\ k \end{Bmatrix}\begin{Bmatrix} \alpha \\ j\ l \end{Bmatrix}$$

$$R_{iklj} = \frac{1}{2}\left(\frac{\partial^2 g_{ji}}{\partial x^l \partial x^k} - \frac{\partial^2 g_{kj}}{\partial x^l \partial x^i} - \frac{\partial^2 g_{li}}{\partial x^j \partial x^k} + \frac{\partial^2 g_{kl}}{\partial x^j \partial x^i}\right)$$
$$+ g_{p\alpha}\begin{Bmatrix} \alpha \\ k\ l \end{Bmatrix}\begin{Bmatrix} p \\ i\ j \end{Bmatrix} - g_{p\alpha}\begin{Bmatrix} p \\ i\ l \end{Bmatrix}\begin{Bmatrix} \alpha \\ k\ j \end{Bmatrix}$$

$$R_{iljk} = \frac{1}{2}\left(\frac{\partial^2 g_{ki}}{\partial x^j \partial x^l} - \frac{\partial^2 g_{lk}}{\partial x^j \partial x^i} - \frac{\partial^2 g_{ji}}{\partial x^k \partial x^l} + \frac{\partial^2 g_{lj}}{\partial x^k \partial x^i}\right)$$
$$+ g_{p\alpha}\begin{Bmatrix} \alpha \\ l\ j \end{Bmatrix}\begin{Bmatrix} p \\ i\ k \end{Bmatrix} - g_{p\alpha}\begin{Bmatrix} p \\ i\ j \end{Bmatrix}\begin{Bmatrix} \alpha \\ l\ k \end{Bmatrix}$$

Adding these three equations, we get
$R_{ijkl} + R_{iklj} + R_{iljk} = 0$, which is known as the cyclic property.

Example 5.3.1. Calculate the components of R_{ijkl} of the Riemannian tensor for the metric

$$ds^2 = (dx^1)^2 - (x^2)^{-2}(dx^2)^2.$$

Solution: Here, $g_{11} = 1, g_{22} = -(x^2)^{-2}, g_{12} = g_{21} = 0$, and

$$g = \begin{vmatrix} 1 & 0 \\ 0 & -(x^2)^{-2} \end{vmatrix} = -(x^2)^{-2}$$

$$g^{11} = 1, g^{12} = g^{21} = 0, g^{22} = (x^2)^2$$

The only non-vanishing Christoffel symbol of 2nd kind is

$$\left\{ \begin{matrix} 2 \\ 2\,2 \end{matrix} \right\} = g^{22}[22,2] = (x^2)^2 \frac{1}{2}\frac{\partial}{\partial x^2}(-(x^2)^{-2})$$

$$= (x^2)^2 \frac{1}{2}(-2)(x^2)^{-3} = -(x^2)^{-1}$$

Using equation (5.6), Riemannian Curvature Tensor R^{α}_{ijk}

$$R^1_{212} = \frac{\partial}{\partial x^2}\left\{ \begin{matrix} 1 \\ 2\,1 \end{matrix} \right\} + \frac{\partial}{\partial x^1}\left\{ \begin{matrix} 1 \\ 2\,2 \end{matrix} \right\}$$

$$+\left\{ \begin{matrix} 1 \\ \beta\,1 \end{matrix} \right\}\left\{ \begin{matrix} \beta \\ 2\,2 \end{matrix} \right\} - \left\{ \begin{matrix} 1 \\ \beta\,2 \end{matrix} \right\}\left\{ \begin{matrix} \beta \\ 2\,1 \end{matrix} \right\} = 0$$

$$R^2_{212} = \frac{\partial}{\partial x^2}\left\{ \begin{matrix} 2 \\ 2\,1 \end{matrix} \right\} + \frac{\partial}{\partial x^1}\left\{ \begin{matrix} 2 \\ 2\,2 \end{matrix} \right\}$$

$$+\left\{ \begin{matrix} 2 \\ \beta\,1 \end{matrix} \right\}\left\{ \begin{matrix} \beta \\ 2\,2 \end{matrix} \right\} - \left\{ \begin{matrix} 2 \\ \beta\,2 \end{matrix} \right\}\left\{ \begin{matrix} \beta \\ 2\,1 \end{matrix} \right\} = 0.$$

We know $R_{ijkl} = g_{i\alpha} R^{\alpha}_{jkl}$

$$R_{1212} = g_{1\alpha} R^{\alpha}_{212} = g_{11} R^1_{212} + g_{12} R^2_{212} = 0.$$

5.3.1 Space of Constant Curvature

An n-dimensional Riemannian space V_n is said to be a space of constant curvature if its Riemann-Christoffel curvature tensor can be expressed as

$R^i_{jkl} = \rho(\delta^i_l g_{jk} - \delta^i_k g_{jl})$, where ρ is a constant.
and $R_{ijkl} = \rho(g_{il} g_{jk} - g_{ik} g_{jl})$.

5.4 Ricci Tensor, Bianchi Identities, Einstein Tensors

5.4.1 Ricci Tensor

We define the *Ricci Tensor as* $R_{ij} = R^{\alpha}_{ij\alpha}$.

We have

$$R^l_{ijk} = \begin{vmatrix} \frac{\partial}{\partial x^j} & \frac{\partial}{\partial x^k} \\ \begin{Bmatrix} l \\ i\ j \end{Bmatrix} & \begin{Bmatrix} l \\ i\ k \end{Bmatrix} \end{vmatrix} + \begin{vmatrix} \begin{Bmatrix} l \\ \alpha\ j \end{Bmatrix} & \begin{Bmatrix} l \\ \alpha\ k \end{Bmatrix} \\ \begin{Bmatrix} \alpha \\ i\ j \end{Bmatrix} & \begin{Bmatrix} \alpha \\ i\ k \end{Bmatrix} \end{vmatrix}$$

$$= \frac{\partial}{\partial x^j}\begin{Bmatrix} l \\ i\ k \end{Bmatrix} - \frac{\partial}{\partial x^k}\begin{Bmatrix} l \\ i\ j \end{Bmatrix} + \begin{Bmatrix} l \\ \alpha\ j \end{Bmatrix}\begin{Bmatrix} \alpha \\ i\ k \end{Bmatrix} - \begin{Bmatrix} l \\ \alpha\ k \end{Bmatrix}\begin{Bmatrix} \alpha \\ i\ j \end{Bmatrix}.$$

Putting $k = l$, we get

$$R^l_{ijl} = \frac{\partial}{\partial x^j} = \begin{Bmatrix} l \\ i\ l \end{Bmatrix} - \frac{\partial}{\partial x^l}\begin{Bmatrix} l \\ i\ j \end{Bmatrix} + \begin{Bmatrix} l \\ \alpha\ j \end{Bmatrix}\begin{Bmatrix} \alpha \\ i\ l \end{Bmatrix} - \begin{Bmatrix} l \\ \alpha\ l \end{Bmatrix}\begin{Bmatrix} \alpha \\ i\ j \end{Bmatrix}$$

$$= \frac{\partial}{\partial x^j}\left(\frac{\partial}{\partial x^i}\log\sqrt{g}\right) - \frac{\partial}{\partial x^l}\begin{Bmatrix} l \\ i\ l \end{Bmatrix} + \begin{Bmatrix} l \\ \alpha\ j \end{Bmatrix}\begin{Bmatrix} \alpha \\ i\ l \end{Bmatrix}$$

$$-\left(\frac{\partial}{\partial x^\alpha}\log\sqrt{g}\right)\begin{Bmatrix} \alpha \\ i\ j \end{Bmatrix} \qquad [\begin{Bmatrix} l \\ i\ l \end{Bmatrix} = \frac{\partial}{\partial x^i}\log\sqrt{g}.]$$

Therefore,

$$R_{ij} = \frac{\partial^2 \log\sqrt{g}}{\partial x^j \partial x^i} - \frac{\partial}{\partial x^\alpha}\begin{Bmatrix} \alpha \\ i\ j \end{Bmatrix} + \begin{Bmatrix} \alpha \\ \beta\ j \end{Bmatrix}\begin{Bmatrix} \beta \\ i\ \alpha \end{Bmatrix} - \begin{Bmatrix} \beta \\ i\ j \end{Bmatrix}\left(\frac{\partial}{\partial x^\beta}\log\sqrt{g}\right) \tag{5.17}$$

(changing dummy index α to β and l to α).

Now, interchanging i and j

$$R_{ji} = \frac{\partial^2 \log\sqrt{g}}{\partial x^i \partial x^j} - \frac{\partial}{\partial x^\alpha}\begin{Bmatrix} \alpha \\ j\ i \end{Bmatrix} + \begin{Bmatrix} \alpha \\ \beta\ i \end{Bmatrix}\begin{Bmatrix} \beta \\ j\ \alpha \end{Bmatrix} - \begin{Bmatrix} \beta \\ j\ i \end{Bmatrix}\left(\frac{\partial}{\partial x^\beta}\log\sqrt{g}\right).$$

Interchanging α and β in the third term,

$$R_{ji} = \frac{\partial^2 \log\sqrt{g}}{\partial x^i \partial x^j} - \frac{\partial}{\partial x^\alpha}\begin{Bmatrix} \alpha \\ j\ i \end{Bmatrix} + \begin{Bmatrix} \beta \\ \alpha\ i \end{Bmatrix}\begin{Bmatrix} \alpha \\ j\ \beta \end{Bmatrix} - \begin{Bmatrix} \beta \\ j\ i \end{Bmatrix}\left(\frac{\partial}{\partial x^\beta}\log\sqrt{g}\right).$$

Therefore,

$$R_{ij} = R_{ji} \tag{5.18}$$

Tensor R_{ij} is symmetric since $R_{ij} = R_{ji}$ is one of the distinct components of R_{ij} is $\frac{1}{2}n(n+1)$. In a four dimensional manifold $n = 4$, so that, if we set $R_{ij} = 0$, we obtain 10 partial differential equations, which Einstein has adopted as his equations of the gravitational field in free space in the general theory of relativity.

5.4.2 Bianchi Identity

In the development of that theory, another tensor introduced by Einstein plays an important role. The tensor is obtained from an identity the called *Bianchi Identity.*

Theorem 5.4.1. $R^i_{jkl,m} + R^i_{jlm,k} + R^i_{jkm,l} = 0$

We know

$$R^i_{jkl} = \frac{\partial}{\partial x^k}\begin{Bmatrix} i \\ j\ l \end{Bmatrix} - \frac{\partial}{\partial x^l}\begin{Bmatrix} i \\ j\ k \end{Bmatrix} + \begin{Bmatrix} i \\ j\ k \end{Bmatrix}\begin{Bmatrix} \alpha \\ j\ l \end{Bmatrix} - \begin{Bmatrix} i \\ j\ l \end{Bmatrix}\begin{Bmatrix} \alpha \\ j\ k \end{Bmatrix}$$

$$\frac{\partial}{\partial x^m} R^i_{jkl} = \frac{\partial}{\partial x^m}\left[\frac{\partial}{\partial x^k}\begin{Bmatrix} i \\ j\ l \end{Bmatrix} - \frac{\partial}{\partial x^l}\begin{Bmatrix} i \\ j\ k \end{Bmatrix} + \begin{Bmatrix} i \\ j\ k \end{Bmatrix}\begin{Bmatrix} s \\ j\ l \end{Bmatrix} - \begin{Bmatrix} i \\ j\ l \end{Bmatrix}\begin{Bmatrix} s \\ j\ k \end{Bmatrix}\right]$$

$$= \frac{\partial^2}{\partial x^m \partial x^k}\begin{Bmatrix} i \\ j\ l \end{Bmatrix} - \frac{\partial^2}{\partial x^m \partial x^l}\begin{Bmatrix} i \\ j\ k \end{Bmatrix} + \begin{Bmatrix} s \\ j\ l \end{Bmatrix}\frac{\partial}{\partial x^m}\begin{Bmatrix} i \\ j\ k \end{Bmatrix} + \begin{Bmatrix} i \\ j\ l \end{Bmatrix}\frac{\partial}{\partial x^m}\begin{Bmatrix} s \\ j\ k \end{Bmatrix}$$

$$- \begin{Bmatrix} s \\ j\ k \end{Bmatrix}\frac{\partial}{\partial x^m}\begin{Bmatrix} i \\ j\ l \end{Bmatrix} - \begin{Bmatrix} i \\ j\ l \end{Bmatrix}\frac{\partial}{\partial x^m}\begin{Bmatrix} s \\ j\ k \end{Bmatrix}$$

From the covariant derivative of a mixed tensor, we have

$$R^i_{jkl,m} = \frac{\partial}{\partial x^m} R^i_{jkl} - R^i_{\alpha kl}\begin{Bmatrix} \alpha \\ j\ m \end{Bmatrix} - R^i_{j\alpha l}\begin{Bmatrix} \alpha \\ k\ m \end{Bmatrix} - R^i_{jk\alpha}\begin{Bmatrix} \alpha \\ l\ m \end{Bmatrix} - R^\alpha_{jkl}\begin{Bmatrix} i \\ \alpha\ m \end{Bmatrix}$$

$$= \frac{\partial^2}{\partial x^m \partial x^k}\begin{Bmatrix} i \\ j\ l \end{Bmatrix} - \frac{\partial^2}{\partial x^m \partial x^l}\begin{Bmatrix} i \\ j\ k \end{Bmatrix} + \begin{Bmatrix} s \\ j\ l \end{Bmatrix}\frac{\partial}{\partial x^m}\begin{Bmatrix} i \\ j\ k \end{Bmatrix}$$

$$+ \begin{Bmatrix} i \\ j\ l \end{Bmatrix}\frac{\partial}{\partial x^m}\begin{Bmatrix} s \\ j\ k \end{Bmatrix} - \begin{Bmatrix} s \\ j\ k \end{Bmatrix}\frac{\partial}{\partial x^m}\begin{Bmatrix} i \\ j\ l \end{Bmatrix} - \begin{Bmatrix} i \\ j\ l \end{Bmatrix}\frac{\partial}{\partial x^m}\begin{Bmatrix} s \\ j\ k \end{Bmatrix}$$

$$- R^i_{\alpha kl}\begin{Bmatrix} \alpha \\ j\ m \end{Bmatrix} - R^i_{j\alpha l}\begin{Bmatrix} \alpha \\ k\ m \end{Bmatrix} - R^i_{jk\alpha}\begin{Bmatrix} \alpha \\ l\ m \end{Bmatrix} - R^\alpha_{jkl}\begin{Bmatrix} i \\ \alpha\ m \end{Bmatrix}$$

Similarly, by changing *k, l, and m* cyclically and adding the resulting equations, we get

$$R^i_{jkl,m} + R^i_{jlm,k} + R^i_{jkm,l} = 0. \tag{5.19}$$

This equation is known as *'Bianchi's Second identity'.*
Multiplying $R^i_{jkl,m}$ by g_{hi} ie, $g_{hi}R^i_{jkl,m} = R_{hjkl,m}$.
Equation (5.19) becomes

$$R_{hjkl,m} + R_{hjlm,k} + R_{hjmk,l} = 0. \tag{5.20}$$

Flat Space

A Riemannian space whose Riemannian-Christoffel tensor R^i_{jkl} vanishes identically is called a *flat space.* Otherwise, it is called non-flat.

Example 5.4.1. The Minkowski Space-Time with metric

$$ds^2 = (dx^1)^2 + (dx^2)^2 + (dx^3)^2 - c^2(dx^4)^2$$

is a flat.

Solution: Here, $ds^2 = (dx^1)^2 + (dx^2)^2 + (dx^3)^2 - c^2(dx^4)^2/$

We know $ds^2 = g_{ij}dx^i dx^j$

$\therefore g_{11} = g_{22} = g_{33} = 1$, $g_{44} = -c^2$. Here, c is the speed of light

and $g_{ij} = 0$ *when* $i \neq j$.

Hence, $\frac{\partial g_{ij}}{\partial x^k} = 0$ *for all* $i, j, k = 1,2,3,4.$

Its implied that all Christoffel symbols are zero and therefore the Riemann-Christoffel curvature tensor also vanishes identically.

⇒Minkowski Space-Time is a flat.

Example 5.4.2. In a V_2, the components of the Ricci tensor are proportional to the components of the metric tensor.

The components of the Ricci tensor are $R_{11}, R_{12}, R_{21}, and\ R_{22}$ and components of metric tensor are $g_{11}, g_{12}, g_{21}, and\ g_{22}$. Both are symmetric with respect to indices.

We know $R_{ij} = g^{kl}R_{lijk} = g^{kl}R_{ilkj}$ (by the property of Ricci tensor)

$$\therefore R_{ij} = g^{11}R_{i11j} + g^{12}R_{i12j} + g^{21}R_{i21j} + g^{22}R_{i22j}$$

$$\therefore R_{11} = g^{11}R_{1111} + g^{12}R_{1121} + g^{21}R_{1211} + g^{22}R_{1221}$$

$$= 0 + 0 + 0 + g^{22}R_{1221}\ [\text{here } R_{1111} = 0\ and\ R_{1121} = -R_{1211}]$$

$$= \frac{g_{11}}{g}(-R_{1212}),\ \text{where } g^{ij} = \frac{cofactor\ of\ g_{ij}\ of\ g}{\det g}$$

$$\therefore \frac{R_{11}}{g_{11}} = -\frac{R_{1212}}{g}$$

Similarly, $R_{22} = g^{11}R_{2112} + g^{12}R_{2122} + g^{21}R_{2212} + g^{22}R_{2222}$

$$= g^{11}R_{2112} + 0 + 0 + 0 = -\frac{g_{22}}{g}R_{1212}$$

$$\frac{R_{22}}{g_{22}} = -\frac{R_{1212}}{g}$$

Lastly, $R_{12} = g^{11}R_{1112} + g^{12}R_{1122} + g^{21}R_{1212} + g^{22}R_{1222}$

$$= g^{21}R_{1212} = -\frac{g_{12}}{g}$$

$$\frac{R_{12}}{g_{12}} = -\frac{R_{1212}}{g}$$

Since both R_{ij} and g_{ij} are symmetric and from above relation we get

$$\frac{R_{12}}{g_{12}} = -\frac{R_{1212}}{g}.$$

Combining all the cases, $\frac{R_{11}}{g_{11}} = \frac{R_{22}}{g_{22}} = \frac{R_{12}}{g_{12}} = \frac{R_{12}}{g_{12}} - \frac{R_{1212}}{g}$

$\therefore$The components of the Ricci tensor are proportional to the components of the metric tensor.

Example 5.4.3. Prove that in a V_2

$$R_{ijkl} = -\frac{R}{2}(g_{ik}g_{jl} - g_{il}g_{jk}).$$

Solution: We know that in V_2 the components of the Ricci tensor are proportional to the components of the metric tensor.

$$\frac{R_{11}}{g_{11}} = \frac{R_{22}}{g_{22}} = \frac{R_{12}}{g_{12}} = \frac{R_{12}}{g_{12}} - \frac{R_{1212}}{g}$$

$\therefore$ or $R_{ij} = -\frac{R_{1212}}{g} g_{ij}$

We know that the covariant curvature tensor vanishes if its three or more indices are same.

$$\therefore R_{1111} = R_{2222} = R_{1112} = R_{1121} = R_{1211} = R_{1121} = R_{2111} = R_{1211} = R_{1222} = R_{2111} = 0$$

Now, we have to show that the given relation $R_{ijkl} = -\frac{R}{2}(g_{ik}g_{jl} - g_{il}g_{jk})$ follows easily.

The non-vanishing components are R_{1212}, R_{2121}, R_{1221}, *and* R_{2112}.

Now, we have
$$\begin{aligned} R &= g^{ij}R_{ij} \\ &= g^{ij}\left(-\frac{R_{1212}}{g} g_{ij}\right) \\ &= -2\frac{1}{g}R_{1212}(i, j = 1, 2), \end{aligned}$$

so
$$R_{1212} = -\frac{R}{2} g$$
$$= -\frac{R}{2}\begin{vmatrix} g_{11} & g_{12} \\ g_{21} & g_{22} \end{vmatrix}$$
$$= -\frac{R}{2}(g_{11}g_{22} - g_{21}g_{12}).$$

Changing indices with symmetric property of g_{ij}, we get

$$R_{2121} = -\frac{R}{2}(g_{22}g_{11} - g_{12}g_{21})$$
$$R_{1221} = -\frac{R}{2}(g_{12}g_{21} - g_{11}g_{22})$$
$$R_{2112} = -\frac{R}{2}(g_{21}g_{12} - g_{22}g_{11}).$$

Combine these relations and we get $R_{ijkl} = -\frac{R}{2}(g_{ik}g_{jl} - g_{il}g_{jk})$.

Example 5.4.4. Prove that in a Riemannian space $div(R^i_{jkl}) = R_{jl,k} - R_{jk,l}$.
Solution: We know the Bianchi Identity

$$R^i_{jkl,m} + R^i_{jlm,k} + R^i_{jkm,l} = 0$$

Contracting in the above relation, we get

$$R^i_{jkl,i} + R^i_{jli,k} + R^i_{jki,l} = 0$$

or
$$R^i_{jkl,i} = -R^i_{jli,k} - R_{jk,l}$$
$$= R_{jl,k} - R_{jk,l}$$
$$\therefore div(R^i_{jkl}) = R_{jl,k} - R_{jk,l}$$

Example 5.4.5. If $R_{ij,k} = 2B_kR_{ij} + B_iR_{kj} + B_jR_{ik}$, prove that $B_k = \frac{\partial}{\partial x^k}(log\sqrt{R})$

Solution: We have $R_{ij,k} = 2B_kR_{ij} + B_iR_{kj} + B_jR_{ik}$.

Using the Associate Tensor Formula, $B^j = g^{ij}B_i$, and the Scalar Curvature. $R = g^{ij}R_{ij}$, we get

$$g^{ij}R_{ij,k} = 2B_kg^{ij}R_{ij} + g^{ij}B_iR_{kj} + g^{ij}B_jR_{ik}$$

$$= 2B_kR + B^jR_{kj} + B^iR_{ik}$$

$$R_{,k} = 2B_kR + B^iR_{ki} + B^iR_{ik} = 2B_kR + 2B^iR_{ik} \text{ ....(i) } [\because R_{ik} = R_{ki}]$$

From the given condition, we get

$$R_{ij,k} - R_{ik,j} = B_kR_{ij} - B_jR_{ik},$$

$$\text{or } g^{ij}R_{ij,k} - g^{ij}R_{ik,j} = B_kg^{ij}R_{ij} - B_jg^{ij}R_{ik},$$

$$\text{or } R_{,k} - \frac{1}{2}R^j_{k,j} = B_kR - B^iR_{ik},$$

$$\text{or } R_{,k} - \frac{1}{2}R_{,k} = B_kR - B^iR_{ik}$$

$$\text{or } R_{,k} = 2B_kR - 2B^iR_{ik} \qquad \text{(ii)}$$

From (i) and (ii) it follows that $B^iR_{ik} = 0$
and by this value, from the equations, we get

$$R_{,k} = 2B_kR$$

$$\therefore B_k = \frac{1}{2R}R_{,k} = \frac{1}{2R}\frac{\partial R}{\partial x^k} = \frac{\partial}{\partial x^k}(\log\sqrt{R}).$$

5.4.3 Einstein Tensor

Theorem 5.4.2. Prove that the tensor $R^i_j - \frac{1}{2}\delta^i_j$ R is divergence free.

Proof: We know Equation (5.20), *Bianchi's Second identity*

$$R_{ijkl,m} + R_{ijlm,k} + R_{ijmk,l} = 0.$$

Multiply it by $g^{il}g^{jk}$ and we get

$$g^{il}g^{jk}R_{ijkl,m} + g^{il}g^{jk}R_{ijlm,k} + g^{il}g^{jk}R_{ijmk,l} = 0,$$

$$\text{or } g^{il}g^{jk}R_{ijkl,m} + g^{il}g^{jk}R_{ijlm,k} + g^{il}g^{jk}R_{ijmk,l} = 0,$$

$$\text{or } g^{jk}R_{jk,m} + g^{il}g^{jk}(-R_{ijml,k}) + g^{il}g^{jk}(-R_{jimk,l}) = 0.$$

Since $R_{ijlm} = -R_{ijml}$, $R_{ijmk} = -R_{jimk,}$

$$\therefore g^{jk}R_{jk,m} - g^{jk}R_{jm,k} - g^{il}R_{im,l} = 0,$$

$$\text{or } R_{,m} - R^k_{m,k} - R^l_{m,l} = 0 \left[\because g^{jk}R_{jk} = \mathrm{R}\right]$$

$$\text{or } R_{,m} - R^k_{m,k} - R^k_{m,k} = 0$$

$$\therefore R_{,m} - 2R^k_{m,k} = 0$$

$$\therefore R^k_{m,k} - \frac{1}{2}R_{,m} = 0,$$

$$\text{or } R^k_{m,k} - \frac{1}{2}\delta^k_m R_{,k} = 0, \left[\because R_{,m} = \delta^k_m R_{,k},\right]$$

$$\text{or } \left(R^k_m - \frac{1}{2}\delta^k_m R\right)_{,k} = 0, \tag{5.21}$$

where $R^k_m = g^{jk}R_{jm}$.

$\Rightarrow R^k_m - \frac{1}{2}\delta^k_m R$ is divergence free.

The tensor of type (1,1) is defined by

$$R^i_j - \frac{1}{2}\delta^i_j R \equiv G^i_j \ and \tag{5.22}$$

is called the *Einstein Tensor*.

Scalar Curvature: The *Scalar Curvature* R of a Riemannian space is defined by

$$R = g^{ij}R_{ij}.$$

Example 5.4.6. In a V_2 where $g_{11} = g_{22} = h > 0$ and $g_{12} = g_{21} = 0$ are functions of x^1 and x^1, show that

$$R_{ij} = \frac{R}{2}g_{ij},$$

where R_{ij} is the Ricci Tensor and R is the Scalar Curvature.

Solution: We know $R_{ij} = g^{kl}R_{lijk}$

$$= g^{11}R_{1ij1} + g^{12}R_{1ij2} + g^{21}R_{2ij1} + g^{22}R_{2ij2} \tag{i}$$

$$det g_{ij} = \begin{vmatrix} h & 0 \\ 0 & h \end{vmatrix} = h^2$$

$$g^{ij} = \frac{cofactor\ of\ g_{ij}\ in\ g}{g} = \frac{cofactor\ of\ g_{ij}\ in\ g}{h^2}$$

$$g^{11} = g^{22} = \frac{h}{h^2} = \frac{1}{h}, g^{12} = g^{21} = 0$$

$$R_{ij} = \frac{1}{h}(R_{1ij1} + R_{2ij2})$$

$$R_{11} = \frac{1}{h}(R_{1111} + R_{2112}) = \frac{1}{h}R_{2112}$$

$$R_{22} = \frac{1}{h}(R_{1221} + R_{2222}) = \frac{1}{h}R_{2112}$$

$$R_{12} = \frac{1}{h}(R_{1121} + R_{2122}) = 0$$

$$R_{21} = \frac{1}{h}(R_{1211} + R_{2212}) = 0$$

Now, we know $R = g^{ij}R_{ij}$

$$= g^{11}R_{11} + g^{12}R_{12} + g^{21}R_{21} + g^{22}R_{22}$$

$$= \frac{1}{h}(R_{11} + R_{22}) = \frac{1}{h}\frac{1}{h}2R_{2112} = \frac{2}{h^2}R_{2112}$$

$$R_{11} = \frac{1}{h}R_{1221} = \frac{1}{h}\frac{R}{2}h^2 = \frac{R}{2}h = \frac{R}{2}g_{11}$$

$$R_{22} = \frac{1}{h}R_{2112} = \frac{1}{h}\frac{R}{2}h^2 = \frac{R}{2}h = \frac{R}{2}g_{22} \text{ and } R_{12} = \frac{R}{2}g_{12} = 0 \; and \; R_{21} = \frac{R}{2}g_{21} = 0$$

$$\therefore R_{ij} = \frac{R}{2}g_{ij}.$$

Example 5.4.7. If in a $V_n(n > 2), R_{ij} - \frac{1}{2}g_{ij}R = 0$, show that $R_{ij} = 0$.

Solution: We have $R_{ij} - \frac{1}{2}g_{ij}R = 0$ (i)

Multiplying the given relation by g^{ij}, we get

$$g^{ij}R_{ij} - \frac{1}{2}g^{ij}g_{ij}R = 0,$$

$$\text{or } R - \frac{1}{2}n\,R = 0,$$

$$\text{or } \frac{2-n}{2}R=0 \text{. Hence, } R = 0.$$

From (i), we get for $R = 0$,

$$R_{ij} = 0.$$

Example 5.4.8. If $g^{ik}R_{kj} = R^i_j$ and $g^{ij}R_{ij} = R$,
show that $R^i_{j,i} = \frac{1}{2}\frac{\partial R}{\partial x}$.
Solution: Using covariant differentiation on both sides of $g^{ik}R_{kj} = R^i_j$, we get

$$g^{ik}R_{kj,l} = R^i_{j,l} \, [\because g^{ik}_{,l} = 0].$$

$$\text{Hence, } R^i_{j,i} = g^{ik}R_{kj,i} = g^{ik}R^l_{kjl,i} \, [\because R^l_{kjl} = R_{kj}]$$

$$= g^{ik}(g^{pl}R_{pkjl}),i \qquad [\because g^{pl}R_{pkjl} = R^l_{kjl}] \quad \text{(i)}$$

$$= g^{ik}g^{pl}R_{pkjl,i}[\because g^{pl}_{,i} = 0]$$

Using a Bianchi Identity, we can write $R_{pkjl,j} + R_{pkli,j} + R_{pkij,l} = 0$. (ii)
Using (ii), we can write (i)

$$R^i_{j,i} = g^{ik}g^{pl}R_{pkjl,i}$$

$$= -g^{ik}g^{pl}(R_{pkli,j} + R_{pkij,l})$$

$$= -g^{pl}(g^{ik}R_{pkli,j} + g^{ik}R_{pkij,l})$$

$$= -g^{pl}\left(-g^{ki}R_{kpli,j} + g^{ik}R_{kpji,l}\right)\left[\because R_{pkli,j} = -R_{kpli,j} \text{ and } R_{pkij,l} = R_{kpji,l}\right]$$

$$-g^{pl}\left(-R^i_{pli,j} + R^i_{pji,l}\right)$$

$$= -g^{pl}(-R_{pl,j} + R_{pj,l})$$

$$= (g^{pl}R_{pl})_{,j} - g^{pl}R_{pj,l}$$

$$= g^{pl}R_{pl,j} - g^{pl}R_{pj,l}$$

$$= R_{,j} - R^i_{j,l} \text{ (replacing dummy index } l \text{ by } i\text{)}$$

Hence, $2R^i_{j,i} = R,j$

$$\text{or } R^i_{j,i} = \frac{1}{2}R_{,j} = \frac{1}{2}\frac{\partial R}{\partial x^j}.$$

5.5 Einstein Space

A Riemannian space V_n is said to be an *Einstein Space* if its Ricci Tensor can be expressed as

$$R_{ij} = \lambda g_{ij}, \tag{5.23}$$

where λ is a scalar.

Now, using inner multiplication by g^{ij}, we get

$$g^{ij}R_{ij} = \lambda g^{ij}g_{ij}.$$

Since $R = g^{ij}R_{ij}$ and $g^{ij}g_{ij} = n$

$$\therefore R = n\lambda$$

or $\lambda = \dfrac{R}{n}$,

from (5.23) we get

$$R_{ij} = \frac{R}{n}g_{ij}. \tag{5.24}$$

The equation $R_{ij} = \dfrac{R}{n}g_{ij} = \rho g_{ij}$, where $\rho = \dfrac{R}{n}$ and $R = g^{ij}R_{ij}$, is known as the Einstein Gravitational Equation at points where matter is present. It corresponds to the Poisson Equation, $\nabla^2 V = \rho$, in the Newtonian Theory of Gravitation.

Theorem 5.5.1. Prove that the Scalar Curvature of an Einstein Space is constant.

Proof: We know from Einstein Space that

$$R_{ij} = \frac{R}{n} g_{ij}.$$

Covariant differentiating, we get $R_{ij,k} = \frac{1}{n} R_{,k} g_{ij} \quad [\because g_{ij,k} = 0].$

Inner multiplying by g^{il}, we get $g^{il} R_{ij,k} = \frac{1}{n} R_{,k} g_{ij} g^{il}$

or $R^{l}_{j,k} = \frac{1}{n} \delta^{l}_{j} R_{,k}.$

Now, contracting *l and k*, we get $R^{l}_{j,l} = \frac{1}{n} \delta^{l}_{j} R_{,l}$

or $\frac{1}{2} R_{,j} = \frac{1}{n} R_{,j}.$

$$\therefore (\frac{1}{n} - \frac{1}{2}) R_{,j} = 0,$$

$$\therefore R_{,j} = 0$$

Hence, for Einstein Spaces, the scalar curvature R is constant.

5.6 Riemannian and Euclidean Spaces

5.6.1 Riemannian Spaces

Let n-dimensional space V_n be in an x-coordinate system and metrize it by prescribing the element of arc ds, such that

$$ds^2 = g_{ij}\, dx^i\, dx^j \tag{5.25}$$

There is a positive definite quadratic form in the differential dx^i. We assume that the functions $g_{ij}(x)$ are to be of class C^1 in V_n.

The space V_n so metrized is called a *Riemannian n-dimensional Space* R_n.

Let a coordinate system Y, defined by the Y-frame, are given by

$$T: y^i = y^i\,(x^1, \ldots\ldots, x^n).$$

Apply a symmetric tensor $g_{ij}(x)$ so that $g_{ij}(x)$ has a constant component, h_{ij}, throughout in R_n.

We note first that the components of $g_{ij}(x)$, when reffered to the Y-frame, are given by

$$h_{ij} = \frac{\partial x^\alpha}{\partial y^i}\frac{\partial x^\beta}{\partial y^j} g_{\alpha\beta}. \tag{5.26}$$

If h_{ij}'s are constants, then Christoffel symbols $\left\{ \begin{matrix} k \\ i\ j \end{matrix} \right\}_y$ vanish identically, $h_{ij,l} = \frac{\partial h_{ij}}{\partial y^l}$, and, since $h_{ij,l} = 0$, by Ricci theorem, we have $\frac{\partial h_{ij}}{\partial y^l} = 0$ in R_n.

Theorem 5.6.1. A necessary and sufficient condition that the metric coefficients $g_{ij}(x)$ reduce to constants h_{ij} in some reference y-frame is that Christoffel symbols $\left\{ \begin{matrix} k \\ i\ j \end{matrix} \right\}_y$ vanish identically.

From this theorem we can deduce a system of differential equations that must be satisfied by the functions $y^i(x^1, \ldots\ldots, x^n)$, if there is to be a y-coordinate system in which the h_{ij}'s are constants.

We know the Law of Transformation is

$$-\frac{\partial y^\alpha}{\partial x^i}\frac{\partial y^\beta}{\partial x^j}\left\{ \begin{matrix} m \\ \alpha\ \beta \end{matrix} \right\}_y = \frac{\partial^2 y^m}{\partial x^i x^j} - \left\{ \begin{matrix} \gamma \\ i\ j \end{matrix} \right\} \frac{\partial y^m}{\partial x^\gamma}$$

and, since $\left\{ \begin{matrix} m \\ \alpha\ \beta \end{matrix} \right\}_y = 0$, we have the system of equations

$$\frac{\partial^2 y^m}{\partial x^i x^j} - \left\{ \begin{matrix} \gamma \\ i\ j \end{matrix} \right\}_x \frac{\partial y^m}{\partial x^\gamma} = 0 \tag{5.27}$$

System (5.27) of second order partial differential equations can be rewritten as a system of 1st order partial differential equations.

$$\begin{cases} \dfrac{\partial y}{\partial x^i} = u_i (i = 1, 2, \dots n) \\ \dfrac{\partial u_i}{\partial x^i} = \begin{Bmatrix} \gamma \\ i \ j \end{Bmatrix} u_\gamma (\gamma = 1, 2, .. n) \end{cases} \tag{5.28}$$

This system, in general, will be incompatible. For existence of the solution, we form these conditions in a symmetric form and consider the system:

$$\frac{\partial f^\alpha}{\partial x^i} = F_i^\alpha (f^1, f^2, \dots . f^m, x^1, x^2, \dots x^n). (\alpha = 1, 2.. m; i = 1, 2, \dots n), \tag{5.29}$$

where F_i^α are known functions of f 's and x's. Equation (5.29) is specialization of (5.28) if we set $f^1 = y, f^2 = u_1, \dots f^m = u_{n,}$. Function F_i^α is defined over the n-dimensional region R. Let us refer to the region of definition of functions as R'. This region consistsof R of variables x^i and $-\infty < f^i < \infty$.

We will suppose that the functions F_i^α are of class C^1 in R'. Since R' is open, we will assume that $\dfrac{\partial F_i^\alpha}{\partial f^i}$ are bounded in R'.

Since F_i^α are of class C^1 in R', it follows that f^α's are of class C^2 and hence,

$$\frac{\partial^2 f^\alpha}{\partial x^i x^j} = \frac{\partial^2 f^\alpha}{\partial x^j x^i}. \tag{5.30}$$

This is the necessary condition for the integrability of system (5.29). Now we obtain

$$\begin{aligned} \frac{\partial^2 f^\alpha}{\partial x^i x^j} &= \frac{\partial F_i^\alpha}{\partial x^j} + \frac{\partial F_i^\alpha}{\partial f^\beta} \frac{\partial f^\beta}{\partial x^j} \\ &= \frac{\partial F_i^\alpha}{\partial x^j} + \frac{\partial F_i^\alpha}{\partial f^\beta} F_{j,}^\beta \quad \text{from (5.29).} \end{aligned}$$

From (5.30), we get a necessary condition of integrability, the set of equations

$$\frac{\partial F_i^\alpha}{\partial x^j}+\frac{\partial F_i^\alpha}{\partial f^\beta}F_{j,}^\beta=\frac{\partial F_j^\alpha}{\partial x^i}+\frac{\partial F_j^\alpha}{\partial f^\beta}F_{i,}^\beta \quad (\alpha,\beta=1,2..m;i,j=1,2,\ldots n). \tag{5.31}$$

The dependent variables in (5.28) are $y, u_1, \ldots. u_n$, and in (5.29) are $f^1, \ldots f^m$, i.e., $f^1 = y, f^2 = u_1, \ldots\ldots f^{n+1} = u_n$. Then, (5.28) gives

$$\frac{\partial f^1}{\partial x^i}=F_i^1=u_i \quad (i=1,2,\ldots n)$$

and $$\frac{\partial f^\alpha}{\partial x^i}=F_i^\alpha=\left\{\begin{matrix}\gamma\\ \alpha-1i\end{matrix}\right\}u_\gamma\begin{pmatrix}\alpha=2,3\ldots n+1\\ i,\gamma=1,2,\ldots n\end{pmatrix}.$$

If we substitute the values of F_i^α in (5.31), we get

$$\begin{cases}\left\{\begin{matrix}\gamma\\ i\ j\end{matrix}\right\}u_\gamma=\left\{\begin{matrix}\gamma\\ j\ i\end{matrix}\right\}u_\gamma\\ R_{kij}^\gamma u_\gamma=0\end{cases} \tag{5.32}$$

The first sets of equations are satisfied identically for the symmetric property of the Christoffel symbol and states that the set equations of (5.28) will have a solution if the Riemann-Christoffel tensor R_{kij}^i vanishes identically, since this tensor vanishes when metric coefficients are constants.

Theorem 5.6.2. A necessary and sufficient condition that a symmetric tensor g_{ij} with $|g_{ij}|\neq 0$ will reduce under a suitable transformation of coordinates to a tensor h_{ij}, where the h_{ij}'s are constants, is that the Riemann-Christoffel tensor formed from the g_{ij}'s be a zero tensor.

5.6.2 Euclidean Spaces

If the quadratic form $Q = h_{ij}y^iy^j$ is a positive definite, there exists a nonsingular linear transformation reducing Q to the canonical form $Q = (y^1)^2 +\cdots.+(y^n)^2$.

Thus, if $g_{ij}(x)$ are the coefficients in the positive definite quadratic differential form

$$ds^2 = g_{ij}dx^i dx^j, \tag{5.33}$$

characterizing metric properties of there exists a real functional transformation

T: $y^i = y^i(x)$, which reduces it to the form

$$ds^2 = (dy^1)^2 + \cdots .. + (dy^n)^2, \tag{5.34}$$

provided that R^i_{jkl} vanishes identically in R_n.

A metric manifold R_n in which it is possible to affect the reduction of $ds^2 = g_{ij}\ dx^i\ dx^j$ to $ds^2 = (dy^1)^2 + \cdots .. + (dy^n)^2$ is called an *Euclidean n-dimensional manifold* (E_n) and we see that

R_n is *Euclidean* if, and only if, the Riemann Tensor of the manifold is a zero tensor.

5.7 Exercises

1. Show that $R_{ijkl} = \frac{\partial}{\partial x^k}[jl,i] - \frac{\partial}{\partial x^l}[jk,i] + \left\{ \begin{matrix} \alpha \\ j\ k \end{matrix} \right\}[il,\alpha] - \left\{ \begin{matrix} \alpha \\ j\ l \end{matrix} \right\}[ik,\alpha]$.
2. Show that

$$R_{ijkl} = R_{ijkl} = \frac{1}{2}\left(\frac{\partial^2 g_{li}}{\partial x^k \partial x^j} - \frac{\partial^2 g_{jl}}{\partial x^k \partial x^i} - \frac{\partial^2 g_{ki}}{\partial x^l \partial x^j} + \frac{\partial^2 g_{jk}}{\partial x^l \partial x^i} \right)$$

$$+ g^{\alpha\beta}([jk,\beta][il,\alpha] - [jl,\beta][ik,\alpha]).$$

3. Show that $R^{\alpha}_{\alpha jk} = 0$.
4. If $R_{ij} = \rho g_{ij}$, then $\rho = \frac{R}{n}$ where $R = g^{ij}R_{ij}$.

   (i) If $n = 2$, show that $\frac{R_{11}}{g_{11}} = \frac{R_{22}}{g_{22}} = \frac{R_{12}}{g_{12}} = \frac{R_{12}}{g_{12}} = -\frac{R_{1212}}{g}$.

   (ii) If $n = 3$, the tensor R_{ijkl} has six distinct components and there are six equations *for* $R_{jk} = g^{il}R_{ijkl}$. Prove that the solutions of these equations for R_{ijkl} are given by

$$R_{ijkl} = g_{il}R_{jk} + g_{jk}R_{il} - g_{ik}R_{jl} - g_{jl}R_{il} + \frac{R}{2}(g_{ik}g_{jl} - g_{il}g_{jk}).$$

5. Find the expression for div. R^{α}_{ijk}.
6. If in a $V_n (n > 2) R_{ij,k} + R_{ki,j} + R_{jk,i} = 0$, prove that the scalar curvature is constant.
7. If $A^i_j = R^i_j + \delta^i_j (aR + b)$, where a and b are constants and notations carry their usual meanings, determine the value of a for which $A^i_{j,i} = 0$.
8. If in a V_n, $R_{ij}\, R^{ij} = \dfrac{R^2}{n}$,
   where $R^{ij} = g^{ip} g^{jq} R_{pq}$ and $R = g^{ij} R_{ij}$,
   prove that $R_{ij} = \dfrac{R}{n} g_{ij}$.
9. If A_i is a covariant vector such that
   $A_{i,j} + A_{j,j} = 0$, show that $A_{i,jk} = -A_r R^r_{kij}$.
10. If $g_{kl} R_{ij} - g_{jk} R_{il} + g_{il} R_{jk} - g_{ij} R_{kl} = 0$, show that the space is an Einstein Space.
11. If in a Riemannian space of dimension n(n>2),

$$R^h_{ijk} = \rho\left(\delta^h_k g_{ij} - \delta^h_j g_{ik}\right),$$

where ρ is a constant, prove that the space is an Einstein Space.

6

The e-Systems and the Generalized Kronecker Deltas

6.1 Introduction

The concept of symmetry and skew-symmetry with respect to pairs of indices can be extended to cover the sets of quantities that are symmetric or skew-symmetric with respect to more than two indices. We consider in this chapter the sets of quantities $A^{i_1 \ldots i_k}$ or $A_{i_1 \ldots i_k}$, depending on k indices, written as subscripts or superscripts, although the quantities of A may not represent tensors.

6.2 e-Systems

Definition 6.2.1. The system of quantities $A^{i_1 \ldots i_k}$ $(or\ A_{i_1 \ldots i_k})$, depending on k indicies, is said to be *completely symmetric* if the value of symbol A is unchanged by any permutation of the indices.

Definition 6.2.2. The system of quantities $A^{i_1 \ldots i_k}$ $(or\ A_{i_1 \ldots i_k})$, depending on k indicies, is said to be *completely skew-symmetric* if the value of symbol A is unchanged by any even permutation of the indices and A merely changes signs after an odd permutations of indices.

For a skew-symmetric system it follows from Definition 6.2.2 that the term containing two of the same indices is not necessarily zero. Thus, if one has a skew-symmetric system of quantities A_{ijk} where *i, j, and k* assume values 1, 2, and 3, then $A_{122} = 0$, $A_{123} = -A_{213}$, $A_{312} = A_{123}$, etc.

In general, the components A_{ijk} of a skew-symmetric system satisfy the relations $A_{ijk} = -A_{ikj} = -A_{jik} = A_{jki} = A_{kij} = -A_{kji}$.

Definition 6.2.3. If the values of $A^{i_1 \ldots i_n}$ $(or\ A_{i_1 \ldots i_n})$ are +1 when $i_1 \ldots i_n$ is an even permutation of the numbers 1,2,...n and -1 when $i_1 \ldots i_n$ is an odd

Dipankar De. Introduction to Differential Geometry with Tensor Applications, (177–192) © 2022 Scrivener Publishing LLC

permutation of numbers 1,2,...n and if it is zero in all other cases, then the system $A^{i_1 \ldots i_n}$ $(or\ A_{i_1 \ldots i_n})$ is called an e-system.

We consider four particular systems, two of which are of second order and another two are of third order. They are represented by e_{ij}, e^{ij} and e_{ijk}, e^{ikj}. The number of components of the first two are each 4 and last two are 27 each.

Now, the components of e_{ij}, e^{ij} *are*:

$$e_{11} = 0,\ e_{22} = 0,\ e_{12} = 1,\ e_{21} = -1$$

$$e^{11} = 0,\ e^{22} = 0,\ e^{12} = 1,\ e^{21} = -1 \tag{6.1}$$

$$e_{123} = e_{231} = e_{312} = 1;$$

$$e_{213} = e_{321} = e_{132} = -1$$

and the remaining 21 components are zero.

$e^{123} = e^{231} = e^{312} = 1$ (if ijk is an even permutation of 123)

$e^{213} = e^{321} = e^{132} = -1$ (if ijk is an odd permutation of 123)

and remaining 21 components are zero.

These systems are called e–systems of the second and third order, respectively.

The covariant ε tensor of second order is defined by

$$\epsilon_{11} = 0,, \epsilon_{12} = \sqrt{g}, \epsilon_{21} = -\sqrt{g}, \epsilon_{22} = 0, g = |g_{ij}| \tag{6.2}$$

This tensor is skew-symmetric. According to the Tensor law of Transformation,

$$\bar{\epsilon}_{ij} = \frac{\partial u^p}{\partial \bar{u}^i}\frac{\partial u^q}{\partial \bar{u}^j}\epsilon_{pq}\ or, \bar{\epsilon}_{ij} = \frac{\partial u^1}{\partial \bar{u}^i}\frac{\partial u^2}{\partial \bar{u}^j}\bar{\epsilon}_{12} + \frac{\partial u^2}{\partial \bar{u}^i}\frac{\partial u^1}{\partial \bar{u}^j}\epsilon_{21}$$

$$= \sqrt{g}\begin{vmatrix} \dfrac{\partial u^1}{\partial \bar{u}^i} & \dfrac{\partial u^2}{\partial \bar{u}^i} \\ \dfrac{\partial u^1}{\partial \bar{u}^j} & \dfrac{\partial u^2}{\partial \bar{u}^j} \end{vmatrix} \text{as } \epsilon_{11} = \epsilon_{22} = 0$$

$$\therefore \bar{\epsilon}_{11} = \bar{\epsilon}_{22} = 0$$

$$\bar{\epsilon}_{22} = \sqrt{g}\,\frac{\partial(u^1,u^2)}{\partial(u^1,u^2)}, = \sqrt{g} \text{ and } \bar{\epsilon}_{21} = \sqrt{g}\,\frac{\partial(u^1,u^2)}{\partial(u^2,u^1)}, = -\sqrt{g}.$$

Let us consider ϵ^{ij},

$\epsilon^{ij} = \epsilon_{rs} g^{ir} g^{js}$

which is called the *contravariant ε tensor of second order.*

Since $\epsilon_{11} = \epsilon_{22} = 0$, we get $\epsilon^{ij} = \epsilon_{rs} g^{ir} g^{js} = \sqrt{g}\left[g^{i1}g^{j2} - g^{i2}g^{j1}\right]$

$\therefore \epsilon^{11} = \epsilon^{22} = 0$ and

$$\epsilon^{12} = \sqrt{g}[g^{11}g^{22} - g^{12}g^{21}] = \sqrt{g}\,\frac{1}{g} = \frac{1}{\sqrt{g}}.$$

and $\epsilon^{21} = \sqrt{g}[g^{21}g^{12} - g^{22}g^{11}] = -\sqrt{g}\,\dfrac{1}{g} = -\dfrac{1}{\sqrt{g}}.$

Hence, we can write $\epsilon^{ij} = \dfrac{1}{\sqrt{g}} e^{ij}, \epsilon_{ij} = \sqrt{g}\, e_{ij}$ (6.3)

The third order e-system e_{ijk}, e^{ijk}: If the e-system depends on three indices, ijk, then

$e_{ijk} = 0$ if any two indices are alike,

$e_{ijk} = e_{123} = 1$, if ijk is an even permutation of 123 and

$e_{ijk} = -e_{123} = -1$, if ijk is an odd permutation of 123.

$$e_{123} = e_{231} = e_{312} = 1 \,;\, e_{213} = e_{321} = e_{132} = -1 \tag{6.4}$$

$$e^{123} = e^{231} = e^{312} = 1 \,;\, e^{213} = e^{321} = e^{132} = -1$$

It can be written that $\epsilon^{ijk} = \frac{1}{\sqrt{g}} e^{ijk}; \epsilon_{ijk} = \sqrt{g} e_{ijk},$ (6.5)

which are respectively contravariant and covariant tensors called permutation tensors in a 3-dimensional space. The formula of permutation tensor for ϵ_{ijk} or ϵ^{ijk} are

$$e_{ijk} = \begin{cases} +1, \text{ when } i, j, k \text{ are in even permutation of } 123. \\ 0, \quad \text{when any two of indices } i, j, k \text{ are alike} \\ -1, \text{ when } i, j, k \text{ are in odd permutation of } 123 \end{cases} \quad (6.6)$$

We shall now establish some results using e-Systems of 2nd and 3rd order.

By means of these systems and Kronecker Delta, proofs of a number of properties of determinants are considerably simplified. They are also useful for writing down briefly different expressions important in the theory of determinants.

Property 6.2.1 e-Systems of Second Order

Let $|a_j^i| = \begin{vmatrix} a_1^1 & a_2^1 \\ a_1^2 & a_2^2 \end{vmatrix} = a_1^1 a_2^2 - a_2^1 a_1^2 = e_{12} a_1^1 a_2^2 + e_{21} a_2^1 a_1^2 \quad [e_{12} = 1, e_{21} = -1]$

$$= e_{ij} a_1^i a_2^j \quad (6.7)$$

Similarly, it can be shown that

$$|a_j^i| = e^{ij} a_i^1 a_j^2. \quad (6.8)$$

Let us now consider the expression $e_{ij} a_p^i a_q^j$, where the indices p and q are free and can be assigned values 1 and 2 at will.

Now we have

$$\begin{aligned} e_{ij} a_1^i a_2^j &= e_{12} a_1^1 a_2^2 + e_{21} a_1^2 a_2^1 \\ &= e_{12} a_1^1 a_2^2 - e_{12} a_1^2 a_2^1 \\ &= e_{12} |a_j^i|. \end{aligned}$$

Thus, $e_{ij}a_2^i a_2^j = e_{21}\left|a_j^i\right|$.

Similarly, $e_{ij}a_2^i a_1^j = e_{21}\left|a_j^i\right|$.

From these relations, we get $e_{ij}a_p^i a_q^j = e_{pq}\left|a_j^i\right|$.

$$\Rightarrow \left|a_j^i\right| e_{pq} = e_{ij}a_p^i a_q^j \tag{6.9}$$

$$\text{Similarly, it can be shown that } \left|a_j^i\right| e^{pq} = e^{ij}a_i^p a_j^q. \tag{6.10}$$

The above definitions of e-systems of second order can obviously be extended to define e-systems of nth order $e^{i_1 \ldots i_n}$ *and* $e_{i_1 \ldots i_n}$ involving n indices and $\left|a_j^i\right| = \begin{vmatrix} a_1^1 & a_2^1 & .. & a_n^1 \\ a_1^2 & a_2^2 & .. & a_n^2 \\ .. & .. & .. & .. \\ a_1^n & a_2^n & .. & a_n^n \end{vmatrix}$.

6.3 Generalized Kronecker Delta

Definition 6.3.1. A symbol $\delta_{j_1 \ldots\ldots j_k}^{i_1 \ldots\ldots i_k}$ depending on k superscripts and k subscripts, each of which runs from 1 to n, is called a *Generalized Kronecker Delta* provided that

(i) it is completely skew-symmetric in superscripts and subscripts;
(ii) if the superscripts are distinct from each other and the subscripts are the same set of numbers as the superscripts, the value of the symbol is +1 *or* −1 according to an even or odd number of transpositions required to arrange the superscripts in the same order as the subscripts;
(iii) in all other cases the value of the symbol is zero.

If we consider δ_{kl}^{ij}, it follows from definition that if $i = j$ *or* $k = l$ or if the set *i*, *j* is not the set *k*, *l*, then $\delta_{kl}^{ij} = 0$. In all other cases, δ_{kl}^{ij} equals to +1 *or* −1 according to whether *kl* is an even or odd permutation of *ij*, i.e.,
$\delta_{kl}^{11} = \delta_{pq}^{22} = \delta_{13}^{23} = \cdots\ldots = 0$ (except odd or even permutation)

and $\delta_{12}^{22} = \delta_{21}^{21} = \delta_{31}^{31} = \delta_{23}^{23} = 1$ (even permutation)
and $\delta_{22}^{12} = \delta_{13}^{31} = \delta_{31}^{13} = \delta_{12}^{21} = -1$ (odd permutation)

From Definition 6.2.3, it follows that the direct product $e^{i_1 \ldots i_n}\, e_{i_1 \ldots i_n}$ of the two systems $e^{i_1 \ldots i_n}$ and $e_{i_1 \ldots i_n}$ is a generalized Kronecker Delta.

For example, $e^{\alpha\beta\gamma}\, e_{ijk}$ has the following values:

(a) 0, if two or more subscripts or superscripts are the same;
(b) +1, if the difference in the number of transpositions of $\alpha\beta\gamma$ and ijk from 123 is an even number.
(c) −1, if the difference in the number of transpositions of $\alpha\beta\gamma$ and ijk from 123 is an odd number.
(b) and (c) can be write another way
(b') +1, if even number of transpositions is required to arrange the subscripts in the same order as the superscripts.
(c') +1 if an odd number of transpositions is required to arrange the subscripts in the same order as the superscripts.

We can write $e^{\alpha\beta\gamma} e_{ijk} = \delta_{ijk}^{\alpha\beta\gamma}$.

It implies that the e-symbols can be defined in terms of the Kronecker Deltas.

$$e^{i_1 \ldots i_n} = \delta_{12 \ldots n}^{i_1 \ldots i_n} \qquad \text{and} \qquad \delta_{i_1 \ldots i_n}^{12 \ldots n} = e_{i_1 \ldots i_n}$$

Since $e = +$ or -1, when the set of distinct integers $i_1 \ldots i_n$ is obtained from the set $1, 2, \ldots n$ by an even or an odd permutation, $e = 0$ in all other cases. The e–systems and generealized Kronecker Deltas prove useful in calculations involving alternating sets of quantities.

Example 6.3.1. Evaluate $e_{ij}e^{ik}$, in e-systems of second order if $i, j = 1,2$.
Solution: By the summation convention, $e_{ij}e^{ik} = e_{1j}e^{1k} + e_{2j}e^{2k}$.

When $j = k = 1$ *or* 2, then

$$e_{i1}e^{i1} = e_{11}e^{11} + e_{21}e^{21} = 1.1 + 0.0 = 1$$

$$e_{i2}e^{i2} = e_{12}e^{12} + e_{22}e^{22} = 0.0 + 1.1 = 1.$$

Thus, when $j = k$, then $e_{ij}e^{ik} = 1$.

Again, when $j \neq k$, say $j = 1, k = 2$ or $j = 2, k = 1$, we get

$$e_{i1}e^{i2} = e_{11}e^{12} + e_{21}e^{22} = 1.0 + 0.1 = 0$$

$$e_{i2}e^{i1} = e_{12}e^{11} + e_{22}e^{21} = 0.1 + 1.0 = 0.$$

Thus, *if* $j \neq k$, then $e_{ij}e^{ik} = 0$.
Therefore, in general, $e_{ij}e^{ik} = \delta_j^k$ for $i, j = 1,2 \ldots. n$.
For particular, when $j = k$

$$e_{ij}e^{ik} = \delta_i^i = \delta_1^1 + \delta_2^2 + \cdots . + \delta_n^n = 1 + 1 + .. + 1 = n.$$

6.4 Contraction of $\delta_{\alpha\beta\gamma}^{ijk}$

Let us contract $\delta_{\alpha\beta\gamma}^{ijk}$ on k and γ. The result for $n = 3$ is

$$\delta_{\alpha\beta k}^{ijk} = \delta_{\alpha\beta 1}^{ij1} + \delta_{\alpha\beta 2}^{ij2} + \delta_{\alpha\beta 3}^{ij3} \equiv \delta_{\alpha\beta}^{ij}.$$

This expression vanishes if $i = j$ or $\alpha = \beta$. If we set $i = 1$ or $j = 2$, we get $\delta_{\alpha\beta 3}^{123} = \delta_{\alpha\beta}^{12}$ and hence, $\delta_{\alpha\beta}^{12} = 0$ unless $\alpha\beta$ is a permutation of 12, and $\delta_{\alpha\beta}^{12} = 1$ if $\alpha\beta$ is an even permutation of 12, $\delta_{\alpha\beta}^{12} = -1$ if $\alpha\beta$ is an odd permutation of 12.

If we contract $\delta_{\alpha\beta}^{ij}$, we obtain a system depending on two indices (first contract it and the multiply the result by $\frac{1}{2}$).

$$\delta_\alpha^i \equiv \frac{1}{2}\delta_{\alpha j}^{ij} = \frac{1}{2}(\delta_{\alpha 1}^{i1} + \delta_{\alpha 2}^{i2} + \delta_{\alpha 3}^{i3})$$

If $\delta_\alpha^i \equiv \frac{1}{2}\delta_{\alpha j}^{ij} = \frac{1}{2}(\delta_{\alpha 1}^{i1} + \delta_{\alpha 2}^{i2} + \delta_{\alpha 3}^{i3})$ in δ_α^i, we get $\delta_\alpha^1 = \frac{1}{2}(\delta_{\alpha 2}^{12} + \delta_{\alpha 3}^{13})$. It vanishes unless $\alpha = 1$, in which $\delta_1^1 = 1$.

Similar results can be obtained by setting $i = 2$ *or* $i = 3$. Thus, δ_α^i has values

(a) 0 if $i \neq \alpha$
(b) 1, if $i = \alpha$.

By counting the number of terms appearing in the sums, it is not difficult to show that, in general,

$$\delta^i_\alpha = \frac{1}{n-1}\delta^{ij}_{\alpha j} \text{ and } \delta^{ij}_{ij} = n(n-1) \tag{6.11}$$

We can also deduce that

$$\delta^{i_1\ldots i_r}_{j_1\ldots j_r} = \frac{(n-k)!}{(n-r)!}\delta^{i_1\ldots i_r i_{r+1}\ldots i_k}_{j_1\ldots j_r i_{r+1}\ldots i_k} \tag{6.12}$$

$$\text{And } \delta^{i_1\ldots i_r}_{i_1\ldots i_r} = n(n-1)(n-2)\ldots(n-r+1) = \frac{n!}{(n-\mathrm{r})!} \tag{6.13}$$

As a special case of (6.13), we have the formula

$$e^{i_1\ldots i_n} e_{i_1\ldots i_n} = n! \tag{6.14}$$

and from (6.12) we can deduce that

$$e^{i_1\ldots i_r i_{r+1}\ldots i_n} e_{j_1\ldots j_r i_{r+1}\ldots i_k} = (n-r)!\,\delta^{i_1\cdots i_r}_{j_1\cdots j_r} \tag{6.15}$$

Example 6.4.1. If a set of n^{p+q} quantities $A^{i_1\ldots i_p}_{j_1\ldots j_q}, i, j = 1,2,..n$, and it symmetric in two or more indices (superscripts or subscripts), then show that

$$\delta^{j_1\cdots j_q}_{i_1\ldots i_q} A^{r_1\ldots r_p}_{j_1\cdots j_q} = 0.$$

Solution: Suppose that $A^{i_1\ldots i_p}_{j_1\cdots j_q}$ is symmetric in j_1 *and* j_2, then

$$\delta^{j_1\cdots j_q}_{i_1\ldots i_q} A^{r_1\ldots r_p}_{j_1\cdots j_2\cdots j_q} = \delta^{j_1 j_2\cdots j_q}_{i_1\ldots i_q} A^{r_1\ldots r_p}_{j_2 j_1\cdots j_q} = -\delta^{j_2 j_1\cdots j_q}_{i_1\ldots i_q} A^{r_1\ldots r_p}_{j_2 j_1\cdots j_q}.$$

However, j_1 *and* j_2 are the dummy indices, hence

$$\delta^{j_1\cdots j_q}_{i_1\ldots i_q} A^{r_1\ldots r_p}_{j_1 j_2\cdots j_q} = -\delta^{j_1\cdots j_q}_{i_1\ldots i_q} A^{r_1\ldots r_p}_{j_1 j_2\cdots j_q}.$$

Therefore, $\delta^{j_1\cdots j_q}_{i_1\ldots i_q} A^{r_1\ldots r_p}_{j_1 j_2\cdots j_q} = 0.$

6.5 Application of e-Systems to Determinants and Tensor Characters of Generalized Kronecker Deltas

The determinant $\left|a_j^i\right|$ of nth order, with elements a_j^i, consists of the sum of the products of the elements where each term in the sum contains one and only one element from each row and each column of the determinant. The sign of each term in the sum is determined by the character of permutation of the indices. Thus, if the superscripts in the product $a_{i_1}^1 a_{i_2}^1 \ldots .a_{i_n}^n$ are arranged in the normal order 1, 2, ... n, then the product will sign if the number of transpositions necessary to arrange the subscripts in the normal order is even. The sign is if the required number of transpositions is odd.

Since $e^{i_1 i_2 \ldots i_n} = \delta_{12\ldots n}^{i_1 \ldots i_n}$ and $e_{i_1 i_2 \ldots i_n} = \delta_{i_1 \ldots i_n}^{12\ldots n}$,

if we take $\left|a_j^i\right| = \begin{vmatrix} a_1^1 & a_2^1 & .. & a_n^1 \\ a_1^2 & a_2^2 & .. & a_n^2 \\ .. & .. & .. & .. \\ a_1^n & a_2^n & .. & a_n^n \end{vmatrix} \equiv a$, we get the following results

$a = \left|a_j^i\right| = e_{i_1 i_2 \ldots i_n} a_1^{i_1} a_2^{i_2} \ldots .a_n^{i_n}$ (by similar arguments as in the cases of (6.7), (6.8), (6.9), and (6.10)

$$\left|a_j^i\right| = e^{i_1 i_2 \ldots i_n} a_{i_1}^1 a_{i_2}^2 \ldots .a_{i_n}^n$$

An example consider:

We take $\left|a_j^i\right| = \begin{vmatrix} a_1^1 & a_2^1 & a_3^1 \\ a_1^2 & a_2^2 & a_3^2 \\ a_1^3 & a_2^3 & a_3^3 \end{vmatrix}$,

If this determinant is expanded to $a = \left|a_j^i\right| = \sum a_1^i a_2^j a_3^k$, where ijk is a permutation of 1, 2, 3 the + *or* –sign is assigned to the term $a_1^i a_2^j a_3^k$ according to whether this permutation is even or odd. Hence, the determinant can be written as $\left|a_j^i\right| = e_{ijk} a_\alpha^i a_\beta^j a_\gamma^k$.

Consider next terms $e_{ijk} a_\alpha^i a_\beta^j a_\gamma^k$ $(i, j, k; \alpha, \beta, \gamma = 1, 2, 3)$.

We will observe that this system is completely skew-symmetric in $\alpha\beta\gamma$. Since ijk are dummy indices, we can change them at will and write

$$e_{ijk}a^i_\alpha a^j_\beta a^k_\gamma = e_{ijk}a^k_\alpha a^j_\beta a^i_\gamma = e_{ijk}a^i_\gamma a^j_\beta a^k_\alpha.$$

If k and i are interchanged, this e–symbol will be changed

$$e_{ijk}a^i_\alpha a^j_\beta a^k_\gamma = -e_{ijk}a^i_\gamma a^j_\beta a^k_\alpha.$$

It implies that an interchange of α and γ changes the sign, so that the system under consideration is skew-symmetric in α and γ and from this study we can write

$$e_{ijk}a^i_\alpha a^j_\beta a^k_\gamma = \left|a^i_j\right| e_{\alpha\beta\gamma}. \tag{6.16}$$

Similarly, we can show that

$$e^{ijk}a^\alpha_i a^\beta_j a^\gamma_k = \left|a^i_j\right| e^{\alpha\beta\gamma}. \tag{6.17}$$

It follows at once from these expressions that an interchange of two columns (or two rows) of the determinant $\left|a^i_j\right|$ changes its sign and if two columns (or two rows) in it are identical, then its value is zero.

This result can be generalized to determinants of the nth order, so that for any permutation of rows we can write

$$e^{\alpha\beta\ldots\ldots\gamma}\left|a^i_j\right| = e^{ij\ldots\ldots k}a^\alpha_i a^\beta_j \ldots\ldots a^\gamma_k \tag{6.18}$$

and for any permutation of columns

$$e_{ij\ldots k}\left|a^i_j\right| = e_{\alpha\beta\ldots\ldots\gamma}a^\alpha_i a^\beta_j \ldots\ldots a^\gamma_k \tag{6.19}$$

The expression of the determinant in terms of the elements of the first column and their cofactors can be written as

$$\left|a^i_j\right| = a^{i_1}_1 e_{i_1 i_2\ldots i_n} a^{i_1}_2 \ldots\ldots a^{i_1}_n$$

$$= a_1^\alpha A_\alpha^1, \tag{6.20}$$

where $A_\alpha^1 = e_{i_1 i_2 \ldots i_n} a_2^{i_1} \ldots\ldots a_n^{i_1}$ is the cofactor of the element a_1^α.

Property 6.5.1 Product of Two Determinant

If $$|a_j^i| = \begin{vmatrix} a_1^1 & a_2^1 & .. & a_n^1 \\ a_1^2 & a_2^2 & .. & a_n^2 \\ .. & .. & .. & .. \\ a_1^n & a_2^n & .. & a_n^n \end{vmatrix} \text{ and } |b_j^i| = \begin{vmatrix} b_1^1 & b_2^1 & .. & b_n^1 \\ b_1^2 & b_2^2 & .. & b_n^2 \\ .. & .. & .. & .. \\ b_1^n & b_2^n & .. & b_n^n \end{vmatrix},$$

then $|a_j^i||b_j^i| = |c_j^i|$, where $c_j^i = a_\alpha^i b_j^\alpha$.

Proof: Since $|b_j^i| = e_{ij\ldots k} b_1^i b_2^j \ldots . b_n^k$,

we know $|a_j^i||b_j^i| = |a_j^i| = e_{ij\ldots k} b_1^i b_2^j \ldots . b_n^k$ (since $e_{ij\ldots k}|a_j^i| = e_{\alpha\beta\ldots\ldots\gamma} a_i^\alpha a_j^\beta \ldots\ldots a_k^\gamma$)

$$= (|a_j^i| e_{ij\ldots k})(b_1^i b_2^j \ldots . b_n^k)$$

$$= (e_{\alpha\beta\ldots\ldots\gamma} a_i^\alpha a_j^\beta \ldots . a_k^\gamma)(b_1^i b_2^j \ldots . b_n^k)$$

$$= \left(e_{\alpha\beta\ldots\ldots\gamma}\right)(a_i^\alpha b_1^i)(a_j^\beta b_2^j)\ldots\ldots(a_k^\gamma b_n^k)$$

$$= \left(e_{\alpha\beta\ldots\ldots\gamma}\right)(c_1^\alpha)(c_2^\beta)\ldots\ldots(c_n^\gamma)$$

$$= |c_j^i|,$$

where $$c_j^i = a_\alpha^i b_j^\alpha = a_1^i b_j^1 + a_2^i b_j^2 \cdots .. + a_n^i b_j^n.$$

Property 6.5.2 Partial Derivative of a Determinant

Show that $\dfrac{\partial a}{\partial x} = \dfrac{\partial a_\beta^\alpha}{\partial x^j} A_\alpha^\beta$,

where $a = |a_j^i|$ and a_j^i are the functions of the variables $x^1, x^2, \ldots x^n$ and A_i^j is the cofactor of element a_j^i of $= |a_j^i|$.
Solution: We know

$$a = e_{i_1 i_2 \ldots i_n} a_1^{i_1} a_2^{i2} \ldots . a_n^{in}.$$

Differentiating this expression, we get

$$\frac{\partial a}{\partial x} = (e_{i_1 i_2 \ldots i_n})(\frac{\partial a_1^{i1}}{\partial x} a_2^{i_2} \ldots a_n^{i_n} + a_1^{i_1} \frac{\partial a_2^{i2}}{\partial x} \ldots . a_n^{i_n} \cdots \ldots + a_1^{i_1} a_2^{i_2} \ldots \frac{\partial a_n^{in}}{\partial x})$$
$$= (\frac{\partial a_1^{i1}}{\partial x} A_{i_1}^1 + \frac{\partial a_2^{i2}}{\partial x} A_{i_1}^2 + \ldots + \frac{\partial a_n^{in}}{\partial x} A_{i_1}^n) \quad [\text{using } A_\alpha^1] = e_{i_1 i_2 \ldots i_n} a_2^{i_1} \ldots . a_n^{i_1}]$$
$$= \frac{\partial a_\beta^\alpha}{\partial x^j} A_\alpha^\beta.$$

Property 6.5.3 *Permutation of* $e_{i_1 i_2 \ldots i_n}$ *and* $e^{i_1 i_2 \ldots i_n}$ is relative tensors of weight −1 *and* +1, respectively.
Proof: Let us consider an admissible transformation

$$T: y^i = y^i(x^1, x^2, \ldots x^n)$$

and its Jacobian $j = \left|\frac{\partial y}{\partial x}\right|$. If we put $a_j^i = \frac{\partial y^i}{\partial x^j}$ in $e^{i_1 i_2 \ldots i_n} |a_j^i| = e^{\alpha_1 \alpha_2 \ldots \alpha_n} a_{i_1}^{\alpha_1} a_{i_2}^{\alpha_2}$ $\ldots . a_{i_n}^{\alpha_n}$ and consider $\frac{1}{j} = \left|\frac{\partial x}{\partial y}\right|$, we obtain

$$e^{i_1 i_2 \ldots i_n} = \left|\frac{\partial x}{\partial y}\right| e^{\alpha_1 \alpha_2 \ldots \alpha_n} a_{\alpha_1}^{i_1} a_{\alpha_2}^{i_2} \ldots . a_{\alpha_n}^{i_n}$$
$$= \left|\frac{\partial x}{\partial y}\right| e^{\alpha_1 \alpha_2 \ldots \alpha_n} \frac{\partial y^{i_1}}{\partial x^{\alpha_1}} \frac{\partial y^{i_2}}{\partial x^{\alpha_2}} \ldots . \frac{\partial y^{i_n}}{\partial x^{\alpha_n}}, \tag{6.21}$$

which is the Law of Transformation of relative tensors of weight +1. In an entirely similar way, we deduce that

$$e_{i_1 i_2 \ldots i_n} = \left|\frac{\partial x}{\partial y}\right|^{-1} e_{\alpha_1 \alpha_2 \ldots \alpha_n} \frac{\partial x^{\alpha_1}}{\partial y^{i_1}} \frac{\partial x^{\alpha_2}}{\partial y^{i_2}} \cdots\cdots \frac{\partial x^{\alpha n}}{\partial y^{i_n}}, \tag{6.22}$$

so that $e_{i_1 i_2 \ldots i_n}$ is a relative tensor of weight -1.

6.5.1 Curl of Covariant Vector

Let A_i be a covariant vector. Then, $A_{i,k}$ is a tensor of type (0,2). The product $\epsilon^{ikl} A_{l,k}$ is then a tensor of type (1,0). If we denote this vector by B^i, then B^i is called the curl of vector A_i.

Thus, in a V_3,

$$\text{Curl } A_i = B^i = \epsilon^{ikl} A_{l,k} = \frac{1}{\sqrt{g}} e^{ikl} A_{l,k}$$

$$\therefore B^i = \frac{1}{\sqrt{g}} e^{ikl} A_{l,k},$$

or $$B^1 = \frac{1}{\sqrt{g}} e^{1kl} A_{l,k} = \frac{1}{\sqrt{g}} e^{123} A_{3,2} + \frac{1}{\sqrt{g}} e^{132} A_{2,3} \quad (\text{other } e^{ikl} = 0)$$
$$= \frac{1}{\sqrt{g}} (A_{3,2} - A_{2,3}),$$

or $$B^2 = \frac{1}{\sqrt{g}} e^{2kl} A_{l,k} = \frac{1}{\sqrt{g}} e^{213} A_{3,1} + \frac{1}{\sqrt{g}} e^{231} A_{1,3} \quad (\text{other } e^{ikl} = 0)$$
$$= \frac{1}{\sqrt{g}} (A_{1,3} - A_{3,1})$$

and $$B^3 = \frac{1}{\sqrt{g}} (A_{2,1} - \mathrm{A}_{1,2}).$$

Thus, components of curl A_i (components of B^i)

$$= \frac{1}{\sqrt{g}}(A_{3,2} - A_{2,3}), \frac{1}{\sqrt{g}}(A_{1,3} - A_{3,1}), \frac{1}{\sqrt{g}}(A_{2,1} - A_{1,2}).$$

Since $A_{i,j} - A_{j,i} = \frac{\partial A_i}{\partial x^j} - \frac{\partial A_j}{\partial x^i}$,

$$A_{3,2} - A_{2,3} = \frac{\partial A_3}{\partial x^2} - \frac{\partial A_2}{\partial x^3}, A_{1,3} - A_{3,1} = \frac{\partial A_1}{\partial x^3} - \frac{\partial A_3}{\partial x^1}, A_{2,1} - A_{1,2} = \frac{\partial A_2}{\partial x^1} - \frac{\partial A_1}{\partial x^2},$$

the components of curl A_i are

$$\frac{1}{\sqrt{g}}\left(\frac{\partial A_3}{\partial x^2} - \frac{\partial A_2}{\partial x^3}\right), \frac{1}{\sqrt{g}}\left(\frac{\partial A_1}{\partial x^3} - \frac{\partial A_3}{\partial x^1}\right), \frac{1}{\sqrt{g}}\left(\frac{\partial A_2}{\partial x^1} - \frac{\partial A_1}{\partial x^2}\right)$$

$$Curl A_i = \left[\frac{1}{\sqrt{g}}\left(\frac{\partial A_3}{\partial x^2} - \frac{\partial A_2}{\partial x^3}\right), \frac{1}{\sqrt{g}}\left(\frac{\partial A_1}{\partial x^3} - \frac{\partial A_3}{\partial x^1}\right), \frac{1}{\sqrt{g}}\left(\frac{\partial A_2}{\partial x^1} - \frac{\partial A_1}{\partial x^2}\right)\right].$$

6.5.2 Vector Product of Two Covariant Vectors

Let A_i and B_i be two covariant vectors of V_3. Then, the product $\epsilon^{ijk}A_jB_k$ is a tensor of type (1,0). If we denote it by C^i, then $C^i = \epsilon^{ijk}A_jB_k$ is called the vector product (cross product) of vectors A_i and B_i.

$$C^i = \epsilon^{ijk}A_jB_k$$

$$C^1 = \epsilon^{1jk}A_jB_k$$

$$= \epsilon^{123}A_2B_3 + \epsilon^{132}A_3B_2 \text{ others } \epsilon^{1jk} = 0$$

$$= \frac{1}{\sqrt{g}}(A_2B_3 - A_3B_2) \quad (\because \epsilon^{ijk} = \frac{1}{\sqrt{g}}e^{ijk} \Rightarrow \epsilon^{123} = \frac{1}{\sqrt{g}}, \epsilon^{132} = -\frac{1}{\sqrt{g}})$$

$$C^2 = \epsilon^{213} A_1 B_3 + \epsilon^{231} A_3 B_1$$

$$= \frac{1}{\sqrt{g}}(-A_1 B_3 + A_3 B_1)$$

and

$$C^3 = \epsilon^{312} A_1 B_2 + \epsilon^{321} A_2 B_1 = \frac{1}{\sqrt{g}}(A_1 B_2 - A_2 B_1).$$

The components of the vector product of two vectors, A_i and B_i, are

$$\frac{1}{\sqrt{g}}(A_2 B_3 - A_3 B_2), \frac{1}{\sqrt{g}}(-A_1 B_3 + A_3 B_1), \frac{1}{\sqrt{g}}(A_1 B_2 - A_2 B_1).$$

Hence,

$$A \times B = \left[\frac{1}{\sqrt{g}}(A_2 B_3 - A_3 B_2), \frac{1}{\sqrt{g}}(-A_1 B_3 + A_3 B_1), \frac{1}{\sqrt{g}}(A_1 B_2 - A_2 B_1) \right].$$

Example 6.5.1. Prove that in a V_3,

$$curl(\phi A) = \phi curl A - A \times grad\phi, \text{ where } \phi \text{ is an invariant.}$$

Solution:

We have $curlA = e^{ijk} A_{k,j}$.

Now, if $\phi A_k = B_k$,

$$\text{then } B_{k,j} = \phi_{,j} A_k + \phi A_{k,j}. \tag{i}$$

Hence, $curl(\phi A) = curlB_k = e^{ijk} B_{k,j}$

$$= e^{ijk}(\phi_{,j} A_k + \phi A_{k,j})$$

$$= e^{ijk}\phi A_{k,j} - e^{ikj}\phi_{,j} A_k \; [e^{ijk} = -e^{ikj}]$$

$$= \phi e^{ijk} A_{k,j} - e^{ikj} A_k \phi_{,j}$$

$$= \phi curl A - A \times grad\phi \; [\because A \times B = e^{ijk} A_j B_k].$$

6.6 Exercises

1. Verify that $\delta^{ijk}_{\alpha\beta\gamma} a^{\alpha\beta\gamma} = a^{ijk} - a^{ikj} + a^{jki} - a^{jik} + a^{kij} + a^{kji}$.
2. Show that
   (a) $e_{ij} e^{ik} = \delta^k_j$
   (b) $e_{ijk} e^{irs} = \delta^r_j \delta^s_k - \delta^r_k \delta^s_j$
   (c) $e_{rij} e_{rkl} = \delta_{ik}\delta_{jl} - \delta_{il}\delta_{jk}$
3. Prove that $\epsilon_{ijk}\sqrt{g}$ is a covariant tensor of rank three where ϵ_{ijk} is the usual permutation tensor.
4. If in a V_3, $C^i = \epsilon^{ipq} A_{pq}$, where A_{pq} is a skew-symmetric tensor, show that $2A_{pq} = \epsilon_{ijt} C^t$.
5. If A^{ij} is a skew-symmetric tensor in V_3, show that $(\sqrt{g}A^{23}, \sqrt{g}A^{23}, and \sqrt{g}A^{23})$ are the components of a covariant vector. Show further that if A_{ij} is a skew-symmetric tensor in V_3, then $(\frac{1}{\sqrt{g}}A_{23}, \frac{1}{\sqrt{g}}A_{31}, and \frac{1}{\sqrt{g}}A_{12})$ are the components of a contravariant vector.
6. Show that the permutation symbols $e_{i_1 i_2 \ldots i_n}$ *and* $e^{i_1 i_2 \ldots i_n}$ are relative tensors of weight -1 *and* $+1$, respectively, and also show that they are associated.
7. Show that the Kroneker Deltas behave as constants in a covariant differentiation.
8. Show that if a set of quantities $A_{i_1 i_2 \ldots i_k}$ is skew-symmetric in the subscripts (k in number), then $\delta^{i_1 i_2 \ldots jk}_{j_1 j_2 \ldots jk} A_{i_1 i_2 \ldots i_k} = k! A_{j_1 j_2 \ldots j_k}$.
9. Show that the value of $e_{i_1 i_2 \ldots i_r i_{r+1} \ldots i_n} e^{j_1 j_2 \ldots j_r j_{r+1} \ldots j_n}$ is $n!$.
10. Prove that ϵ_{ij} is a covariant tensor of rank 2.
11. Show that $\epsilon_{ijk}\epsilon^{ist} = \begin{vmatrix} \delta^s_j & \delta^t_j \\ \delta^s_k & \delta^t_k \end{vmatrix}$.

    Hence, deduce that

    $$\text{(i) } \epsilon_{ijk}\epsilon^{ijt} = 2\delta^t_k \quad \text{(ii) } \epsilon_{ijk}\epsilon^{ijk} = 3!$$

12. Show that $e_{imn} e^{irs} a^{mn} = a^{rs} - a^{sr}$.
13. Show that $e_{ijk} e^{ijk}$, where e_{ijk} *and* e^{ijk} are e –systems of third order are 6.
14. Show that $\delta^{ij}_{\alpha\beta} = \begin{vmatrix} \delta^i_\alpha & \delta^i_\beta \\ \delta^j_\alpha & \delta^j_\beta \end{vmatrix}$ and $\delta^{ijk}_{\alpha\beta\gamma} = \begin{vmatrix} \delta^i_\alpha & \delta^i_\beta & \delta^i_\gamma \\ \delta^j_\alpha & \delta^j_\beta & \delta^k_\gamma \end{vmatrix}$.

Part II
DIFFERENTIAL GEOMETRY

7
Curvilinear Coordinates in Space

7.1 Introduction

What is the pure idea of coordinates? The pure idea is in representing points of space by triple numbers. This means that we should have a one to one map $P(x^1, x^2, x^3) \leftrightarrow (y^1, y^2, y^3)$ in the whole space, or at least in some domain, where we are going to use coordinates. In Cartesian coordinates, this map is constructed by means of vectors and bases. The triplets of numbers (x^1, x^2, *and* x^3) are curvilinear coordinates of point *P in R*. Arranging other coordinate systems, one can use other methods. For example, in spherical coordinates, $y^1 = r$ is a distance from the point P to the center of a sphere and, $y^2 = \theta$, $y^3 = \phi$ are two angles. The discussion of this chapter is confined mainly in curvilinear coordinates in E_3 and studies reciprocal base systems and the meaning of covariant derivatives.

7.2 Length of Arc

Consider that n–dimensional space R can be covered by an x-coordinate system and that a one-dimensional subspace of R is determined by a curve, C, so that

$$C: x^i =: x^i(t),\ i = 1,2 \ldots n, \tag{7.1}$$

where t is a real parameter in $t_1 \leq t \leq t_2$. The one-dimensional manifold C is called an *arc of a curve*.

We deal in this book only with those curves for which $x^i(t)$ and $\dot{x}^i(t) \equiv \frac{dx^i}{dt}$ are continuous functions in $t_1 \leq t \leq t_2$.

Let $\mathrm{F}(x^1, x^2, \ldots . x^n, \dot{x}^1, \dot{x}^2, \ldots . \dot{x}^n)$ be a continuous function in the interval $t_1 \leq t \leq t_2$. We suppose that $\mathrm{F}(\mathrm{x}, \dot{x}) > 0$ unless every $\dot{x} = 0$ and for every positive number k,

Dipankar De. Introduction to Differential Geometry with Tensor Applications, (195–220) © 2022 Scrivener Publishing LLC

$$F(x^1, x^2, \ldots\ldots x^n, k\dot{x}^1, k\dot{x}^2, \ldots\ldots k\dot{x}^n) = k\,\mathrm{F}(x^1, x^2, \ldots\ldots x^n, \dot{x}^1, \dot{x}^2, \ldots\ldots \dot{x}^n).$$

$$\text{The integral } s = \int_{t_1}^{t_2} \mathrm{F}(\mathrm{x}, \dot{x})\mathrm{dt} \tag{7.2}$$

is called the *length* of C and space R is called *metrized* by (7.2).

We choose to define the length of arc by the formula

$$s = \int_{t_1}^{t_2} \sqrt{g_{\alpha\beta}(\mathrm{x}) \frac{dx^\alpha}{dt} \frac{dx^\beta}{dt}}\, \mathrm{dt} \quad ,(\alpha, \beta = 1, 2, \ldots n,) \tag{7.3}$$

where $g_{\alpha\beta}(\mathrm{x})$ is a positive definite quadratic form in variable $\dot{x}^\alpha$.

This resulting geometry is called *Riemannian geometry* and the space is R-metrized and, in this way, is the *Riemannian n-dimensional space* R_n.

Consider the admissible transformation of coordinates T: $y^i = y^i(, x^2, \ldots\ldots x^n)$, such that the square of arc element ds

$$ds^2 = g_{\alpha\beta} dx^\alpha dx^\beta \tag{7.4}$$

can be reduced to the form

$$ds^2 = dy^i dy^i. \tag{7.5}$$

Then, the Riemannian manifold R_n is called an *n –dimensional Euclidean manifold* E_n and the reference frame Y in which arc ds in E_n is given by (7.5) is an orthogonal Cartesian reference frame. Here, E_n is the generalization of a Euclidian plane determined by the totality of the pairs of real values (y^1, y^2). The square of the element of arc ds assumes the familiar form

$$ds^2 = (dy^1)^2 + (dy^2)^2.$$

A function $\mathrm{F}(\mathrm{x}, \dot{x})$, satisfying the condition $\mathrm{F}(\mathrm{x}, k\dot{x}) = k\mathrm{F}(\mathrm{x}, \dot{x})$ for every $k > 0$, is called *positively homogenous of degree 1 in the* $\dot{x}^i$.

Theorem 7.2.1. A function, $F(x,\dot{x})$ satisfies the condition $F(x,k\dot{x}) = kF(x,\dot{x})$ for every $k > 0$. This condition is both necessary and sufficient to ensure independence of the value of' integral $s = \int_{t_1}^{t_2} F(x,\dot{x})dt$ of a particular mode of parametrization of C. Thus, if t in C : $x^i =: x^i(t)$ is replaced by some function, $t = \phi(s)$, and we denote $x^i(\phi(s))$ by $\xi^i(\lambda)$ so that: $x^i(t) = \xi^i(s)$ and we have equality $\int_{t_1}^{t_2} F(x,\dot{x})dt = \int_{t_1}^{t_2} F(\xi,\xi')ds,$

where $\xi'^i = \dfrac{dx^i}{ds}$ and $t_1 = \phi(s_1)$ and $t_2 = \phi(s_2)$.

Proof: Suppose that k is an arbitrary positive number and put $t = ks$ so that $t_1 = ks_1$ and $t_2 = ks_2$.

Then, by (7.1), $x^i(t) = x^i(ks)$

and $\xi'^i = \dfrac{dx^i}{ds} = \dfrac{dx^i(ks)}{ds} = \dfrac{dx^i(ks)}{dt}\dfrac{dt}{ds} = k\dot{x}^i(ks).$ $\quad \left[\because \dfrac{dt}{ds} = k\right]$

Putting these values in $s = \int_{t_1}^{t_2} F(x,\dot{x})dt$, we get

$$s = \int_{ks_1}^{ks_2} F[x(ks),\dot{x}(ks)]dt = \int_{ks_1}^{ks_2} F[x(ks),\dot{x}(ks)]kds \qquad [\because dt = kds]$$

$$= \int_{s_1}^{s_2} F[\xi^i(s),\xi'^i(s)]ds. \qquad \left[\because x(ks) = \xi^i(s)\right]$$

We must have the relation $F(\xi,\xi') = F(x,k\dot{x}) = kF(x,\dot{x})$

Conversely, if this relation is true for every line element of C and each $k > 0$, then the equality of integrals is assured for every choice of parameters $t = \phi(s)$, $\phi'(s) > 0$, *and* $s_1 \le s \le s_2$ with $t_1 = \phi(s_1)$ *and* $t_2 = \phi(s_2)$.

Example 7.2.1. Consider a sphere, S, of radius a, immersed in a 3-dimensional Euclidean manifold E_3, with a center at origin (0,0,0) of a set of orthogonal Cartesian axes $0 - X^1X^2X^3$.

Solution: Let T be a plane tangent to S at (0,0, $-a$) and let the points on this plane be referred to as a set of orthogonal Cartesian axes $O'-Y^1Y^2$, as shown in Figure (7.1).

If we draw from $O(0,0,0)$ a radial line OM, intersect sphere S at $M(x^1, x^2, x^3)$, and the plane T at $O(y^1, y^2, -a)$, then the points M on the lower half of the sphere S are in 1 – 1 correspondence with points (y^1, y^2) of the tangent plane T.

If $M(x^1, x^2, x^3)$ is any point on a radial line OM, then the symmetric equation of this line is:

$$\frac{x^1-0}{y^1-0}=\frac{x^2-0}{y^2-0}=\frac{x^3-0}{-a-0}=\lambda$$

$$\text{or } x^1=\lambda y^1,\ x^2=\lambda y^2,\ x^3=-\lambda a \tag{7.6}$$

These are general points of sphere, so it satisfies the sphere with equation:

$$(x^1)^2+(x^2)^2+(x^3)^2=a^2,$$

$$\text{or } (\lambda y^1)^2+(\lambda y^2)^2+(-\lambda a)^2=a^2,$$

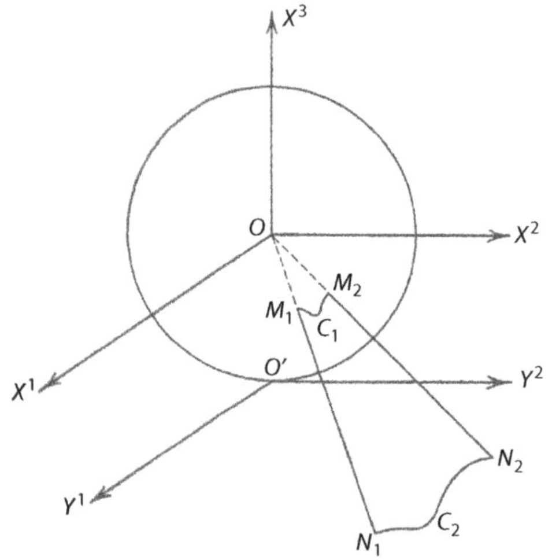

Figure 7.1

$$\text{or } \lambda[(y^1)^2 + (y^2)^2 + (a)^2] = a^2$$

$$\therefore \lambda^2 = \frac{a^2}{(y^1)^2 + (y^2)^2 + (a)^2} \Rightarrow \lambda = \frac{a}{\sqrt{(y^1)^2 + (y^2)^2 + (a)^2}}$$

$$x^1 = \frac{ay^1}{\sqrt{(y^1)^2 + (y^2)^2 + (a)^2}}, x^2 = \frac{ay^2}{\sqrt{(y^1)^2 + (y^2)^2 + (a)^2}}, x^3$$
$$= -\frac{a^2}{\sqrt{(y^1)^2 + (y^2)^2 + (a)^2}}. \tag{7.7}$$

These are the equations giving the analytical 1 – 1 correspondence of the points of N *on* T and S.

Let $M_1(x^1, x^2, x^3)$ and $M_2(x^1 + dx^1, x^2 + dx^2, x^3 + dx^3)$ be two close points on curve C lying on S.

The Euclidean distance $\overline{M_1M_2}$, along C is given by

$$ds^2 = dx^i dx^i \quad (i = 1,2,3) \tag{7.8}$$

and $dx^i = \frac{\partial x^i}{\partial y^\alpha} dy^\alpha \quad (\alpha = 1,2)$.

(7.8) becomes $ds^2 = dx^i dx^i = \frac{\partial x^i}{\partial y^\alpha} dy^\alpha \frac{\partial x^i}{\partial y^\beta} dy^\beta = \frac{\partial x^i}{\partial y^\alpha} \frac{\partial x^i}{\partial y^\beta} dy^\alpha dy^\beta$

$$ds^2 = g_{\alpha\beta}(y) dy^\alpha dy^\beta,$$

where $g_{\alpha\beta}(y) = \frac{\partial x^i}{\partial y^\alpha} \frac{\partial x^i}{\partial y^\beta}$ are functions of y computed from (7.7).

If the image C_2 of C *on* T is given by

$$C_2: \begin{cases} y^1 = y^1(t) \\ y^2 = y^2(t) \end{cases} \quad t_1 \le t \le t_2,$$

then the length of C can be computed from integral

$$s=\int_{t_1}^{t_2}\sqrt{g_{\alpha\beta}(y)dy^{\alpha}dy^{\beta}}=\int_{t_1}^{t_2}\sqrt{g_{\alpha\beta}(y)\dot{y}^{\alpha}\dot{y}^{\beta}}\,dt$$

$$\therefore\ ds^2=\frac{(dy^1)^2+dy^2)^2+\frac{1}{a^2}(y^1-\dot{y}^2-\dot{y}^1y^1)}{1+\frac{1}{a^2}[(y^1)^2+(y^2)^2]}. \tag{7.9}$$

$$s=\int_{t_1}^{t_2}\sqrt{\frac{(dy^1)^2+dy^2)^2+\frac{1}{a^2}(y^1\dot{y}^2-\dot{y}^1y^1)}{1+\frac{1}{a^2}[(y^1)^2+(y^2)^2]}}\,dt$$

We see that the resulting formulas refer to a 2-dimensional manifold determined by the variables (y^1, y^2) in a Cartesian plane T and that the geometry of the surface of the sphere imbedded in a 3-dimensional Euclidean manifold can be visualized on a 2-dimensional manifold R_2 with a metric determined by (7.9).

If the radius of S is very large, in (7.9), implying $\frac{1}{a^2}\to 0$ and the geometry of the surface of the sphere is then approximated by the Euclidean metric

$$ds^2=(dy^1)^2+dy^2)^2. \tag{7.10}$$

For a large value of radius a, metric properties of the sphere S are indistinguishable from those of the Euclidean plane and the sum of the angles of the curvilinear triangle drawn on S will be nearly equal to 180° by Euclidean geometry.

The main point of this example is to indicate that the geometry of a sphere, imbedded in a Euclidean 3-dimensional space with the element of arc in the form of (7.8) is indistinguishable from the Riemannian geometry of a 2-dimensional manifold R_2 with metric (7.9).

The latter manifold, referred to as Cartesian frame Y, is not Euclidean since (7.9) cannot be reduced by admissible transformation to (7.10).

7.3 Curvilinear Coordinates in E_3

The apparatus of tensor analysis was developed initially as a tool of the analytic study of geometries. Since dynamics, mechanics of its continuous media, and relativity lean heavily on geometrical properties of the 3-dimensional space of physical experience, we devote most of the chapters to an investigation of properties of curves and surfaces imbedded in E_3.

Let $P(y)$ be the point in a Euclidean 3-space E_3, referred to a set of orthogonal Cartesian axes Y. Consider a coordinate transformation

$$T: x^i = x^i(y^1, y^2, y^3), \ (i = 1,2,3)$$

such that the x^i are of class C^1 and $J = \left|\frac{\partial x^i}{\partial y^j}\right| \neq 0$ in the same region R of E_3.

The inverse transformation T^{-1}: $y^i = y^i(x^1, x^2, x^3)$, $(i = 1,2,3)$ will be single valued and the transformation T and T^{-1} establishes a 1 – 1 correspondence between the sets of values (x^1, x^2, x^3) and (y^1, y^2, y^3).

The triplets of numbers (x^1, x^2, x^3) are called *curvilinear coordinates* of the points P in R.

7.3.1 Coordinate Surfaces

Let x^1 be kept fixed, i.e.

$$x^1 = \textit{constant} \text{ in } T, \text{ then}$$

$$x^1 = x^1(y^1, y^2, y^3) = \textit{constant}, \tag{7.11}$$

defines a surface. If the constant is now to assume different values, we get a one-parameter family of surfaces. Similarly, other families

$$x^2 = x^2(y^1, y^2, y^3) = \textit{constant} \text{ and } x^3 = x^3(y^1, y^2, y^3) = \textit{constant}.$$

The condition is that the Jacobian, $j \neq 0$, in the region under consideration that the surfaces

$$x^1 = c_1, \quad x^2 = c_1, \quad x^3 = c_1, \tag{7.12}$$

intersects at one, and only one, point. The surfaces of (7.12) are called *coordinate surfaces* and their intersection pair by pair are called *coordinate lines* because, along this line, the variable x^3 is the only one that is changing.

7.3.2 Coordinate Curves

Let two of the coordinates x^1, x^2, x^3 be kept fixed in T, i.e. $x^2 = c^2$, *constant and* $x^3 = c^3$, *constant* in T, and x^1 be allowed to vary. Then, $P(y^1, y^2y^3)$ will satisfy the relations:

$$y^i = y^i(x^1, c^2, c^3).$$

Since y^1, y^2, *and* y^3 are the functions of one variable, it follows that $P(y^1, y^2y^3)$ will lie on a curve, called a coordinate curve. This coordinate curve

$$x^2 = c^2, x^3 = c^3$$

is called the x^1 –curve. Thus, the line of intersection of $x^2 = c^2$, $x^3 = c^3$ is the x^1 –coordinate line. Therefore, the x^1 –curve lies on both the surfaces of $x^2 = c^2$, $x^3 = c^3$. Similarly, we can define the x^2 – *and* x^3 –curves. It should also be noted that through a given point $P(y^1, y^2y^3)$, 3 coordinate curves corresponding to fixed values c^1, c^2, *and* c^3 *are passed through.*

Example 7.3.1. Consider a coordinate system defined by the transformation

$$y^1 = x^1 sinx^2 cosx^3$$

$$y^2 = x^1 sinx^2 sinx^3$$

$$y^3 = x^1 cosx^2.$$

The surface x^1 = constant are spheres, x^2 = constant are circular cones, and x^3 = constant are planes passing through the Y^3-axis (Figure 7.2).
Solution: Squaring and adding the equation, we get

$$(y^1)^2 + (y^2)^2 + (y^3)^2 = (x^1 sinx^2 cosx^3)^2 + (x^1 sinx^2 sinx^3)^2 + (x^1 cosx^2)^2$$

$$= (x^1 sinx^2)^2 + (x^1 cosx^2)^2 = (x^1)^2 \qquad \text{(i)}$$

Squaring and adding its two equations, $(y^1)^2 + (y^2)^2 = (x^1 sinx^2)^2$

$$\text{or} \qquad x^1 sinx^2 = \sqrt{(y^1)^2 + (y^2)^2} \qquad \text{(ii)}$$

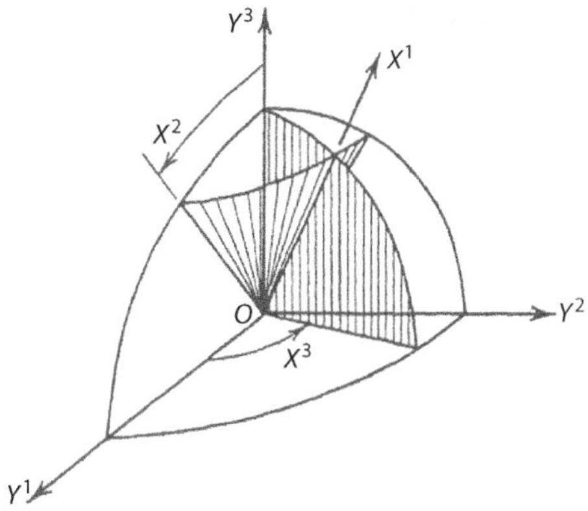

Figure 7.2

Dividing (ii) by the last given equation, we get

$$tan x^2 = \frac{\sqrt{(y^1)^2 + (y^2)^2}}{y^3} \quad \text{(iii)}$$

Dividing the first two equations, we get

$$tan x^3 = \frac{y^2}{y^1}. \quad \text{(iv)}$$

From (i), $x^1 = \sqrt{(y^1)^2 + (y^2)^2 + (y^3)^2}$,

from (iii) $x^2 = tan^{-1} \dfrac{\sqrt{(y^1)^2 + (y^2)^2}}{y^3}$,

and from (iv) $x^3 = tan^{-1} \dfrac{y^2}{y^1}$.

So, the inverse transformation is given by the above equations.

If $x^1 > 0, 0 < x^2 < \pi$, $0 \le x^3 < 2\pi$, this is the familiar spherical coordinate system.

Example 7.3.2. The transformation

$$y^1 = x^1 cos x^2$$

$$y^2 = x^1 sin x^2$$

$$y^3 = x^3$$

defines a cylindrical coordinate system.
Solution: The Jacobian of transformation is given by

$$J = \begin{vmatrix} \cos x^2 & -x^1 sin x^2 & 0 \\ sin x^2 & x^1 cos x^2 & 0 \\ 0 & 0 & 1 \end{vmatrix} = x^1 \neq 0.$$

Hence, the inverse transformation exists and is given by

$$T^{-1} : x^1 = \sqrt{(y^1)^2 + (y^2)^2}$$

$$\tan x^2 = \frac{y^2}{y^1} \text{ or } x^2 = tan^{-1} \frac{y^2}{y^1}$$

$$\text{and } x^3 = y^3$$

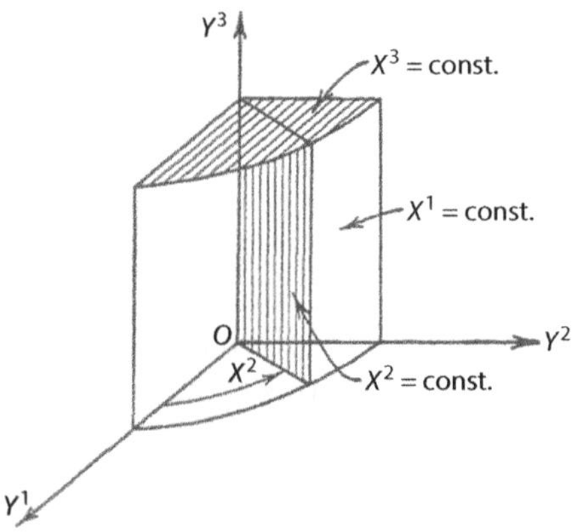

Figure 7.3

if $x^1 > 0, 0 \le x^2 < 2\pi, -\infty < x^3 < \infty$, this coordinate system defines a cylindrical coordinate system, as shown in Figure (7.3).

The coordinates of the surface are

$$x^1 = constant = \sqrt{c_1} \Rightarrow (y^1)^2 + (y^2)^2 = c_1,$$

which is circle.

$x^2 = tan^{-1}\dfrac{y^2}{y^1} = c_2 \Rightarrow y^2 = y^1 tan c_2$, which is a straight line.

and $y^3 = c_3$, which is a plane parallel to *the* y^1y^2 *– plane*

7.3.3 Line Element

Here, we have to obtain the line element of E_3 in curvilinear coordinates. Let $P(y^1, y^2, y^3)$ and $Q(y^1 + dy^1, y^2 + dy^2, y^3 + dy^3)$ be two neighboring points in R. The Euclidean distance between these points is determined by the quadratic form

$$ds^2 = (dy^1)^2 + dy^2)^2 + dy^3)^2$$

$$= dy^i dy^i$$

and since $dy^i = \dfrac{\partial y^i}{\partial x^\alpha} dx^\alpha$, we get

$$\begin{aligned} ds^2 = dy^i dy^i &= \frac{\partial y^i}{\partial x^\alpha} dx^\alpha \frac{\partial y^i}{\partial x^\alpha} dx^\alpha \\ &= \frac{\partial y^i}{\partial x^\alpha}\frac{\partial y^i}{\partial x^\alpha} dx^\alpha dx^\alpha = \frac{\partial y^\alpha}{\partial x^i}\frac{\partial y^\alpha}{\partial x^j} dx^i dx^j \\ &= g_{ij} dx^i dx^j, \end{aligned} \tag{7.13}$$

where $g_{ij} = \dfrac{\partial y^\alpha}{\partial x^i}\dfrac{\partial y^\alpha}{\partial x^j}$ $(\alpha, \beta = 1,2,3)$.

Obviously, g_{ij} is symmetric. Moreover, it is a tensor, since ds^2 is an invariant and the vector dx^i is arbitray.

Denote by g the determinant $|g_{ij}|$; this is positive in R. Since $g_{ij}dx^i dx^j$ is a. positive definite form, we can introduce the conjugate symmetric tensor g^{ij}, defined by $g^{ij} = \dfrac{G^{ij}}{g}$, where G^{ij} is the cofactor of the element g_{ij} in g.

7.3.4 Length of a Vector

Consider a contravariant vector A^i in a curvilinear coordinate system. Now, we form the invariant

$$A = (g_{ij}A^iA^j)^{\frac{1}{2}} \tag{7.14}$$

In orthogonal Cartesian coordinate $g_{ij} = \delta_{ij}$ and *when* $i = j$, $g_{ij} = 1$,

$$A = (A^iA^i)^{\frac{1}{2}}.$$

Therefore, in the orthogonal Cartesian frame, (7.14) assumes the form

$$A = [(A^1)^2 + (A^2)^2 + (A^3)^2]^{\frac{1}{2}}.$$

We see that A represents the length of the vector A^i.

Similarly, the length of the covariant vector A_i is defined by the formula

$$A = (g^{ij}A_iA_j)^{\frac{1}{2}}, \tag{7.15}$$

a vector whose length is 1 and is called a *unit vector*. From (7.13) we get

$$ds^2 = g_{ij}dx^idx^j$$

$$\text{or} \qquad 1 = g_{ij}\frac{dx^i}{ds}\frac{dx^j}{ds},$$

so that $\frac{dx^i}{ds} \equiv \lambda^i$ is a unit vector. It is a contravariant vector.

If $x^i = y^i$ and the coordinate system is Cartesian,

then $\frac{dx^1}{ds} \equiv \lambda^1, \frac{dx^2}{ds} \equiv \lambda^2, \frac{dx^3}{ds}\lambda^3$ are direction cosines of the displacement vector (dx^1, dx^2, dx^3). Accordingly, we take this vector λ^i to define the direction in space relative to a curvilinear coordinate system X.

7.3.5 Angle Between Two Vectors

Consider two unit vectors of different directions λ^i and μ^i at same point P, as shown in Figure (7.4). Since the manifold under consideration is Euclidean, the cosine law with Pythagoras's formula gives

$$\overline{ST^2} = \overline{RS^2} + \overline{RT^2} - 2\overline{RS}\,\overline{RT}\ Cos\theta$$

and since λ^i and μ^i are unit vectors, $RT = RS = 1$, and hence

$$\overline{ST}^2 = 2(1 - \cos\theta) \qquad (7.16)$$

and the position vector $\overline{ST} = \lambda^i - \mu^i$.

Using the formula (7.15), we get $\overline{ST}^2 = g_{ij}(\lambda^i - \mu^i)(\lambda^j - \mu^j)$

$$= g_{ij}\lambda^i\lambda^j + g_{ij}\,\mu^i\,\mu^j - 2g_{ij}\lambda^i\,\mu^j$$

$$= 1 + 1 - 2g_{ij}\lambda^i\,\mu^j$$

$$= 2(1 - g_{ij}\lambda^i\,\mu^j) \qquad (7.17)$$

From (7.16) and (7.17), we get

$$cos\theta = g_{ij}\lambda^i\,\mu^j \qquad (7.18)$$

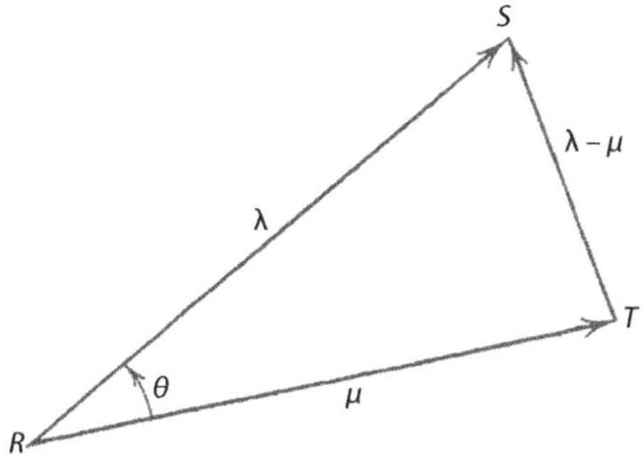

Figure 7.4

We can use this formula to define the angle θ between two directions λ^i *and* μ^j.

If A^i and B^i are any two vectors, then from definition of the length of a vector, it is clear

$$\cos\theta = \frac{g_{ij}A^i B^j}{\sqrt{g_{ij}A^i A^j}\sqrt{g_{ij}B^i B^j}}.$$

This leads to the formula $AB\cos\theta = g_{ij}A^iB^j$, defining an invariant which is a "scalar product" AB of elementary vector analysis.

Example 7.3.3. Prove that the angles θ_{12}, θ_{23}, *and* θ_{31} between the coordinate curves in a three-dimensional coordinate system are given by

$$\cos\theta_{12} = \frac{g_{12}}{\sqrt{g_{11}}\sqrt{g_{22}}}, \cos\theta_{23} = \frac{g_{23}}{\sqrt{g_{22}}\sqrt{g_{33}}}, \text{and } \cos\theta_{31} = \frac{g_{31}}{\sqrt{g_{33}}\sqrt{g_{11}}}.$$

Solution:

We know the expression

$$ds^2 = g_{ij}dx^idx^j$$

for the square of the element of arc (ds) between P_1 (x^1, x^2, x^3) and $P_2(x^1 + dx^1, x^2 + dx^2, x^3 + dx^3)$ and the lengths of the elements of arc measured along the coordinate lines of a curvilinear system X are $ds_{(1)}^2 = g_{11}dx^1dx^1 = g_{11}(dx^1)^2 \Rightarrow ds_{(1)} = \sqrt{g_{11}}(dx^1)$, where $s_{(1)}$ denotes the arc length along *the* x^1 curve.

$$\text{Similarly}, ds_{(2)}^2 = \sqrt{g_{22}}(dx^2)\ ,\ ds_{(3)}^2 = \sqrt{g_{33}}(dx^3). \tag{7.19}$$

Thus, the length of displacement vector (dx^1, 0,0) is given by $\sqrt{g_{11}}(dx^1)$ and that of (0, dx^2, 0) is $\sqrt{g_{22}}(dx^2)$ and the length of vector (0,0, dx^1) is given by $\sqrt{g_{33}}(dx^3)$, as shown in Figure (7.5).

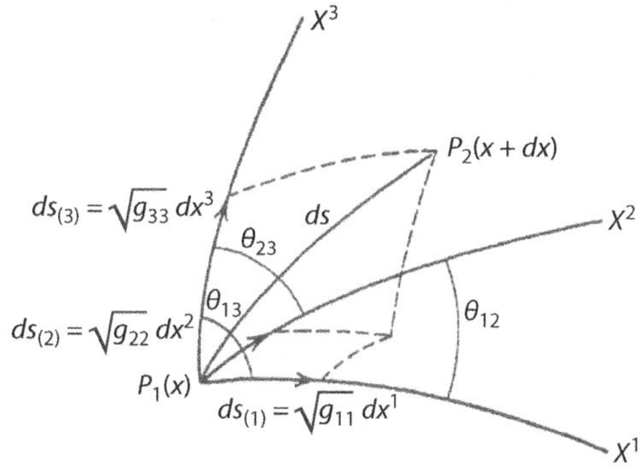

Figure 7.5

Angles θ_{12}, θ_{23}, *and* θ_{31} are between the coordinate curves in a three-dimensional coordinate system. Its cosines are

$$cos\theta_{12} = \frac{g_{12}dx^1dx^2}{\sqrt{g_{11}}(dx^1)\sqrt{g_{22}}(dx^2)} = \frac{g_{12}}{\sqrt{g_{11}g_{22}}}, cos\theta_{23} = \frac{g_{22}}{\sqrt{g_{22}g_{33}}},$$

$$and\ cos\theta_{31} = \frac{g_{31}}{\sqrt{g_{33}g_{11}}}.$$

Theorem 7.3.1. A necessary and sufficient condition that a given curvilinear coordinate system X be orthogonal is that $g_{ij} = 0$ for $i \neq j$ at every point of the region *R*.
Proof: We know that if θ_{ij} is the angle between λ^i *and* μ^j, they are displacement vectors along coordinate lines and

$$cos\theta_{ij} = \frac{g_{ij}\lambda^i \mu^j}{\sqrt{g_{ij}\lambda^i\lambda^i}\sqrt{g_{ij}\mu^j\mu^j}}.$$

Angles θ_{12}, θ_{23}, *and* θ_{31} are between the coordinate curves in a three-dimensional coordinate system. Its cosines are

$$cos\theta_{12} = \frac{g_{12}dx^1dx^2}{\sqrt{g_{11}}(dx^1)\sqrt{g_{22}}(dx^2)} = \frac{g_{12}}{\sqrt{g_{11}g_{22}}}, cos\theta_{23} = \frac{g_{22}}{\sqrt{g_{22}g_{33}}}, and\ cos\theta_{31}$$

$$= \frac{g_{31}}{\sqrt{g_{33}g_{11}}},$$

since curvilinear coordinate system X is orthogonal, i.e., $\theta_{ij} = 90°$.

Therefore, $\frac{g_{ij}}{\sqrt{g_{ii}g_{jj}}} = 0 \Rightarrow g_{ij} = 0.$

Conversely, let $g_{ij} = 0 \Rightarrow \frac{g_{ij}}{\sqrt{g_{ii}g_{jj}}} = 0$

$\therefore cos\theta_{ij} = 0.$

Therefore, curvilinear coordinate system X is orthogonal.

7.4 Reciprocal Base Systems

Here, we interpret the main results of curvilinear coordinate systems in notation of elementary vector analysis, as shown Figure (7.6). Let a Cartesian coordinate system be determined by a set of orthonormal base vectors $\boldsymbol{b}_1$, $\boldsymbol{b}_2$, ***and*** $\boldsymbol{b}_3$. Then, the position vector **r** of any point $P(y^1, y^2, y^3)$ can be represented in the form

$$\boldsymbol{r} = \boldsymbol{b}_i y^i \quad (i = 1,2,3). \tag{7.20}$$

Since the base vectors $\boldsymbol{b}_i$ are independent of the position of the point,

$$\boldsymbol{dr} = \boldsymbol{b}_i dy^i. \tag{7.21}$$

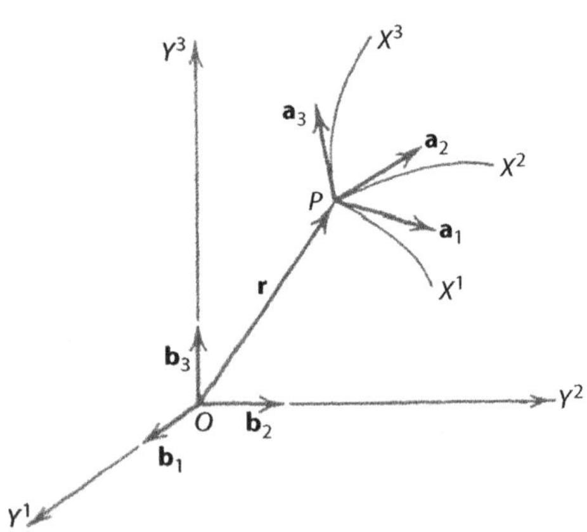

Figure 7.6

By definition, the square of the length of the arc element (ds) is between points

$$P(y^1, y^2, y^3) \text{ and } Q(y^1 + dy^1, y^2 + dy^2, y^3 + dy^3)$$

$$ds^2 = \boldsymbol{dr}.\boldsymbol{dr} \tag{7.22}$$

$$= \boldsymbol{b}_i.\boldsymbol{b}_j dy^i dy^j$$

$$= \delta^i_j dy^i dy^j = dy^i dy^i,$$

a familiar expression for the square of the element of arc in orthogonal Cartesian coordinates.

Let a set of equations of transformation

$$x^i = x^i(y^1, y^2 y^3) \quad (i = 1,2,3)$$

define a curvilinear coordinate system X. The position vector $\boldsymbol{r}$ can now be regarded as a function of coordinates x^i and we can write

$$\boldsymbol{dr} = \frac{\partial r}{\partial x^i} dx^i \tag{7.23}$$

$$\text{and } ds^2 = \boldsymbol{dr}.\boldsymbol{dr} = \frac{\partial r}{\partial x^i}\frac{\partial r}{\partial x^j} dx^i dx^j$$

$$= g_{ij} dx^i dx^j,$$

$$\text{where } g_{ij} = \frac{\partial r}{\partial x^i}\frac{\partial r}{\partial x^j} \tag{7.24}$$

The geometrical interpretation of $\frac{\partial r}{\partial x^i}$ is a base vector directed tangentially to the X^i coordinate curve, so that we denote $\frac{\partial r}{\partial x^i} = \boldsymbol{a}_i$.

(7.22) and (7.23) become

$$\boldsymbol{dr} = \boldsymbol{a}_i dx^i \tag{7.25}$$

$$g_{ij} = \boldsymbol{a}_i . \boldsymbol{a}_j.$$

We observe that the base vector a_i is no longer independent of the coordinates (x^1, x^2, x^3).

Result 7.4.1. Here $dr = \boldsymbol{a}_i dx^i$ and $dr = \boldsymbol{b}_i dy^i$ and we get $\boldsymbol{a}_j dx^j = \boldsymbol{b}_i dy^i = \boldsymbol{bi} \frac{\partial y^i}{\partial x^j} dx^j$,

$$\Rightarrow \boldsymbol{a}_j = \boldsymbol{b}_i \frac{\partial y^i}{\partial x^j} \text{ (since } dx^j \text{ } are\ arbitrary)$$

We see that the base vector $\boldsymbol{a}_j$ transforms according to the Law of Transformation of components of covariant vectors.

Result 7.4.2. The components of base vectors $\boldsymbol{a}_i$, when referred to in X^i coordinate systems, are

$$\boldsymbol{a}_1\text{: } (a_1, 0,0) \quad \boldsymbol{a}_2\text{: } (0, a_2, 0), \text{ and } \boldsymbol{a}_3\text{: } (0,0, a_3).$$

They are not necessarily unit vectors because

$$g_{11} = \boldsymbol{a}_1 . \boldsymbol{a}_1 = (a_1)^2 \neq 1, \text{ similarly } g_{22} = \boldsymbol{a}_2 . \boldsymbol{a}_2 \neq 1,\ g_{33} = \boldsymbol{a}_3 . \boldsymbol{a}_3 \neq 1.$$

Again, $g_{ij} = \boldsymbol{a}_i . \boldsymbol{a}_j = |\boldsymbol{a}_i||\boldsymbol{a}_j| \cos\theta = 0 \ if\ i \neq j$ implies that a curvilinear coordinate system is orthogonal if $g_{ij} = 0$.

Result 7.4.3. Any vector can be written in the form $\boldsymbol{A} = k\boldsymbol{dr}$.

We know $\boldsymbol{dr} = \frac{\partial r}{\partial x^i} dx^i$

or $\boldsymbol{A} = kd\boldsymbol{r} = k\frac{\partial r}{\partial x^i} dx^i = \frac{\partial r}{\partial x^i}(kdx^i) = \boldsymbol{a}_i A^i$, where $A^i = kdx^i$.

Here, A^i is the contravariant component of a vector of A.

Result 7.4.4. The vectors $\boldsymbol{a}_1 A^1$, $\boldsymbol{a}_2 A^2$, *and* $\boldsymbol{a}_3 A^3$ form the edges of the parallel pipes, whose diagonal will be $\boldsymbol{A}$. Since the $\boldsymbol{a}_i$ are not unit vectors in general, we see that the length of edges of Parallelepiped are

$$\sqrt{g_{11}}A^1, \sqrt{g_{22}}A^2, \sqrt{g_{33}}A^3 \text{ since } g_{11} = \boldsymbol{a}_1 . \boldsymbol{a}_1 = (a_1)^2.$$

Result 7.4.5. Consider three non-coplanar vectors

$$a^1 = \frac{a_2 \times a_3}{[a_1 \boldsymbol{a}_2 a_3]} \quad a^2 = \frac{a_3 \times a_1}{[a_1 \boldsymbol{a}_2 a_3]} \quad a^3 = \frac{a_1 \times a_2}{[a_1 \boldsymbol{a}_2 a_3]}, \tag{7.26}$$

where $a_2 \times a_3$ etc. Denote that the vector product of a_2, a_3 and $[a_1 \boldsymbol{a}_2 a_3]$ is a scalar triple product $a_{1.} \boldsymbol{a}_2 \times a_3$.

$$\therefore a^1 . a_1 = a_1 . \frac{a_2 \times a_3}{[a_1 \boldsymbol{a}_2 a_3]} = \frac{[a_1 \boldsymbol{a}_2 a_3]}{[a_1 \boldsymbol{a}_2 a_3]} = 1$$

$$a^1 . a_2 = a_2 . \frac{a_2 \times a_3}{[a_1 \boldsymbol{a}_2 a_3]} = \frac{[a_2 \boldsymbol{a}_2 a_3]}{[a_1 \boldsymbol{a}_2 a_3]} = 0.$$

It is obvious that $a^i . a_j = \delta^i_j$

Example 7.4.1. Show that $[a_1 \boldsymbol{a}_2 a_3] = \sqrt{g}$ and $[a^1 a^2 a^3] = \frac{1}{\sqrt{g}}$, where $g = |g_{ij}|$.

Solution: Let the components of base vectors a_i be

$$\boldsymbol{a}_1: (a_1, 0, 0) \quad \boldsymbol{a}_2: (0, a_2, 0), \boldsymbol{a}_3: (0, 0, a_3).$$

$$\text{Then, } [\boldsymbol{a}_1 \boldsymbol{a}_2 \boldsymbol{a}_3] = \begin{vmatrix} a_1 & 0 & 0 \\ 0 & a_2 & 0 \\ 0 & 0 & a_3 \end{vmatrix} = a_1 \boldsymbol{a}_2 a_3$$

[$[\boldsymbol{a}_1 \boldsymbol{a}_2 \boldsymbol{a}_3] = \boldsymbol{a}_1 . \boldsymbol{a}_2 \times \boldsymbol{a}_3$ is numerically equal to the volume of parallelepiped made by its edges a_1, $\boldsymbol{a}_2$, a_3,]

$$\text{and } g = |g_{ij}| = \begin{vmatrix} g_{11} & g_{12} & g_{13} \\ g_{21} & g_{22} & g_{23} \\ g_{31} & g_{32} & g_{33} \end{vmatrix}$$

$$g_{11} = \boldsymbol{a}_1 . \boldsymbol{a}_1 = (a_1)^2.$$

Similarly, $g_{22} = (a_2)^2$ *and* $g_{33} = (a_3)^2$ and $g_{12} = g_{23} = g_{13} = 0$

$$\therefore\ g = \begin{vmatrix} (a_1)^2 & 0 & 0 \\ 0 & (a_2)^2 & 0 \\ 0 & 0 & (a_3)^2 \end{vmatrix} = (a_1)^2 (a_2)^2 (a_3)^2$$

$$\text{or } a_1 \boldsymbol{a}_2 a_3 = \sqrt{g}.$$

Therefore, $[\boldsymbol{a}_1 \boldsymbol{a}_2 \boldsymbol{a}_3] = a_1 \boldsymbol{a}_2 a_3 = \sqrt{g}$ and $[\boldsymbol{a}^1 \boldsymbol{a}^2 \boldsymbol{a}^3] = \frac{1}{\sqrt{g}}$, since $a^i . a_j = \delta^i_j$.

Definition 7.4.1. The triple products $[\boldsymbol{a}^1 \boldsymbol{a}^2 \boldsymbol{a}^3]$ and $[\boldsymbol{a}_1 \boldsymbol{a}_2 \boldsymbol{a}_3]$ are reciprocally related. Moreover,

$$a_1 = \frac{\boldsymbol{a}^2 \times \boldsymbol{a}^3}{[\boldsymbol{a}^1 \boldsymbol{a}^2 \boldsymbol{a}^3]}, \quad a_2 = \frac{\boldsymbol{a}^3 \times \boldsymbol{a}^1}{[\boldsymbol{a}^1 \boldsymbol{a}^2 \boldsymbol{a}^3]}, \textit{and } a_3 = \frac{\boldsymbol{a}^1 \times \boldsymbol{a}^2}{[\boldsymbol{a}^1 \boldsymbol{a}^2 \boldsymbol{a}^3]}, \tag{7.27}$$

The system of vectors $\boldsymbol{a}^1$, $\boldsymbol{a}^2$, $\boldsymbol{a}^3$ are called the *reciprocal base system.*

Hence, if the vectors $\boldsymbol{a}^1$, $\boldsymbol{a}^2$, ***and*** $\boldsymbol{a}^3$ are unit vectors associated with orthogonal Cartesian, coordinates then the reciprocal system of the vectors defines the same system of coordinates.

The differential of a vector **r** in the reciprocal base system is $d\boldsymbol{r} = \boldsymbol{a}^i dx_i$, where dx_i are the components of $d\boldsymbol{r}$. Then,

$$ds^2 = d\boldsymbol{r}.d\boldsymbol{r} = (\boldsymbol{a}^i dx_i)(\boldsymbol{a}^j dx_j)$$

$$= \boldsymbol{a}^i \boldsymbol{a}^j dx_i dx_j$$

$$\therefore ds^2 = g^{ij} dx_i dx_j,$$

$$\text{where } g^{ij} = \boldsymbol{a}^i \boldsymbol{a}^j = g^{ji}. \tag{7.28}$$

The system of base vectors determined by Equation (7.25) can be used to represent an arbitrary vector A in the form of $\boldsymbol{A} = \boldsymbol{a}^i A_i$, where A_i are the covariant components of $\boldsymbol{A}$.

The scalar product of the vector $\boldsymbol{a}^i A_i$, with the base vector a_j, noted that the latter is directed along the X^j-coordinate line and we get

$$a_j.\boldsymbol{a}^i A_i = \delta_j^i A_i = A_j.$$

Thus, $\frac{A_j}{\sqrt{g_{jj}}}$ is the length of the orthogonal projection of vector $\boldsymbol{A}$ on the tangent to the $\boldsymbol{x}^j$ coordinate curve at P, whereas $\frac{A_j}{\sqrt{g_{jj}}}$ is the length of the edge of the parallel pipes whose diagonal is vector $\boldsymbol{A}$.

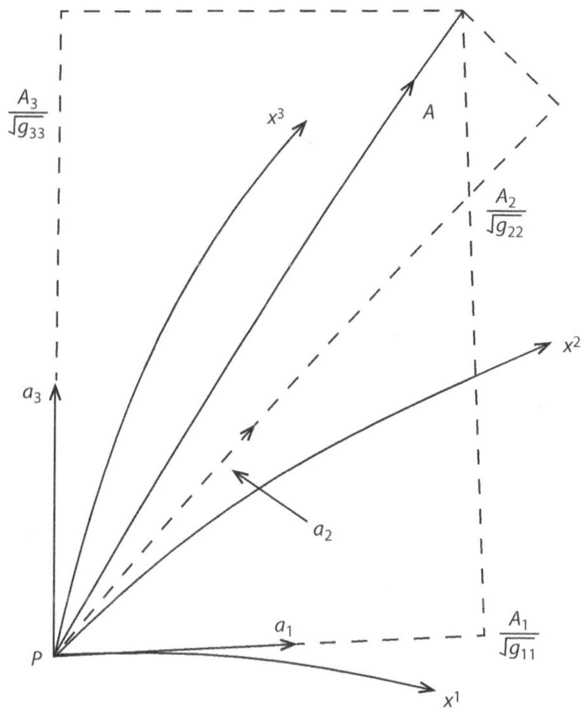

Figure 7.7

$$\text{Since } A = \boldsymbol{a}_i A^i =. \boldsymbol{a}^i A_{i,}$$

$$\text{or } \boldsymbol{a}_i.\ \boldsymbol{a}_j A^i =. \boldsymbol{a}^i.\ \boldsymbol{a}_j A_i,$$

$$\text{or } g_{ij}A^i = \delta^i_j.A_i = A_j,$$

we see that the vector obtained by lowering the index in A^i is precisely the covariant vector A_i and thus, it is seen to represent the same vector A referred to by two different base vectors. Hence, the distinction between the covariant and contrvariant components of A disappears whenever the base vectors are orthogonal.

7.5 Partial Derivative

Theorem 7.5.1. If $\boldsymbol{A}$ is a vector along the curve in E_3, prove that $\dfrac{\partial A}{\partial x^j} = A^{\alpha}_{,j} a_{\alpha}$.

Proof: We suppose that the components of A are continuously differentiable functions of y^i in R and, if we introduce a curvilinear coordinate system X by means of transformation

$$\text{T}: x^i = x^i(y^1, y^2, y^3),$$

then, the corresponding components $A^i(x)$ will be continuously differentiable functions of point $(x^1\ x^2, x^3)$ determined by the position vector $\boldsymbol{r}(x^1\ x^2, x^3)$.

A vector $\boldsymbol{A}$ can be expressed by base vector $\boldsymbol{a}_i = \dfrac{\partial \boldsymbol{r}}{\partial \boldsymbol{x}^i}$ as

$$\boldsymbol{A} = A^i \boldsymbol{a}_i, \tag{7.29}$$

where A^i is the component of $\boldsymbol{A}$,

Change $\Delta \boldsymbol{A}$ in $\boldsymbol{A}$ at point $P(x^1\ x^2, x^3)$ assumes a different position

$$P'(x^1 + \Delta\, x^1, x^2 + \Delta\, x^2, x^3 + \Delta\, x^3).$$

$$\text{Then, } \Delta \boldsymbol{A} = (A^i + \Delta A^i)(\boldsymbol{a}_i + \Delta \boldsymbol{a}_i) - A^i \boldsymbol{a}_i$$

$$= \Delta A^i \boldsymbol{a}_i + \Delta \boldsymbol{a}_i A^i + \Delta A^i \Delta \boldsymbol{a}_i$$

In ordinary calculus, the principal part of the change by dA

$$dA = dA^i \boldsymbol{a}_i + d\boldsymbol{a}_i A^i \tag{7.30}$$

$$(\text{neglecting } \Delta A^i \Delta \boldsymbol{a}_i).$$

This formula states that the change in A arises from two sources:

(i) change in the components A^i as the values $(x^1\ x^2, x^3)$ are changed,
(ii) change in the base vectors $\boldsymbol{a}_i$ as the position of *point* $(x^1\ x^2, x^3)$ is altered.

The partial derivative of $\boldsymbol{A}$, with respect to x^i, is defined as the limit

$$\lim_{\Delta x^i \to 0} \frac{\Delta \mathbf{A}}{\Delta x^j} = \frac{\partial \mathbf{A}}{\partial x^j}$$

$$\therefore \frac{\partial \mathbf{A}}{\partial x^j} = \frac{\partial A^i}{\partial x^j}\boldsymbol{a}_i + \frac{\partial \boldsymbol{a}_i}{\partial x^j}A^i. \tag{7.31}$$

Now, we have to show that $\dfrac{\partial \mathbf{A}}{\partial x^j}$ is a covariant derivative of A^i.

$$\text{Now, we establish } \frac{\partial \boldsymbol{a}_i}{\partial x^j} = \begin{Bmatrix} a \\ i\ j \end{Bmatrix} \boldsymbol{a}_\alpha. \tag{7.32}$$

We know $g_{ij} = \boldsymbol{a}_i . \boldsymbol{a}_j$, differentiating with respect to x^k

$$\frac{\partial}{\partial x^k} g_{ij} = \frac{\partial \boldsymbol{a}_i}{\partial x^k}.\boldsymbol{a}_j + \frac{\partial \boldsymbol{a}_j}{\partial x^k}.\boldsymbol{a}_i.$$

Similarly, $\dfrac{\partial}{\partial x^j} g_{ik} = \dfrac{\partial \boldsymbol{a}_i}{\partial x^j}.\boldsymbol{a}_k + \dfrac{\partial \boldsymbol{a}_k}{\partial x^j}.\boldsymbol{a}_i$

$$\frac{\partial}{\partial x^i} g_{jk} = \frac{\partial \boldsymbol{a}_j}{\partial x^i}.\boldsymbol{a}_k + \frac{\partial \boldsymbol{a}_k}{\partial x^i}.\boldsymbol{a}_j$$

and we assume T is of class C^2, therefore, $\dfrac{\partial \boldsymbol{a}_i}{\partial x^j} = \dfrac{\partial \boldsymbol{a}_j}{\partial x^i}$

$$\therefore [ij,k] = \frac{1}{2}\left(\frac{\partial}{\partial x^j} g_{ik} + \frac{\partial}{\partial x^i} g_{jk} - \frac{\partial}{\partial x^k} g_{ij} \right),$$

$$\text{or} \quad [ij,k] = \frac{\partial \boldsymbol{a}_i}{\partial x^j}.\boldsymbol{a}_k, \tag{7.33}$$

$$\text{or}\quad \frac{\partial \boldsymbol{a}_i}{\partial x^j} = [ij,k]a^k,$$

$$\text{or}\quad \frac{\partial \boldsymbol{a}_i}{\partial x^j}.\, a^\alpha = [ij,k]a^k.\, a^\alpha$$

$$= [ij,k]g^{k\alpha} = \left\{ \begin{matrix} \alpha \\ i\ j \end{matrix} \right\},$$

implying that $\frac{\partial \boldsymbol{a}_i}{\partial x^j} = \left\{ \begin{matrix} a \\ i\ j \end{matrix} \right\} \boldsymbol{a}_\alpha$. This is established in (7.31).

Inserting these values in (7.30), we get

$$\frac{\partial \mathbf{A}}{\partial x^j}\frac{\partial A^i}{\partial x^j}\boldsymbol{a}_i + \frac{\partial \boldsymbol{a}_i}{\partial x^j}A^i = \frac{\partial A^i}{\partial x^j}\boldsymbol{a}_i + \left\{ \begin{matrix} \alpha \\ i\ \ j \end{matrix} \right\}\boldsymbol{a}_\alpha A^i$$

$$= \left[\frac{\partial A^\alpha}{\partial x^j} + \left\{ \begin{matrix} \alpha \\ i\ \ j \end{matrix} \right\} A^i \right] \boldsymbol{a}_\alpha.$$

The expression in the bracket is $A^i_{,j}$, thus

$$\frac{\partial \mathbf{A}}{\partial x^j} = A^\alpha_{,j}\boldsymbol{a}_\alpha. \tag{7.34}$$

Theorem 7.5.2. If $\boldsymbol{A}$ is a vector along the curve in E_3, prove that $\frac{\partial \mathbf{A}}{\partial x^k} = A_{j,k}a^j$.

Proof: We know $\boldsymbol{A} = A_i a^i$

$$\frac{\partial \mathbf{A}}{\partial x^k} = \frac{\partial A_i}{\partial x^k}a^i + \frac{\partial a^i}{\partial x^k}A_i \tag{i}$$

$$a^i a_j = \delta^i_j$$

Differentiating with respect to x^k, we get

$$\frac{\partial \boldsymbol{a}^i}{\partial x^k}.\boldsymbol{a}_j+\frac{\partial \boldsymbol{a}_j}{\partial x^k}.\boldsymbol{a}^i=0$$

$$\text{or}\quad \frac{\partial \boldsymbol{a}^i}{\partial x^k}.\boldsymbol{a}_j=-\frac{\partial \boldsymbol{a}_j}{\partial x^k}.\boldsymbol{a}^i$$

$$=-\boldsymbol{a}^i\begin{Bmatrix} a \\ j\quad k\end{Bmatrix}\boldsymbol{a}_\alpha.\ \text{(from 7.31)}$$

Since $\boldsymbol{a}^i\boldsymbol{a}_\alpha=\delta^i_\alpha$,

$$\text{therefore,}\ \frac{\partial \boldsymbol{a}^i}{\partial x^k}.\boldsymbol{a}_j=-\delta^i_\alpha\begin{Bmatrix} a \\ j\quad k\end{Bmatrix}=-\begin{Bmatrix} i \\ j\quad k\end{Bmatrix}$$

$$\frac{\partial \boldsymbol{a}^i}{\partial x^k}=-\begin{Bmatrix} i \\ j\quad k\end{Bmatrix}\boldsymbol{a}^j. \tag{7.35}$$

Substituting this value in (i), we get

$$\frac{\partial \mathbf{A}}{\partial x^k}=\frac{\partial A_i}{\partial x^k}a^i+\frac{\partial a^i}{\partial x^k}A_i=\frac{\partial A_i}{\partial x^k}a^i-\begin{Bmatrix} i \\ k\quad j\end{Bmatrix}\boldsymbol{a}^jA_i$$

$$=\frac{\partial A_j}{\partial x^k}a^j-\begin{Bmatrix} i \\ j\quad k\end{Bmatrix}\boldsymbol{a}^jA_i$$

$$=\left(\frac{\partial A_j}{\partial x^k}-\begin{Bmatrix} i \\ j\quad k\end{Bmatrix}A_i\right)\boldsymbol{a}^j=A_{j,k}\boldsymbol{a}^j.$$

7.6 Exercises

1. If $A_i=g_{ij}A^j$, show that $A_{i,k}=g_{i\alpha}A^\alpha_{,k}$.
2. Prove that if A is the magnitude of A^i, then $A_j=\dfrac{A_{i,j}A^i}{A}$.
3. If y^i are the rectangular Cartesian coordinates, show that in E_3

$$[\alpha\beta,\gamma]=\frac{\partial^2 y^i}{\partial x^\alpha\partial x^\beta}\frac{\partial y^i}{\partial x^\gamma}\ \text{and}\begin{Bmatrix} \gamma \\ \alpha\quad \beta\end{Bmatrix}=\frac{\partial^2 y^i}{\partial x^\alpha\partial x^\beta}\frac{\partial y^\gamma}{\partial x^i}.$$

4. Show that the area of the parallelogram constructed on the base vectors a_2, *and* a_3 is $\sqrt{gg_{11}}$, where g^{ij} and g_{ij} are the conjugate and metric tensors in a curvilinear coordinate system and $g = |g_{ij}|$.
5. Show that $\frac{d(g_{ij}A^i)}{dx^k} = A_{i,k}B^i + B_{i,k}A^i$.

8

Curves in Space

8.1 Introduction

In this chapter we describe the geometry of space curve and the Serret-Frenet formulas analogous to the derivatives of the tangent, normal, and binormal unit vectors in terms each other. The formulas are named after two French mathematicians who independently discovered them: J. F. Frenet, in his thesis of 1847 and J. A. Serret in 1851. A set of three remarkable formulas are generally known as Frenet's formulas, which characterized, in small, all essential geometric properties of space curves. We also deal with Helix, an equation of straight lines, and intrinsic differentiation.

8.2 Intrinsic Differentiation

We are now going to introduce another kind of differentiation which may be regarded as a generalization of ordinary differentiation in Euclidean spaces with rectangular Cartesian coordinates.

Let a vector field $\boldsymbol{A}(x)$ be defined in some region of E_3 and let

$$C: x^i = x^i(t)\ ,\ t_1 \le t \le t_2$$

be a curve in that region. The vector $\boldsymbol{A}(x)$ is defined over the one dimensional manifold C, depending on the parameter t, and if $\boldsymbol{A}(x)$ is a differentiable vector and the $x^i(t)$ belong to the class C^1, then

$$\begin{aligned} \frac{d\boldsymbol{A}}{dt} &= \frac{\partial A}{\partial x^j}\frac{dx^j}{dt} \\ &= A^{\alpha}_{,j} a_{\alpha} \frac{dx^j}{dt}. \end{aligned}$$

Dipankar De. Introduction to Differential Geometry with Tensor Applications, (221–264) © 2022 Scrivener Publishing LLC

We know

$$\frac{\partial A}{\partial x^j} = A^{\alpha}_{,j} a_{\alpha} = \left[\frac{\partial A^{\alpha}}{\partial x^j} + \left\{ \begin{matrix} \alpha \\ i \ j \end{matrix} \right\} A^i \right] a_{\alpha},$$

so

$$\frac{dA}{dt} = \left[\frac{\partial A^{\alpha}}{\partial x^j} + \left\{ \begin{matrix} \alpha \\ i \ j \end{matrix} \right\} A^i \right] a_{\alpha} \frac{dx^j}{dt}$$

$$= \left[\frac{dA^{\alpha}}{dt} + \left\{ \begin{matrix} \alpha \\ i \ j \end{matrix} \right\} A^i \frac{dx^j}{dt} \right] a_{\alpha}.$$

The formula $\dfrac{\delta A^{\alpha}}{\delta t} \equiv \dfrac{dA^{\alpha}}{dt} + \left\{ \begin{matrix} \alpha \\ i \ j \end{matrix} \right\} A^i \dfrac{dx^j}{dt} \quad (\alpha = 1,2,3)$ (8.1)

is called the *absolute or intrinsic derivative* of A^{α}, with respect to parameter t, and is denoted by $\dfrac{\delta A^{\alpha}}{\delta t}$.

We observe the following properties of intrinsic derivatives.

1. We will make free use of intrinsic differentiation in the treatment of geometry of curves and surfaces.
2. It also follows the familiar rules for differentiation of sums, products, etc. and remains valid for the process of intrinsic differentiation.
3. If the vector field A^{α} is defined in the neighborhood of C, as well as on C, we can write

$$\frac{\delta A^{\alpha}}{\delta t} = A^{\alpha}_{,\beta} \frac{dA^{\beta}}{dt}.$$

4. If A is scalar, then obviously, $\frac{\delta A^\alpha}{\delta t} = \frac{dA^\alpha}{dt}$.
5. $\frac{\delta}{\delta t} g_{ij} = g_{ij,l} \frac{dx^l}{dt} = 0$ since $g_{ij,l} = 0$. Hence, show that the intrinsic derivative of $g_{ij} = 0$.
   Also, the intrinsic derivative of $\delta^i_j = 0$.w

With the extension of the process of intrinsic differentiation to tensors of rank greater than one, we can write

(i) If A_i is a covariant vector,

$$\frac{\delta A_i}{\delta t} = \frac{dA_i}{dt} - \begin{Bmatrix} \alpha \\ i \ \beta \end{Bmatrix} A_\alpha \frac{dx^\beta}{dt}$$

(ii) $$\frac{\delta A^{ij}}{\delta t} \equiv \frac{dA^{ij}}{dt} + \begin{Bmatrix} i \\ \alpha \ \beta \end{Bmatrix} A^{\alpha j} \frac{dx^\beta}{dt} + \begin{Bmatrix} j \\ \alpha \ \beta \end{Bmatrix} A^{i\alpha} \frac{dx^\beta}{dt}$$

(iii) $$\frac{\delta A^i_j}{\delta t} = \frac{dA^i_j}{dt} + \begin{Bmatrix} i \\ \alpha \ \beta \end{Bmatrix} A^\alpha_j \frac{dx^\beta}{dt} - \begin{Bmatrix} \alpha \\ i \ \beta \end{Bmatrix} A^i_\alpha \frac{dx^\beta}{dt}$$

(iv)

$$\frac{\delta A^i_{jk}}{\delta t} = \frac{dA^i_{jk}}{dt} + \begin{Bmatrix} i \\ \alpha \ \beta \end{Bmatrix} A^\alpha_{jk} \frac{dx^\beta}{dt} - \begin{Bmatrix} i \\ j \ \beta \end{Bmatrix} A^i_{\alpha k} \frac{dx^\beta}{dt} - \begin{Bmatrix} \alpha \\ k \ \beta \end{Bmatrix} A^i_{j\alpha} \frac{dx^\beta}{dt}$$

We observe that $\frac{\delta}{\delta t} g_{ij} = 0$ and the fundamental tensors g_{ij} and g^{ij} can be taken outside the sign of intrinsic differentiation.

Example 8.2.1. Prove that $\frac{d(g_{ij} A^i A^j)}{dt} = 2 g_{ij} A^i \frac{\delta A^j}{\delta t}$.

Solution: Since $g_{ij} A^i A^j$ is scalar,

then $$\frac{d(g_{ij} A^i A^j)}{dt} = \frac{\delta(g_{ij} A^i A^j)}{\delta t}$$
$$= g_{ij} \frac{\delta(A^i A^j)}{\delta t},$$

and since g_{ij} is independent of t and g_{ij}, $A^i A^j$ is invariant.

$$= g_{ij} \left[A^j \frac{\delta A^i}{\delta t} + A^i \frac{\delta A^j}{\delta t} \right].$$

Interchanging *i and j* in first term, we get

$$\frac{d(g_{ij}A^iA^j)}{dt} = g_{ij}\left[A^i\frac{\delta A^j}{\delta t} + A^i\frac{\delta A^j}{\delta t}\right] = 2g_{ij}A^i\frac{\delta A^j}{\delta t} \quad \text{(since } g_{ij} \text{ is symmetric)}$$

Example 8.2.2. If g_{ij} are components of metric tensors, show that $\frac{\delta g_{ij}}{\delta t} = 0.$
Solution: The intrinsic derivative of g_{ij} is

$$\frac{\delta g_{ij}}{\delta t} = \frac{dg_{ij}}{dt} - \begin{Bmatrix} \alpha \\ i \quad \beta \end{Bmatrix} g_{\alpha j}\frac{dx^\beta}{dt} - \begin{Bmatrix} \alpha \\ \beta \quad j \end{Bmatrix} g_{i\alpha}\frac{dx^\beta}{dt}$$

$$= \frac{\partial g_{ij}}{\partial x^\beta}\frac{dx^\beta}{dt} - \begin{Bmatrix} \alpha \\ i \quad \beta \end{Bmatrix} g_{\alpha j}\frac{dx^\beta}{dt} - \begin{Bmatrix} \alpha \\ \beta \quad j \end{Bmatrix} g_{i\alpha}\frac{dx^\beta}{dt}$$

$$= \left[\frac{\partial g_{ij}}{\partial x^\beta} - \begin{Bmatrix} \alpha \\ i \quad \beta \end{Bmatrix} g_{\alpha j} - \begin{Bmatrix} \alpha \\ \beta \quad j \end{Bmatrix} g_{i\alpha}\right]\frac{dx^\beta}{dt}.$$

$$\frac{\delta g_{ij}}{\delta t} = \left\{\frac{\partial g_{ij}}{\partial x^\beta} - [i\beta, j] - [\beta j, i]\right\}\frac{dx^\beta}{dt}, \qquad \left(as \begin{Bmatrix} \alpha \\ i \quad \beta \end{Bmatrix} g_{\alpha j} = [i\beta, j]\right.$$

$$\left. and \begin{Bmatrix} \alpha \\ \beta \quad j \end{Bmatrix} g_{i\alpha} = [\beta j, i]\right)$$

$$\text{but } \frac{\partial g_{ij}}{\partial x^\beta} = [i\beta, j] + [\beta j, i]$$

$$\Rightarrow \frac{\delta g_{ij}}{\delta t} = 0$$

Example 8.2.3. Show that $\frac{\delta}{\delta t}\left(\frac{dx^i}{dt}\right) = \frac{d^2x^i}{dt^2} + \begin{Bmatrix} i \\ j \quad k \end{Bmatrix} A^i \frac{dx^j}{dt}\frac{dx^k}{dt}.$

We know that $\frac{dx^i}{dt}$ are the components of a contravaiant vector.

$$\text{Now } \frac{\delta}{\delta t}\left(\frac{dx^i}{dt}\right)=\left(\frac{dx^i}{dt}\right)_{,k}\left(\frac{dx^k}{dt}\right)=\left[\frac{\partial\left(\frac{dx^i}{dt}\right)}{\partial x^k}+\begin{Bmatrix} i \\ j\ k \end{Bmatrix}\frac{dx^j}{dt}\right]\frac{dx^k}{dt}$$

$$=\frac{\partial\left(\frac{dx^i}{dt}\right)}{\partial xk}\frac{dx^k}{dt}+\begin{Bmatrix} i \\ j\ k \end{Bmatrix}\frac{dx^j}{dt}\frac{dx^k}{dt}$$

$$=\frac{d^2x^i}{dt^2}+\begin{Bmatrix} i \\ j\ k \end{Bmatrix}\frac{dx^j}{dt}\frac{dx^k}{dt}.$$

Example 8.2.4. Show that the intrinsic derivative of a tensor of type (p, q) is again a tensor of type (p, q).
Solution: Let us consider a tensor of type (p, q) whose components $A^{i_1 \ldots i_p}_{j_1 \ldots j_q}$. are functions of a parameter, t. Then, the intrinsic derivative of the tensor, with respect to parameter t, is denoted by $\frac{\delta}{\delta t} A^{i_1 \ldots i_p}_{j_1 \ldots j_q}$. and is defined as

$$\frac{\delta}{\delta t} A^{i_1 \ldots i_p}_{j_1 \ldots j_q} = A^{i_1 \ldots i_p}_{j_1 \ldots j_q, l} \frac{dx^l}{dt},$$

where a comma (,) denotes covariant differentiation.

Since $\frac{dx^l}{dt}$ is a contravariant vector, a tensor of type (1,0) is along the curve $x^i = \phi^i(t)$ and $A^{i_1 \ldots i_p}_{j_1 \ldots j_q, l}$ is a tensor of type $(p, q + 1)$ along the same curve and $A^{i_1 \ldots i_p}_{j_1 \ldots j_q, l}. \frac{dx^l}{dt}$ is a tensor of type (p,q) along the curve, i.e. $\frac{\delta}{\delta t} A^{i_1 \ldots i_p}_{j_1 \ldots j_q}$. is a tensor of type (p, q) along the curve.

Thus, the intrinsic derivative of a tensor of type (p, q) is again a tensor of type (p, q).

Example 8.2.5. If the intrinsic derivative of a contravariant vector vanishes identically, show that its magnitude is constant.
Solution: Let A^i be the components of a cotravariant vector such that

$$\frac{\delta A^i}{\delta t} = 0.$$

If the square of magnitude of vector A^i is $g_{ij}A^iA^j$,

$$\text{then } \frac{\delta (g_{ij}A^iA^j)}{\delta t} = 2g_{ij}A^i\frac{\delta A^j}{\delta t} = 0.$$

$\Rightarrow g_{ij}A^iA^j$ is constant
$\Rightarrow$The square of magnitude of the vector A^i is constant.

8.3 Parallel Vector Fields

Consider a curve, as shown in Figure (8.1).

$$C: x^i = x^i(t), \quad t_1 \le t \le t_2 \text{ and } i = 1,2,3,$$

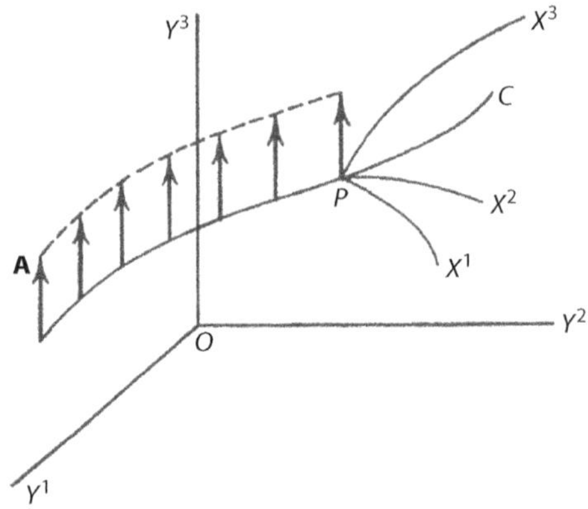

Figure 8.1

in some region of E_3 with a vector **A** at point P of C. We suppose that the functions $x^i(t)$ are of class C^1. If we construct at every point of C a vector equal to **A** in magnitude and parallel to it in direction, we obtain what is known as a *parallel field* of vectors along curve C.

If $\boldsymbol{A}$ is a parallel field along C, then the vectors of A do not change along the curve and we can write $\frac{dA}{dt}=0.$

It follows that components A^i of A satisfy a set of simultaneous differential equations $\frac{\delta A^i}{\delta t}=0$

$$\text{or} \qquad \frac{dA^i}{dt}+\left\{\begin{matrix} i \\ \alpha\ \beta \end{matrix}\right\}A^\alpha \frac{dx^\beta}{dt}=0 \tag{8.2}$$

Every solution of Equation (8.2) satisfies the initial conditions and must form a parallel field along C.

Example 8.3.1. If A^i and B^i are two vectors of constant magnitudes and undergo parallel displacements along a given curve, then show that they are inclined at a constant angle.

Let $A^i(t)$ and $B^i(t)$ be any two solutions of (8.2). We verify that the lengths of these two vectors indeed do not change as we move along the curve. Moreover, the angle θ between vectors A^i and B^i remains fixed as parameter t is allowed to change.

Solution: prove this

We have $A.B = AB\ cos\theta = g_{ij}A^iB^j$ and, if $g_{ij}A^iB^j$ is to remain constant along C, then $\frac{d}{dt}g_{ij}A^iB^j=0.$

But $g_{ij}A^iB^j$ is an invariant, and, since g_{ij} behave like constant in the process of covariant differentiation. We can write

$$\frac{d}{dt}(g_{ij}A^iB^j)=\frac{\delta}{\delta t}(g_{ij}A^iB^j)=g_{ij}\frac{\delta}{\delta t}(A^i)B^j+g_{ij}\frac{\delta}{\delta t}(B^j)A^i.$$

Since the field A^i and B^j satisfy (8.2), $\frac{\delta A^i}{\delta t}=0\ and\ \frac{\delta B^i}{\delta t}=0$, we conclude that $g_{ij}A^iB^j$ is constant along C.

$$\Rightarrow \frac{d}{dt}(g_{ij}A^iB^j)=0, \text{or}$$

$$\Rightarrow \frac{d}{dt}(AB\ cos\theta)=0$$

$$\therefore\ cos\theta = constant.$$

If $A^i = B^i$, *then* $g_{ij}A^iB^j = A^2$ is a constant along C and this implies that $\theta =$ constant.

Since vectors A^i are defined at every point (x^i) of the manifold, we can write

$$\frac{dA^i}{dt} = \frac{\partial A^i}{\partial x^k}\frac{dx^k}{dt},$$

so that Equation (8.2) assumes the form

$$\left(\frac{\partial A^i}{\partial x^k} + \begin{Bmatrix} i \\ \alpha\ k \end{Bmatrix} A^\alpha\right)\frac{dx^k}{dt} = 0.$$

The parallel vector field in E_3 satisfies the system of equations

$$\frac{\partial A^i}{\partial x^k} + \begin{Bmatrix} i \\ \alpha\ k \end{Bmatrix} A^\alpha = 0 \ \ or \ \ A^i_{,k} = 0.$$

Similarly, the condition for parallel displacement of covariant vector A_i is

$$\frac{\partial A_i}{\partial x^k} + \begin{Bmatrix} i \\ \alpha\ k \end{Bmatrix} A_\alpha = 0 \ \ or \ \ A_{i,k} = 0.$$

8.4 Geometry of Space Curves

Let the parametric equations of the curve C in E_3 be

$$C: x^i = x^i(t), \quad t_1 \le t_2 \ and \ i = 1,2,3.$$

The square of the length of an element of C is given by

$$ds^2 = g_{ij}dx^i dx^j \tag{8.3}$$

and the length of arc s of C is defined by the integral

$$s = \int_{t1}^{t2} \sqrt{g_{ij} \frac{dx^i}{dt} \frac{dx^j}{dt}} dt \tag{8.4}$$

$$\text{From (8.3), we get } 1 = g_{ij} \frac{dx^i}{ds} \frac{dx^j}{ds} \tag{8.5}$$

and if we put $\frac{dx^i}{ds} = \lambda^i$, Equation (8.5) can be written as

$$g_{ij}\lambda^i\lambda^j = 1. \tag{8.6}$$

Thus, the vector λ with components λ^i is a unit vector. Moreover, λ is tangent to C since its components λ^i, when the curve C is referred to a rectangular Cartesian reference frame Y, become $\lambda^i = \frac{dy^i}{ds}$. These are the direction cosines of the tangent vector to curve C.

Consider a pair of unit vectors λ and μ with components λ^i and μ^i at any point P *of* C, as shown in Figure (8.2). We suppose that λ is tangent to C *at* P. The cosine of the angle θ between λ and μ is given by

$$\cos\theta = g_{ij}\lambda^i\mu^j \tag{8.7}$$

and if λ and μ are orthogonal, we have

$$g_{ij}\lambda^i\mu^j = 0 \tag{8.8}$$

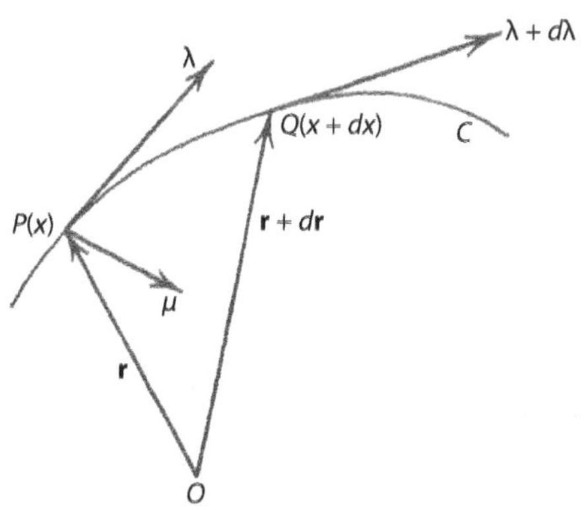

Figure 8.2

Any vector μ satisfying Equation (8.7) is said to be *normal* to *C at P*.

If we take the intrinsic derivative with respect to the arc parameter *s* of quadratic relation (8.6), g_{ij} behaves in covariant differentiation like constants and we obtain

$$g_{ij}\frac{\delta\lambda^i}{\delta s}\lambda^j + g_{ij}\frac{\delta\lambda^j}{\delta s}\lambda^i = 0.$$

Since g_{ij} is symmetric, $g_{ij}\frac{\delta\lambda^i}{\delta s}\lambda^j = 0.$

Here, the vector $\frac{\delta\lambda^j}{\delta s}$ either vanishes or is normal to *C* (for orthogonality) and if it does not vanish, we denote that the unit vector is co-directional with $\frac{\delta\lambda^j}{\delta s}$ by μ^j and write

$$\mu^j = \frac{1}{\chi}\frac{\delta\lambda^j}{\delta s}, \tag{8.9}$$

where $\chi > 0$ is so chosen to make μ^j a unit vector.

The vector μ^j, determined by Formula (8.9), is called the *principal normal vector* to curve C at point *P* and χ is the *curvature* of C at P.

The plane determined by the tangent vector λ and the principal normal vector μ is called the osculating plane to curve *C at P*.

Since μ is a unit vector

$$g_{ij}\mu^i\mu^j = 1. \tag{8.10}$$

Differentiating intrinsically the orthogonality relation (8.10), we get

$$g_{ij}\frac{\delta\lambda^i}{\delta s}\mu^j + g_{ij}\frac{\delta\mu^j}{\delta s}\lambda^i = 0$$

$$g_{ij}\frac{\delta\mu^j}{\delta s}\lambda^i = -g_{ij}\frac{\delta\lambda^i}{\delta s}\mu^j$$

$= -g_{ij}\chi\mu^i\mu^j$ [since $\chi\mu^j = \frac{\delta\lambda^j}{\delta s}$ and $g_{ij}\mu^i\mu^j = 1$ from (8.9) & (8.10)]

$= -\chi.1 = -\chi$

$$\text{Therefore, } g_{ij}\frac{\delta\mu^j}{\delta s}\lambda^i = -\chi \tag{8.11}$$

and since $g_{ij}\lambda^i\lambda^j = 1$ and we multiply $g_{ij}\lambda^i\lambda^j$ on the right hand side of Equation (8.11),

$$g_{ij}\frac{\delta\mu^j}{\delta s}\lambda^i = -\chi g_{ij}\lambda^i\lambda^j.$$

$$g_{ij}\lambda^i\left(\frac{\delta\mu^j}{\delta s} + \chi\lambda^j\right) = 0$$

shows that vector $\frac{\delta\mu^j}{\delta s} + \chi\lambda^j$ is orthogonal to λ^i.

We define a unit vector ν, with components ν^j, by formula

$$\nu^j = \frac{1}{\tau}\left(\frac{\delta\mu^j}{\delta s} + \chi^{\lambda j}\right), \tag{8.12}$$

where $\tau > 0$ is chosen to make ν^j a unit vector.

The vector ν will be orthogonal to both λ and μ.

The number τ in (8.12) is called the *torsion* of C at P and the vector ν is *binormal.*

8.4.1 Plane

If a curve lies on a plane, then it is called a *plane curve,* otherwise it is called a skew curve.

The plane determined by λ and μ is called the *osculating plane* at $P(x_0^i)$ and its equation is given by

$$g_{ij}(x^i - x_0^i)\nu^j = 0,$$

where x^i is any point on the plane.

The plane determined by ν and μ is called the *normal plane* at $P(x_0^i)$ and its equation is given by

$$g_{ij}(x^i - x_0^i)\lambda^j = 0$$

and that of the plane determined by ν and λ is called the *rectifying plane* at $P(x_0^i)$ and its equation is given by

$$g_{ij}(x^i - x_0^i)\mu^j = 0.$$

Result 8.4.1. Choose the sign of τ in such a way that

$$\sqrt{g}\,e_{ijk}\lambda^i\mu^j\nu^k = 1, \tag{8.13}$$

so that the triad of unit vectors λ, μ, ν and forms, at each point *P of* C, a right–handed system of axes.

Result 8.4.2. From the application of e on determinant and triple scalar products, we get

$$e_{ijk}\,\lambda^i\mu^j\nu^k = \begin{vmatrix} \lambda^1 & \mu^1 & \nu^1 \\ \lambda^2 & \mu^2 & \nu^2 \\ \lambda^3 & \mu^3 & \nu^3 \end{vmatrix} = \frac{1}{\sqrt{g}}\lambda.(\mu\times\nu).$$

Result 8.4.3. Since e_{ijk} is a relative tensor of weight −1 and $g = \left|\frac{\partial y^i}{\partial x^j}\right|^2$, it follows that

$$\epsilon_{ijk} = \sqrt{g}\,e_{ijk} \text{ and } \epsilon^{ijk} = \frac{1}{\sqrt{g}}e^{ijk}$$

is an *absolute tensor*, the left hand side of (8.13) is an invariant, and ν^k in (8.13) is determined by the formula

$$\nu^k = \epsilon^{ijk}\lambda_i\mu_j, \tag{8.14}$$

where λ_i *and* μ_j are associated vectors $g_{i\alpha}\lambda^\alpha$ and $g_{i\alpha}\mu^\alpha$.

The number τ in Equation (8.12) is called the *torsion of C at P and the vector* **v** is *binormal.*

We recall formula $\frac{\partial \boldsymbol{A}}{\partial x^i} = A^\alpha_{,i} a_\alpha$.

If vector field A is defined along C, we can write

$$\frac{\partial \boldsymbol{A}}{\partial x^i}\frac{dx^i}{ds} = A^\alpha_{,i}\frac{dx^i}{ds}\boldsymbol{a}_\alpha \tag{8.15}$$

Using intrinsic derivative $\frac{\delta A^\alpha}{\delta s} = A^\alpha_{,i}\frac{dA^i}{ds}$, we can write the above equation as

$$\frac{dA}{ds} = \frac{\delta A^\alpha}{ds}\boldsymbol{a}_\alpha.$$

Let r be the position vector of point P *on* C, then tangent vector $\boldsymbol{\lambda}$ is determined by

$$\frac{dr}{ds} = \lambda^i a_i = \boldsymbol{\lambda}.$$

From (8.15) and the above equation, we get

$$\frac{d^2r}{ds^2} = \frac{d\boldsymbol{\lambda}}{ds} = \frac{\delta\lambda^\alpha}{\delta s}a_\alpha = \boldsymbol{c},$$

where $\boldsymbol{c}$ is perpendicular to λ.

8.5 Serret-Frenet Formula

The set of three formulas known as Serret-Frenet formulas characterize all essential geometry properties of space curves. Two of these formulas have already been derived. These are

$$\text{(i) } \mu^i = \frac{1}{\chi}\frac{\delta\lambda^i}{\delta s} \quad \chi > 0 \tag{8.9}$$

$$\text{(ii) } \nu^i = \frac{1}{\tau}\left(\frac{\delta\mu^i}{\delta s} + \chi\lambda^i\right), \text{or } \frac{\delta\mu^i}{\delta s} = \tau\nu^i - \chi\lambda^i. \tag{8.12}$$

The third is expressed as

$$\text{(iii)} \frac{\delta\nu^i}{\delta s} = -\tau\mu^i. \tag{8.16}$$

The first gives the rate of turning of tangent vector λ as the points move along the curve, the second gives that of the principal normal μ, and the third one specifies the rate of turning of the binormal ν as point P moves along the curve.

$$\text{We know,} \qquad \nu^k = \epsilon^{ijk}\lambda_i\mu_j,$$

where λ_i and μ_j are the associated vectors $g_{i\alpha}\mu^\alpha$ and $g_{i\alpha}\lambda^\alpha$ and $\epsilon^{ijk} \equiv \frac{1}{\sqrt{g}}e^{ijk}$ is an absolute tensor.

If we differentiate the above equation intrinsically, we get

$$\frac{\delta\nu^k}{\delta s} = \epsilon^{ijk}\frac{\delta\lambda_i}{\delta s}\mu_j + \epsilon^{ijk}\lambda_i\frac{\delta\mu_j}{\delta s}, \tag{8.17}$$

since the covariant derivatives of ϵ^{ijk} are zero.

Using (8.9) as $\frac{\delta\lambda_i}{\delta s} = \chi\mu_i$ and (8.12) as $\frac{\delta\mu_i}{\delta s} = \tau\nu_i - \chi\lambda_i$ and putting them in (8.17),

$$\frac{\delta\nu^k}{\delta s} = \epsilon^{ijk}\chi\mu_i\mu_j + \epsilon^{ijk}\lambda_i(\tau\nu_j - \chi\lambda_j).$$

Here, $\epsilon^{ijk}\chi\mu_i\mu_j = 0$ and $\epsilon^{ijk}\chi\lambda_i\lambda_j = 0$ since ϵ^{ijk} is skew symmetric and $\epsilon^{ijk}\lambda_i\nu_j = -\mu_k$

$$\therefore \frac{\delta v^k}{\delta s} = \epsilon^{ijk} \lambda_i \tau v_j$$

$= -\tau\ \mu_k$, established in (8.16).

The Serret-Frenet formula can be written in Christoffel symbols as

$$\frac{d\lambda^i}{ds} + \left\{ \begin{matrix} i \\ j \quad k \end{matrix} \right\} \lambda^j \frac{dx^k}{ds} = \chi \mu^i$$

$$\frac{d\mu^i}{ds} + \left\{ \begin{matrix} i \\ j \quad k \end{matrix} \right\} \mu^j \frac{dx^k}{ds} = \tau v^i - \chi \lambda^i \tag{8.18}$$

$$\frac{dv^i}{ds} + \left\{ \begin{matrix} i \\ j \quad k \end{matrix} \right\} v^j \frac{dx^k}{ds} = \tau \mu^i$$

Equation (8.18) can be written in matrix form as

$$\frac{\delta}{\delta s} \begin{pmatrix} \lambda^i \\ \mu^i \\ v^i \end{pmatrix} = \begin{pmatrix} 0 & \chi & 0 \\ -\chi & 0 & \tau \\ 0 & -\tau & 0 \end{pmatrix} \begin{pmatrix} \lambda^i \\ \mu^i \\ v^i \end{pmatrix} \tag{8.19}$$

8.5.1 Bertrand Curves

The curves whose principal normals are also the principal normals of another curve are called Bertrand curves.

If the two curves are. C_1 *and* C_2:

$$C_1 : x^i = x^i(s)\, and\, C_2 : \overline{x}^i = \overline{x}^i(s)$$

have common principal normals at any of their points and one is called the Bertrand of another or one is called Bertrand associate of another.

Example 8.5.1. Show that a twisted curve ($\tau \neq 0$) is a Bertrand curve if, and only if, its curvature χ and torsion τ are connected by a linear equation of the form $a\chi + b\tau = 1$, where a and b are non-zero constants, hence showing that a circular helix is a Bertrand curve.

Solution: Let Γ: $x^i = x^i(s)$

$$\text{and} \quad \bar{\Gamma}: \bar{x}^i = \bar{x}^i(s) = x^i(s) + a(s)\mu^i \tag{i}$$

Differentiating intrinsically with respect to s, we get

$$\frac{\delta \bar{x}^i}{\delta s} = \frac{\delta x^i}{\delta s} + a'(s)\mu^i + a(s)\frac{\delta \mu^i}{\delta s} \qquad \left[\text{Here } a' = \frac{\delta a}{\delta s}\right]$$

$$= \lambda^i + a'(s)\mu^i + a(s)\left(\tau v_1^i - \chi\lambda^i\right)$$

$$\text{or } \bar{\lambda}^i \frac{\delta \bar{s}^i}{\delta s} = \lambda^i + a'(s)\mu^i + a(s)(\tau v^i - \chi\lambda^i) \tag{ii}$$

since the two curves are Bertrand, μ^i *is* parallel to $\bar{\mu}^i$. Taking the inner product of (ii), we get

$$\bar{\mu}^i \bar{\lambda}^i \frac{\delta \bar{s}^i}{\delta s} = \lambda^i \mu^i + a'(s)\mu^i\mu^i + a(s)(\tau v^i \mu^i - \chi\lambda^i\mu^i)$$

$$\text{or } 0 = 0 + a'(s).\ 1 + a(s)(\tau.0 - \chi.0)$$

$$\Rightarrow a'(s) = 0,$$

$$\therefore a(s) = constant,$$

which is the necessary condition so that two curves are Bertrand curves.

Now, (ii) becomes

$$\bar{\lambda}^i \frac{\delta \bar{s}^i}{\delta s} = \lambda^i + a(s)(\tau v^i - \chi\lambda^i)$$

$$\text{or} \quad \bar{\lambda}^i \frac{\delta \bar{s}^i}{\delta s} = \lambda^i(1 - a\chi) + a\tau v^i. \tag{iii}$$

For curve $\bar{\Gamma}$, we have

$$\frac{d}{ds}(\bar{\lambda}^i\lambda^i)=\frac{d\lambda^i}{ds}\bar{\lambda}^i+\frac{d\bar{\lambda}^i}{ds}\lambda^i$$

$$=\chi\mu^i\bar{\lambda}^i+(\bar{\chi}\bar{\mu}^i)\lambda^i\frac{d\bar{s}}{ds}=0.$$

$$\Rightarrow \bar{\lambda}^i\lambda^i=constant$$

This implies that the tangents to the two curves are inclined at a constant angle, but their principal normals coincide, and, therefore, the binormals of these curves are inclined at the same constant angle.

Let θ be the the inclination of v^i to $\bar{v}^i$, then θ is constant.

If the curve Γ is not plane, i.e., $\tau \neq 0$ (twisted curve), then from (iii)

$$\bar{\lambda}^i=\lambda^i(1-a\chi)\frac{ds}{d\bar{s}}+a\tau v^i\frac{ds}{d\bar{s}}. \tag{iv}$$

$\bar{\lambda}^i$ lies in the plane $\lambda^i v^i$, i.e., on the rectifying plane.

$$\text{Also, we can write } \bar{\lambda}^i=\cos\theta\,\lambda^i+sin\theta v^i. \tag{v}$$

Comparing Equations (iv) and (v), we have

$$\frac{1-a\chi}{\cos\theta}=\frac{a\tau}{sin\theta}=\frac{d\bar{s}}{ds}$$

$$\Rightarrow 1-a\chi=a\tau\ cot\theta. \tag{vi}$$

Differentiating Equation (v) with respect to s, we get

$$\bar{\chi}\bar{\mu}^i\frac{d\bar{s}}{ds}=-sin\theta\lambda^i\frac{d\theta}{ds}+\cos\theta\chi\mu^i+\cos\theta v^i\frac{d\theta}{ds}-\tau\mu^i sin\theta$$

$$=\mu^i(\chi\cos\theta-\tau sin\theta)+\frac{d\theta}{ds}(\cos\theta v^i-sin\theta\lambda^i).$$

Since the curves are Bertrand curves, $\bar{\mu}^i$ and μ^i are parallel and equal.

Therefore, $$\frac{d\theta}{ds}(\cos\theta v^i - sin\theta\lambda^i) = 0$$

$$\Rightarrow \frac{d\theta}{ds} = 0 \Rightarrow \theta = constant$$

$$1 - a\chi = a\tau\, cot\theta = b\tau, \text{ where } b = a\, cot\theta = constant.$$

Hence, we get a linear equation in τ and χ with constant coefficients as

$$a\chi + b\tau = 1.$$

This is the necessary condition so that the two curves are Bertrand curves when the curves are not plane.

Example 8.5.2. Find the curvature and torsion at any point of the curve: The equations of curve C in cylindrical coordinates

$$x^1 = a$$

$$x^2 = \theta(s)$$

$$x^3 = 0.$$

Solution: This curve is a circle of radius a. The square of the elements of s in cylindrical coordinate is

$$ds^2 = (dx^1)^2 + (dx^2)^2 + (dx^3)^2$$

$$= g_{ij}dx^i dx^j \qquad i, j = 1,2,3$$

where, $x^1 = a$, $x^2 = \theta(s)$, $x^3 = 0$; $y^1 = x^1 \cos x^2$, $y^2 = x^1 \sin x^2$, $y^3 = x^3$

$ds^2 = (dy^1)^2 + (dy^2)^2 + (dy^3)^2 = (dx^1)^2 + (x^1)^2(dx^2)^2 + (dx^3)^2)$, so that $g_{11} = 1$, $g_{22} = (x^1)^2$, $g_{33} = 1$, *and* $g_{ij} = 0$ *where* $i \neq j$, $g = (x^1)^2 = a^2$, and this verifies that the non-vanishing Christoffel symbols are

$$\begin{Bmatrix} 1 \\ 22 \end{Bmatrix} = -x^1, \begin{Bmatrix} 2 \\ 12 \end{Bmatrix} = \frac{1}{x^1}.$$

The components of tangent vector λ to circle C are

$$\lambda^i = \frac{dx^i}{ds} : \lambda^1 = \frac{dx^1}{ds} = 0, \lambda^2 = \frac{dx^2}{ds} = \frac{d\theta}{ds}, \textit{and } \lambda^3 = 0.$$

Since λ^i is a unit vector,

$$\therefore g_{ij}\lambda^i\lambda^j = 1 \text{ at all points of } C,$$

$$g_{11}(\lambda^1)^2 + g_{22}(\lambda^2)^2 + g_{33}(\lambda^3)^2 = 1,$$

$$or\ 1.0 + (x_1)^2\left(\frac{d\theta}{ds}\right)^2 + 1.0 = 1,$$

$$or\ a^2.\left(\frac{d\theta}{d(s}\right)^2 = 1, \therefore \left(\frac{d\theta}{d(s}\right) = \frac{1}{a}.$$

Now, using the 1st Serret-Frenet formula,

$$\frac{d\lambda^i}{ds} + \left\{ \begin{matrix} i \\ j\ k \end{matrix} \right\} \lambda^j \frac{dx^k}{ds} = \chi\mu^i, \textit{ for } i = 1$$

$$\chi\mu^1 = \frac{d\lambda^1}{ds} + \left\{ \begin{matrix} 1 \\ j\ k \end{matrix} \right\} \lambda^i \frac{dx^k}{ds}$$

$$= 0 + \left\{ \begin{matrix} 1 \\ 2\ \ 2 \end{matrix} \right\} \lambda^2 \frac{dx^2}{ds} = 0 + (-x^1)\lambda^2 \frac{d\theta}{ds}$$

$$= 0 + (-x^1)\lambda^2 \frac{d\theta}{ds} = -a.(\lambda^2)^2 = -a.\left(\frac{1}{a}\right)^2 = -a.\frac{1}{a^2} = -\frac{1}{a} \tag{8.20a}$$

$$for\ i=2,\ \chi\mu^2=\frac{d\lambda^2}{ds}+\left\{\begin{matrix}2\\ j\ k\end{matrix}\right\}\lambda^j\frac{dx^k}{ds}=\frac{d}{ds}\left(\frac{d\theta}{ds}\right)+\left\{\begin{matrix}2\\ 2\ 1\end{matrix}\right\}\lambda^2\frac{dx^1}{ds}$$

$$=\frac{d}{ds}\left(\frac{1}{a}\right)+\frac{1}{x^1}\frac{d\theta}{ds}\frac{dx^1}{ds}=0+\frac{1}{a}\frac{1}{a}.0=0 \qquad (8.20b)$$

$$for\ i=3,\ \chi\mu^3=\frac{d\lambda^3}{ds}+\left\{\begin{matrix}3\\ j\ k\end{matrix}\right\}\lambda^j\frac{dx^k}{ds}=0+\left\{\begin{matrix}3\\ 2\ 2\end{matrix}\right\}\lambda^2\frac{dx^2}{ds}=0+0=0. \qquad (8.20c)$$

For μ as a unit vector, we have

$$\therefore g_{ij}\mu^i\mu^j = 1 \text{ at all points of C}$$

$$\text{or } g_{11}(\mu^1)^2+g_{22}(\mu^2)^2+g_{33}(\mu^3)^2=1,\ or\ 1.(\mu^1)^2+0+0=1$$

or,

$$(\mu^1)^2=1 \qquad (8.20d)$$

(Here,

$$[22,1]=\frac{1}{2}\left[\frac{\partial g_{21}}{\partial x^2}+\frac{\partial g_{21}}{\partial x^2}-\frac{\partial g_{22}}{\partial x^1}\right]=--\frac{1}{2}\frac{\partial g_{22}}{\partial x^1}=--\frac{1}{2}\frac{\partial(x^1)^2}{\partial x^1}=-\frac{1}{2}.2x^1=-x^1.$$

$$\left\{\begin{matrix}1\\ 22\end{matrix}\right\}=g^{1l}[22,l]=g^{11}[22,1]=\frac{1}{g_{11}}.(-x^1)=-x^1.$$

$$[12,2]=\frac{1}{2}\left[\frac{\partial g_{22}}{\partial x^1}+\frac{\partial g_{12}}{\partial x^2}-\frac{\partial g_{12}}{\partial x^2}\right]=\frac{1}{2}\frac{\partial(x^1)^2}{\partial x^1}=\frac{1}{2}.2x^1=x^1\left\{\begin{matrix}2\\ 21\end{matrix}\right\}$$

$$=\left\{\begin{matrix}2\\ 12\end{matrix}\right\}=g^{2l}[12,l]=g^{22}[12,2]=\frac{1}{g_{22}}(x^1)=\frac{1}{(x^1)^2}x^1=\frac{1}{x^1},\left\{\begin{matrix}2\\ 22\end{matrix}\right\}$$

$$=g^{2l}[22,l]=g^{22}[22,2]=\frac{1}{(x^1)^2}.\frac{1}{2}\frac{\partial(x^1)^2}{\partial x^2}=0.$$

All others Christoffel symbols are zero.)

From above,

$$\chi\mu^1 = -\frac{1}{a}$$

$$\text{or } (\chi\mu^1)^2 = \left(-\frac{1}{a}\right)^2$$

$$\therefore \chi^2(\mu^1)^2 = \frac{1}{a^2}$$

$$\Rightarrow \chi^2.1 = \frac{1}{a^2}\chi = \frac{1}{a} \text{ (nonnegative).}$$

From (8.20a), $\chi\mu^1 = -\frac{1}{a}$

$$\therefore \frac{1}{a}\mu^1 = -\frac{1}{a}$$

$$\Rightarrow \mu^1 = -1.$$

We get, $\mu^1 = -1$, $\mu^2 = \mu^3 = 0$, *and* $\chi = \frac{1}{a}$

2nd Part : Using the 2nd Serret-Frenet formula:

$$\text{for}, i=1, \quad \frac{d\mu^i}{ds}\begin{Bmatrix} i \\ j\ k \end{Bmatrix}\mu^j\frac{dx^k}{ds} = \tau\nu^i - \chi\lambda^i,$$

$$\text{or } 0 + \begin{Bmatrix} 1 \\ 1\ 2 \end{Bmatrix}\mu^1\frac{dx^2}{ds} = \tau\nu^1 - \chi\lambda^1,$$

$$\text{or } 0 + 0.\mu^1\frac{d\theta}{ds} = \tau\nu^1 - \frac{1}{a}.0$$

$$\therefore \tau\nu^1 = 0 \qquad (8.20e)$$

$$\text{for } i=2, \frac{d\mu^2}{ds}+\begin{Bmatrix} 2 \\ j\ k \end{Bmatrix}\mu^j\frac{dx^k}{ds}=\tau\nu^2-\chi\lambda^2,$$

$$\text{or } 0+\begin{Bmatrix} 2 \\ 1\ 2 \end{Bmatrix}\mu^1\frac{dx^2}{ds}=\tau\nu^2-\chi\lambda^2,$$

$$\text{or } 0+\frac{1}{x^1}\mu^1\frac{d\theta}{ds}=\tau\nu^2-\frac{1}{a}\frac{1}{a},$$

$$\text{or } \frac{1}{a}(-1)\frac{1}{a}=\tau\nu^2-\frac{1}{a}\frac{1}{a}$$

$$\therefore \tau\nu^2=0 \tag{8.20f}$$

For ν is a unit vector, we have

$$g_{ij}\nu^i\nu^j = 1 \quad or\, g_{11}(\nu^1)^2 + g_{22}(\nu^2)^2 + g_{33}(\nu^3)^2 = 1$$

$$\text{or } 1.\,(\nu^1)^2 + (x^1)^2(\nu^2)^2 + 1.(\nu^3)^2 = 1.$$

Multiplying τ^2, we get $\tau^2(\nu^1)^2 + \tau^2\,(x^1)^2(\nu^2)^2 + \tau^2\,.\,(\nu^3)^2 = \tau^2$

$$\text{or, by (8.20e) and (8.20f), } (\nu^3)^2 = 1 \tag{8.20g}$$

Using the S.F. 2nd formula,

$$\frac{d\mu^3}{ds}+\begin{Bmatrix} 3 \\ j\ k \end{Bmatrix}\mu^j\frac{dx^k}{ds}=\tau\nu^3-\chi\lambda^3$$

$$\text{or } \ 0+0=\tau\nu^3-\frac{1}{a}.0, \therefore \tau\nu^3=0 \tag{8.20h}$$

By (8.20g) in (8.20h), we get $\tau = 0$ and we can easily find ν^i as $\nu^1 = \nu^2 = 0$, $\nu^3 = 1$.

To determine the components of ν, we know

$$\nu^i=\epsilon^{irs}\lambda_r\mu_s.$$

$$i=1 \text{ gives } v^1 = \epsilon^{1rs}\lambda_r\mu_s = \epsilon^{123}\lambda_2\mu_3 + \epsilon^{132}\lambda_3\mu_2 = 0.$$

$$i=2 \text{ gives } v^2 = \epsilon^{2rs}\lambda_r\mu_s = \epsilon^{213}\lambda_1\mu_3 + \epsilon^{231}\lambda_3\mu_1 = 0.$$

$$i=3 \text{ gives } v^3 = \epsilon^{3rs}\lambda_r\mu_s = \epsilon^{312}\lambda_1\mu_2 + \epsilon^{321}\lambda_2\mu_1 = \epsilon^{321}\lambda_2\mu_1$$

$$= \frac{1}{\sqrt{g}} e^{321}\lambda_2\mu_1.$$

$$\text{Here, } \lambda_2 = g_{22}\lambda^2 = a^2\frac{1}{a} = a$$

$$\text{and } \mu_1 = g_{11}\mu^1 = 1 \text{ and } (-1) = -1$$

$$\text{Therefore, } v^3 = \frac{1}{\sqrt{g}} e^{321}\lambda_2\mu_1 = \frac{1}{a}(-1).a.(-1) = 1$$

$$\therefore v = (0,0,1)$$

Example 8.5.3.

(a) Find the curvature and torsion at any point of the circular helix where the equation in cylindrical coordinates is: C: $x^1 = a$, $x^2 = \theta(s)$, $x^3 = k\theta$.

(b) Show that the tangent vector λ at every point of C makes a constant angle with the direction of x^3-axis.

Solution:

Here, $ds^2 = (dx^1)^2 + (dx^2)^2 + (dx^3)^2$.

$= g_{ij}\, dx^i dx^j$, $i, j = 1,2,3$.

This curve is a circle of radius a. The square of the element of ds in cylindrical coordinates is

$$ds^2 = (dx^1)^2 + (dx^2)^2 + (dx^3)^2$$

$$= g_{ij} dx^i dx^j \quad i, j = 1,2,3$$

here, $x^1 = a$, $x^2 = \theta(s)$, $x^3 = k\theta$; $y^1 = x^1 \cos x^2$, $y^2 = x^1 \sin x^2$, $y^3 = x^3$

$$ds^2 = (dy^1)^2 + (dy^2)^2 + (dy^3)^2 = (dx^1)^2 + (x^1)^2(dx^2)^2 + (dx^3)^2,$$

so that $g_{11} = 1$, $g_{22} = (x^1)^2$, $g_{33} = 1$ *and* $g_{ij} = 0$ *where* $i \neq j$, and $g = (x^1)^2 = a^2$,
so that $g_{11} = 1$, $g_{22} = (x^1)2$, $g_{33} = 1$ *and* $g_{ij} = 0$, *where* $i \neq j$,

$$and \quad \lambda^1 = 0, \quad \lambda^2 = \frac{dx^2}{ds} = \frac{d\theta}{ds} \quad \lambda^3 = \frac{dx^3}{ds} = k\frac{d\theta}{ds}.$$

We know $g_{ij}\lambda^i \lambda^j = 1 \Rightarrow g_{11}(\lambda^1)^2 + g_{22}(\lambda^2)^2 + g_{33}(\lambda^3)^2 = 1$

$$0 + a^2.\left(\frac{d\theta}{ds}\right)^2 + k^2\left(\frac{d\theta}{ds}\right)^2 = 1$$

$$\left(\frac{d\theta}{ds}\right)^2 = \frac{1}{k^2 + a^2}$$

$$\left(\frac{d\theta}{ds}\right) = \frac{1}{\sqrt{k^2 + a^2}} = A(let) \tag{8.21a}$$

Using the 1st Serret-Frenet formula,

$$\chi\mu^i = \frac{d\lambda^i}{ds} + \begin{Bmatrix} i \\ j \quad k \end{Bmatrix} \lambda^j \frac{dx^k}{ds}$$

for $i = 1$

$$\chi\mu^1 = \frac{d\lambda^i}{ds} + \begin{Bmatrix} i \\ j \quad k \end{Bmatrix} \lambda^j \frac{dx^k}{ds} = 0 + \begin{Bmatrix} 1 \\ 2 \quad 2 \end{Bmatrix} \lambda^2 \frac{dx^2}{ds}$$

$$= (-x^1)\lambda^2.\frac{d\theta}{ds} = -a\left(\frac{d\theta}{ds}\right)^2 \tag{8.21b}$$

$$\chi\mu^2 = \frac{d\lambda^2}{ds} + \left\{ \begin{matrix} 2 \\ j \quad k \end{matrix} \right\} \lambda^j \frac{dx^k}{ds} = \frac{d}{ds}\left(\frac{d\theta}{ds}\right) + \left\{ \begin{matrix} 2 \\ 2 \quad 2 \end{matrix} \right\} \lambda^2 \frac{dx^2}{ds}$$

$$+ \left\{ \begin{matrix} 2 \\ 2 \quad 1 \end{matrix} \right\} \lambda^2 \frac{dx^1}{ds} = 0.\left(\frac{d\theta}{ds}\right)^2 + \frac{1}{a}\frac{d\theta}{ds}.0 = 0 \qquad (8.21c)$$

$$\chi\mu^3 = \frac{d\lambda^3}{ds} + \left\{ \begin{matrix} 3 \\ j \quad k \end{matrix} \right\} \lambda^j \frac{dx^k}{ds} = 0 + \left\{ \begin{matrix} 3 \\ 3 \quad 2 \end{matrix} \right\} \lambda^3 \frac{dx^2}{ds} = 0 + 0 = 0.$$

(8.21d)

We know, $g_{ij}\mu^i \mu^j = 1$,

or $g_{11}\mu^1 \mu^1 + g_{22}\mu^2 \mu^2 + g_{33}\mu^3 \mu^3 = 1$,

or $\chi^2(\mu^1)^2 + 0 + 1.0 = \chi^2$,

or $a^2A^4 = \chi^2$

$\chi^2 = a^2A^{4,}$

$$\text{or } \chi = aA^2 = \frac{a}{a^2 + k^2} \qquad (8.21e)$$

(Here,

$$[22,1] = \frac{1}{2}\left[\frac{\partial g_{21}}{\partial x^2} + \frac{\partial g_{21}}{\partial x^2} - \frac{\partial g_{22}}{\partial x^1}\right] = -\frac{1}{2}\frac{\partial g_{22}}{\partial x^1} = -\frac{1}{2}\frac{\partial (x1)2}{\partial x^1} = -\frac{1}{2}.2x^1 = -x^1 .$$

$$\left\{ \begin{matrix} 1 \\ 22 \end{matrix} \right\} = g^{1l}[22,l] = g^{11}[22,1] = \frac{1}{g_{11}}.(-x^1) = -x^1.$$

$$[12,2] = \frac{1}{2}\left[\frac{\partial g_{22}}{\partial x^1} + \frac{\partial g_{12}}{\partial x^2} - \frac{\partial g_{12}}{\partial x^2}\right] = \frac{1}{2}\frac{\partial (x1)2}{\partial x^1} = \frac{1}{2}.2x^1 = x^1$$

$$\left\{ \begin{matrix} 2 \\ 21 \end{matrix} \right\} = \left\{ \begin{matrix} 2 \\ 12 \end{matrix} \right\} = g^{2l}[12,l] = g^{22}[12,2] = \frac{1}{g_{22}}(x^1) = \frac{1}{(x^1)^2}x^1 = \frac{1}{x^1}$$

All other Christoffel symbols are zero.)

$$\mu^1 = \frac{-aA^2}{aA^2} = -1,$$

$$\therefore \mu^1 = -1, \mu^2 = \mu^3 = 0, \chi = \frac{a}{a^2 + k^2},$$

Using the 2nd Serret-Frenet formula,

$\frac{d\mu^1}{ds} + \left\{ \begin{matrix} 1 \\ j\ k \end{matrix} \right\} \mu^j \frac{dx^k}{ds} = \tau \nu^1 - \chi \lambda^1$ and we can easily find τ.

For $i = 1$, $\frac{d\mu^i}{ds} + \left\{ \begin{matrix} i \\ j\ k \end{matrix} \right\} \mu^j \frac{dx^k}{ds} = \tau \nu^i - \chi \lambda^i$

$$\frac{d\mu^1}{ds} + \left\{ \begin{matrix} 1 \\ j\ k \end{matrix} \right\} \mu^j \frac{dx^k}{ds} = \tau \nu^1 - \chi \lambda^1,$$

$$\text{or } \tau \nu^1 - \chi \lambda^1 = 0 + \left\{ \begin{matrix} 1 \\ 2\ 2 \end{matrix} \right\} \mu^2 \frac{dx^2}{ds},$$

$$\text{or } \tau \nu^1 - \chi .0 = 0 + \left\{ \begin{matrix} 1 \\ 2\ 2 \end{matrix} \right\} \mu^2 \frac{dx^2}{ds} = a.0. \frac{dx^2}{ds} = 0$$

$$\left[\left\{ \begin{matrix} 1 \\ 2\ \ 2 \end{matrix} \right\} = g^{11}[22,1] = g^{11} \frac{1}{2} \left(-\frac{\partial g_{22}}{\partial x^1} \right) = -x^1 = -a, \left\{ \begin{matrix} 2 \\ 1\ \ 2 \end{matrix} \right\} \right.$$

$$\left. = g^{22}[12,2] = \frac{1}{a^2} \frac{1}{2} \left(\frac{\partial g_{22}}{\partial x^1} \right) = \frac{1}{a} \right]$$

$$\therefore \tau \nu^1 = 0.$$

For $i=2$, $\tau v^2 - \chi\lambda^2 = \frac{d\mu^2}{ds} + \left\{ \begin{matrix} 2 \\ j\ k \end{matrix} \right\} \mu^j \frac{dx^k}{ds}$

$$= 0 + \left\{ \begin{matrix} 2 \\ 1\ 2 \end{matrix} \right\} \mu^1 \frac{dx^2}{ds} \left\{ \begin{matrix} 2 \\ 2\ 1 \end{matrix} \right\} \mu^2 \frac{dx^1}{ds}$$

$$+ \left\{ \begin{matrix} 2 \\ 3\ 2 \end{matrix} \right\} \mu^3 \frac{dx^2}{ds} + \left\{ \begin{matrix} 2 \\ 2\ 3 \end{matrix} \right\} \mu^2 \frac{dx^3}{ds}$$

$$= \left\{ \begin{matrix} 2 \\ 1\ 2 \end{matrix} \right\} \mu^1 \frac{dx^2}{ds} + \left\{ \begin{matrix} 2 \\ 2\ 1 \end{matrix} \right\} \mu^2 \frac{dx^1}{ds} = \frac{1}{a}(-1)\frac{d\theta}{ds} + \frac{1}{a}.0.0$$

$$= \left(-\frac{1}{a}\right)\frac{d\theta}{ds} = -\frac{1}{a\sqrt{k^2+a^2}}$$

$$\left[\left\{ \begin{matrix} 2 \\ 1\ 2 \end{matrix} \right\} = g^{22}[12,2] = g^{22}\frac{1}{2}\left(\frac{\partial g_{22}}{\partial x^1}\right) \right.$$

$$= \frac{1}{2}.2\frac{x^1}{a^2} = \frac{1}{a} \ \& \left\{ \begin{matrix} 2 \\ 3\ 2 \end{matrix} \right\} = g^{22}[32,2]$$

$$\left. = g^{22}\frac{1}{2}\left(-\frac{\partial g_{22}}{\partial x^3}\right) = 0 \ \right]$$

$$\tau v^2 = \chi\lambda^2 - \frac{1}{a\sqrt{k^2+a^2}}$$

$$= \frac{a}{a^2+k^2}\frac{1}{\sqrt{k^2+a^2}} - \frac{1}{a\sqrt{k^2+a^2}} = -\frac{k^2}{a(a^2+k^2)^{\frac{3}{2}}}.$$

For $i=3$, $\tau v^3 - \chi\lambda^3 = \frac{d\mu^3}{ds} + \left\{ \begin{matrix} 3 \\ j\ \ k \end{matrix} \right\} \mu^j \frac{dx^k}{ds} = 0$

$$\left[\left\{ \begin{matrix} 3 \\ 3\ \ 2 \end{matrix} \right\} = g^{33}[23,3] = 1.\frac{1}{2}\left(\frac{\partial g_{33}}{\partial x^2}\right) = 0 \ \& \left\{ \begin{matrix} 3 \\ 3\ \ 1 \end{matrix} \right\} = g^{33}[13,3] = 0 \right]$$

$$\text{Or } \tau v^3 = \chi\lambda^3 = \frac{\alpha k}{(a^2+k^2)^{\frac{3}{2}}}.$$

We can find ν by the following relations: $\nu^t = \epsilon^{trs}\lambda_r \mu_s$.

$$\nu^1 = \epsilon^{123}\lambda_2\mu_3 + \epsilon^{132}\lambda_3\mu_2 = 0$$

$$\nu^2 = \epsilon^{213}\lambda_1\mu_3 + \epsilon^{231}\lambda_3\mu_1 = 0 + \frac{1}{\sqrt{g}}e^{231}(-1)k\frac{d\theta}{ds} = \frac{1}{\sqrt{g}}1.(-1)k\frac{1}{\sqrt{k^2+a^2}}$$

$$= \frac{-k}{a\sqrt{k^2+a^2}}$$

and $\nu^3 = \epsilon^{312}\lambda_1\mu_2 + \epsilon^{321}\lambda_2\mu_1 = 0 + \frac{1}{\sqrt{g}}e^{321}\frac{d\theta}{ds}(-1) = \frac{1}{\sqrt{g}}(-1)\frac{d\theta}{ds}(-1)$

$$= \frac{1}{a\sqrt{k^2+a^2}}.$$

We know $g_{ij}\nu^i \nu^j = 1$,

or $g_{11}\nu^1 \nu^1 + g_{22}\nu^2 \nu^2 + g_{33}\nu^3 \nu^3 = 1$,

or $\tau^2 g_{22}\nu^2 \nu^2 + \tau^2 g_{33}\nu^3 \nu^3 = \tau^2$,

$$\text{or } a^2\left(-\frac{k^2}{a(a^2+k^2)^{\frac{3}{2}}}\right)^2 + \left(-\frac{ak}{(a^2+k^2)^{\frac{3}{2}}}\right)^2 = \tau^2,$$

$$\tau^2 = \frac{k^4}{(a^2+k^2)^3} + \frac{a^2k^2}{(a^2+k^2)^3} = \frac{k^2(a^2+k^2)}{(a^2+k^2)^3} = \frac{k^2}{(a^2+k^2)^2},$$

$$\text{or } \tau^2 = \frac{k^2}{(a^2+k^2)^2}$$

$$\tau = \frac{k}{(a^2+k^2)}.$$

We can verify $\tau\nu^i$ with these results.

Example 8.5.4. Using the results of Problem 1, show that the ratio of the curvature χ to the torsion τ is constant. Show using Frenet's formulas that whenever $\frac{\tau}{\chi} = \text{constant}$ and the coordinates are Cartesian, $\nu i = c\lambda^i + b^i$, where c and b^i are constants.

$$\text{Here, } \tau = \frac{k}{(a^2+k^2)} \text{ and } \chi = \frac{a}{(a^2+k^2)}.$$

$$\text{Therefore, } \frac{\tau}{\chi} = \frac{\dfrac{k}{(a^2+k^2)}}{\dfrac{a}{(a^2+k^2)}} = \frac{k}{a} \text{ is constant}$$

$$\text{and } \lambda^1 = 0, \quad \lambda^2 = \frac{dx^2}{ds} = \frac{d\theta}{ds} = \frac{1}{(a^2+k^2)^{\frac{1}{2}}} \qquad \lambda^3 = \frac{dx^3}{ds} = k\frac{d\theta}{ds} = \frac{k}{(a^2+k^2)^{\frac{1}{2}}}$$

$$v^1 = 0 \ , \ v^2 = \frac{-k}{a\sqrt{k^2+a^2}} v^3 = \frac{1}{a\sqrt{k^2+a^2}}.$$

$$v^i = c\lambda^i$$

Example 8.5.5. Show that

$$(i)\ \chi = \left(g_{ij} \frac{\delta\lambda^i}{\delta s} \frac{\delta\lambda^j}{\delta s} \right)^{\frac{1}{2}}$$

$$(ii)\ \tau = \epsilon_{ijk} \lambda^i \mu^j \frac{\delta\mu^k}{\delta s}$$

Solution: We know the 1st Serret-Frenet formula

$$\mu^i = \frac{1}{\chi} \frac{\delta\lambda^i}{\delta s} \quad \text{or} \quad \frac{\delta\lambda^i}{\delta s} = \chi\mu^i,$$

where $\chi > 0$ is chosen to make μ^i a unit vector.

$$g_{ij} \frac{\delta\lambda^i}{\delta s} \mu^j = g_{ij} \chi \mu^i \mu^j$$

$$= \chi \qquad\qquad [\because g_{ij}\mu^i\mu^j = 1 \ ],$$

$$\text{or } \chi = g_{ij} \frac{\delta\lambda^i}{\delta s} \frac{1}{\chi} \frac{\delta\lambda^j}{\delta s} \quad (\text{as } \chi > 0),$$

$$\text{or } \chi^2 = g_{ij} \frac{\delta\lambda^i}{\delta s} \frac{\delta\lambda^j}{\delta s}$$

$$\therefore \chi = \left(g_{ij} \frac{\delta\lambda^i}{\delta s} \frac{\delta\lambda^j}{\delta s} \right)^{\frac{1}{2}}.$$

(ii) We know from the 2nd Serret-Frenet formula,

$$\frac{\delta\mu^i}{\delta s} = \tau\nu^i - \chi\lambda^i$$

$$\begin{aligned}
\epsilon_{ijk}\lambda^i\mu^j\frac{\delta\mu^k}{\delta s} &= \epsilon_{ijk}\lambda^i\mu^j(\tau\nu^k - \chi\lambda^k) \\
&= \epsilon_{ijk}\lambda^i\mu^j\tau\nu^k - \epsilon_{ijk}\lambda^i\mu^j\chi\lambda^k \\
&= \left(\epsilon_{ijk}\lambda^i\mu^j\right)\tau\nu^k - 0 = \nu_k\nu^k\tau = \tau \qquad \left(as\ \nu_k\nu^k = 1 \text{ and } \epsilon_{ijk}\lambda^i\mu^j\lambda^k = 0\right).
\end{aligned}$$

Example 8.5.6. Show that $\tau = \frac{1}{\chi^2}\epsilon_{ijk}\lambda^i\frac{\delta\lambda^i}{\delta s}\frac{\delta^2\lambda k}{\delta s^2}$, where notations have their usual meaning.

Solution: We know that $\mu^i = \frac{1}{\chi}\frac{\delta\lambda^i}{\delta s}$ or, $\frac{\delta\lambda^i}{\delta s} = \chi\mu^i$.

Differentiating intrinsically with respect to s, we get

$\frac{\delta^2\lambda^i}{\delta s^2} = \frac{\delta\chi}{\delta s}\mu^i + \frac{\delta\mu^i}{\delta s}\chi = \frac{\delta\chi}{\delta s}\mu^i + (\tau\nu^i - \chi\lambda^i)\chi$ (using the 2nd Serret-Frenet formula).

$$\begin{aligned}
\text{Now, } \frac{1}{\chi^2}\epsilon_{ijk}\lambda^i\frac{\delta\lambda^i}{\delta s}\frac{\delta^2\lambda^k}{\delta s^2} &= \frac{1}{\chi^2}\epsilon_{ijk}\lambda^i\chi\mu^i\left[\frac{\delta\chi}{\delta s}\mu^k + (\tau\nu^k - \chi\lambda^k)\chi\right] \\
&= \frac{1}{\chi}\left[\frac{\delta\chi}{\delta s}\epsilon_{ijk}\lambda^i\mu^j\mu^k + \tau\chi\epsilon_{ijk}\lambda^i\mu^j\nu^k - \chi^2\epsilon_{ijk}\lambda^i\mu^i\lambda^k\right] \\
&= \frac{1}{\chi}(0 + \tau\chi - 0) = \tau.
\end{aligned}$$

Example 8.5.7.

Show that

$$\frac{\delta^2\lambda^i}{\delta s^2} = \frac{d\chi}{ds}\mu^i + \chi(\tau\nu^i - \chi\lambda^i)$$

$$\frac{\delta^2\lambda^i}{\delta s^2} = \frac{d\tau}{ds}\nu^i - (\chi^2 + \tau^2)\mu^i - \frac{d\chi}{ds}\lambda^i$$

$$\frac{\delta^2\nu^i}{\delta s^2} = \tau(\chi\lambda^i - \tau\nu^i) - \frac{d\tau}{ds}\mu^i,$$

where symbols have their usual meanings.
Solution: We know from the first Serret-Frenet formula

$$\frac{\delta\lambda^i}{\delta s} = \chi\mu^i.$$

Differentiating intrinsically with respect to s, we get

$$\frac{\delta^2\lambda^i}{\delta s^2} = \frac{\delta}{\delta s}(\chi\mu^i) = \frac{d(\chi\mu^i)}{ds} + \left\{ \begin{matrix} i \\ j \;\; k \end{matrix} \right\}(\chi\mu^j)\frac{dx^k}{ds}$$

$$= \frac{d(\chi)}{ds}\mu^i + \frac{d(\mu^i)}{ds}\chi + \chi\mu^j \left\{ \begin{matrix} i \\ j \;\; k \end{matrix} \right\}\frac{dx^k}{ds}$$

$$= \frac{d(\chi)}{ds}\mu^i + \chi\left[\frac{d(\mu^i)}{ds} + \mu^j \left\{ \begin{matrix} i \\ j \;\; k \end{matrix} \right\}\frac{dx^k}{ds}\right]$$

$$= \frac{d(\chi)}{ds}\mu^i + \chi\frac{\delta\mu^i}{\delta s}$$

$$= \frac{d(\chi)}{ds}\mu^i + \chi(\tau\nu^i - \chi\lambda^i) \text{ (from the Serret-Frenet 2}^{\text{nd}}\text{ formula)}$$

$$\frac{\delta^2\lambda^i}{\delta s^2} = \frac{d(\chi)}{ds}\mu^i + \chi(\tau\nu^i - \chi\lambda^i)$$

(ii) Using the Serret-Frenet formula,

$$\frac{\delta\mu^i}{\delta s} = \tau\nu^i - \chi\lambda^i$$

$$\frac{\delta^2\mu^i}{\delta s^2} = \frac{\delta(\tau\nu^i)}{\delta s} - \frac{\delta(\chi\lambda^i)}{\delta s}$$

$$= \frac{d(\tau\nu^i)}{ds} + \tau\nu^i \left\{ \begin{matrix} i \\ j \;\; k \end{matrix} \right\}\frac{dx^k}{ds} - \left[\frac{d(\chi\lambda^i)}{ds} + \chi\lambda^i \left\{ \begin{matrix} i \\ j \;\; k \end{matrix} \right\}\frac{dx^k}{ds}\right]$$

$$= \frac{d(\tau v^i)}{ds} + \tau v^i \left\{ \begin{matrix} i \\ j\ k \end{matrix} \right\} \frac{dx^k}{ds} - \left[\frac{d(\chi\lambda^i)}{ds} + \chi\lambda^i \left\{ \begin{matrix} i \\ j\ k \end{matrix} \right\} \frac{dx^k}{ds} \right]$$

$$= \frac{d(\tau)}{ds} v^i + \frac{d(v^i)}{ds}\tau + \tau v^i \left\{ \begin{matrix} i \\ j\ k \end{matrix} \right\} \frac{dx^k}{ds} - \left[\frac{d(\chi)}{ds}\lambda^i + \frac{d(\lambda^i)}{ds}\chi + \chi\lambda^i \left\{ \begin{matrix} i \\ j\ k \end{matrix} \right\} \frac{dx^k}{ds} \right]$$

$$= \frac{d(\tau)}{ds} v^i + \tau \left[\frac{d(v^i)}{ds} + v^i \left\{ \begin{matrix} i \\ j\ k \end{matrix} \right\} \frac{dx^k}{ds} \right] - \frac{d(\chi)}{ds}\lambda^i - \chi \left[\frac{d(\lambda^i)}{ds} + \lambda^i \left\{ \begin{matrix} i \\ j\ k \end{matrix} \right\} \frac{dx^k}{ds} \right]$$

$$= \frac{d(\tau)}{ds} v^i + \tau \frac{\delta v^i}{\delta s} - \frac{d(\chi)}{ds}\lambda^i - \chi \frac{\delta \lambda^i}{\delta s}$$

$$= \frac{d(\tau)}{ds} v^i + \tau(-\tau\mu^i) - \frac{d(\chi)}{ds}\lambda^i - \chi(\chi\mu^i) \qquad \left[\frac{\delta v^i}{\delta s} = -\tau\mu^i, \frac{\delta\lambda^i}{\delta s} = \chi\mu^i \right]$$

$$= \frac{d(\tau)}{ds} v^i - (\tau^2 + \chi^2)\mu^i - \frac{d(\chi)}{ds}\lambda^i.$$

(iii) $$\frac{\delta^2 v^i}{\delta s^2} = \frac{\delta}{\delta s}(-\tau\mu^i) = -\left[\frac{d(\tau\mu^i)}{ds} + \tau\mu^i \left\{ \begin{matrix} i \\ j\ k \end{matrix} \right\} \frac{dx^k}{ds} \right]$$

$$= -\frac{d\tau}{ds}\mu^i - \frac{d\mu^i}{ds}\tau - \tau\mu^i \left\{ \begin{matrix} i \\ j\ k \end{matrix} \right\} \frac{dx^k}{ds}$$

$$= -\frac{d\tau}{ds}\mu^i - \tau \left[\frac{d\mu^i}{ds} + \mu^i \left\{ \begin{matrix} i \\ j\ k \end{matrix} \right\} \frac{dx^k}{ds} \right]$$

$$= -\frac{d\tau}{ds}\mu^i - \tau \frac{\delta\mu^i}{\delta s}$$

$$= -\frac{d\tau}{ds}\mu^i - \tau(\tau v^i - \chi\lambda^i) \qquad \left[\Theta \ \frac{\delta\mu^i}{\delta s} = (\tau v^i - \chi\lambda^i) \right]$$

$$= \tau(\chi\lambda^i - \tau v^i) - \frac{d\tau}{ds}\mu^i.$$

8.6 Equations of a Straight Line

Let A^i be a vector field defined along a curve C in E_3, where C is given parametrically as

$$C{:}x^i = x^i(s) \quad , \quad s_1 \le s \le s_2 \, , \, (i = 1,2,3.),$$

s being the arc parameter.

If the vector field A^i is parallel, then it follows from (8.3) that

$$\frac{\delta A^i}{\delta s} = 0$$

$$\text{or } \frac{dA^i}{ds} + \left\{ \begin{matrix} i \\ \alpha\,\beta \end{matrix} \right\} A^\alpha \frac{dx^\beta}{ds} = 0 \tag{8.22}$$

We shall use this equation to obtain the equations of a straight line in general curvilinear coordinates. The characteristic property of a straight line is that the tangent vector **λ** to a straight lines is directed along the straight line, so that the totality of tangent vectors **λ** forms a parallel vector field. Thus, the field of tangent vector $\lambda^i = \dfrac{dx^i}{ds}$ must satisfy (8.22) and we have

$$\frac{\delta \lambda^i}{\delta s} = \frac{d^2x^i}{ds^2} + \left\{ \begin{matrix} i \\ \alpha\,\beta \end{matrix} \right\} \frac{dx^\alpha}{ds} \frac{dx^\beta}{ds} = 0.$$

The equation $\dfrac{d^2x^i}{ds^2} + \left\{ \begin{matrix} i \\ \alpha\,\beta \end{matrix} \right\} \dfrac{dx^\alpha}{ds} \dfrac{dx^\beta}{ds} = 0$ is a straight line equation.

In Cartesian coordinates, the Christoffel symbols vanish and we obtain the familiar straight line equation $\dfrac{d^2x^i}{ds^2} = 0.$

In view of geometrical interpretation of curvature χ as a measure of the rate of turning of the tangent vector to the curve will be zero since the angle is constant.

Example 8.6.1. Show that a space curve is a straight line if, and only if, its curvature is zero at all points of it.

Solution: Suppose that the curvature $\chi = 0$ at all points of a space curve Γ.

Then, by Serret-Frenet's 1st formula,

$$\frac{\delta\lambda^i}{\delta s} = \chi\mu^i$$

$$\frac{\delta\lambda^i}{\delta s} = 0 \quad \text{for all s,} \quad [\text{since } \chi = 0]$$

$\Rightarrow \lambda^i$ = Constant (with respect to s)
This means that λ^i is fixed direction
$\therefore$ Space curve is a straight line.
Hence Γ is a straight line
Conversely, Space curve is a straight line.
$\Rightarrow$ direction of the tangent vector of space is fixed.
$\Rightarrow \chi\mu^i = 0$ (since $\mu^i = 0$)
It follows that $\chi = 0$.

$$\text{or } \frac{d\lambda^i}{ds} + \begin{Bmatrix} i \\ j\ k \end{Bmatrix} \lambda^j \frac{dx^k}{ds} = 0$$

$$\Rightarrow, \ \frac{d\lambda^i}{ds} = 0 \ , or \ \lambda^i = ks + c,$$

which is a straight line. Hence Γ is a straight line.

8.7 Helix

A space curve is called a helix if the tangent vector at every point to helix makes a constant angle with a fixed direction.

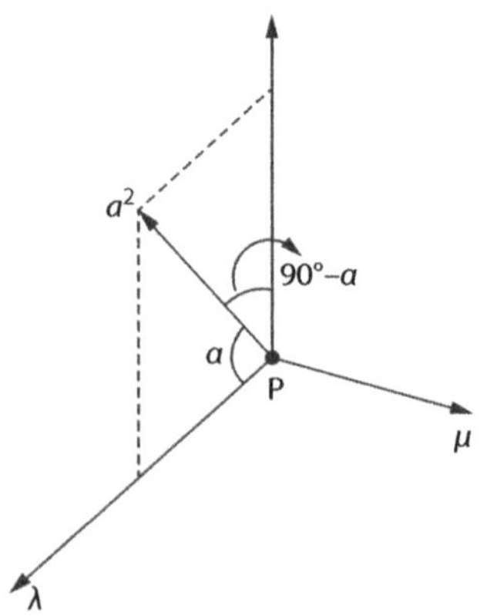

Figure 8.3

Let C: $x^i = x^i(s)$ be an equation of a helix. By definition, if a^i is the fixed direction and α is the angle between a^i and the tangent vector λ^j, as shown in Figure (8.3), then

$$\cos\alpha = g_{ij}a^j\lambda^i \quad , \; 0<|\alpha|\leq\frac{\pi}{2} \tag{8.23}$$

Differentiating intrinsically with respect to s, we get

$$g_{ij}a^j\frac{\delta\lambda^i}{\delta s}+g_{ij}\frac{\delta a^j}{\delta s}\lambda^i=0$$

$$\text{or } g_{ij}a^j\frac{\delta\lambda^i}{\delta s}=0,$$

$$\text{or } g_{ij}a^j\chi\mu^i=0, \qquad \left[\because\frac{\delta\lambda^i}{\delta s}=\chi\mu^i\right]$$

$$\text{or } g_{ij}a^j\mu^i=0 \quad , \quad as\ \chi\neq 0 \tag{8.24}$$

This shows that a^j is orthogonal to μ^i.

Since the curvature χ of a helix does not vanish, the principal normal μ^i is perpendicular everywhere to the generators. Hence, the fixed direction of the generators is parallel to the plane of μ^i and ν^i and, since it makes a constant angle α with λ^i, it makes a constant angle of $90° - \alpha$ with ν^i.

We know from the definition of a helix that the curvature and torsion are in a constant ratio. Since the angle between a^j and ν^i is $90° - \alpha$,

$$\text{Cos}(90° - \alpha) = g_{ij}a^j\nu^i \qquad \text{or} \qquad sin\alpha = g_{ij}a^j\nu^i \tag{8.25}$$

Differentiating intrinsically (8.24) with respect to s, we get

$$g_{ij}a^j\frac{\delta\mu^i}{\delta s}=0.$$

Using Serret-Frenet's 2nd formula,

$$g_{ij}a^j(\tau\nu^i - \chi\lambda^i) = 0.$$

Using (8.23) and (8.25), we have τ

$$\tau g_{ij}a^j\nu^i - \chi g_{ij}a^j\lambda^i = 0$$

$$\text{or } \tau\sin\alpha - \chi\cos\alpha = 0.$$

$$\text{Therefore, } \frac{\chi}{\tau} = \tan\alpha = constant,$$

but a^i is parallel to the plane of λ^i and ν^i and must be parallel to the vector $\tau\lambda^i + \chi\nu^i$, which is inclined to λ^i at an angle $\tan^{-1}\frac{\chi}{\tau}$. This angle is constant. Therefore, the curvature and torsion are in a constant ratio.

Conversely, we can prove that a curve whose curvature and torsion are in a constant ratio is a helix.

We suppose that a curve different from a straight line is

$$\frac{\chi}{\tau} = \frac{1}{c}, constant \ \ ie, \tau = c\chi.$$

Using Serret-Frenet's 3rd formula, we get

$$\frac{\delta\nu^i}{\delta s} = -\tau\mu^i = -c\chi\mu^i$$

$$= -c\frac{\delta\lambda^i}{\delta s} \qquad \left(using \text{ 1st } frenet \ formula, \frac{\delta\lambda^i}{\delta s} = \chi\mu^i\right)$$

$$\text{or } \frac{\delta}{\delta s}(\nu^i + c\lambda^i) = 0.$$

Therefore, $\nu^i + c\lambda^i$ is a constant vector, say a^i, i.e., $\nu^i + c\lambda^i = a^i$.

$$\text{Now, } g_{ij}a^i\lambda^j = g_{ij}(\nu^i + c\lambda^i)\lambda^j$$
$$= g_{ij}\nu^i\lambda^j + cg_{ij}\lambda^i\lambda^j = 0 + c.1 = c.$$

This shows that the tangent vector λ^i makes a constant angle with a fixed direction a^i and the curve is therefore a general helix.

Theorem 8.7.1. The necessary and sufficient condition for a given curve to be a helix is that the ratio of the curvature to the torsion is constant.

8.7.1 Cylindrical Helix

The cylindrical helix is a curve upon a cylinder which cuts the generator of the cylinder at a constant angle α with a fixed direction. Let us consider

x^3 as the axis of the cylinder. Then, the equation of cylinder C: $x^i = f^i(t)$, $i = 1,2,3$ (Figure 8.4).

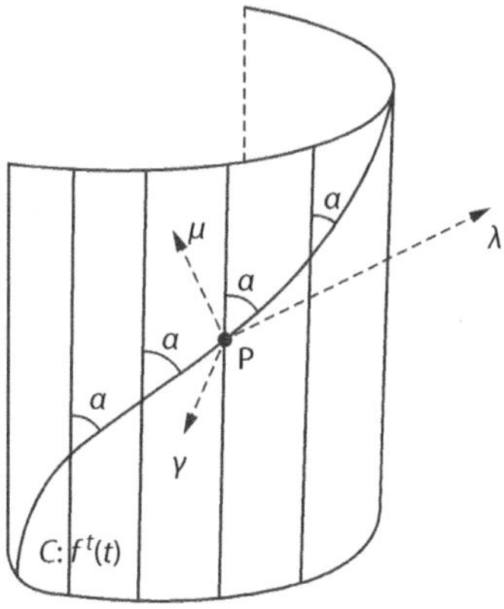

Figure 8.4

Let α be the angle at any point P on the curve between tangent a and the generator of the cylinder passing through P. The unit vector along the generator is

$$\boldsymbol{k} = (0,0,1)$$

and the unit tangent vector to the curve at P is

$$\lambda^i = \frac{\delta x^i}{\delta s} = \frac{\delta x^i/\delta t}{\delta s/\delta t} = \frac{\delta x^i/\delta t}{|\delta x^i/\delta f|}$$

$$= \frac{(\dot{f}^1, \dot{f}^2, \dot{f}^3)}{\left[\{\dot{f}^1(t)\}^2 + \{\dot{f}^2(t)\}^2 + \{\dot{f}^3(t)\}^2\right]^{\frac{1}{2}}}$$

$$\therefore \cos\alpha = g_{ij} k^j \lambda^i = \frac{(\dot{f}^1, \dot{f}^2, \dot{f}^3).(\dot{0},0,1)}{\left[\{\dot{f}^1(t)\}^2 + \{\dot{f}^2(t)\}^2 + \{\dot{f}^3(t)\}^2\right]^{\frac{1}{2}}}$$

$$= \frac{\dot{f}^3}{\left[\{\dot{f}^1(t)\}^2 + \{\dot{f}^2(t)\}^2 + \{\dot{f}^3(t)\}^2\right]^{\frac{1}{2}}},$$

$$\text{or } (\dot{f}^3)^2 = \cos^2\alpha\left[\{\dot{f}^1(t)\}^2 + \{\dot{f}^2(t)\}^2 + \{\dot{f}^3(t)\}^2\right]$$

$$(\dot{f}^3(t))^2 - \cos^2\alpha\{\dot{f}^3(t)\}^2 = \cos^2\alpha\left[\{\dot{f}^1(t)\}^2 + \{\dot{f}^2(t)\}^2\right],$$

$$\text{or } \sin^2\alpha\{\dot{f}^3(t)\}^2 = \cos^2\alpha\left[\{\dot{f}^1(t)\}^2 + \{\dot{f}^2(t)\}^2\right]$$

$$\dot{f}^3(t)=\cot\alpha\left[\{\dot{f}^1(t)\}^2+\{\dot{f}^2(t)\}^2\right]^{\frac{1}{2}}$$

$$\therefore f^3(t)=\cot\alpha\int\left[\{\dot{f}^1(t)\}^2+\{\dot{f}^2(t)\}^2\right]^{\frac{1}{2}}dt+constant,$$

$$\text{or, } f^3(t)=x^3(t)=\cot\alpha\int\left[\{\dot{x}^1\}^2+\{\dot{x}^2\}^2\right]^{\frac{1}{2}}dt+constant.$$

Therefore, the equation of the helix is

$$x^1(t)=f^1(t),x^2(t)=f^2(t),\ x^3(t)=\cot\alpha\int\left[\{\dot{x}^1\}^2+\{\dot{x}^2\}^2\right]^{\frac{1}{2}}dt+constant. \tag{8.26}$$

8.7.2 Circular Helix

It follows from the definition that a helix can always be drawn on the surface of a cylinder.

If a space curve Γ lies on a circular cylinder and the tangent at each point of it makes a constant angle with the axis of the cylinder, then the curve is a helix. Such helixes are said to be Circular Helixes.

A particular case of the cylindrical helix is the circular helix.

Let $x^1 = a\cos\theta$, $x^2 = a\sin\theta$, and

$$x^3=\cot\alpha\int_0^\theta\sqrt{a^2(\cos\theta+\sin^2\theta)}\,d\alpha=\cot\alpha\int_0^\theta a\,d\alpha$$

$$=a\cot\alpha.\theta=b\theta\ ,\text{ where } b=a\cot\alpha=constant.$$

The parametric equation of a circular helix is

$$x^1 = a\cos\theta,\ x^2 = a\sin\theta,\ x^3 = b\theta,\ b\neq 0. \tag{8.27}$$

Example 8.7.1. Prove that

$$\epsilon_{ijk}\frac{\delta\lambda^i}{\delta s}\frac{\delta^2\lambda^j}{\delta s^2}\frac{\delta^3\lambda^k}{\delta s^3}=\chi^5\frac{d}{dx}\left(\frac{\tau}{\chi}\right).$$

Hence, a space curve is a helix if, and only if,

$$\epsilon_{ijk}\frac{\delta\lambda^i}{\delta s}\frac{\delta^2\lambda^j}{\delta s^2}\frac{\delta^3\lambda^k}{\delta s^3}=0.$$

Solution:

Using Serret-Frenet's 1st formula,

$$\frac{\delta\lambda^i}{\delta s} = \chi\mu^{i.}$$

Differentiating intrinsically with respect to s, we get

$$\frac{\delta^2\lambda^i}{\delta s^2} = \chi\frac{\delta\mu^i}{\delta s} + \mu^i\frac{\delta\chi}{\delta s}$$
$$= \mu^i\frac{\delta\chi}{\delta s} + \chi\left(\tau\nu^i - \chi\lambda^i\right).$$

$$\text{Now, } \frac{\delta^3\lambda^i}{\delta s^3} = \mu^i\frac{\delta^2\chi}{\delta s^2} + \frac{\delta\mu^i}{\delta s}\frac{\delta\chi}{\delta s} + \frac{\delta\chi}{\delta s}\left(\tau\nu^i - \chi\lambda^i\right) + \chi\frac{\delta}{\delta s}\left(\tau\nu^i - \chi\lambda^i\right)$$
$$= \mu^i\frac{\delta^2\chi}{\delta s^2} + \frac{\delta\mu^i}{\delta s}\frac{\delta\chi}{\delta s} + \frac{\delta\chi}{\delta s}\left(\tau\nu^i - \chi\lambda^i\right) + \chi\left[\tau\frac{\delta\nu^i}{\delta s} + \nu^i\frac{\delta\tau}{\delta s} - \chi\frac{\delta\lambda^i}{\delta s} - \frac{\delta\chi}{\delta s}\lambda^i\right]$$
$$= \mu^i\frac{\delta^2\chi}{\delta s^2} + \frac{\delta\chi}{\delta s}\left(\tau\nu^i - \chi\lambda^i\right) + \chi\tau(-\tau\mu^i) + \chi\nu^i\frac{\delta\tau}{\delta s} - \chi^2(\chi\mu^i)$$
$$-\chi\frac{\delta\chi}{\delta s}\lambda^i] + (\tau\nu^i - \chi\lambda^i)\frac{\delta\chi}{\delta s} = \mu^i\frac{\delta^2\chi}{\delta s^2} + \frac{\delta\chi}{\delta s}\left(\tau\nu^i - \chi\lambda^i\right) + \chi\tau(-\tau\mu^i)$$
$$+\nu^i\left(\chi\frac{\delta\tau}{\delta s} + \tau\frac{\delta\chi}{\delta s}\right) - \chi^2(\chi\mu^i) - 2\chi\frac{\delta\chi}{\delta s}\lambda^i = \left(\frac{\delta^2\chi}{\delta s^2} - \chi\tau^2 - \chi^3\right)\mu^i$$
$$+\left(\chi\frac{\delta\tau}{\delta s} + 2\tau\frac{\delta\chi}{\delta s}\right)\nu^i - 3\chi\frac{\delta\chi}{\delta s}\lambda^{i.}$$

Therefore,

$$\epsilon_{ijk}\frac{\delta\lambda^i}{\delta s}\frac{\delta^2\lambda^j}{\delta s^2}\frac{\delta^3\lambda^k}{\delta s^3}$$

$$= \epsilon_{ijk}\chi\mu^i\left[\mu^j\frac{\delta\chi}{\delta s} + \chi(\tau\nu^j - \chi\lambda^j)\right]\left[\left(\frac{\delta^2\chi}{\delta s^2} - \chi\tau^2 - \chi^3\right)\mu^k\right.$$
$$\left. + \left(\chi\frac{\delta\tau}{\delta s} + 2\tau\frac{\delta\chi}{\delta s}\right)\nu^k - 3\chi\frac{\delta\chi}{\delta s}\lambda^k\right]$$

$$=\left(\chi\epsilon_{ijk}\mu^i\mu^j\frac{\delta\chi}{\delta s}+\chi^2\tau\epsilon_{ijk}\mu^i v^j-\chi^3\epsilon_{ijk}\mu^i\lambda^j\right)$$

$$\left[\left(\frac{\delta^2\chi}{\delta s^2}-\chi\tau^2-\chi^3\right)\mu^k+\left(\chi\frac{\delta\tau}{\delta s}+2\tau\frac{\delta\chi}{\delta s}\right)v^k-3\chi\frac{\delta\chi}{\delta s}\lambda^k\right]$$

$$=(0+\chi^2\tau\lambda^k-\chi^3v^k)\left[\left(\frac{\delta^2\chi}{\delta s^2}-\chi\tau^2-\chi^3\right)\mu^k+\left(\chi\frac{\delta\tau}{\delta s}+2\tau\frac{\delta\chi}{\delta s}\right)v^k-3\chi\frac{\delta\chi}{\delta s}\lambda^k\right]$$

$$=\left(\frac{\delta^2\chi}{\delta s^2}-\chi\tau^2-\chi^3\right)(\chi^2\tau\lambda^k\mu^k+\chi^3v^k\mu^k)+\left(\chi\frac{\delta\tau}{\delta s}+2\tau\frac{\delta\chi}{\delta s}\right)(\chi^2\tau\lambda^k v^k+\chi^3v^k v^k)$$

$$-3\chi\frac{\delta\chi}{\delta s}(\chi^2\tau\lambda^k\lambda^k+\chi^3v^k\lambda^k)$$

$$=0+\chi^3\left(\chi\frac{\delta\tau}{\delta s}+2\tau\frac{\delta\chi}{\delta s}\right)-3\chi\frac{\delta\chi}{\delta s}(\chi^2\tau)$$

$$=\chi^3\left(\chi\frac{\delta\tau}{\delta s}-\tau\frac{\delta\chi}{\delta s}\right)=\chi^5\frac{1}{\chi^2}\left(\chi\frac{\delta\tau}{\delta s}-\tau\frac{\delta\chi}{\delta s}\right)=\chi^5\frac{d}{d^s}\left(\frac{\tau}{\chi}\right).$$

Second Part: If a curve is a helix, then $\frac{\tau}{\chi}$ is constant.

Hence, $\frac{d}{ds}\left(\frac{\tau}{\chi}\right)=0.$

Therefore, $\epsilon_{ijk}\frac{\delta\lambda^i}{\delta s}\frac{\delta^2\lambda^j}{\delta s^2}\frac{\delta^3\lambda^k}{\delta s^3}=0.$

Conversely, let $\epsilon_{ijk}\frac{\delta\lambda^i}{\delta s}\frac{\delta^2\lambda^j}{\delta s^2}\frac{\delta^3\lambda^k}{\delta s^3}=0,$

$$\text{or } \epsilon_{ijk}(\chi\mu^i)\chi(\tau v^j-\chi\lambda^j)(-\chi\tau^2-\chi^3)\mu^k=0,$$

$$\text{or } \epsilon_{ijk}(\chi\mu^i)\chi(\tau v^j-\chi\lambda^j)(\tau^2+\chi^2)\mu^k=0,$$

$$\text{or } -\epsilon_{ijk}\chi^3\mu^i\mu^k(\tau v^j-\chi\lambda^j)(\tau^2+\chi^2)=0,$$

or $-\epsilon_{ijk}\chi^3\mu^k\tau v^j+\epsilon_{ijk}\chi^3\mu^k\chi\lambda^j=0,$ as $(\tau^2+\chi^2)\neq 0$ and μ^i are an arbitrary vector,

$$or - \epsilon_{ijk} \tau \mu^k \nu^j + \epsilon_{ijk} \chi \mu^k \lambda^j = 0,$$

$$\text{or} \qquad (\tau\lambda_i + \chi\nu_i) = 0$$

$$g_{ij}(\tau\lambda^j + \chi\nu^j) = 0.$$

If a^i is the fixed direction, then, by hypothesis, $g_{ij}a^i\lambda^j = \cos\alpha$ and $g_{ij}a^i\nu^j = \sin\alpha$.

$$\text{Therefore, } g_{ij}a^i(\tau\lambda^j + \chi\nu^j) = 0$$

$$\therefore \tau\cos\alpha + \chi\sin\alpha = 0$$

$$\text{or } \frac{\chi}{\tau} = -tan\alpha = constant,$$

hence, the curve is a helix.

Example 8.7.2. If the tangent and the binormal to a space curve make angles α and β with a fixed direction, then show that $\frac{\chi}{\tau} = -\frac{\sin\alpha \, d\alpha}{\sin\beta \, d\beta}$.

Let a^i be the fixed direction and the tangent vector with component λ^i to a space curve C and make an angle α with a^i. The binormal μ^i makes an angle β with a^i. Then, by the hypothesis

$$g_{ij}a^i\lambda^j = \cos\alpha \text{ and } g_{ij}a^i\nu^j = cos\beta.$$

since α and β are invariants, $\frac{\delta\alpha}{\delta s} = \frac{d\alpha}{ds}$ and $\frac{\delta\beta}{\delta s} = \frac{d\beta}{ds}$

Differentiating intrinsically, we get

$$g_{ij}a^i \frac{\delta\lambda^j}{\delta s} = \frac{\delta}{\delta s}\cos\alpha = \frac{d}{ds}(cos\alpha) = -sin\alpha\frac{d\alpha}{ds}$$

$$g_{ij}a^i \frac{\delta\nu^j}{\delta s} = \frac{\delta}{\delta s}\cos\beta = \frac{d}{ds}(cos\beta) = -sin\alpha\frac{d\beta}{ds}.$$

Using Serret-Frenet formulas, we get

$$g_{ij}a^i \frac{\delta\lambda^j}{\delta s} = -sin\alpha\frac{d\alpha}{ds} \quad \text{or, } g_{ij}a^i\chi\mu^j = -sin\alpha\frac{d\alpha}{ds}$$

$$\text{and } g_{ij}a^i \frac{\delta v^j}{\delta s} - sin\beta \frac{d\beta}{ds} \text{ or, } g_{ij}a^i(-\tau\mu i) = -sin\beta \frac{d\beta}{ds}.$$

Dividing these two, we get

$$\frac{g_{ij}a^i \chi \mu^j}{g_{ij}a^i \tau \mu^i} = \frac{-sin\alpha \frac{d\alpha}{ds}}{sin\beta \frac{d\beta}{ds}}$$

$$\text{or } \frac{\chi}{\tau} = -\frac{\sin\alpha \, d\alpha}{\cos\beta \, d\beta}.$$

8.8 Exercises

1. If for a contravariant vector A^i, $\frac{\delta A^i}{\delta t} = 0$, prove that $\frac{\delta A^i}{\delta t} = 0$ where A_i is the associate tensor of A^i.
2. Prove that $\frac{d(g_{ij}A^iB^j)}{dt} = g_{ij}\frac{\delta A^i}{\delta t}B^j + g_{ij}A^i\frac{\delta B^j}{\delta t}$.
3. If the intrinsic derivative of a vector **A** along a curve C vanishes at all points of C, show that the magnitude of **A** is constant along curve C.
4. If two unit vectors are such that at all points of a given curve C their intrinsic derivatives along C are zero, show that they are inclined at a constant angle along C.
5. Find the curvature and torsion at any point of a given curve, namely the circle C, where

$$C: x^1 = a, x^2 = t, x^3 = 0, \text{ and } ds^2 = (dx^1)^2 + (x^1)^2(dx^2)^2 + (dx^3)^2.$$

6. Show from the Serret-Frenet formula that whenever $\frac{\chi}{\tau}$ = constant and the coordinates are Cartesian,

$$v^j = c\lambda^i + b^i,$$

where c and b^i are constants and the tangent vector makes a constant angle with a fixed direction.

7. Let C be a cylindrical helix determined by

$$C: \begin{cases} y^1 = \phi(\sigma) \\ y^2 = \varphi(\sigma) \\ y^3 = k\sigma, k = constant, \end{cases}$$

where σ is the arc parameter of the directrix curve C' in the y^1y^2 −plane, so that
$(d\sigma)^2 = (dy^1)^2 + (dy^2)^2$ and $(ds)^2 = (1 + k^2)(d\sigma)^2$ and show that

$$\tau = \frac{\begin{vmatrix} \phi' \; \varphi' \; k \\ \phi'' \; \varphi'' \; 0 \\ \phi''' \; \varphi''' \; 0 \end{vmatrix}}{\chi^2} \cdot \frac{1}{(1+k^2)^3}$$

$$\chi = \frac{\phi'\varphi'' - \varphi'\phi''}{1+k^2}.$$

Also, verify that $\frac{\chi}{\tau} = \text{k}$.

8. Show that a space curve is a straight line if, and only if, its curvature is zero.
9. Show that a space curve is a plane curve if, and only if, its torsion is zero.
10. Show that $\tau = \frac{1}{\chi^2} \epsilon_{ijk} \lambda^i \frac{\delta\lambda^i}{\delta s} \frac{\delta^2 \lambda^k}{\delta s^2}$.

9

Intrinsic Geometry of Surfaces

9.1 Introduction

We will discuss the differential geometry of surfaces by means of Tensor Calculus. In this chapter we study the properties of surfaces imbedded in three-dimensional Euclidean spaces. It will be shown that the properties can be phrased independently of the space in which the surface is immersed and that they are concerned solely with the structure of the differential quadratic form for the element of arc of a curve drawn on the surface.

All such properties of surfaces are termed as *intrinsic* properties and the geometry based on the study of differential quadratic forms is called the *intrinsic geometry* of the surface.

9.2 Curvilinear Coordinates on a Surface

We study the surface in which the surface is imbedded to a set of orthogonal Cartesian axes Y and the locus of points are satisfying the equation

$$F(y^1, y^1, y^1) = 0 \tag{9.1}$$

A surface S is defined, in general, due to Gauss as the set of points whose coordinates are functions of two independent parameters. Thus, the equation of a surface is of the form

$$S{:}y^i = y^i\,(u^1, u^2), \tag{9.2}$$

where $u_1^1 \le u^1 \le u_2^1$ *and* $u_1^2 \le u^2 \le u_2^2$ and y^i are real functions of class C^1 in the region of definition of the independent parameters u^1 *and* u^2. If we assign to u^1 in (9.2) some fixed value $u^1 = c$, we obtain Figure (9.1) as a locus of the one–dimensional manifold

Dipankar De. Introduction to Differential Geometry with Tensor Applications, (265–320) © 2022 Scrivener Publishing LLC

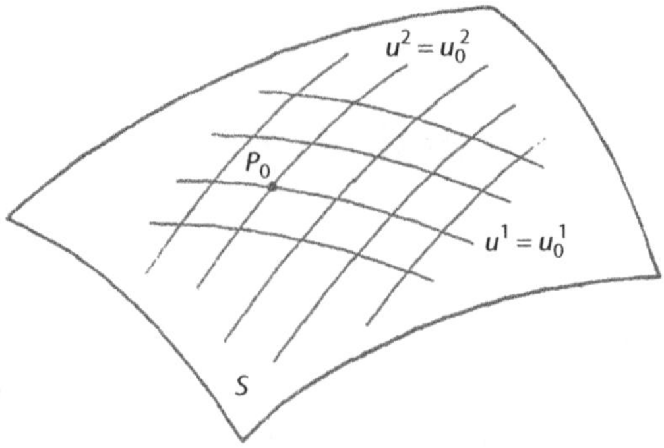

Figure 9.1

$$y^i = y^i(c, u^2), \quad (i = 1,2,3),$$

which is a curve lying on the surface S defined by Equation (9.2). We shall call it a u^2–curve. Similarly, u^2 = constant in (9.2) defines a u^1– curve, along which only u^1 varies. By assigning to u^1 *and* u^2 a succession of fixed values, we obtain a *net of curves*, on the surface which are termed *coordinate curves*.

The intersection of a pair of coordinate curves is obtained by setting $u^1 = \overset{1}{u_0}, u^2 = \overset{2}{u_0}$ determines P_0.

The variables $u^1$and u^2 determining the point P on S are called the *curvilinear or Gaussian coordinates* on the surface. If one introduces a transformation

$$\left\{\begin{array}{l} u^1 = u^1\left(\bar{u}^1, \bar{u}^2\right) \\ u^2 = u^2\left(\bar{u}^1, \bar{u}^2\right) \end{array}\right\}, \tag{9.3}$$

where $u^\alpha\left(\bar{u}^1, \bar{u}^2\right)$ are of class C^1 and Jacobian $J = \dfrac{\partial\left(u^1, u^2\right)}{\partial\left(\bar{u}^1, \bar{u}^2\right)}$ does not vanish in the region of variables $\bar{u}^i$, then one can insert the values from (9.3) in (9.2) to obtain a different set of parametric equations

$$y^i = f^i\left(\bar{u}^1, \bar{u}^2\right) \quad , i = 1,2,3. \tag{9.4}$$

The parametric representation of a surface is not unique and there are infinitely many curvilinear coordinates systems which can be used to locate points on a given surface *S*.

For example, the equations

$$x^1 = u^1 + u^2 \qquad x^1 = v^1 \cosh v^2$$

$$x^2 = u^1 - u^2 \qquad \text{and } x^2 = v^1 \sinh v^2$$

$$x^3 = 4u^1 - u^2 \qquad x^3 = (v^1)^2$$

represent the same surface: $(x^1)^2 - (x^2)^2 = x^3$.

The two representations may be related by parametric transformation $v^1 = 2\sqrt{u^1 u^2}$,

$$v^2 = \frac{1}{2} \log \frac{u^1}{u^2}.$$

9.3 Intrinsic Geometry: First Fundamental Quadratic Form

In the last section, we saw that the properties of surfaces that can be described without reference to the space in which the surface is imbedded are called *intrinsic* properties. A study of intrinsic properties is made to depend on a certain quadratic differential forms describing the metric character of the surface.

Here, we will be dealing with two distinct sets of variables: those referring to the space E_3 in which the surface is imbedded and two curvilinear coordinats u^1 *and* u^2, referring to the two dimensional manifold S. We use Latin letters for the indices referring to space variables and Greek letters for the surface variables.

A transformation *T* of space coordinates from one system (x^i) to another system $(\bar{x}^i)$ will be written as

$$\mathrm{T}: x^i = x^i\left(\bar{x}^1, \bar{x}^2, \bar{x}^3\right)$$

and a transformation of Gaussian surface coordinates will be denoted by

$$u^{\alpha} = u^{\alpha}(\bar{u}^1, \bar{u}^2).$$

Consider a surface S defined by

$$y^i = y^i(u^1, u^2), \tag{9.5}$$

where y^i are the orthogonal Cartesian coordinates covering space E_3 in which the surface S is imbedded and a curve *C on S* is defined by

$$u^{\alpha} = u^{\alpha}(t), \quad t_1 \leq t \leq t_2, \tag{9.6}$$

where u^{α}'s are the Gausian coordinates covering S. The curve defined by (9.6) is a curve in a three-dimensional Euclidean space and its element of arc is given by the formula

$$ds^2 = dy^i dy^i. \tag{9.7}$$

From (9.5), we get

$$dy^i = \frac{\partial y^i}{\partial u^{\alpha}} du^{\alpha} \tag{9.8}$$

and from (9.6),

$$du^{\alpha} = \frac{\partial u^{\alpha}}{\partial t} dt.$$

In (9.7), we have

$$\begin{aligned} ds^2 &= \frac{\partial y^i}{\partial u^{\alpha}} \frac{\partial y^i}{\partial u^{\beta}} du^{\alpha} du^{\beta} \\ &= a_{\alpha\beta} du^{\alpha} du^{\beta}, \end{aligned}$$

Where
$$a_{\alpha\beta} = \frac{\partial y^i}{\partial u^\alpha}\frac{\partial y^i}{\partial u^\beta}. \tag{9.9}$$

The expression of ds^2

$$ds^2 = a_{\alpha\beta}du^\alpha du^\beta \tag{9.10}$$

Is the square of the linear element of C lying on the surface S and $a_{\alpha\beta}du^\alpha du^\beta$ is called the *first fundamental quadratic form of the surface.*

The length of the arc of the curve defined by Equation (9.6) is given by the formula

$$s = \int_{t_1}^{t_2} \sqrt{a_{\alpha\beta}\dot{u}^\alpha \dot{u}^\beta}\, dt, \text{ where } \dot{u}^\alpha = \frac{du^\alpha}{dt}.$$

From (9.10), it follows that since $ds^2>0$, at once upon setting u^α = *constant* and u^1 = *constant* in turn, $ds^2_{(1)} = a_{11}(du^1)^2$ and $ds^2_{(2)} = a_{22}(du^2)^2$. Thus, a_{11} and a_{22} are positive functions of u^1 and u^2.

Example 9.3.1. Show that $a_{\alpha\beta}$ is a covariant metric tensor of the surface.

Consider a transformation of surface coordinates

$$u^\alpha = u^\alpha\left(\bar{u}^1, \bar{u}^2\right) \tag{9.11}$$

with a non-vanishing Jacobian $J = \left|\frac{\partial u^\alpha}{\partial \bar{u}^\alpha}\right|$.

We have $u^\alpha = u^\alpha\left(\bar{u}^1, \bar{u}^2\right)$, that

$$du^\alpha = \frac{\partial u^\alpha}{\partial \bar{u}^\gamma} d\bar{u}^\gamma.$$

We have

$$\begin{aligned} ds^2 &= a_{\alpha\beta}du^\alpha du^\beta \\ &= a_{\alpha\beta}\frac{\partial u^\alpha}{\partial \bar{u}^\gamma}\frac{\partial u^\beta}{\partial \bar{u}^\delta} d\bar{u}^\gamma d\bar{u}^\delta \end{aligned}$$

If we set
$$\bar{a}_{\gamma\delta} = a_{\alpha\beta}\frac{\partial u^\alpha}{\partial \bar{u}^\gamma}\frac{\partial u^\beta}{\partial \bar{u}^\delta},$$

we see that the set of quantities $a_{\alpha\beta}$ represents a covariant tensor of rank two.

Now, we have to show that $a_{\alpha\beta}$ is a symmetric tensor.

Let

$$a_{\alpha\beta} = \frac{1}{2}\left(a_{\alpha\beta} + a_{\beta\alpha}\right) + \frac{1}{2}\left(a_{\alpha\beta} - a_{\beta\alpha}\right) = A_{\alpha\beta} + B_{\alpha\beta}.\cdot$$

It is clear to us that $A_{\alpha\beta}$ is symmetric and $B_{\alpha\beta}$ is a skew-symmetric tensor.

Now, $a_{\alpha\beta}du^\alpha du^\beta = A_{\alpha\beta}du^\alpha du^\beta + B_{\alpha\beta}du^\alpha du^\beta$,

$$\text{or } \left(a_{\alpha\beta} - A_{\alpha\beta}\right)du^\alpha du^\beta = B_{\alpha\beta}du^\alpha du^\beta, \tag{i}$$

or $\left(a_{\alpha\beta} - A_{\alpha\beta}\right)du^\alpha du^\beta = -B_{\beta\alpha}du^\alpha du^\beta$, since $B_{\alpha\beta}$ is skew-symmetric,
or $\left(a_{\alpha\beta} - A_{\alpha\beta}\right)du^\alpha du^\beta = -B_{\alpha\beta}du^\beta du^\alpha$, with an interchanging dummy index,

$$\text{or } \left(a_{\alpha\beta} - A_{\alpha\beta}\right)du^\alpha du^\beta = -B_{\alpha\beta}du^\alpha du^\beta. \tag{ii}$$

Thus, we get $2\, B_{\alpha\beta}du^\alpha du^\beta = 0$ from (i) and (ii).

$\Rightarrow B_{\alpha\beta} = 0$ (Since du^α *and* du^β are arbitrary)

Therefore, $a_{\alpha\beta} = \frac{1}{2}\left(a_{\alpha\beta} + a_{\beta\alpha}\right) = A_{\alpha\beta}$

Hence, the set of quantities $a_{\alpha\beta}$ represents a symmetric covariant tensor of rank 2.

The tensor $a_{\alpha\beta}$ is called the *covariant metric tensor* of the surface.

9.3.1 Contravariant Metric Tensor

Since $ds^2 = a_{\alpha\beta}du^\alpha du^\beta$ is positive definite, the determinant

$$a = \begin{vmatrix} a_{11} a_{12} \\ a_{21}\ a_{22} \end{vmatrix} > 0.$$

If we can define the reciprocal tensor by the formula $a^{\alpha\beta} = \dfrac{cofactor\ of\ a_{\alpha\beta}\ in\ a}{a}$, then $a^{\alpha\beta}a_{\beta\gamma} = \delta^\alpha_\gamma$.

We have $a^{11} = \frac{a_{22}}{a}$, $a^{12} = a^{21} = -\frac{a_{12}}{a}$, *and* $a^{22} = \frac{a_{11}}{a}$

The contravariant tensor $a^{\alpha\beta}$ is called the *contravariant metric tensor.*

The direction of a linear element on the surface can be specified either by *direction cosine or direction parameters.*

$$\lambda^{\alpha} = \frac{du^{\alpha}}{ds} \tag{9.12}$$

and

$$\frac{dy^i}{ds} = \frac{\partial y^i}{\partial u^{\alpha}} \frac{du^{\alpha}}{ds}.$$

We define the length of the *surface vector* A^{α}. That is, the vector determined by $A^1(u^1,u^2)$ and $A^2(u^1,u^2)$, by the formula

$$A = \sqrt{a_{\alpha\beta} A^{\alpha} A^{\alpha}}.$$

We have $ds^2 = a_{\alpha\beta} du^{\alpha} du^{\beta}$

or

$$1 = a_{\alpha\beta} \frac{du^{\alpha}}{ds} \frac{du^{\beta}}{ds}$$

$$= a_{\alpha\beta}\, \lambda^{\alpha}\, \lambda^{\beta}.$$

This shows that λ^{α} are components of a unit vector.

The covariant vector

$$\lambda_{\beta} = a_{\alpha\beta}\, \lambda^{\alpha} \tag{9.13}$$

is sometimes called the *direction moment* and from (9.13) we get

$$a^{\gamma\beta}\, \lambda_{\beta} = a^{\gamma\beta} a_{\alpha\beta}\, \lambda^{\alpha} = \delta^{\gamma}_{\alpha}\, \lambda^{\alpha} = \lambda^{\gamma} \text{ and } a^{\gamma\beta}\, \lambda_{\beta}\, \lambda_{\gamma} = \lambda^{\gamma}\, \lambda_{\gamma} = |\lambda|^2, \tag{9.14}$$

so that

$$\lambda^{\alpha}\, \lambda_{\alpha} = a_{\alpha\beta}\, \lambda^{\alpha}\, \lambda^{\beta}.$$

If we interchange the dummy index β and γ in (9.14), we get $a^{\beta\gamma}\ \lambda_\gamma\ \lambda_\beta = |\lambda|^2$
Therefore, $a^{\gamma\beta}$ is a symmetric tensor.

9.4 Angle Between Two Intersecting Curves on a Surface

We have to find the angle between two intersecting curves on a surface. The equation of curve C drawn on the surface S can be written in the form

$$C\ : u^\alpha = u^\alpha(t).$$

Since the $u^\alpha(t)$ are assumed to be class C^2, curve C has a continuously turning tangent. Let C_1 and C_2 be two such curves intersecting at point *P of* S, as shown in Figure (9.2). If λ^α and μ^α are two unit vectors in the direction of the tangents to C_1 and C_2 at P, respectively, then $\lambda^\alpha = \frac{d_1u^\alpha}{ds_1}, \mu^\alpha = \frac{d_2\mu^\alpha}{ds_2}$, where s_1 and s_2 denote the arc length of C_1 and C_2 respectively. We denote the direction cosine of the tangent lines to C_1 and C_2 at P by ξ^i *and* η^i, respectively. The cosine of angle θ between C_1 and C_2 is

$$cos\theta = \xi^i\eta^i, \tag{9.15}$$

where $\xi^i = \frac{\partial y^i}{\partial u^\alpha}\frac{d_1u^\alpha}{ds_1}, \eta^i = \frac{\partial y^i}{\partial u^\beta}\frac{du^\beta}{ds_2}$ *and*

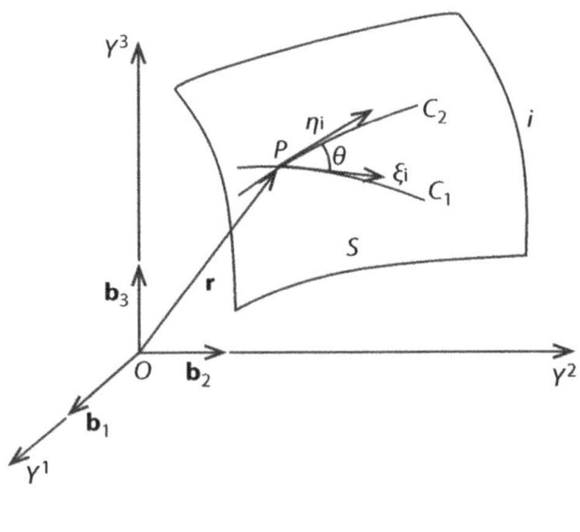

Figure 9.2

$$\xi^i = \frac{\partial y^i}{\partial u^\alpha}\lambda^\alpha, \eta^i = \frac{\partial y^i}{\partial u^\beta}\mu^\beta \tag{9.16}$$

Putting these values in (9.15), we get

$$cos\theta = \frac{\partial y^i}{\partial u^\alpha}\frac{\partial y^i}{\partial u^\beta}\mu^\beta\lambda^\alpha$$

and since $a_{\alpha\beta} = \frac{\partial y^i}{\partial u^\alpha}\frac{\partial y^i}{\partial u^\beta}$,

Therefore,

$$cos\theta = a_{\alpha\beta}\mu^\beta\lambda^\alpha, \tag{9.17}$$

If curves C_1 and C_2 are orthogonal,

$$\theta = \frac{\pi}{2}, \Rightarrow cos\theta = 0.$$

From (9.17), we get

$$a_{\alpha\beta}\lambda^\alpha\mu^\beta = 0 \tag{9.18}$$

Example 9.4.1. The coordinate curves will form an orthogonal net if a_{12} = 0 at every point of the surface.
Solution: The surface vectors λ^α and μ^β are taken along the coordinate curves, that is, if λ^α and μ^β are orthogonal, then

$$\lambda^1 = \frac{1}{\sqrt{a_{11}}}, \lambda^2 = 0 \text{ and } \mu^1 = 0, \ \mu^2 = \frac{1}{\sqrt{a_{22}}}.$$

We know $a_{\alpha\beta}\lambda^\alpha\mu^\beta = 0$ and substituting these values, we get

$$a_{11}\mu^1\lambda^1 + a_{22}\mu^2\lambda^2 + a_{12}\lambda^1\mu^2 + a_{21}\lambda^2\mu^1 = 0$$

or $0 + 0 + a_{12}\frac{1}{\sqrt{a_{11}}}\frac{1}{\sqrt{a_{22}}} + 0 = 0$

$$\therefore\ a_{12} = 0$$

Conversely, it can be easily shown.

Example 9.4.2. If θ is the angle between the parametric curves, show that $cos\theta = \frac{a_{12}}{\sqrt{a_{11}a_{22}}}$ and $sin\theta = \frac{a}{\sqrt{a_{11}a_{22}}}$.

Solution: We know $cos\theta = a_{\alpha\beta}\lambda^{\alpha}\mu^{\beta} = a_{\alpha\beta}\delta_1^{\alpha}\delta_2^{\beta}\frac{1}{\sqrt{a_{11}a_{22}}} = \frac{a_{12}}{\sqrt{a_{11}a_{22}}}$

$$sin^2\theta = 1 - cos^2\theta = 1 - \frac{a^2_{12}}{a_{11}a_{22}} = \frac{a_{11}a_{22} - a^2_{12}}{a_{11}a_{22}} = \frac{a}{a_{11}a_{22}}$$

$$\therefore\ sin\theta = \frac{a}{\sqrt{a_{11}a_{22}}}.$$

9.4.1 Pictorial Interpretation

Here, $\boldsymbol{r}$ denotes the position vector of any point P on the surface and *if* $\boldsymbol{b}_i$ are the unit vectors directed along the orthogonal coordinate axes Y, then the equation $y^i = y^i(u^1,u^2)$ of the surface S can be written in vector form, as shown in Figure (9.3).

$$\boldsymbol{r}(u^1,u^2) = \boldsymbol{b}_i y^i(u^1,u^2)$$

We can write

$$ds^2 = dr.dr = \frac{\partial \boldsymbol{r}}{\partial u^{\alpha}}\frac{\partial \boldsymbol{r}}{\partial u^{\alpha}}du^{\alpha}du^{\alpha}$$

$$= a_{\alpha\beta}\, du^{\alpha}\, du^{\alpha},$$

where $a_{\alpha\beta} = \frac{\partial \boldsymbol{r}}{\partial u^{\alpha}}\frac{\partial \boldsymbol{r}}{\partial u^{\alpha}}$.

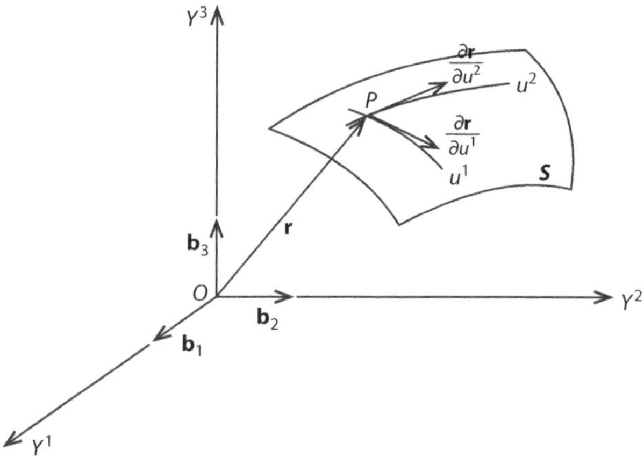

Figure 9.3

Putting $\frac{\partial \boldsymbol{r}}{\partial u^\alpha} = \boldsymbol{a}_\alpha$ where $\boldsymbol{a}_1$ *and* $\boldsymbol{a}_2$ are tangent vectors to the coordinate curves, we get $\boldsymbol{a}_{11} = \boldsymbol{a}_1.a_1, a_{12} = \boldsymbol{a}_1.\boldsymbol{a}_2, a_{22} = \boldsymbol{a}_2.\boldsymbol{a}_2$.

The space components of $\boldsymbol{a}_1$ *and* $\boldsymbol{a}_2$ are ξ^i *and* η^i, respectively

We can define an element area dσ of the surface S by the formula

$$d\sigma = |\boldsymbol{a}_1 \times \boldsymbol{a}_2| du^1 du^2$$

$$= \sqrt{a_{11}a_{11} - a_{12}{}^2} du^1 du^2 = \sqrt{a} du^1 du^2.$$

In a two-dimensional manifold, the skew-symmetric e-systems can be defined by the formulas

$$e_{11} = e_{22} = e^{11} = e^{22} = 0 \; and \; e^{12} = -e^{21} = e_{12} = -e_{21} = 1$$

and since these systems are relative tensors, the expression $\epsilon^{\alpha\beta} = \frac{1}{\sqrt{a}} e^{\alpha\beta}$ $and \epsilon_{\alpha\beta} = \sqrt{a} e_{\alpha\beta}$ are absolute tensors. Using $\epsilon-$ symbols, let θ be the angle between the unit vectors λ^α and μ^α, then

$$Sin^2\theta = 1 - cos^2\theta = (\xi^i\xi^i)(\eta^i\eta^i) - (\xi^i\eta^i)^2$$

$$= \begin{vmatrix} \xi^1 & \eta^1 \\ \xi^2 & \eta^2 \end{vmatrix}^2 + \begin{vmatrix} \xi^1 & \eta^1 \\ \xi^3 & \eta^3 \end{vmatrix}^2 + \begin{vmatrix} \xi^2 & \eta^2 \\ \xi^3 & \eta^3 \end{vmatrix}^2$$

$$= (J_1^2 + J_2^2 + J_3^2) \begin{vmatrix} \lambda^1 & \mu^1 \\ \lambda^2 & \mu^2 \end{vmatrix}^2 = a(\lambda^1\mu^2 - \lambda^2\mu^1)^2$$

$$sin\theta = \sqrt{a}(\lambda^1\mu^2 - \lambda^2\mu^1) = \sqrt{a}e_{\alpha\beta}\lambda^\alpha\mu^\beta,$$

which is numerically equal to the area of the parallelogram constructed by unit vectors λ^α *and* μ^α.

It follows from the result that a necessary and sufficient condition for the orthogonality of two surface unit vectors λ^α *and* μ^α is $|\epsilon_{\alpha\beta}\lambda^\alpha\mu^\beta| = 1$.

Example 9.4.3. Prove that the parametric curves on a surface given by $x^1 = a \sin u \cos v$, $x^2 = a \sin u \sin v$, *and* $x^3 = a\cos u$ form an orthogonal system.
Solution: Here, the symmetric covariant tensors $\alpha_{\alpha\beta}$ are

$$a_{11} = \left(\frac{\partial x^1}{\partial u}\right)^2 + \left(\frac{\partial x^2}{\partial u}\right)^2 + \left(\frac{\partial x^3}{\partial u}\right)^2 = (a \cos u \cos v)^2 + (a \cos u \sin v)^2$$
$$+ (-a \sin u)^2$$
$$= a^2$$

$$a_{22} = \left(\frac{\partial x^1}{\partial v}\right)^2 + \left(\frac{\partial x^2}{\partial v}\right)^2 + \left(\frac{\partial x^3}{\partial v}\right)^2 = (-a \sin u \sin v)^2 + (a \sin u \cos v)^2$$
$$= a^2 \sin^2 u$$

$$a_{12} = \frac{\partial x^1}{\partial u}\frac{\partial x^1}{\partial v} + \frac{\partial x^2}{\partial u}\frac{\partial x^2}{\partial v} + \frac{\partial x^3}{\partial u}\frac{\partial x^3}{\partial v}$$

$$= (a \cos u \cos v)(-a \sin u \sin v) + (a \cos u \sin v)(a \sin u \cos v) = 0.$$

Here, $a_{12} = 0$, implying that the coordinate curves are orthogonal.
Let λ^α and μ^β be taken along the parametric curves.
For u–curve, $dv = 0$ and for v– curve, $du = 0$.
Thus, unit vector λ_1^α along the u–curve is

$$\lambda_1^\alpha = \left(\frac{du}{ds_1}, \frac{dv}{ds_2}\right) = \left(\frac{1}{\sqrt{a_{11}}}, 0\right) = \left(\frac{1}{a}, 0\right)$$

$$\lambda_2^\beta = \left(\frac{du}{ds_2}, \frac{dv}{ds_2}\right) = \left(0, \frac{1}{\sqrt{a_{22}}},\right) = \left(0, \frac{1}{a \sin u}\right)$$

$$\therefore \cos\theta = a_{\alpha\beta}\lambda_1^\alpha \lambda_2^\beta = 0$$

$$\Rightarrow \cos\theta = 0$$

$$\Rightarrow \theta = \frac{\pi}{2}.$$

Thus, the given parametric curves on a surface form an orthogonal system.

9.5 Geodesic in R^n

We know that if a Riemannian Space V_n is in a Euclidean Space, E_n, a coordinate system exists in which the components $g_{ij}(i,j = 1,2,..n)$ of the metric tensor are constant throughout the space. In this case, all the Christoffel symbols vanish.

If V_n is not Euclidean, then the Christoffel symbols do not vanish at all points of V_n, but it is possible to find a coordinate system. In fact infinitely many Christoffel symbols are at a given point P of V_n. Such coordinates are *geodesic coordinates* for that particular point, P. Point P is called the *pole or origin* of the geodesic coordinate system.

We are now in a position to discuss the problem of finding curves of minimum length joining a pair of given points on the surface. Such a curve is called a *geodesic* through P and Q. We are going to deduce the equation of geodesic for the case of the n-dimensional Riemannian manifolds.

Let the metric properties of the n–dimensional manifold R_n be determined by

$$ds^2 = g_{ij}\, dx^i\, dx^j\,,\ (i,j = 1,2,3) \tag{9.19}$$

The length of a curve C, is represented in R_n by equations

$$C: x^i = x^i(t)\,, \quad t_1 \le t \le t_2,$$

given by

$$s = \int_{t_1}^{t_2} \sqrt{g_{\alpha\beta}\dot{x}^{\alpha}\dot{x}^{\beta}}\, dt, \quad (\alpha,\beta = 1,\dots n) \tag{9.20}$$

The extremals of the function of (9.20) are geodesics in R_n.
Let

$$F = \sqrt{g_{\alpha\beta}\dot{x}^{\alpha}\dot{x}^{\beta}} \tag{9.21}$$

and to form Euler's equations, we compute F_{x^i} and $F_{\dot{x}^i}$.

$$F_{x^j} = \frac{1}{2}\left(g_{\alpha\beta}\dot{x}^{\alpha}\dot{x}^{\beta}\right)^{\frac{-1}{2}} \frac{\partial g_{\alpha\beta}}{\partial x^j}\dot{x}^{\alpha}\dot{x}^{\beta},$$

$$F_{\dot{x}^j} = \left(g_{\alpha\beta}\dot{x}^{\alpha}\dot{x}^{\beta}\right)^{\frac{-1}{2}} g_{\alpha j}\dot{x}^{\alpha}$$

Substituting these values in Euler's equations gives

$$\frac{d}{dt}\left[\frac{g_{\alpha j}\dot{x}^{\alpha}}{\sqrt{\left(g_{\alpha\beta}\dot{x}^{\alpha}\dot{x}^{\beta}\right)}}\right]-\frac{\frac{\partial g_{\alpha\beta}}{\partial x^{j}}\dot{x}^{\alpha}\dot{x}^{\beta}}{2\sqrt{\left(g_{\alpha\beta}\dot{x}^{\alpha}\dot{x}^{\beta}\right)}}=0 \tag{9.22}$$

Since $\frac{ds}{dt}=\sqrt{g_{\alpha\beta}\dot{x}^{\alpha}\dot{x}^{\beta}}$,
Equation (9.22) can be written in the form

$$\frac{d}{dt}\left[\frac{g_{\alpha j}\dot{x}^{\alpha}}{\frac{ds}{dt}}\right]-\frac{\frac{\partial g_{\alpha\beta}}{\partial x^{j}}\dot{x}^{\alpha}\dot{x}^{\beta}}{2\frac{ds}{dt}}=0,$$

or $$g_{\alpha j}\ddot{x}^{\alpha}+\frac{\partial g_{\alpha j}}{\partial x^{\beta}}\dot{x}^{\alpha}\dot{x}^{\beta}-\frac{1}{2}\frac{\partial g_{\alpha\beta}}{\partial x^{j}}\dot{x}^{\alpha}\dot{x}^{\beta}=\frac{\frac{g_{\alpha j}\dot{x}^{\alpha}d^{2}s}{dt^{2}}}{\frac{ds}{dt}},$$

or $$g_{\alpha j}\ddot{x}^{\alpha}+\frac{1}{2}\left(\frac{\partial g_{\alpha j}}{\partial x^{\beta}}+\frac{\partial g_{\beta j}}{\partial x^{\alpha}}-\frac{\partial g_{\alpha\beta}}{\partial x^{j}}\right)\dot{x}^{\alpha}\dot{x}^{\beta}=\frac{g_{\alpha j}\dot{x}^{\alpha}d^{2}s/dt^{2}}{\frac{ds}{dt}},$$

$$g_{\alpha j}\ddot{x}^{\alpha}+[\alpha\beta,j]\dot{x}^{\alpha}\dot{x}^{\beta}=g_{\alpha j}\dot{x}^{\alpha}\left(\frac{d^{2}s}{dt^{2}}/\frac{ds}{dt}\right) \tag{9.23}$$

These are the desired equations of geodesics.
If we chose parameter t to be the arc length *s* of the curve, we get

$$\frac{ds}{dt}=\sqrt{g_{\alpha\beta}\dot{x}^{\alpha}\dot{x}^{\beta}}=1.$$

Then, $\frac{d^{2}s}{dt^{2}}=0$ *and* from (9.23),

$$g_{\alpha j}\ddot{x}^{\alpha}+[\alpha\beta,j]\dot{x}^{\alpha}\dot{x}^{\beta}=0. \tag{9.24}$$

Here, the symbol '.' denotes the differentiation w.r.t arc parameter s. If we multiply by tensor g^{ij} and sum,

$$g^{ij}g_{\alpha j}\ddot{x}^{\alpha}+g^{ij}[\alpha\beta,j]\dot{x}^{\alpha}\dot{x}^{\beta}=0$$

$$\text{or } \delta^{i}_{\alpha}\ddot{x}^{\alpha}+\begin{Bmatrix} i \\ \alpha\ \beta \end{Bmatrix}\dot{x}^{\alpha}\dot{x}^{\beta}=0$$

$$\text{or } \ddot{x}^{i}+\begin{Bmatrix} i \\ \alpha\beta \end{Bmatrix}\dot{x}^{\alpha}\dot{x}^{\beta}=0 \quad (i=1,2,\dots n) \tag{9.25}$$

$$(\alpha,\beta=1,2,\dots n),$$

which is a simple form of *geodesics* in R_n.

We observe that these above equations are the same with equations of a straight line in E_3.

If the manifold R_n is Euclidean and a coordinate system exists in which the Christoffel symbols vanish, then in (9.25), the 2nd part of it vanishes.

Therefore,

$$\ddot{x}^{i}=\frac{d^2x^i}{ds^2}=0.$$

The general solution of the equation is

$$x^i=A^i s+B^i,$$

that is, the geodesic in E_n are straight lines.

If we assume a given surface S as a Riemannian two-dimensional manifold R_2, covered by Gaussian coordinates u^{α}, then (9.25) takes the form

$$\frac{d^2u^\gamma}{ds^2}+\begin{Bmatrix}\gamma\\ \alpha\ \beta\end{Bmatrix}\frac{du^\alpha}{ds}\frac{du^\beta}{ds}=0\quad (\alpha,\beta,,\gamma=1,2). \tag{9.26}$$

Hence, at each point of S there exists a unique geodesic with arbitrary prescribed direction $\lambda^\alpha=\frac{du^\alpha}{ds}$. Thus, if there exists a unique solution to $u^\alpha(s)$ passing through two given points on S, then the curve *of* $u^\alpha(s)$ is the curve of shortest length joining these points.

Example 9.5.1. Find the differential equation for the geodesics on an arbitrary cylinder immersed in E_3
Solution: We choose the Y^3- axis parallel to the generators of the cylinder and let the trace of the generators of the cylinder on aY^1Y^2- plane be given as (Figure (9.4)).

$$C: y^1=\phi(\sigma),$$

$$y^2=\psi(\sigma),$$

where $\boldsymbol{\sigma}$ is the arc length of C.

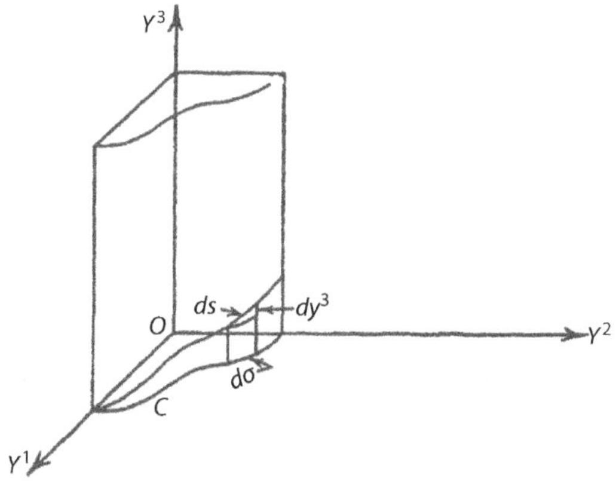

Figure 9.4

Since $(d\sigma)^2 = (dy^1)^2 + (dy^2)^2$ and the element of arc (ds) of the geodesic is

$$(ds)^2 = (d\sigma)^2 + (dy^3)^2,$$

we get $a_{11} = 1. a_{22} = 1, a_{12} = a_{21} = 0.$

We know, $\frac{d^2u^\gamma}{ds^2} + \begin{Bmatrix} \gamma \\ \alpha\ \beta \end{Bmatrix} \frac{du^\alpha}{ds}\frac{du^\beta}{ds} = 0$

for $\gamma = 1, \quad u^\gamma = u^1 = \sigma$, and the second part of above equation is zero.

$$\frac{d^2\sigma}{ds^2} = 0$$

and $\gamma = 2$, $u^2 = y^3$

$$\frac{d^2y^3}{ds^2} = 0.$$

We get $\sigma = As + B$

$$y^3 = A_1 s + B_1.$$

If $A \neq 0$, we can write $y^3 = C_1\sigma + C_2$, where C_1 *and* C_2, are arbitrary constants. Therefore, the equations of the geodesics are

$$y^1 = \phi(\sigma),$$

$$y^2 = \psi(\sigma)$$

$$y^3 = C_1\sigma + C_2.$$

Hence, the curve is a helix whose pitch is determined by C_1 and C_2 determining the origin for the arc parameter σ.

Example 9.5.2. Find the differential equation for geodesic in spherical coordinates:

$$y^1 = acosu^1cosu^2$$

$$y^2 = acosu^1sinu^2$$

$$y^3 = asinu^1.$$

Solution: Here, $(ds)^2 = a^2(\mathrm{d}u^1)^2 + a^2(cosu^1)^2(\mathrm{d}u^2)^2$

$$a_{11} = a^2,\ a_{22} = a^2(cosu^1)^2,\ a_{12} = a_{21} = 0.$$

$$a = \begin{vmatrix} a_{11} & a_{12} \\ a_{21} & a_{22} \end{vmatrix} = \begin{vmatrix} a^2 & 0 \\ 0 & a^2(cosu^1)^2 \end{vmatrix} = a^4(cosu^1)^2,$$

$$a^{11} = \frac{a^2(cosu^1)^2}{a^4(cosu^1)^2} = \frac{1}{a^2}, a^{22} = \frac{a^2}{a^4(cosu^1)^2} = \frac{1}{a^2(cosu^1)^2}, a^{12} = a^{21} = 0,$$

and $s = \int_{u_0^1}^{u_1^1} \sqrt{(1+(cos^2u^1(\dot{u}^2)^2}\, du^1.$

We know, $\frac{d^2u^\gamma}{ds^2} + \begin{Bmatrix} \gamma \\ \alpha\ \beta \end{Bmatrix} \frac{du^\alpha}{ds}\frac{du^\beta}{ds} = 0.$

For $\gamma = 1$,

$$\frac{d^2u^1}{ds^2} + \begin{Bmatrix} 1 \\ 2\ 2 \end{Bmatrix} \frac{du^2}{ds}\frac{du^2}{ds} + \begin{Bmatrix} 1 \\ 1\ 2 \end{Bmatrix} \frac{du^1}{ds}\frac{du^2}{ds} + \begin{Bmatrix} 1 \\ 1\ 1 \end{Bmatrix} \frac{du^1}{ds}\frac{du^1}{ds} = 0$$

Here,

$$\begin{Bmatrix} 1 \\ 2\ 2 \end{Bmatrix} = a^{11}[22,1] = \frac{1}{a^2}\left(-\frac{1}{2}\frac{\partial a_{22}}{\partial u^1}\right) = -\frac{1}{2a^2}2a^2cosu^1sinu^1 = -cosu^1sinu^1$$

$$\begin{Bmatrix} 1 \\ 1\ 2 \end{Bmatrix} = a^{11}[12,1] = 0, \begin{Bmatrix} 1 \\ 1\ 1 \end{Bmatrix} = a^{11}[11,1] = 0$$

$$\therefore \frac{d^2u^1}{ds^2}+\begin{Bmatrix}1\\2\ 2\end{Bmatrix}\frac{du^2}{ds}\frac{du^2}{ds}+\begin{Bmatrix}1\\2\ 2\end{Bmatrix}\frac{du^1}{ds}\frac{du^2}{ds}+\begin{Bmatrix}1\\2\ 2\end{Bmatrix}\frac{du^1}{ds}\frac{du^1}{ds}=0,$$

$$\text{or } \frac{d^2u^1}{ds^2}+(-cosu^1 sinu^1)\left(\frac{du^2}{ds}\right)^2+0+0=0,$$

$$\text{or } \frac{d^2u^1}{ds^2}-cosu^1 sinu^1\left(\frac{du^2}{ds}\right)^2=0. \tag{i}$$

For $\gamma=2$,

$$\frac{d^2u^2}{ds^2}+\begin{Bmatrix}2\\2\ 2\end{Bmatrix}\frac{du^2}{ds}\frac{du^2}{ds}+\begin{Bmatrix}2\\1\ 2\end{Bmatrix}\frac{du^1}{ds}\frac{du^2}{ds}+\begin{Bmatrix}2\\1\ 1\end{Bmatrix}\frac{du^1}{ds}\frac{du^1}{ds}=0$$

$$\begin{Bmatrix}2\\2\ 2\end{Bmatrix}=0,\ \begin{Bmatrix}2\\1\ 2\end{Bmatrix}=a^{22}[12,2]=\frac{1}{a^2(cosu^1)^2}\frac{1}{2}.\left(-2a^2\,cosu^1 sinu^1\right)=-tanu^1$$

$$\begin{Bmatrix}2\\1\ 1\end{Bmatrix}=a^{22}[11,2]=0$$

$$\frac{d^2u^2}{ds^2}+\begin{Bmatrix}2\\2\ 2\end{Bmatrix}\frac{du^2}{ds}\frac{du^2}{ds}+\begin{Bmatrix}2\\1\ 2\end{Bmatrix}\frac{du^1}{ds}\frac{du^2}{ds}+0=0$$

$$\text{or } \frac{d^2u^2}{ds^2}-tanu^1\frac{du^1}{ds}\frac{du^2}{ds}=0 \tag{ii}$$

(i) and (ii) are required equations of the geodesic for the surface.

Example 9.5.3. In orthogonal Cartesian frame *Y*, if curve C is to lie on the sphere of radius *a* show that the geodesic are great circles.
Solution: We consider the function

$$J=\int_{t_1}^{t_2}\sqrt{\dot{x}^i\dot{x}^i}\,dt\ \ (i=1,2,3) \tag{i}$$

If C is to lie on a sphere of radius *a*, the constraining relations are in the form

$$\phi = x^i x^i - a^2 = 0, \tag{ii}$$

when the center of the sphere is at origin, the function

$$G = \sqrt{\dot{x}^i \dot{x}^i} - \lambda(t)\left(x^i x^i - a^2\right) \tag{iii}$$

Now, the differential equation for the extremal is

$$\frac{dG_{\dot{x}^i}}{dt} - G_{x^i} = 0$$

or
$$\frac{d}{dt}\frac{dx^i}{ds} - 2\lambda(t)x^i = 0\,(i = 1,2,3), \tag{iv}$$

where $ds = \sqrt{\dot{x}^i \dot{x}^i}$, $G_{\dot{x}^i} = \frac{dx^i}{ds}$ *and* $G_{x^i} = -2\lambda(t)x^i$.

Eliminating $\lambda(t)$ from two sets of equations of (iv), we get

$$x^2 d\left(\frac{dx^1}{ds}\right) - x^1 d\left(\frac{dx^2}{ds}\right) = 0$$

or
$$d\left(x^2\left(\frac{dx^1}{ds}\right) - x^1\left(\frac{dx^2}{ds}\right)\right) = 0$$

$$\therefore\; x^2\left(\frac{dx^1}{ds}\right) - x^1\left(\frac{dx^2}{ds}\right) = constant = C_1. \tag{v}$$

Similarly, eliminating of $\lambda(t)$ from last two sets of equations of (iv), we get

$$x^3\left(\frac{dx^1}{ds}\right)-x^1\left(\frac{dx^3}{ds}\right)=constant=C_2. \quad \text{(vi)}$$

From (v) and (vi), we get

$$C_1\frac{x^3dx^1-x^1dx^3}{(x^1)^2}=C_2\frac{x^2dx^1-x^1dx^2}{(x^1)^2}$$

or

$$C_1d\left(\frac{x^3}{x^1}\right)=C_2d\left(\frac{x^2}{x^1}\right).$$

Integrating both sides, we get

$$C_1\left(\frac{x^3}{x^1}\right)+C_3=C_2\left(\frac{x^2}{x^1}\right)$$

or

$$C_3x^1-C_2x^2+C_1x^3=0, \quad \text{(vii)}$$

which is the plane passing through the origin.

Equation (vii) with equation ii shows that the geodesic on the surface of a sphere are an arc of great circle.

Example 9.5.4. Find the geodesic on the surface

$$y^1 = u^1 cosu^2$$

$$y^2 = u^1 sinu^2$$

$$y^3 = 0$$

imbedded in E_3. The coordinate y^i are orthogonal Cartesian.
Solution: Here, $ds^2=(du^1)^2+(u^1)^2(du^2)^2$

$$a_{11}=1,\ a_{22}=(u^1)^2, a_{12}=a_{21}=0\ ,$$

so that $a = (u^1)^2$ and

$$a^{11} = 1, a^{22} = \frac{1}{(u^1)^2}, a^{12} = a^{21} = 0 \text{ and}$$

$$\left\{ \begin{matrix} 1 \\ 2\ 2 \end{matrix} \right\} = a^{11}[22,1] = 1.\left(-\frac{1}{2}\frac{\partial a_{22}}{\partial u^1} \right) = -\frac{1}{2}2u^1 = -u^1$$

$$\left\{ \begin{matrix} 2 \\ 1\ 2 \end{matrix} \right\} = \left\{ \begin{matrix} 2 \\ 2\ 1 \end{matrix} \right\} = a^{22}[12,2] = \frac{1}{(u^1)^2}\frac{1}{2}\frac{\partial a_{22}}{\partial u^1} = \frac{1}{(u^1)^2}\frac{1}{2}2u^1 = \frac{1}{u^1},$$

all others

$$\left\{ \begin{matrix} i \\ j\ k \end{matrix} \right\} = 0$$

The differential equations of a geodesic for the surface are

$$\frac{d^2u^\gamma}{ds^2} + \left\{ \begin{matrix} \gamma \\ \alpha\ \beta \end{matrix} \right\} \frac{du^\alpha}{ds}\frac{du^\beta}{ds} = 0 \cdot$$

For $\gamma = 1$ $\dfrac{d^2u^1}{ds^2} + \left\{ \begin{matrix} 1 \\ \alpha\ \beta \end{matrix} \right\} \dfrac{du^\alpha}{ds}\dfrac{du^\beta}{ds} = 0,$

$$\text{or } \frac{d^2u^1}{ds^2} + \left\{ \begin{matrix} 1 \\ 2\ 2 \end{matrix} \right\} \frac{du^2}{ds}\frac{du^2}{ds} = 0,$$

$$\text{or } \frac{d^2u^1}{ds^2} - u^1\left(\frac{du^2}{ds} \right)^2 = 0. \tag{i}$$

For $\gamma = 2$ $\dfrac{d^2u^2}{ds^2} + \left\{ \begin{matrix} 2 \\ \alpha\ \beta \end{matrix} \right\} \dfrac{du^\alpha}{ds}\dfrac{du^\beta}{ds} = 0,$

$$\text{or } \frac{d^2u^2}{ds^2}+2\begin{Bmatrix} 1 \\ 1\ 2 \end{Bmatrix}\frac{du^1}{ds}\frac{du^2}{ds}=0,$$

$$\text{or } \frac{d^2u^2}{ds^2}+2\frac{1}{u^1}\frac{du^1}{ds}\frac{du^2}{ds}=0 \qquad \text{(ii)}$$

(i) and (ii) are the differential equations of geodesic on the surfaces.

Example 9.5.5. Find the differential equations of the geodesic for the metric

$$ds^2=(dx^1)^2+\{(x^2)^2-(x^1)^2\}(dx^2)^2.$$

Solution: $a_{11}=1,\ a_{22}=(x^2)^2-(x^1)^2\ \ a_{12}=a_{21}=0,$

$$a=(x^2)^2-(x^1)^2$$

$$a^{11}=1, a^{22}=\frac{1}{(x^2)^2-(x^1)^2}, a^{12}=a^{21}=0.$$

Thus, $\dfrac{\partial a_{22}}{\partial x^1}=-2x^1,\ \dfrac{\partial a_{22}}{\partial x^2}=2x^2$.

The non-zero Christoffel symbols are

$$\begin{Bmatrix} 1 \\ 2\ 2 \end{Bmatrix}=a^{11}[22,1]=1.-\frac{1}{2}\frac{\partial a_{22}}{\partial x^1}=\frac{1}{2}\left(-2x^1\right)=x^1$$

$$\begin{Bmatrix} 2 \\ 1\ 2 \end{Bmatrix}=\begin{Bmatrix} 2 \\ 2\ 1 \end{Bmatrix}=\frac{1}{2}a^{22}\,[21,2]=\frac{1}{2\{(x^2)^2-(x^1)^2\}}\frac{\partial g_{22}}{\partial x^1}$$

$$=\frac{-2x^1}{2\{(x^2)^2-(x^1)^2}=\frac{-x^1}{(x^2)^2-(x^1)^2}$$

$$\begin{Bmatrix} 2 \\ 2\ 2 \end{Bmatrix} = a^{22}[22,2] = \frac{1}{2\{(x^2)^2 - (x^1)^2\}}\frac{\partial a_{22}}{\partial x^2} = \frac{x^2}{(x^2)^2 - (x^1)^2}.$$

Hence, the desired geodesic equations are

$$\ddot{x}^i + \begin{Bmatrix} i \\ \alpha\ \beta \end{Bmatrix} \dot{x}^\alpha \dot{x}^\beta = 0 \text{ are:}$$

for $\gamma = 1$

$$\frac{d^2x^1}{ds^2} + \begin{Bmatrix} 1 \\ 2\ 2 \end{Bmatrix}\left(\frac{dx^2}{ds}\right)^2 = 0$$

$$\frac{d^2x^1}{ds^2} + x^1\left(\frac{dx^2}{ds}\right)^2 = 0 \qquad \text{(i)}$$

and for $\gamma = 2$

$$\frac{d^2x^2}{ds^2} + \begin{Bmatrix} 2 \\ 2\ 2 \end{Bmatrix}\left(\frac{dx^2}{ds}\right)^2 + 2\begin{Bmatrix} 2 \\ 1\ 2 \end{Bmatrix}\frac{dx^1}{ds}\frac{dx^2}{ds} = 0$$

$$\frac{d^2x^2}{ds^2} + \frac{x^2}{(x^2)^2 - (x^1)^2}\left(\frac{dx^2}{ds}\right)^2 + \frac{-2x^1}{(x^2)^2 - (x^1)^2}\frac{dx^1}{ds}\frac{dx^2}{ds} = 0 \qquad \text{(ii)}$$

(i) and (ii) are required equations of geodesic.

9.6 Geodesic Coordinates

If Riemannian space R_n is Euclidean, a coordinate system exists in which the components g_{ij} of metric tensors are constants throughout the space, which implies that $\frac{\partial g_{ij}}{\partial x^k} = 0$. Consequently, the vanishing of $\frac{\partial g_{ij}}{\partial x^k}$ is equivalent to the vanishing of all Christoffel symbols, i.e. $[ij,k]=\frac{1}{2}\left(\frac{\partial g_{ik}}{\partial x^j}+\frac{\partial g_{jk}}{\partial x^i}-\frac{\partial g_{ij}}{\partial x^k}\right)=0$. If R_n is not Euclidean, then the Christoffel symbols do not vanish at all points in R_n, but it is possible to find infinitely many coordinate systems in which they vanish at any given point of R_n. Such coordinates systems are called *geodesic coordinate system.*

Let us consider a surface net S whose curvilinear coordinates are and also consider a point $P(u_0^1, u_0^2)$ on *S*. Let v^α be coordinates of some other net on *S*.

Then,

$$u^\alpha = u^\alpha(v^1, v^2) \quad (\alpha = 1,2). \tag{9.27}$$

The second derivative formula of the transformation of Christoffel symbols is

$$\frac{\partial^2 u^\alpha}{\partial v^\lambda \partial v^\mu} + \left\{ \begin{matrix} \alpha \\ \beta\,\gamma \end{matrix} \right\}_u \frac{\partial u^\beta}{\partial v^\lambda}\frac{\partial u^\gamma}{\partial v^\mu} = \left\{ \begin{matrix} \nu \\ \lambda\,\mu \end{matrix} \right\}_v \frac{\partial u^\alpha}{\partial v^\nu}. \tag{9.28}$$

If there exists a transformation of coordinates of (9.27) for which $\left\{ \begin{matrix} \nu \\ \lambda\,\mu \end{matrix} \right\}$ vanishes at P, then for that particular point,.

$$\frac{\partial^2 u^\alpha}{\partial v^\lambda \partial v^\mu} + \left\{ \begin{matrix} \alpha \\ \beta\,\gamma \end{matrix} \right\}_u \frac{\partial u^\beta}{\partial v^\lambda}\frac{\partial u^\gamma}{\partial v^\mu} = 0 \tag{9.29}$$

We exhibit the next solution of Equation (9.29), yielding a transformation of (9.27) to a coordinate system v^α in which Christoffel symbols vanish at *P*.

Let us take a second degree polynomial

$$u^\alpha = u_P^\alpha + v^\alpha - \frac{1}{2}\left\{ \begin{matrix} \alpha \\ \lambda\,\mu \end{matrix} \right\}_P v^\lambda v^\mu, \tag{9.30}$$

where u_P^α is the value of u^α at *P* and $\left\{ \begin{matrix} \alpha \\ \lambda\,\mu \end{matrix} \right\}_P$ are the values of the Christoffel symbols at *P*.

For verification, if Equation (9.30) satisfies (9.29), we get

$$\frac{\partial u^{\alpha}}{\partial v^{\mu}} = \delta^{\alpha}_{\mu} - \left\{ \begin{matrix} \alpha \\ \lambda\,\mu \end{matrix} \right\}_{P} v^{\lambda} \tag{9.31}$$

and

$$\frac{\partial^2 u^{\alpha}}{\partial v^{\lambda}\, \partial v^{\mu}} = -\left\{ \begin{matrix} \alpha \\ \lambda\,\mu \end{matrix} \right\}_{P}. \tag{9.32}$$

From this computation we conclude that the coordinates are geodesic at all points of an arbitrarily prescribed analytical curve.

9.7 Parallel Vectors on a Surface

The concept of parallel vector fields along a curve C imbedded in E_3 was generalized by Levi-Ciita to curves imbedded in $n-$ dimensional Riemannian manifolds.

For the usefulness of this concept we consider:

A surface S imbedded in E_3 and a curve C *on* S.

We take equations of C in the form:

$$C:\ u^{\alpha} = u^{\alpha}(t) \quad , \quad t_1 \leq t \leq t_2$$

and suppose that the metric properties of S are governed by the tensor $a_{\alpha\beta}$.

If A^{α} is a surface vector field defined along C, we can calculate the surface intrinsic derivative

$$\frac{\delta A^{\alpha}}{\delta t} = \frac{dA^{\alpha}}{dt} + \left\{ \begin{matrix} \alpha \\ \beta\,\gamma \end{matrix} \right\} A^{\beta} \frac{du^{\gamma}}{dt}. \tag{9.33}$$

It is identical with parallel vector fields along a space curve.

Accordingly, we take the differential equation

$$\frac{dA^\alpha}{dt} + \begin{Bmatrix} \alpha \\ \beta\,\gamma \end{Bmatrix} A^\beta \frac{du^\gamma}{dt} = 0 \tag{9.34}$$

If parameter t is chosen as the arc length, s, Equation (9.34) can be written as

$$\frac{dA^\alpha}{ds} + \begin{Bmatrix} \alpha \\ \beta\,\gamma \end{Bmatrix} A^\beta \frac{du^\gamma}{ds} = 0 \tag{9.35}$$

and if A^α is taken to be a unit tangent vector to C so that $A^\alpha = \frac{du^\alpha}{ds} \equiv \lambda^\alpha$, with $a_{\alpha\beta}\lambda^\alpha\lambda^\beta = 1$, then

$$\frac{d^2u^\alpha}{ds^2} + \begin{Bmatrix} \alpha \\ \beta\,\gamma \end{Bmatrix} \frac{du^\alpha}{ds}\frac{du^\gamma}{ds} = 0\,. \tag{9.36}$$

This is an equation of geodesic on S.

9.8 Isometric Surface

The properties of surface (i.e., the lengths of curves and angle between intersecting curves) with which we have been concerned, depend entirely on the study of the first fundamental quadratic form

$$ds^2 = a_{\alpha\beta}du^\alpha du^\beta. \tag{9.37}$$

These properties constitute a body of what is known as the *Intrinsic geometry of surfaces*.

We have seen that the intrinsic property of a surface depends on the metric tensor of the surface and its derivatives. For example, two surfaces, a cylinder and a cone, appear to be entirely different when viewed from the enveloping space and yet their intrinsic geometries are completely indistinguishable since the metric properties of cylinders and cones can be described by identical expressions for the square of the elements of arc.

If a coordinate system exists on each of the two surfaces such that the elements on them are characterized by the same metric coefficients, $a_{\alpha\beta}$, the surfaces are called *isometric.*

The surfaces of the cylinder and cone are isometric with the Euclidean plane since these surfaces can be rolled out or developed on the plane without changing the length of arc elements and the measurements of angles and areas.

9.8.1 Developable

A surface which is isometric with a Euclidean plane E^2 is called a *developable.*

We introduce an important scalar invariant, known as the Gaussian Curvature, which will enable us to determine the circumstances under which a given surface is *developable*, that is, isometric with a Euclidean plane.

A surface S is called a developable surface or simply a developable if the Gaussian Curvature vanishes at every point of it.

Example 9.8.1. Consider Catenoid

$$S_1 : y^1 = v^1 cos v^2$$

$$y^2 = v^1 sin v^2$$

$$y^3 = a cosh^{-1} \frac{v^1}{a}$$

obtained by revolving the catenary $y^2 = cosh\left(\frac{y^3}{a}\right)$ about *the* y^3-axis.

Show that surface S_1 is isometric with the surface of Helicoid, defined by

$$S_2 : y^1 = u^1 cos u^2$$

$$y^2 = u^1 sin u^2$$

$$y^3 = a u^2.$$

Solution: For S_1: The 1st fundamental form

$ds^2 = dy^i dy^j$ is easily found to be

$$ds^2 = a_{\alpha\beta} dv^\alpha dv^\beta = \frac{(v^1)^2}{(v^1)^2 - a^2}(dv^1)^2 + (v^1)^2 (dv^2)^2 \qquad \text{(i)}$$

$$[ds^2 = (dy^1)^2 + (dy^2)^2 + (dy^3)^2 = (dv^1)^2 + (v^1)^2(dv^2)^2 + \frac{a^2}{(v^1)^2 - a^2}(dv^1)^2$$

$$= \frac{(v^1)^2}{(v^1)^2 - a^2}(dv^1)^2 + (v^1)^2(dv^2)^2 \text{ and } v^1 = a\cosh\frac{y^3}{a},$$

$$\therefore dv^1 = \text{a sinh}\left(\frac{y^3}{a}\right)\frac{1}{a}dy^3, dy^3 = \frac{1}{\sinh\left(\frac{y^3}{a}\right)}dv^1, \sin^2\text{h}\left(\frac{y^3}{a}\right) = cos^2h\frac{y^3}{a} - 1$$

$$= \frac{(v^1)^2}{a^2} - 1 = \frac{(v^1)^2 - a^2}{a^2}],$$

so that $a_{11} = \dfrac{(v^1)^2}{(v^1)^2 - a^2}, a_{22} = (v^1)^2 \equiv a^2 + \left[(v^1)^2 - a^2\right],\ a_{12} = a_{21} = 0.$

For surface S_2

$$ds^2 = a_{\alpha\beta} dv^\alpha dv^\beta = (du^1)^2 + \left[(u^1)^2 + a^2\right](du^2)^2 \qquad \text{(ii)}$$

$$a_{11} = 1, a_{12} = 0,\ a_{22} = (u^1)^2 + a^2$$

Now, if we set in (i), $(v^1)^2 - a^2 = (u^1)^2, v^2 = u^2$ and $\left[(dv^1)^2 = \dfrac{(u^1)^2}{(v^1)^2}(du^1)^2\right]$, we obtain, (i) to become (ii).

Therefore, $ds^2 = (du^1)^2 + \left[(u^1)^2 + a^2\right](du^2)^2$.

Since this is identical with (ii), the surfaces of S_1 and S_2 are isometric.

9.9 The Riemannian–Christoffel Tensor and Gaussian Curvature

With the first fundamental quadratic form of the surface

$$ds^2 = a_{\alpha\beta} du^\alpha du^\beta, \tag{9.10}$$

we can form the Christoffel symbols with respect to this surface and the corresponding Riemann tensor

$$R_{\alpha\beta\gamma\delta} = \begin{vmatrix} \dfrac{\partial}{\partial x^\gamma} & \dfrac{\partial}{\partial x^\delta} \\ [\beta\gamma,\alpha] & [\beta\delta,\alpha] \end{vmatrix} + \begin{vmatrix} \left\{ \begin{matrix} \lambda \\ \beta\,\gamma \end{matrix} \right\} & \left\{ \begin{matrix} \lambda \\ \beta\,\delta \end{matrix} \right\} \\ [\alpha\gamma,\lambda] & [\alpha\delta,\lambda] \end{vmatrix}.$$

This tensor is skew-symmetric in the first two and last two indices, so that, for the surface, S

$$R_{\alpha\alpha\beta\gamma} = R_{\alpha\beta\gamma\gamma} = 0, \; R_{1212} = R_{2121} = -\, R_{2112} = -R_{1221}. \tag{9.38}$$

Hence, all non-vanishing components of Riemann-Christoffel tensors are equal to R_{1212} or its negative.

We define the quantity κ

$$\kappa = \frac{R_{1212}}{a}, \text{ where } a = |a_{\alpha\beta}| \tag{9.39}$$

and it is called *Gaussian Curvature or the total Curvature* of the surface S.

If we introduce the two dimensional $\epsilon-$ tensors,

$$\epsilon_{\alpha\beta} = \sqrt{a}\; e_{\alpha\beta} \qquad \epsilon^{\alpha\beta} = \frac{1}{\sqrt{a}} e^{\alpha\beta}$$

we can write (9.39) using (9.38) as

$$R_{1212} = a\kappa = \sqrt{a}\; e_{12} \sqrt{a}\; e_{12} \kappa = \epsilon_{12}\epsilon_{12}\kappa$$

and generalize it as

$$R_{\alpha\beta\gamma\delta} = \epsilon_{\alpha\beta}\epsilon_{\gamma\delta}\kappa. \tag{9.40}$$

Since $\epsilon^{\alpha\beta}\epsilon_{\alpha\beta} = 2,$

$$[\epsilon^{\alpha\beta}\epsilon_{\alpha\beta} = \epsilon^{11}\epsilon_{11} + \epsilon^{22}\epsilon_{22} + \epsilon^{12}\epsilon_{12} + \epsilon^{21}\epsilon_{21} = 0 + 0 + \sqrt{a}\frac{1}{\sqrt{a}}1.1$$

$$+\sqrt{a}\frac{1}{\sqrt{a}}(-1).(-1) = 2]$$

we can solve (9.40) for κ and obtain

$$\epsilon^{\alpha\beta}\epsilon^{\gamma\delta}R_{\alpha\beta\gamma\delta} = \epsilon^{\alpha\beta}\epsilon^{\gamma\delta}\kappa\epsilon_{\alpha\beta}\epsilon_{\gamma\delta} = 4\kappa$$

$$\therefore \ \kappa = \frac{1}{4}R_{\alpha\beta\gamma\delta}\epsilon^{\alpha\beta}\epsilon^{\gamma\delta}. \tag{9.41}$$

These equations show that Gaussian Curvature is an invariant.

Theorem 9.9.1. A necessary and sufficient condition that a surface S is isometric with a Euclidean plane is that the Riemannian tensor (or Gaussian curvature of S) be identically zero.
Proof: When a surface S is isometric with the Euclidean plane, there exists on S a coordinate system with respect to which $a_{11} = a_{22} = 1, a_{12} = 0$. It is obvious that in this case $R_{\alpha\beta\gamma\delta} = 0$ in this particular coordinate system and since $R_{\alpha\beta\gamma\delta}$ is a tensor, it must vanish in every coordinate system.

Conversely, if the Riemannian tensor vanishes at all points of the surface, Theorem of §5.6.1 states that there exists a coordinate system on the surface such that $a_{11} = a_{22} = 1, a_{12} = 0$.

9.9.1 Einstein Curvature

We consider an invariant

$$R = a^{\mu\nu,}R_{\mu\nu,} \tag{9.42}$$

where

$$R_{\mu\nu} = R^{\alpha}_{\mu\nu\alpha} = a^{\lambda\alpha}R_{\lambda\mu\nu\alpha} \tag{9.43}$$

are the Ricci tensors (introduced in § 5.4).

If we multiply (9.43) by $a^{\mu\nu}$

$$R = a^{\mu\nu} R_{\mu\nu} = a^{\mu\nu} a^{\lambda\alpha} R_{\lambda\mu\nu\alpha} \tag{9.44}$$

Using (9.38), (9.44) is equivalent to

$$R = -2R_{1212}\left(a^{11}a^{22} - a^{12}a^{12}\right).$$

Since $a^{11} = \frac{a_{22}}{a}, a^{22} = \frac{a_{11}}{a}, a^{12} = -\frac{a_{12}}{a}$,

$$\therefore\ R = -2R_{1212}\frac{a}{a^2} = -2\frac{R_{1212}}{a} \tag{9.45}$$

Comparing this with $\kappa = \frac{R_{1212}}{a}$, we get

$$\kappa = -\frac{R}{2}$$

$$\therefore\ R = -2\kappa$$

The invariant R is called the *Einstein Curvature* of S.

A surface which holds

$$R_{\alpha\beta\gamma\delta} = \rho\left(a_{\alpha\delta}a_{\beta\gamma} - a_{\alpha\gamma}a_{\beta\delta}\right),$$

where ρ is scalar, is called a *surface of constant curvature.*

Geometrical properties which are expressible in terms of the first fundamental form may be called *intrinsic properties.* Since only metric coefficients $a_{\alpha\beta}$ are involved in this definition, the properties described by κ are intrinsic properties of the surface, S.

Example 9.9.1. If the coordinate system is orthogonal, show that

$$\kappa = -\frac{1}{2\sqrt{a}}\left[\frac{\partial}{\partial u^1}\left(\frac{1}{\sqrt{a}}\frac{\partial a_{22}}{\partial u^1}\right) + \frac{\partial}{\partial u^2}\left(\frac{1}{\sqrt{a}}\frac{\partial a_{11}}{\partial u^2}\right)\right]$$

Solution: If the system is orthogonal,

$$a_{12} = a_{21} = 0 \qquad \therefore\ a = \begin{vmatrix} a_{11} & 0 \\ 0 & a_{22} \end{vmatrix} = a_{11}a_{22}$$

$$\therefore\ a^{11} = \frac{1}{a_{11}},\ a^{22} = \frac{1}{a_{22}}.$$

We know

$$R_{\alpha\beta\gamma\delta} = \begin{vmatrix} \dfrac{\partial}{\partial x^{\gamma}} & \dfrac{\partial}{\partial x^{\delta}} \\ [\beta\gamma,\alpha] & [\beta\delta,\alpha] \end{vmatrix} + \begin{vmatrix} \left\{ \begin{matrix} \lambda \\ \beta\,\gamma \end{matrix} \right\} & \left\{ \begin{matrix} \lambda \\ \beta\,\delta \end{matrix} \right\} \\ [\alpha\gamma,\lambda] & [\alpha\delta,\lambda] \end{vmatrix}$$

$\therefore$

$$R_{1212} = \begin{vmatrix} \dfrac{\partial}{\partial u^{1}} & \dfrac{\partial}{\partial u^{2}} \\ [21,1] & [22,1] \end{vmatrix} + \begin{vmatrix} \left\{ \begin{matrix} \lambda \\ 2\,1 \end{matrix} \right\} & \left\{ \begin{matrix} \lambda \\ 2\,2 \end{matrix} \right\} \\ [11,\lambda] & [12,\lambda] \end{vmatrix}$$

$$[11,1] = \frac{1}{2}\frac{\partial a_{11}}{\partial u^{1}},\ [12,1] = \frac{1}{2}\frac{\partial a_{11}}{\partial u^{2}},\ [11,2] = -\frac{1}{2}\frac{\partial a_{11}}{\partial u^{2}},\ [12,2] = \frac{1}{2}\frac{\partial a_{22}}{\partial u^{1}},$$

$$[21,1] = \frac{1}{2}\frac{\partial a_{11}}{\partial u^{2}},\ [22,1] = -\frac{1}{2}\frac{\partial a_{22}}{\partial u^{1}},\ [22,2] = \frac{1}{2}\frac{\partial a_{22}}{\partial u^{2}}$$

$$\left\{ \begin{matrix} \lambda \\ 2\,1 \end{matrix} \right\} : \left\{ \begin{matrix} 1 \\ 2\,1 \end{matrix} \right\} = a^{11}[21,1] = \frac{1}{2a_{11}}\frac{\partial a_{11}}{\partial u^{2}}\ and \left\{ \begin{matrix} 2 \\ 2\,1 \end{matrix} \right\} = a^{22}[21,2] = \frac{1}{2a_{22}}\frac{\partial a_{22}}{\partial u^{1}}$$

$$\left\{ \begin{matrix} \lambda \\ 2\,2 \end{matrix} \right\} : \left\{ \begin{matrix} 1 \\ 2\,2 \end{matrix} \right\} = a^{11}[22,1] = -\frac{1}{2a_{11}}\frac{\partial a_{22}}{\partial u^{1}} \text{ and } \left\{ \begin{matrix} 2 \\ 2\,2 \end{matrix} \right\} = a^{22}[22,2] = \frac{1}{2a_{22}}\frac{\partial a_{22}}{\partial u^{2}}$$

$$\therefore R_{1212} = \frac{\partial}{\partial u^1}[22,1] - \frac{\partial}{\partial u^2}[21,1] + \left\{ \begin{matrix} \lambda \\ 2\ 1 \end{matrix} \right\}[12,\lambda] - \left\{ \begin{matrix} \lambda \\ 2\ 2 \end{matrix} \right\}[11,\lambda]$$

$$= \frac{\partial}{\partial u^1}[22,1] - \frac{\partial}{\partial u^2}[21,1] + \left\{ \begin{matrix} 1 \\ 2\ 1 \end{matrix} \right\}[12,1] + \left\{ \begin{matrix} 2 \\ 2\ 1 \end{matrix} \right\}[12,2]$$
$$- \left\{ \begin{matrix} 1 \\ 2\ 2 \end{matrix} \right\}[11,1] - \left\{ \begin{matrix} 2 \\ 2\ 2 \end{matrix} \right\}[11,2]$$

$$= -\frac{1}{2}\frac{\partial}{\partial u^1}\frac{\partial a_{22}}{\partial u^1} - \frac{1}{2}\frac{\partial}{\partial u^2}\frac{\partial a_{11}}{\partial u^2} + \frac{1}{2a_{11}}\frac{\partial a_{11}}{\partial u^2}\frac{1}{2}\frac{\partial a_{11}}{\partial u^2} + \frac{1}{2a_{22}}\frac{\partial a_{22}}{\partial u^1}\frac{1}{2}\frac{\partial a_{22}}{\partial u^1}$$
$$+ \frac{1}{2a_{11}}\frac{\partial a_{22}}{\partial u^1}\frac{1}{2}\frac{\partial a_{11}}{\partial u^1} + \frac{1}{2a_{22}}\frac{\partial a_{22}}{\partial u^2}\frac{1}{2}\frac{\partial a_{11}}{\partial u^2}$$

$$= -\frac{1}{2}\left[\frac{\partial^2 a_{22}}{(\partial u^1)^2} + \frac{\partial^2 a_{11}}{(\partial u^2)^2}\right] + \frac{1}{4a_{11}}\left[\left(\frac{\partial a_{11}}{\partial u^2}\right)^2 + \frac{\partial a_{22}}{\partial u^1}\frac{\partial a_{11}}{\partial u^1}\right]$$
$$+ \frac{1}{4a_{22}}\left[\left(\frac{\partial a_{22}}{\partial u^1}\right)^2 + \frac{\partial a_{22}}{\partial u^2}\frac{\partial a_{11}}{\partial u^2}\right]$$

$$= -\frac{1}{2}\sqrt{a_{11}a_{22}}\left[\frac{\partial}{\partial u^1}\left(\frac{1}{\sqrt{a_{11}a_{22}}}\frac{\partial a_{22}}{\partial u^1}\right) + \frac{\partial}{\partial u^2}\left(\frac{1}{\sqrt{a_{11}a_{22}}}\frac{\partial a_{11}}{\partial u^2}\right)\right]$$

$$= -\frac{1}{2}\sqrt{a}\left[\frac{\partial}{\partial u^1}\left(\frac{1}{\sqrt{a}}\frac{\partial a_{22}}{\partial u^1}\right) + \frac{\partial}{\partial u^2}\left(\frac{1}{\sqrt{a}}\frac{\partial a_{11}}{\partial u^2}\right)\right] \text{ [since } a = a_{11}\, a_{22}]$$

$$= a\left(-\frac{1}{2}\frac{1}{\sqrt{a}}\left[\frac{\partial}{\partial u^1}\left(\frac{1}{\sqrt{a}}\frac{\partial a_{22}}{\partial u^1}\right) + \frac{\partial}{\partial u^2}\left(\frac{1}{\sqrt{a}}\frac{\partial a_{11}}{\partial u^2}\right)\right]\right).$$

Therefore, the Gaussian Curvature is

$$\kappa = \frac{R_{1212}}{a} = -\frac{1}{2}\frac{1}{\sqrt{a}}\left[\frac{\partial}{\partial u^1}\left(\frac{1}{\sqrt{a}}\frac{\partial a_{22}}{\partial u^1}\right) + \frac{\partial}{\partial u^2}\left(\frac{1}{\sqrt{a}}\frac{\partial a_{11}}{\partial u^2}\right)\right].$$

Example 9.9.2. Calculate the total curvature (Gaussian) of the manifold whose quadratic form is

$$ds^2 = a^2 \sin^2 u^1 (du^2)^2 + a^2 (du^1)^2.$$

Solution: Here, $a_{11} = a^2, a_{22} = a^2 sin^2 u^1, a_{12} = a_{21} = 0$

$$\therefore\ a = \begin{vmatrix} a_{11} & 0 \\ 0 & a_{22} \end{vmatrix} = a_{11}a_{22} = a^4 sin^2 u^1$$

$$a^{11} = \frac{a^2 sin^2 u^1}{a^4 sin^2 u^1} = \frac{1}{a^2},\quad a^{22} = \frac{1}{a_{22}} = \frac{1}{a^2 sin^2 u^1}$$

$$R_{1212} = a_{1\alpha} R^{\alpha}_{212} = a_{11} R^1_{212} + a_{12} R^2_{212} = a_{11} R^1_{212}.$$

We know

$$R^i_{jkl} \equiv \begin{vmatrix} \dfrac{\partial}{\partial x^k} & \dfrac{\partial}{\partial x^l} \\ \begin{Bmatrix} i \\ j\ k \end{Bmatrix} & \begin{Bmatrix} i \\ j\ l \end{Bmatrix} \end{vmatrix} + \begin{vmatrix} \begin{Bmatrix} i \\ \alpha\ k \end{Bmatrix} & \begin{Bmatrix} i \\ \alpha\ l \end{Bmatrix} \\ \begin{Bmatrix} \alpha \\ j\ k \end{Bmatrix} & \begin{Bmatrix} \alpha \\ j\ l \end{Bmatrix} \end{vmatrix}$$

$$R_{1212} = a_{11} R^1_{212} = a_{11}\left[\frac{\partial}{\partial x^1}\begin{Bmatrix} 1 \\ 2\ 2 \end{Bmatrix} - \frac{\partial}{\partial x^2}\begin{Bmatrix} 1 \\ 2\ 1 \end{Bmatrix}\right.$$
$$\left. + \begin{Bmatrix} 1 \\ \alpha\ 2 \end{Bmatrix}\begin{Bmatrix} \alpha \\ 2\ 2 \end{Bmatrix} - \begin{Bmatrix} 1 \\ 2\ 2 \end{Bmatrix}\begin{Bmatrix} 1 \\ \alpha\ 2 \end{Bmatrix}\begin{Bmatrix} \alpha \\ 2\ 1 \end{Bmatrix}\right]$$

Here,

$$\begin{Bmatrix} 1 \\ 2\ 2 \end{Bmatrix} = a^{11}[22,1] = \frac{1}{a^2}\left(-\frac{1}{2}\frac{\partial a_{22}}{\partial u^1}\right) = -\frac{1}{2a^2} a^2 2 sinu^1 cosu^1 = -\frac{1}{2} sin 2u^1$$

$$\begin{Bmatrix} 1 \\ 2\ 1 \end{Bmatrix} = a^{11}[21,1] = \frac{1}{a^2}\left(\frac{1}{2}\frac{\partial a_{11}}{\partial u^2}\right) = 0,$$

$$\begin{Bmatrix} 2 \\ 2\ 1 \end{Bmatrix} = a^{22}\left[21,2\right] = \frac{1}{a^2 sin^2 u^1}\left(\frac{1}{2}\frac{\partial a_{22}}{\partial u^1}\right) = \frac{1}{a^2 sin^2 u^1}\frac{1}{2}a^2 2 sinu^1 cosu^1$$

$$= cotu^1$$

$$\begin{Bmatrix} 2 \\ 2\ 2 \end{Bmatrix} = 0$$

$$R_{1212} = a_{11}\left[\frac{\partial}{\partial u^1}\left(-\frac{1}{2}sin2u^1\right) + \begin{Bmatrix} 1 \\ \alpha\ 1 \end{Bmatrix}\begin{Bmatrix} \alpha \\ 2\ 2 \end{Bmatrix} - \begin{Bmatrix} 1 \\ \alpha\ 2 \end{Bmatrix}\begin{Bmatrix} 1 \\ 2\ 1 \end{Bmatrix}\right]$$

$$= a^2\left[\frac{\partial}{\partial u^1}\left(-\frac{1}{2}sin2u^1\right) + \begin{Bmatrix} 1 \\ 1\ 1 \end{Bmatrix}\begin{Bmatrix} 1 \\ 2\ 2 \end{Bmatrix} - \begin{Bmatrix} 1 \\ 1\ 2 \end{Bmatrix}\begin{Bmatrix} 1 \\ 2\ 1 \end{Bmatrix}\right.$$

$$\left. + \begin{Bmatrix} 1 \\ 2\ 1 \end{Bmatrix}\begin{Bmatrix} 2 \\ 2\ 2 \end{Bmatrix} - \begin{Bmatrix} 1 \\ 2\ 2 \end{Bmatrix}\begin{Bmatrix} 2 \\ 2\ 1 \end{Bmatrix}\right]$$

$$= a^2\left[-\frac{1}{2}2cos2u^1 + 0 - 0 + 0 + \frac{1}{2}sin2u^1 cotu^1\right]$$

$$= a^2\left[-cos2u^1 + sinu^1 cosu^1 \frac{cosu^1}{sinu^1}\right]$$

$$= a^2\left[sin^2u^1 - cos^2u^1 + cos^2u^1\right] = a^2 sin^2 u^1$$

$$\kappa = \frac{R_{1212}}{a^4 sin^2 u^1} = \frac{a^2 sin^2 u^1}{a^4 sin^2 u^1} = \frac{1}{a^2}.$$

Example 9.9.3. Show a surface of revolution defined by

$$y^1 = u^1 cos u^2.$$

$$y^2 = u^1 sin u^2$$

$$y^3 = f(u^1)$$

$$\kappa = \frac{f' f''}{u^1\left[1+(f')^2\right]^2}, \text{ when } f \text{ is class } C^2.$$

Solution:

$$ds^2 = (dy^1)^2 + (dy^2)^2 + (dy^3)^2 = (du^1)^2 + (u^1)^2(du^2)^2 + \left(\frac{\partial f}{\partial u^1}\right)^2 (du^1)^2$$

$$= \left(1+\left(\frac{\partial f}{\partial u^1}\right)^2\right)(du^1)^2 + (u^1)^2(du^2)$$

$$\therefore a_{11} = 1+\left(\frac{\partial f}{\partial u^1}\right)^2 = 1+(f')^2, a_{22} = (u^1)^2, a_{12} = a_{21} = 0$$

$$\therefore a = \begin{vmatrix} a_{11} & 0 \\ 0 & a_{22} \end{vmatrix} = a_{11}a_{22} = (u^1)^2\left[1+(f')^2\right], \text{where } f' = \frac{\partial f}{\partial u^1}$$

$$a^{11} = \frac{a_{22}}{a_{11}a_{22}} = \frac{1}{a_{11}} = \frac{1}{1+(f')^2},\quad a^{22} = \frac{1}{a_{22}} = \frac{1}{(u^1)^2}, a^{12} = a^{21} = 0,$$

$$R_{1212} = a_{11}R^1_{212} = a_{11}\begin{bmatrix} \frac{\partial}{\partial u^1}\begin{Bmatrix} 1 \\ 2\ 2 \end{Bmatrix} - \frac{\partial}{\partial u^2}\begin{Bmatrix} 1 \\ 2\ 1 \end{Bmatrix} \\ + \begin{Bmatrix} 1 \\ \alpha\ 1 \end{Bmatrix}\begin{Bmatrix} \alpha \\ 2\ 2 \end{Bmatrix} - \begin{Bmatrix} 1 \\ \alpha\ 2 \end{Bmatrix}\begin{Bmatrix} \alpha \\ 2\ 1 \end{Bmatrix} \end{bmatrix}$$

$$= a_{11}\left[\frac{\partial}{\partial u^1}\begin{Bmatrix}1\\2\ 2\end{Bmatrix} - \frac{\partial}{\partial u^2}\begin{Bmatrix}1\\2\ 1\end{Bmatrix} + \begin{Bmatrix}1\\1\ 1\end{Bmatrix}\begin{Bmatrix}1\\2\ 2\end{Bmatrix} - \begin{Bmatrix}1\\1\ 2\end{Bmatrix}\begin{Bmatrix}1\\2\ 1\end{Bmatrix} + \begin{Bmatrix}1\\2\ 1\end{Bmatrix}\begin{Bmatrix}2\\2\ 2\end{Bmatrix} - \begin{Bmatrix}1\\2\ 2\end{Bmatrix}\begin{Bmatrix}2\\2\ 1\end{Bmatrix}\right]$$

$$\begin{Bmatrix}1\\1\ 1\end{Bmatrix} = a^{11}[11,1] = a^{11}\left(\frac{1}{2}\frac{\partial a_{11}}{\partial u^1}\right) = \frac{1}{2}.\frac{1}{1+(f')^2}(2f'f'') = \frac{f'f''}{1+(f')^2}$$

$$\begin{Bmatrix}1\\2\ 2\end{Bmatrix} = a^{11}[22,1] = a^{11}\left(-\frac{1}{2}\frac{\partial a_{22}}{\partial u^1}\right) = -\frac{u^1}{1+(f')^2},$$

$$\begin{Bmatrix}1\\2\ 1\end{Bmatrix} = a^{11}[21,1] = a^{11}\left(-\frac{1}{2}\frac{\partial a_{11}}{\partial u^2}\right) = 0$$

$$\begin{Bmatrix}2\\2\ 1\end{Bmatrix} = a^{22}[21,2] = a^{22}\frac{1}{2}\frac{\partial a_{22}}{\partial u^1} = \frac{1}{(u^1)^2}\frac{1}{2}\ 2u^1 = \frac{1}{u^1}$$

$$\therefore R_{1212} = a_{11}\left[\frac{\partial}{\partial u^1}\begin{Bmatrix}1\\22\end{Bmatrix} - \frac{\partial}{\partial u^2}\begin{Bmatrix}1\\21\end{Bmatrix} + \begin{Bmatrix}1\\11\end{Bmatrix}\begin{Bmatrix}1\\22\end{Bmatrix} - \begin{Bmatrix}1\\12\end{Bmatrix}\begin{Bmatrix}1\\21\end{Bmatrix} + \begin{Bmatrix}1\\21\end{Bmatrix}\begin{Bmatrix}2\\22\end{Bmatrix} - \begin{Bmatrix}1\\22\end{Bmatrix}\begin{Bmatrix}2\\21\end{Bmatrix}\right]$$

$$= [1+(f')^2]\left[\frac{\partial}{\partial u^1}\left(-\frac{u^1}{1+(f')^2}\right) - 0 + \frac{f'f''}{1+(f')^2}\frac{u^1}{1+(f')^2} - 0 + 0 + \frac{u^1}{1+(f')^2}\frac{1}{u^1}\right]$$

$$= [1+(f')^2]\left[\left(-\frac{1}{1+(f')^2}\right) + \frac{f'f''u^1}{[1+(f')^2]^2} + \frac{1}{1+(f')^2}\right]$$

$$= [1+(f')^2]\frac{f'f''u^1}{[1+(f')^2]^2} = \frac{f'f''u^1}{[1+(f')^2]}$$

$$\kappa = \frac{R_{1212}}{a} = \frac{\dfrac{ff''u^1}{\left[1+(f')^2\right]}}{(u^1)^2\left[1+(f')^2\right]} = \frac{ff''}{u^1\left[1+(f')^2\right]^2}.$$

Example 9.9.4. Also show that surfaces of the above problem (Example 9.8.1) are non-developable.

Solution: Now we calculate the Gaussian Curvature of S_1

$$a_{11} = \frac{(v^1)^2}{(v^1)^2 - a^2}, a_{22} = (v^1)^2 \text{ and } a_{12} = a_{21} = 0$$

$$a = \begin{vmatrix} a_{11} & 0 \\ 0 & a_{22} \end{vmatrix} = a_{11}a_{22} = \frac{(v^1)^4}{(v^1)^2 - a^2}$$

$$a^{11} = \frac{1}{a_{11}} = \frac{(v^1)^2 - a^2}{(v^1)^2}, \quad a^{22} = \frac{1}{a_{22}} = \frac{1}{(v^1)^2}$$

$$R_{1212} = a_{11}\left[\frac{\partial}{\partial v^1}\begin{Bmatrix} 1 \\ 2\,2 \end{Bmatrix} - \frac{\partial}{\partial v^2}\begin{Bmatrix} 1 \\ 2\,1 \end{Bmatrix} + \begin{Bmatrix} 1 \\ 1\,1 \end{Bmatrix}\begin{Bmatrix} 1 \\ 2\,2 \end{Bmatrix}\right.$$
$$\left. - \begin{Bmatrix} 1 \\ 1\,2 \end{Bmatrix}\begin{Bmatrix} 1 \\ 2\,1 \end{Bmatrix} + \begin{Bmatrix} 1 \\ 2\,1 \end{Bmatrix}\begin{Bmatrix} 2 \\ 2\,2 \end{Bmatrix} - \begin{Bmatrix} 1 \\ 2\,2 \end{Bmatrix}\begin{Bmatrix} 2 \\ 2\,1 \end{Bmatrix}\right]$$

so

$$\begin{Bmatrix} 1 \\ 1\,1 \end{Bmatrix} = a^{11}[11.1] = \frac{(v^1)^2 - a^2}{(v^1)^2}\left(\frac{1}{2}\frac{\partial a_{11}}{\partial v^1}\right) = \frac{1}{2}\frac{(v^1)^2 - a^2}{(v^1)^2}.$$
$$\frac{(v^1)^2 2v^1 - 2v^1\left[(v^1)^2 - a^2\right]}{\left[(v^1)^2 - a^2\right]^2} = \frac{(v^1)^2 - a^2}{(v^1)^2}\cdot\frac{v^1 a^2}{\left[(v^1)^2 - a^2\right]^2} = \frac{a^2}{v^1\left[(v^1)^2 - a^2\right]}$$

$$\begin{Bmatrix} 1 \\ 2\,2 \end{Bmatrix} = a^{11}[22,1] = a^{11}\left(-\frac{1}{2}\frac{\partial a_{22}}{\partial v^1}\right) = -\frac{1}{2}\frac{(v^1)^2 - a^2}{(v^1)^2} 2v^1$$

$$= -\frac{(v^1)^2 - a^2}{v^1}, \begin{Bmatrix} 1 \\ 2\,1 \end{Bmatrix} = 0$$

$$\begin{Bmatrix} 2 \\ 2\,1 \end{Bmatrix} = a^{22}[21,2] = a^{22}\frac{1}{2}\frac{\partial a_{22}}{\partial v^1} = \frac{1}{2}\frac{1}{(v^1)^2} 2v^1 = \frac{1}{v^1}$$

$$R_{1212} = \frac{(v^1)^2}{(v^1)^2 - a^2}\left[\frac{(v^1)^2 + a^2}{(v^1)^2} - 0 - \frac{a^2}{v^1\left[(v^1)^2 - a^2\right]}\frac{(v^1)^2 - a^2}{v^1} - 0 + 0 + \frac{(v^1)^2 - a^2}{v^1}\frac{1}{v^1}\right]$$

$$= \frac{(v^1)^2}{(v^1)^2 - a^2}\left[\frac{(v^1)^2 + a^2}{(v^1)^2} - \frac{a^2}{(v^1)^2} + \frac{(v^1)^2 - a^2}{(v^1)^2}\right]$$

$$= \frac{(v^1)^2}{(v^1)^2 - a^2}\frac{2(v^1)^2 - a^2}{(v^1)^2} = \frac{2(v^1)^2 - a^2}{v^1)^2 - a^2}$$

$$\text{Start here } \kappa = \frac{R_{1212}}{\text{a}} = \frac{\dfrac{2(v^1)^2 - a^2}{(v^1)^2 - a^2}}{\dfrac{(v^1)^4}{(v^1)^2 - a^2}} = \frac{2(v^1)^2 - a^2}{(v^1)^4} \neq 0.$$

Thus, S_1 is not developable.
For S_2 surface: $a_{11} = 1, a_{12} = a_{21} = 0, a_{22} = (u^1)^2 + a^2$ and $a = (u^1)^2 + a^2$

$$a^{11} = \frac{1}{a_{11}} = 1,\ a^{22} = \frac{1}{a_{22}} = \frac{1}{(u^1)^2 + a^2}$$

$$R_{1212}=a_{11}\left[\frac{\partial}{\partial v^1}\begin{Bmatrix}1\\2\ 2\end{Bmatrix}-\frac{\partial}{\partial v^2}\begin{Bmatrix}1\\2\ 1\end{Bmatrix}+\begin{Bmatrix}1\\1\ 1\end{Bmatrix}\begin{Bmatrix}1\\2\ 2\end{Bmatrix}-\begin{Bmatrix}1\\1\ 2\end{Bmatrix}\begin{Bmatrix}1\\2\ 1\end{Bmatrix}\right.$$
$$\left.+\begin{Bmatrix}1\\2\ 1\end{Bmatrix}\begin{Bmatrix}2\\2\ 2\end{Bmatrix}-\begin{Bmatrix}1\\2\ 2\end{Bmatrix}\begin{Bmatrix}2\\2\ 1\end{Bmatrix}\right]$$

$$\begin{Bmatrix}1\\2\ 2\end{Bmatrix}=a^{11}[22,1]=a^{11}\left(-\frac{1}{2}\frac{\partial a_{22}}{\partial u^1}\right)=1.\left(-\frac{1}{2}\right)2u^1=-u^1$$

$$\begin{Bmatrix}2\\2\ 1\end{Bmatrix}=a^{22}[21,2]=a^{22}\frac{1}{2}\frac{\partial a_{22}}{\partial u^1}=\frac{1}{(u^1)^2+a^2}\frac{1}{2}2u^1=\frac{u^1}{(u^1)^2+a^2}$$

$$R_{1212}=\left[-1-0+0-0+0+u^1\frac{u^1}{(u^1)^2+a^2}\right]=\frac{(u^1)^2}{(u^1)^2+a^2}-1=\frac{-a^2}{(u^1)^2+a^2}$$

$$\kappa=\frac{R_{1212}}{a}=\frac{\frac{-a^2}{(u^1)^2+a^2}}{(u^1)^2+a^2}=\frac{-a^2}{\left[(u^1)^2+a^2\right]^2}\neq 0$$, implying that the surface S_2 is not developable.

Example 9.9.5. Determine whether the surface with the metric

$$ds^2=(u^2)^2(du^1)^2+(u^1)^2\,du^2)^2$$

is developable or not

Solution: Here, $a_{11}=(u^2)^2, a_{22}=(u^1)^2, a_{12}=a_{21}=0, a=(u^1)^2(u^2)^2$.

We know

$$\kappa=-\frac{1}{2}\frac{1}{\sqrt{a}}\left[\frac{\partial}{\partial u^1}\left(\frac{1}{\sqrt{a}}\frac{\partial a_{22}}{\partial u^1}\right)+\frac{\partial}{\partial u^2}\left(\frac{1}{\sqrt{a}}\frac{\partial a_{11}}{\partial u^2}\right)\right]$$

$$=-\frac{1}{2}\frac{1}{u^1u^2}\left[\frac{\partial}{\partial u^1}\left(\frac{1}{u^1u^2}\frac{\partial(u^1)^2}{\partial u^1}\right)+\frac{\partial}{\partial u^2}\left(\frac{1}{u^1u^2}\frac{\partial(u^2)^2}{\partial u^2}\right)\right]$$

$$= -\frac{1}{2}\frac{1}{u^1u^2}\left[\frac{\partial}{\partial u^1}\left(\frac{1}{u^1u^2}2u^1\right)+\frac{\partial}{\partial u^2}\left(\frac{1}{u^1u^2}2u^2\right)\right]$$

$$= -\frac{1}{2}\frac{1}{u^1u^2}\left[\frac{\partial}{\partial u^1}\left(\frac{2}{u^2}\right)+\frac{\partial}{\partial u^2}\left(\frac{2}{u^1}\right)\right]$$

$$= -\frac{1}{2}\frac{1}{u^1u^2}[0+0]=0.$$

Since the Gaussian Curvature $\kappa = 0$, the surface is developable.

Example 9.9.6. *Show that the surface is defined by*

$$y^1 = f_1(u^1)$$

$$y^2 = f_2(u^1)$$

$$y^3 = u^2,$$

where f_1 and f_2 are differentiable functions that are developable.
Solution:

$$ds^2 = (dy^1)^2 + (dy^2)^2 + (dy^3)^2 = \left[\left(\frac{\partial f_1}{\partial u^1}\right)^2 + \left(\frac{\partial f_2}{\partial u^1}\right)^2\right](du^1)^2 + (du^2)^2$$

$$a_{11} = \left(\frac{\partial f_1}{\partial u^1}\right)^2 + \left(\frac{\partial f_2}{\partial u^1}\right)^2 = (f_2')^2 + (f_2')^2, a_{12} = a_{21} = 0, a_{22} = 1$$

and $a = (f_1')^2 + (f_2')^2$, where $\frac{\partial f_1}{\partial u^1} = f_1', \frac{\partial f_1}{\partial u^1} = f_1', \frac{\partial f_2}{\partial u^1} = f_2'$

$$a^{11} = \frac{1}{(f_1')^2 + (f_2')^2}, a^{22} = \frac{1}{a_{22}} = 1$$

$$R_{1212} = a_{11}\left[\frac{\partial}{\partial u^1}\begin{Bmatrix}1\\2\,2\end{Bmatrix} - \frac{\partial}{\partial u^2}\begin{Bmatrix}1\\2\,1\end{Bmatrix} + \begin{Bmatrix}1\\1\,1\end{Bmatrix}\begin{Bmatrix}1\\2\,2\end{Bmatrix}\right.$$

$$\left. - \begin{Bmatrix}1\\1\,2\end{Bmatrix}\begin{Bmatrix}1\\2\,1\end{Bmatrix} + \begin{Bmatrix}1\\2\,1\end{Bmatrix}\begin{Bmatrix}2\\2\,2\end{Bmatrix} - \begin{Bmatrix}1\\2\,2\end{Bmatrix}\begin{Bmatrix}2\\2\,1\end{Bmatrix}\right]$$

$$= \left[(f_1')^2 + (f_2')^2\right]\left[\frac{\partial}{\partial u^1}\begin{Bmatrix}1\\2\;2\end{Bmatrix} - \frac{\partial}{\partial u^2}\begin{Bmatrix}1\\2\;1\end{Bmatrix} + \begin{Bmatrix}1\\1\;1\end{Bmatrix}\begin{Bmatrix}1\\2\;2\end{Bmatrix}\right.$$

$$\left. - \begin{Bmatrix}1\\1\;2\end{Bmatrix}\begin{Bmatrix}1\\2\;1\end{Bmatrix} + \begin{Bmatrix}1\\2\;1\end{Bmatrix}\begin{Bmatrix}2\\2\;2\end{Bmatrix} - \begin{Bmatrix}1\\2\;2\end{Bmatrix}\begin{Bmatrix}2\\2\;1\end{Bmatrix}\right]$$

Here,

$$\begin{Bmatrix}1\\1\;1\end{Bmatrix} = a^{11}[11,1] = \frac{1}{(f_1')^2 + (f_2')^2}\frac{1}{2}\frac{\partial a_{11}}{\partial u^1} = \frac{1}{2}\frac{1}{(f_1')^2 + (f_2')^2}\left[2f_1'f_1'' + 2f_2'f_2''\right],$$

$$all\begin{Bmatrix}i\\j\;k\end{Bmatrix} = 0$$

,

$$\therefore\; R_{1212} = \left[(f_1')^2 + (f_2')^2\right][0 - 0 + 0 - 0 + 0 - 0] = 0$$

$$\kappa = \frac{R_{1212}}{a} = 0.$$

Therefore, the surface is developable.

9.10 The Geodesic Curvature

We study the intrinsic geometry of surfaces with a derivation of a formula describing the behavior of the tangent vector to a surface curve, which is analogous to the Serret-Frenet formula.

$$\text{Let } C: u^\alpha = u^\alpha(s) \tag{9.46}$$

be a curve on S whose metric is given by

$$a_{\alpha\beta}\frac{du^\alpha}{ds}\frac{du^\beta}{ds}=1. \tag{9.47}$$

Here, the tangent vector λ^α to C:

$$\lambda^\alpha=\frac{du^\alpha}{ds} \tag{9.48}$$

is a unit vector. If we differentiate the quadratic relation $a_{\alpha\beta}\lambda^\alpha\lambda^\beta=1$ intrinsically with respect to s, we obtain $a_{\alpha\beta}\lambda^\alpha\frac{\delta\lambda^\beta}{\delta s}=0$, from which it follows that surface vector $\frac{\delta\lambda^\alpha}{\delta s}$ is orthogonal to λ^α.

Now, we introduce a unit surface vector η^α normal to λ^α, so that

$$\frac{\delta\lambda^\alpha}{\delta s}=\chi_g\eta^\alpha, \tag{9.49}$$

where χ_α is a suitable scalar.

We choose the sense of η^α in such a manner that $(\lambda^\alpha, \eta^\alpha)$ is positive, i.e.,

$$\epsilon_{\alpha\beta}\lambda^\alpha\eta^\alpha=+1$$

(the choice of orientation of λ *and* η uniquely determines the sign of χ_g and the sine of the angle between λ & η is + 1).

The vector η^α is the unit surface vector orthogonal to curve C and the scalar χ_g is called the *Geodesic Curvature* of C.

We know $\epsilon^{\alpha\beta}\lambda_\alpha=\eta^\beta$ and $\epsilon^{\alpha\beta}\eta_\beta=\lambda^\alpha$.

Now, $\frac{\delta\eta^\beta}{\delta s}=\epsilon^{\alpha\beta}\frac{\delta\lambda_\alpha}{\delta s}=\epsilon^{\alpha\beta}\frac{\delta}{\delta s}\left(a_{\alpha\beta}\lambda^\beta\right)=\epsilon^{\alpha\beta}a_{\alpha\beta}\frac{\delta}{\delta s}\lambda^\beta=\epsilon^{\alpha\beta}a_{\alpha\beta}\chi_g\eta^\beta$

$$=\chi_g\epsilon^{\alpha\beta}\eta_\alpha=-\chi_g\epsilon^{\beta\alpha}\eta_\alpha=-\chi_g\lambda^\beta$$

$$\therefore\ \frac{\delta\eta^\alpha}{\delta s}=-\chi_g\lambda^\alpha. \tag{9.50}$$

We may refer to (9.48) and (9.49) as Serret-Frenet formulas for curve C lying on S.

The transversion of $\frac{\delta\lambda_\alpha}{\delta s} = \chi_g \eta_\alpha$ and by η_α,

we get $\chi_g = \frac{\delta\lambda_\alpha}{\delta s}\lambda^\alpha = \frac{\delta\lambda_\alpha}{\delta s}\left(\epsilon^{\alpha\beta}\lambda_\beta\right) = \epsilon^{\alpha\beta}\lambda_\beta \frac{\delta\lambda_\alpha}{\delta s}$

$$= \frac{1}{\sqrt{a}} e^{\alpha\beta}\lambda_\alpha \frac{\delta\lambda_\alpha}{\delta s} = \sqrt{a}\begin{vmatrix} \lambda_1 & \lambda_2 \\ \frac{\delta\lambda_1}{\delta s} & \frac{\delta\lambda_2}{\delta s} \end{vmatrix}.$$

Alternatively, using $\frac{\delta\lambda^\alpha}{\delta s} = \chi_g \eta^\alpha$,

$$\chi_g = \epsilon_{\alpha\beta}\lambda^\alpha \frac{\delta\lambda^\beta}{\delta s} = \sqrt{a}\begin{vmatrix} \lambda^1 & \lambda^2 \\ \frac{\delta\lambda^1}{\delta s} & \frac{\delta\lambda^2}{\delta s} \end{vmatrix}$$

$$= \sqrt{a}\begin{vmatrix} \frac{du^1}{ds} & \frac{du^2}{ds} \\ \frac{d^2u^1}{ds^2} + \begin{Bmatrix} 1 \\ \beta\gamma \end{Bmatrix}\frac{du^\beta}{ds}\frac{du^\gamma}{ds} & \frac{d^2u^2}{ds^2} + \begin{Bmatrix} 2 \\ \beta\gamma \end{Bmatrix}\frac{du^\beta}{ds}\frac{du^\gamma}{ds} \end{vmatrix}.$$

These expressions are known as *Beltrami's Formula* for Geodesic Curvature.

Theorem 9.10.1. A necessary and sufficient condition that a curve on a surface *S* is a geodesic is that its geodesic curvature is zero.
Proof: The differential equation of a geodesic is

$$\frac{d^2u^\alpha}{ds^2} + \begin{Bmatrix} \alpha \\ \beta\ \gamma \end{Bmatrix}\frac{du^\beta}{ds}\frac{du^\gamma}{ds} = 0,$$

$$\text{or } \frac{d}{ds}\left(\frac{du^\alpha}{ds}\right)+\begin{Bmatrix}\alpha \\ \beta\gamma\end{Bmatrix}\frac{du^\beta}{ds}\frac{du^\gamma}{ds}=0,$$

$$\text{or } \frac{\partial}{\partial u^\beta}\left(\frac{du^\alpha}{ds}\right)\frac{du^\beta}{ds}+\begin{Bmatrix}\alpha \\ \beta\gamma\end{Bmatrix}\frac{du^\beta}{ds}\frac{du^\gamma}{ds}=0,$$

$$\text{or } \left[\frac{\partial}{\partial u^\beta}\left(\frac{du^\alpha}{ds}\right)+\begin{Bmatrix}\alpha \\ \beta\gamma\end{Bmatrix}\frac{du^\gamma}{ds}\right]\frac{du^\beta}{ds}=0,$$

$$\text{or } \left(\frac{du^\alpha}{ds}\right)_{,\beta}\frac{du^\beta}{ds}=0$$

$$\therefore\ \frac{\delta}{\delta s}\left(\frac{du^\alpha}{ds}\right)=0,$$

$$\text{or } \frac{\partial}{\partial s}\lambda^\alpha=0$$

$$\therefore\ \chi_g\eta^\alpha=0.$$

Since η^α is a unit vector, we must have $\chi_g=0$
and conversely, let $\chi_g=0$

$$\frac{\delta}{\delta s}\left(\frac{du^\alpha}{ds}\right)=\frac{\partial}{\partial s}\lambda^\alpha=\chi_g\eta^\alpha=0.$$

$$\Rightarrow \frac{du^\alpha}{ds}=constant,\ \mathrm{c}$$

$$\text{or } u^\alpha = cs+d.$$

This shows that geodesic of a surface is therefore analogous to a straight line in a plane. Therefore, a curve C on a surface S is geodesic if its geodesic curvature χ_g vanishes identically. A straight line on any surface is geodesic.

Example 9.10.1. Find the geodesic curvature of a small circle:

$$C: u^1 = constant = u_0^1 \neq 0,\ u^2 = u^2$$

on the surface of the sphere

$$S: y^1 = a\cos u^1 \cos u^2$$
$$y^2 = a\cos u^1 \sin u^2$$
$$y^3 = a\sin u^1$$

(or show that the geodesic curvature $\chi_g = \dfrac{\tan u_0^1}{a}$).

Solution: If the arc-length s of C is measured from the plane $u_1^2 = 0$, we have

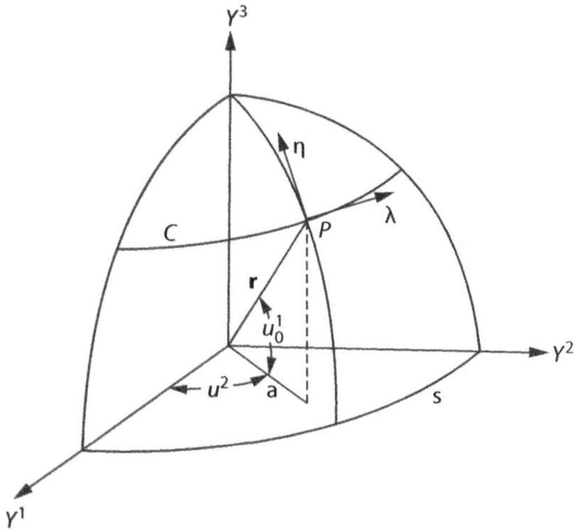

Figure 9.5

$$\left[\frac{r}{a}=\cos u_0^1,\quad s=r\theta=a\cos u_0^1 u^2;\ \therefore u^2=\frac{s}{a\cos u_0^1}\right].$$

$$u^2=\frac{s}{a\cos u_0^1}$$

and the equation of C can be written in the form

$$u^1=u_0^1 \text{ and } u^2=\frac{s}{a\cos u_0^1} \tag{i}$$

from i, we find the components of the unit tangent vector $\lambda^\alpha=\frac{du^\alpha}{ds}$ along C:

$$\lambda^1=0,\lambda^2=\frac{du^2}{ds}=\frac{1}{a\cos u_0^1}, \tag{ii}$$

so that

$$\frac{\delta\lambda^1}{\delta s}=\frac{d\lambda^1}{ds}+\begin{Bmatrix}1\\ \alpha\ \beta\end{Bmatrix}\lambda^\alpha\frac{du^\beta}{ds}=0+\begin{Bmatrix}1\\ 2\ 2\end{Bmatrix}\lambda^2\frac{du^2}{ds}$$

$$=\cos u^1\sin u^1\left(\frac{1}{a\cos u_0^1}\right)^2=\frac{1}{a^2}\tan u_0^1$$

$$\frac{\delta\lambda^2}{\delta s}=\frac{d\lambda^2}{ds}+\begin{Bmatrix}2\\ \alpha\ \beta\end{Bmatrix}\lambda^\alpha\frac{du^\beta}{ds}=0+\begin{Bmatrix}2\\ 2\ 2\end{Bmatrix}\lambda^2\frac{du^2}{ds}=0$$

$$\left[\text{since}\frac{d\lambda^2}{ds}=\frac{d}{ds}\left(\frac{1}{a\cos u_0^1}\right)=0\right].$$

We know $\frac{\delta\lambda^{\alpha}}{\delta s} = \chi_g \eta^{\alpha} \therefore \frac{\delta\lambda^{1\alpha}}{\delta s} = \chi_g \eta^{1}$

$$\therefore \frac{1}{a^2} tan\ u_0^1 = \chi_g \eta^1 \quad \text{(iii)}$$

$$\frac{\delta\lambda^2}{\delta s} = \chi_g \eta^2 = 0 \quad \text{(iv)}$$

Since C is not a geodesic, $\chi_g \neq 0$, *from* (iv) $\eta^2 = 0$, *and*

ηi is a unit vector, we have $a_{ij}\,\eta^i\eta^j = 1 \Rightarrow a^2\,(\eta)^2 = 1, \therefore \eta^1 = \frac{1}{a}$.

From (iii), we get $\therefore \frac{1}{a^2} tan\ u_0^1 = \chi_g \eta^1$ *or* $\chi_g = \frac{1}{a^2} tan\ u_0^1 . a = \frac{1}{a} tan\ u_0^1$.

Example 9.10.2. Consider the surface of the right circular cone

$$S: : x^1 = u^1 cos u^2$$

$$x^2 = u^1 sin u^2$$

$$x^3 = u^1$$

the curve $C: u^1 = a, u^2 = \frac{s}{a}$.

Show that the geodesic curvature is $\frac{\sqrt{2}}{2a}$.

Solution: Here, $ds^2 = 2(du^1)^2 + (u^1)^2(du^2)^2$

$$a_{11} = 2, a_{22} = (u^1)^2,\ a_{12} = a_{21} = 0,$$

$$\lambda^{\alpha} = \frac{du^{\alpha}}{ds}, \lambda^1 = \frac{du^1}{ds} = 0, \lambda^2 = \frac{du^2}{ds} = \frac{1}{a}$$

$$\begin{Bmatrix} 1 \\ 2\ 2 \end{Bmatrix} = g^{1\alpha}[22,\alpha] = g^{11}[22,1] = \frac{1}{g_{11}}\left(-\frac{1}{2}\frac{\partial g_{22}}{\partial u^1}\right) = -\frac{1}{2}\frac{1}{2}2u^1 = -\frac{u^1}{2}$$

$$\begin{Bmatrix} 2 \\ 2\ 2 \end{Bmatrix} = g^{2\alpha}[22,\alpha] = g^{22}[22,2] = g^{22}\frac{1}{2}\frac{\partial g_{22}}{\partial u^2} = 0,$$

so that $\dfrac{\delta\lambda^1}{\delta s} = \dfrac{d\lambda^1}{ds} + \begin{Bmatrix} 1 \\ \alpha\ \beta \end{Bmatrix}\lambda^\alpha \dfrac{du^\beta}{ds} = 0 + \begin{Bmatrix} 1 \\ 2\ 2 \end{Bmatrix}\lambda^2 \dfrac{du^2}{ds} = -\dfrac{u^1}{2}\left(\dfrac{1}{a}\right)^2 = -\dfrac{1}{2a}$

$$\frac{\delta\lambda^2}{\delta s} = \frac{d\lambda^2}{ds} + \begin{Bmatrix} 2 \\ \alpha\ \beta \end{Bmatrix}\lambda^\alpha \frac{du^\beta}{ds} = 0 + \begin{Bmatrix} 2 \\ 2\ 2 \end{Bmatrix}\lambda^2 \frac{du^2}{ds} = 0.$$

Again, by the geodesic curvilinear formula,

$$\frac{\delta\lambda^\alpha}{\delta s} = \chi_g\eta^\alpha, \frac{\delta\lambda^1}{\delta s} = \chi_g\eta^1 \quad or, \chi_g\eta^1 = -\frac{1}{2a} \tag{i}$$

and $\chi_g\eta^2 = 0\, since\, C\, is\, not\, a\, geodesic, \chi_g \neq 0, \eta^2 = 0.$ (ii)

Since C is not a geodesic and

ηi is a unit vector, we have $a_{ij}\eta^i\eta^j = 1 \Rightarrow 2.\ (\eta^1)^2 = 1, \therefore\ \eta^1 = -\dfrac{1}{\sqrt{2}}$.

From (ii), $\chi_g\eta^1 = -\dfrac{1}{2a}$, or $\chi_g\left(-\dfrac{1}{\sqrt{2}}\right) = -\dfrac{1}{2a}$

$$\therefore\ \chi_g = \frac{\sqrt{2}}{2a}\ Ans.$$

[$ds^2 = a^2(-\cos u^1 \sin u^2 du^2 - \cos u^2 \sin u^1 du^1)^2 + a^2(\cos u^1 \cos u^2 du^2 - \sin u^2 \sin u^1 du^1)^2 + a^2\cos^2 u^1 (du^1)^2$

$= a^2[\cos^2 u^1 (du^2)^2 + \sin^2 u^1 (du^1)^2 + \cos^2 u^1 (du^1)^2] = a^2(du^1)^2 + a^2\cos^2 u^1 (du^2)^2$. Here, $a_{11} = a^2, a_{12} = 0, a_{22} = a^2\cos^2 u^1$

$\begin{Bmatrix} 1 \\ 2\ 2 \end{Bmatrix} = g^{1\alpha}[22,\alpha] = g^{11}[22,1] = \dfrac{1}{g_{11}}.\dfrac{1}{2}\left(-\dfrac{\partial g_{22}}{\partial u^1}\right) = \dfrac{1}{a^2}\dfrac{1}{2}\left(-\dfrac{\partial}{\partial u^1}(a^2\cos^2 u^1)\right)$

$= -\dfrac{a^2}{2a^2} 2\cos u^1(-\sin u^1) = \cos u^1 \sin u^1$

$$\left\{ \begin{matrix} 2 \\ 2\,2 \end{matrix} \right\} = g^{2\alpha}[22,\alpha] = g^{22}[22,2] = g22\,\frac{1}{2}\frac{\partial g_{22}}{\partial u^2} = 0.\ ]$$

Example 9.10.3. Show the geodesic curvature of the curve $u = c$ on a surface with metric

$$\phi^2(du)^2 + \mu^2(dv)^2 = \frac{1}{\phi\mu}\frac{\partial \mu}{\partial u}.$$

Solution: Here, $ds^2 = \phi^2(du)^2 + \mu^2(dv)^2$

$$a_{11} = \phi^{2,}a_{22} = \mu^2 \text{ and } a_{12} = a_{21} = 0$$

$$a = \begin{vmatrix} a_{11} & 0 \\ 0 & a_{22} \end{vmatrix} = a_{11}a_{22} = \phi^2\mu^2$$

$$a^{11} = \frac{1}{a_{11}} = \frac{1}{\phi^2},\ a^{22} = \frac{1}{a_{22}} = \frac{1}{\mu^2}, a^{12} = a^{21} = 0.$$

Applying the Beltrami Formula to find geodesic curvature χ_g,

$$\chi_g = \sqrt{a}\begin{vmatrix} \dfrac{du^1}{ds} & \dfrac{du^2}{ds} \\ \dfrac{d^2u^1}{ds^2} + \left\{ \begin{matrix} 1 \\ \beta\gamma \end{matrix} \right\}\dfrac{du^\beta}{ds}\dfrac{du^\gamma}{ds} & \dfrac{d^2u^2}{ds^2} + \left\{ \begin{matrix} 2 \\ \beta\gamma \end{matrix} \right\}\dfrac{du^\beta}{ds}\dfrac{du^\gamma}{ds} \end{vmatrix}$$

$$= \frac{du^1}{ds}\left(\frac{d^2u^2}{ds^2} + \left\{ \begin{matrix} 2 \\ \beta\gamma \end{matrix} \right\}\frac{du^\beta}{ds}\frac{du^\gamma}{ds}\right) - \frac{du^2}{ds}\left(\frac{d^2u^1}{ds^2} + \left\{ \begin{matrix} 1 \\ \beta\gamma \end{matrix} \right\}\frac{du^\beta}{ds}\frac{du^\gamma}{ds}\right)$$

$$= \frac{du^1}{ds}\left(\frac{d^2u^2}{ds^2} + \begin{Bmatrix} 2 \\ 1\,1 \end{Bmatrix}\frac{du^1}{ds}\frac{du^1}{ds} + \begin{Bmatrix} 2 \\ 2\,2 \end{Bmatrix}\frac{du^2}{ds}\frac{du^2}{ds} + 2\begin{Bmatrix} 2 \\ 1\,2 \end{Bmatrix}\frac{du^1}{ds}\frac{du^2}{ds}\right)$$

$$-\frac{du^2}{ds}\left(\frac{d^2u^1}{ds^2} + \begin{Bmatrix} 1 \\ 1\,1 \end{Bmatrix}\frac{du^1}{ds}\frac{du^1}{ds} + \begin{Bmatrix} 1 \\ 2\,2 \end{Bmatrix}\frac{du^2}{ds}\frac{du^2}{ds} + 2\begin{Bmatrix} 1 \\ 1\,2 \end{Bmatrix}\frac{du^1}{ds}\frac{du^2}{ds}\right)$$

$$\begin{Bmatrix} 2 \\ 1\,1 \end{Bmatrix} = a^{22}[11,2] = \frac{1}{\mu^2}\left(-\frac{1}{2}\frac{\partial a_{11}}{\partial v}\right) = -\frac{1}{2\mu^2}2\phi\frac{\partial \phi}{\partial v} = -\frac{\phi}{\mu^2}\frac{\partial \phi}{\partial v}$$

$$\begin{Bmatrix} 2 \\ 2\,2 \end{Bmatrix} = a^{22}[22,2] = \frac{1}{\mu^2}\left(\frac{1}{2}\frac{\partial a_{22}}{\partial v}\right) = \frac{1}{2}\frac{1}{\mu^2}\,2\mu\frac{\partial \mu}{\partial v} = \frac{1}{\mu}\frac{\partial \mu}{\partial v}$$

$$\begin{Bmatrix} 2 \\ 1\,2 \end{Bmatrix} = a^{22}[12,2] = \frac{1}{\mu^2}\left(\frac{1}{2}\frac{\partial a_{22}}{\partial u}\right) = \frac{1}{2\mu^2}2\mu\frac{\partial \mu}{\partial u} = \frac{1}{\mu}\frac{\mu}{\partial u}$$

$$\begin{Bmatrix} 1 \\ 1\,1 \end{Bmatrix} = a^{11}[11,1] = \frac{1}{\phi^2}\left(\frac{1}{2}\frac{\partial a_{11}}{\partial u}\right) = \frac{1}{2\phi^2}2\phi\frac{\partial \phi}{\partial u} = \frac{1}{\phi}\frac{\partial \phi}{\partial u}$$

$$\begin{Bmatrix} 1 \\ 2\,2 \end{Bmatrix} = a^{11}[22,1] = \frac{1}{\phi^2}\left(-\frac{1}{2}\frac{\partial a_{22}}{\partial u}\right) = -\frac{1}{2\phi^2}2\mu\frac{\partial \mu}{\partial u} = -\frac{\mu}{\phi^2}\frac{\partial \mu}{\partial u}$$

$$\begin{Bmatrix} 1 \\ 1\,2 \end{Bmatrix} = a^{11}[12,1] = \frac{1}{\phi^2}\left(\frac{1}{2}\frac{\partial a_{11}}{\partial v}\right) = \frac{1}{2\phi^2}2\phi\frac{\partial \phi}{\partial v}$$

Since for the parametric curve $u = c$ *is* constant,

$$\chi_g = -\sqrt{a}\frac{du^2}{ds}\begin{Bmatrix} 1 \\ 2\,2 \end{Bmatrix}\frac{du^2}{ds}\frac{du^2}{ds} = -\sqrt{a}\begin{Bmatrix} 1 \\ 2\,2 \end{Bmatrix}\left(\frac{dv}{ds}\right)^3$$

$$= -\phi\mu\frac{\mu}{\phi^2}\frac{\partial \mu}{\partial u}\left(\frac{dv}{ds}\right)^3 = \phi\mu\frac{\mu}{\phi^2}\frac{1}{\mu^3}\frac{\partial \mu}{\partial u} = \frac{1}{\phi\mu}\frac{\partial \mu}{\partial u}$$

$$\left[1=\phi^2\left(\frac{du}{ds}\right)^2+\mu^2\left(\frac{dv}{ds}\right)^2,\left(\frac{dv}{ds}\right)^2=\frac{1}{\mu^2},\ since\ u=c\right]$$

Example 9.10.4. Show the condition that the $u^1-curve\ and\ u^2$–curve geodesics are $\left\{\begin{matrix}2\\1\ 1\end{matrix}\right\}=0\ and\left\{\begin{matrix}1\\2\ 2\end{matrix}\right\}=0$, respectively.

Solution: We have

$$\chi_g=\epsilon_{\alpha\beta}\lambda^\alpha\frac{\delta\lambda^\beta}{\delta s}.$$

Now, for the $u^1-curve$, we have

$$\lambda^\alpha_{(1)}=\left(\frac{1}{\sqrt{a_{11}}},0\right)$$

$$\lambda_g^{(1)}=\epsilon_{12}\lambda^1\frac{\delta\lambda^2}{\delta s}\ \text{and others are zero}$$

$$=\epsilon_{12}\lambda^1\left[\frac{d\lambda^2}{ds}+\left\{\begin{matrix}2\\1\ 1\end{matrix}\right\}\lambda^1\frac{du^1}{ds}\right]\ \text{and all other Christoffel symbols are zero,}$$

$$=\sqrt{a}e_{12}\lambda^1\left[\frac{d\lambda^2}{ds}+\left\{\begin{matrix}2\\1\ \ 1\end{matrix}\right\}\lambda^1\lambda^1\right]=\sqrt{a}.1\frac{1}{\sqrt{a_{11}}}\left[0+\left\{\begin{matrix}2\\1\ 1\end{matrix}\right\}\frac{1}{\sqrt{a_{11}}}\frac{1}{\sqrt{a_{11}}}\right]$$

$$\lambda_g^{(1)}=\sqrt{a}\frac{1}{(a_{11})^{\frac{3}{2}}}\left\{\begin{matrix}2\\1\ 1\end{matrix}\right\}.$$

Similarly,

$$\lambda_g^{(2)}=\epsilon_{21}\lambda^2\frac{\delta\lambda^1}{\delta s},\ \text{where}\ \lambda^\alpha_{(2)}=\left(0,\frac{1}{\sqrt{a_{22}}}\right)$$

$$=\sqrt{a}e_{21}\lambda^2\left[\frac{d\lambda^1}{ds}+\left\{\begin{matrix}1\\2\ 2\end{matrix}\right\}\lambda^2\lambda^2\right]\ \text{and all other Christoffel symbols are zero}$$

$$= \sqrt{a}(-1)(\lambda^2)^3\left[\frac{d\lambda^1}{ds} + \left\{\begin{matrix} 1 \\ 2\ 2 \end{matrix}\right\}\right]$$

$$= -\sqrt{a}(\lambda^2)^3\left[\frac{d\lambda^1}{ds} + \left\{\begin{matrix} 1 \\ 2\ 2 \end{matrix}\right\}\right]$$

$$= -\sqrt{a}\frac{1}{(a_{22})^{\frac{3}{2}}}\left[0 + \left\{\begin{matrix} 1 \\ 2\ 2 \end{matrix}\right\}\right] = -\sqrt{a}\frac{1}{(a_{22})^{\frac{3}{2}}}\left\{\begin{matrix} 1 \\ 2\ 2 \end{matrix}\right\}.$$

A curve will be geodesic if the geodesic curvature is zero. Therefore, $\lambda_g^{(1)} = 0$ *and* $\lambda_g^{(2)} = 0$.

This implies that $\left\{\begin{matrix} 2 \\ 1\ 1 \end{matrix}\right\} = 0$ *and* $\left\{\begin{matrix} 1 \\ 2\ 2 \end{matrix}\right\} = 0$, respectively.

9.11 Exercises

1. Find the element of area of the surface of radius r if the equations of the surface are given in the form $x^1 = a \sin u \cos v$, $x^2 = a \sin u \sin v$, *and* $x^3 = a\cos u$, where x^i are the orthogonal Cartesian coordinates.
2. Find the differential equations for the geodesic in cylindrical coordinates:

$$x^1 = u^1 \cos u^2$$

$$x^2 = u^1 \sin u^2$$

$$x^3 = u^3.$$

3. Show that the formula for the Gaussian Curvature K can be written in the form

$$K = \frac{1}{2\sqrt{a}}\left\{\left[\frac{\partial}{\partial u^1}\left(\frac{a_{12}}{a_{11}\sqrt{a}}\frac{\partial a_{11}}{\partial u^2} - \frac{1}{\sqrt{a}}\frac{\partial a_{22}}{\partial u^1}\right)\right]\right.$$

$$\left.+\left[\frac{\partial}{\partial u^2}\left(\frac{2}{\sqrt{a}}\frac{\partial a_{12}}{\partial u^1} - \frac{a_{12}}{a_{11}\sqrt{a}}\frac{\partial a_{11}}{\partial u^1} - \frac{1}{\sqrt{a}}\frac{\partial a_{11}}{\partial u^2}\right]\right]\right\}$$

4. Prove that the catenoid

$$ds^2 = a^2\, cosh^2\, x^1\, (dx^1)^2 + a^2\, cosh^2\, x^2\, (dx^2)^2$$

and the right helicoid

$$ds^2 = (dy^1)^2 + [(y^1)^2 + a^2](dy^2)^2$$

are locally isometric.

5. Determine whether the surface of a helicoid given by

$$x^1 = u^1\, cosu^2$$
$$x^2 = u^1\, sinu^2$$
$$x^3 = cu^1$$

is developable, where is a constant and u^1 *and* u^2 are curvilinear coordinates of the surface.

6. Show that the surface with metric

(i) $ds^2 = (du^1)^2 + [(u^1)^2 + a^2](du^2)^2$ is not developable.

(ii) $ds^2 = (u^2)^2(\mathrm{d}u^1)^2 + [(u^1)^2(du^2)^2$ is developable.

7. Show that for the sphere of radius c, with the equation of the form

$$x^1 = c\, cosu^1\, cosu^2$$
$$x^2 = c\, sinu^1\, sinu^2$$
$$x^3 = c\, cosu^1, \text{ where c is a constant,}$$

the total curvature is $\kappa = \frac{1}{c^2}$.

10

Surfaces in Space

10.1 Introduction

Our study of the geometry of surfaces was carried out from the point of view of a two dimensional being whose universe is determined by the surface parameters u^1 and u^2. The study of surfaces was based entirely on the first quadratic differential form. In the discussion of isometric surfaces in the previous chapter, we remarked that a pair of isometric surfaces, for example, a cone and a cylinder which are indistinguishable in intrinsic geometry, appear to be quite distinct to an observer examining them from a reference frame located in the space in which the surfaces are imbedded. An entity that provides a characterization of the shape of the surface as it appears from enveloping space is the normal line to the surface.

In this chapter, the geometric shape and properties of a surface with the aid of the quadratic form that depends fundamentally on the behavior of the normal line is discussed.

10.2 The Tangent Vector

A surface S imbedded in E_3 was defined by three parametric equations

$$y^i = y^i\,(u^1, u^2)\,,\ (i=1,2,3), \tag{10.1}$$

so that

$$dy^i = \frac{\partial y^i}{\partial u^\alpha} du^\alpha,$$

where the y^i are orthogonal Cartesian coordinates of the reference frame located in the space surrounding S. An element of arc ds of a curve on S is determined by the formula

Dipankar De. Introduction to Differential Geometry with Tensor Applications, (321–348) © 2022 Scrivener Publishing LLC

$$ds^2=a_{\alpha\beta}\, du^\alpha\, du^\beta, \tag{10.2}$$

where $$a_{\alpha\beta}=\frac{\partial y^i}{\partial u^\alpha}\frac{\partial y^i}{\partial u^\beta}.$$

The choice of Cartesian variables y^i in the space enveloping the surface is clearly not essential and we could have equally referred the points of E_3 to a curvilinear coordinate system X related to Y by the transformation

$$x^i=x^i\,(y^1,y^2,y^3).$$

Now, relative to the frame X, the line element in E_3 is given by

$$ds^2=g_{ij}\, dx^i\, dx^j, \tag{10.3}$$

where $$g_{ij}=\frac{\partial y^k}{\partial x^i}\frac{\partial y^k}{\partial x^j}.$$

Then, Equation (10.1) for the surface S can be written as

$$S{:}x^i=x^i\,(u^1,u^2), \tag{10.4}$$

so that $$dx^i=\frac{\partial x^i}{\partial u^\alpha}du^\alpha. \tag{10.5}$$

Hence, Equation (10.3) can be formed as

$$ds^2=g_{ij}\, dx^i\, dx^j$$

$$=g_{ij}\frac{\partial x^i}{\partial u^\alpha}\frac{\partial x^j}{\partial u^\beta}du^\alpha du^\beta.$$

Comparing this to Equation (10.2), we get

$$a_{\alpha\beta} = g_{ij}\frac{\partial x^i}{\partial u^\alpha}\frac{\partial x^j}{\partial u^\beta}, (i,j=1,2,3),(\alpha,\beta=1,2), \tag{10.6}$$

where dx^i *and* g_{ij}'s are tensors with respect to the transformations induced on the space variablesx^i, whereas du^α and $a_{\alpha\beta}$ are tensors with respect to the transformation of Gaussian surface coordinates u^α.

Since both $a_{\alpha\beta}$ *and* g_{ij} in Equation (10.6) are tensors, this formula suggests that $\frac{\partial x^i}{\partial u^\alpha}$ can be regarded either as a contravariant space vector or as a covariant surface vector.

From Equation (10.5), $dx^i = \frac{\partial x^i}{\partial u^\alpha} du^\alpha$.

The left member dx^i is invariant relative to a change of the surface coordinate u^α. Since du^α is an arbitrary surface vector, we conclude that

$$\frac{\partial x^i}{\partial u^\alpha} \tag{10.7}$$

is a covariant surface vector. On the other hand, if we change the space coordinates, du^α, being a surface vector, is invariant relative to this change, so that Equation (10.7) must be a contravariant space vector. Hence, we can write Equation (10.7) as

$$x^i_\alpha = \frac{\partial x^i}{\partial u^\alpha} \tag{10.8}$$

Again, $$dx^i = \frac{\partial x^i}{\partial u^\alpha} du^\alpha.$$

We have $$\frac{dx^i}{ds} = \frac{\partial x^i}{\partial u^\alpha}\frac{du^\alpha}{ds}$$

or $$\lambda^i = x^i_\alpha \lambda^\alpha \tag{10.9}$$

This equation tells us that any surface vector A^α (that is, a vector lying in a tangent plane to S) can be viewed as a space vector with components A^i determined by

$$A^i = x^i_\alpha A^\alpha. \tag{10.10}$$

We shall refer to a vector A^i determined by this formula as a *tangent vector to surface S.*

10.3 The Normal Line to the Surface

Let A *and* B be a pair of surface vectors drawn at point P *of* S, as shown in Figure (10.1). According to Formula (10.10),

$$A^i = x^i_\alpha A^\alpha \quad (A^\alpha \text{ is lying in the tangent plane to } S)$$

$$\left[x^i_\alpha = \left(\frac{\partial x^1}{\partial u^\alpha}, \frac{\partial x^2}{\partial u^\alpha}, \frac{\partial x^3}{\partial u^\alpha}\right)\right].$$

They can be represented in the form

$$A^i = x^i_\alpha A^\alpha \quad B^i = x^i_\alpha B^\alpha \tag{10.11}$$

$$\left[\frac{\partial x^i}{\partial \bar{u}^\beta} = \bar{x}^i_\beta, \quad x^i_\alpha = \frac{\partial \bar{u}^\beta}{\partial u^\alpha}\bar{x}^i_\beta\right]$$

The vector product:

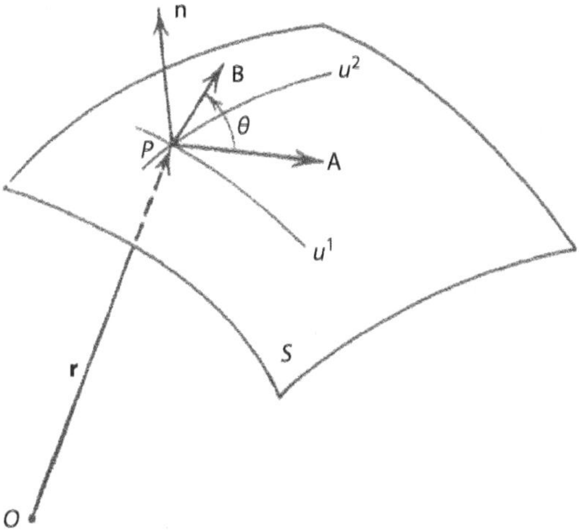

Figure 10.1

$A \times B$ is the vector normal to the tangent plane determined by the vectors A *and* B and the unit vector $\boldsymbol{n}$ perpendicular to the tangent plane, so oriented that $A,B,$ *and* $\boldsymbol{n}$ from a right handed system, are

$$\boldsymbol{n} = \frac{\mathbf{A} \times \boldsymbol{B}}{|\mathbf{A} \times \boldsymbol{B}|} = \frac{\mathbf{A} \times \boldsymbol{B}}{AB|sin\theta|}, \tag{10.12}$$

where θ is the angle between A *and* B.

n is called the unit normal vector to the surface S at P. Clearly, n is a function of coordinates (u^1,u^2) and, as point P (u^1,u^2) is displaced to a new position $P(u^1+du^1,u^2+du^2)$, the vector $\boldsymbol{n}$ undergoes a change:

$$d\boldsymbol{n} = \frac{\partial \boldsymbol{n}}{\partial u^\alpha} du^\alpha, \tag{10.13}$$

where the position vector r is changed by the amount $dr = \frac{\partial \boldsymbol{r}}{\partial u^\beta} du^\beta$.

Let us use the scalar product $dr.d\boldsymbol{n} = \frac{\partial \boldsymbol{n}}{\partial u^\alpha} \frac{\partial \boldsymbol{r}}{\partial u^\beta} du^\alpha du^\beta$. (10.14)

If we define

$$b_{\alpha\beta} = -\frac{1}{2}\left(\frac{\partial \boldsymbol{n}}{\partial u^\alpha} \frac{\partial \boldsymbol{r}}{\partial u^\beta} + \frac{\partial \boldsymbol{n}}{\partial u^\beta} \frac{\partial \boldsymbol{r}}{\partial u^\alpha} \right),$$

(10.14) becomes

$$\boldsymbol{dr.dn} = -b_{\alpha\beta}\, du^\alpha\, du^\beta. \tag{10.15}$$

The left hand of (10.15), being the scalar product of two vectors, is obviously an invariant. Moreover, from symmetry with regards to α and β, it is clear that the coefficient of $du^\alpha\, du^\beta$ in the r.h.s. of (10.15) is defined as a covariant tensor of rank two.

The quadratic form $$\mathcal{B} \equiv b_{\alpha\beta}\, du^\alpha\, du^\beta \tag{10.16}$$

is just like the first fundamental quadratic form $\mathcal{A} \equiv dr.dr$.

$$\mathcal{A} \equiv a_{\alpha\beta}\, du^\alpha\, du^\beta,$$

which is in the study of intrinsic properties of surface.

Differential form (10.16) was introduced by Gauss and is called the *second fundamental quadratic form of the surface.*

This differential equation plays an important role in the study of the geometry of a surface when it is viewed from the surrounding space.

We write (10.12) in terms of the components of x^i_α of the base vectors a_α.

$$\boldsymbol{n}_i = \frac{\epsilon_{ijk} A^j B^k}{AB sin\theta} \tag{10.17}$$

Using the formula $\epsilon_{\alpha\beta}\, \lambda^\alpha\, \mu^\beta = sin\theta$,

$$AB sin\theta = \epsilon_{\alpha\beta}\, A^\alpha\, B^\beta. \tag{10.18}$$

Substitute this value and from (10.11) in (10.17), we get

$$\boldsymbol{n}_i \epsilon_{\alpha\beta} A^\alpha B^\beta = \epsilon_{ijk}\ x^i_\alpha x^i_\beta A^\alpha A^\beta$$

$$\text{or } (\boldsymbol{n}_i \epsilon_{\alpha\beta} - \epsilon_{ijk}\ x^i_\alpha x^i_\beta) A^\alpha A^\beta = 0.$$

Since this relation is valid for all surface vectors, we can write

$$\boldsymbol{n}_i \epsilon_{\alpha\beta} - \epsilon_{ijk}\ x^i_\alpha x^i_\beta = 0 \tag{10.19}$$

Multiplying (10.19) by $\epsilon^{\alpha\beta}$ and putting the value of $\epsilon^{\alpha\beta}\, \epsilon_{\alpha\beta} = 2$,

$$\boldsymbol{n}_i = \frac{1}{2}\epsilon^{\alpha\beta}\epsilon_{ijk}\ x^i_\alpha x^i_\beta \tag{10.20}$$

It is clear from the formula that $\boldsymbol{n}_i$ is a space vector which does not depend on the choice of the surface coordinates.

Example 10.3.1. Find the Second Fundamental quadratic form of right Helicoid

$$r = (ucosv, usinv, cv).$$

Solution:

$$r_u = \frac{\partial r}{\partial u} = (cosv, sinv, 0)$$

$$r_v = \frac{\partial r}{\partial v} = (-usinv, ucosv, c)$$

$$r_u \times r_v = \begin{vmatrix} i & j & k \\ cosu & sinv & 0 \\ -usinv & ucosv & c \end{vmatrix} = (csinv, -c\ cosv, u)$$

$$\therefore |r_u \times r_v| = (u^2 + c^2)^{\frac{1}{2}}$$

$$\eta = \frac{(csinv, -ccosv,\ u)}{(u^2 + c^2)^{\frac{1}{2}}}$$

$$\eta_u = \frac{\partial \eta}{\partial u} = \left[\frac{1}{\sqrt{u^2 + c^2}}(0,0,1) - \frac{u}{(u^2 + c^2)^{\frac{3}{2}}}(csinv, -ccosv,\ u)\right].$$

$$\eta_v = \frac{\partial \eta}{\partial v} = \frac{1}{\sqrt{u^2 + c^2}}(c.cosv, c.sinv,\ 0)$$

$$b_{\alpha\beta} = -\frac{1}{2}(\eta_\alpha r_\beta + \eta_\beta r_\alpha)$$

$$b_{11} = -\frac{1}{2}(\eta_1 r_1 + \eta_1 r_1)$$

$$= -(\eta_u r_u) = -\left[\frac{1}{\sqrt{u^2 + c^2}}(0,0,1) - \frac{u}{(u^2 + c^2)^{\frac{3}{2}}}(\text{csinv}, -\text{ccosv}, u).\right].$$
$$(\text{cosv}, \text{sinv}, 0) = (0,0,0)$$

$$b_{12} = -\frac{1}{2}(\eta_1 r_2 + \eta_2 r_1) = \frac{c}{\sqrt{u^2 + c^2}}$$

$$b_{22} = -\frac{1}{2}(\eta_2 r_2 + \eta_2 r_2) = -\eta_2 r_2 = -\eta_v r_v = 0$$

$$\mathcal{B} \equiv b_{\alpha\beta} du^\alpha du^\beta = -2b_{12} du dv = \frac{-2c}{\sqrt{u^2 + c^2}} du dv.$$

Example 10.3.2. Find the second fundamental form for the paraboloid given by

$$r = (u, v, u^2 - v^2).$$

Solution: Here, $r = (u, v, u^2 - v^2)$

So $\mathbf{A} = \frac{\partial r}{\partial u^1} = (1,0,2u)$, $\mathbf{B} = \frac{\partial r}{\partial u^2} = (0,1,-2v)$

$$\mathbf{A} \times \mathbf{B} = \begin{vmatrix} i & j & k \\ 1 & 0 & 2u \\ 0 & 1 & -2v \end{vmatrix} = -2ui + 2vj + k$$

$$n = \frac{\mathbf{A} \times \mathbf{B}}{|\mathbf{A} \times \mathbf{B}|} = \frac{-2ui + 2vj + k}{\sqrt{4u^2 + 4v^2 + 1}}$$

$$b_{11} = -n_u . r_u = \frac{\left(2\{4v^2+1\},-8uv,4u\right).(1,o,2u)}{(4u^2+4v^2+1)^{\frac{3}{2}}} = \frac{2}{\sqrt{4u^2+4v^2+1}}$$

$$b_{22} = -n_v . r_v = \frac{\left(-8uv,-2\{4u^2+1\},4v\right).(0,1,-2v)}{(4u^2+4v^2+1)^{\frac{3}{2}}} = -\frac{2}{\sqrt{4u^2+4v^2+1}}$$

$$b_{12} = b_{21} = 0$$

$$\mathcal{B} \equiv b_{\alpha\beta} du^{\alpha} du^{\beta} = b_{11}(du^1)^2 + b_{22}(du^2)^2 + b_{12} du^1 du^2$$

$$= \frac{2}{\sqrt{4u^2+4v^2+1}}\left[(du^1)^2 - (du^2)^2\right].$$

10.4 Tensor Derivatives

We are introducing the tensor differentiation of tensor fields which are functions of both surface and space coordinates.

Let us consider a curve C lying on a given surface S and vector A^i defined along C, whose parameter is t:

$$\frac{\delta A^i}{\delta t} = \frac{dA^i}{dt} + \left\{ \begin{matrix} i \\ j \quad k \end{matrix} \right\}_g A^j \frac{dx^k}{dt}, \qquad (10.21)$$

where $\left\{ \begin{matrix} i \\ j \quad k \end{matrix} \right\}_g$ refers to space coordinates x^i and are formed from metric coefficient g_{ij}.

Again, if we consider a surface vector A^{α} defined along the same curve C, we can form the intrinsic derivative with respect to the surface variables

$$\frac{\delta A^{\alpha}}{\delta t} = \frac{dA^{\alpha}}{dt} + \left\{ \begin{matrix} \alpha \\ \beta \quad \gamma \end{matrix} \right\}_a A^{\beta} \frac{dx^{\gamma}}{dt}. \qquad (10.22)$$

$\left\{ \begin{matrix} \alpha \\ \beta\gamma \end{matrix} \right\}_a$ refer to Gaussian surface coordinates u^α and are formed from metric coefficient $a_{\alpha\beta}$.

A geometric interpretation of these formulas is at hand when the fields A^i *and* A^α are such that $\frac{\delta A^i}{\delta t} = 0$ and $\frac{\delta A^\alpha}{\delta t} = 0$. In the first equation, the vectors A^i form a parallel field with respect to C, considered a *space curve*, whereas $\frac{\delta A^\alpha}{\delta t} = 0$ defines a parallel field with respect to C as a *surface curve*.

Intrinsic derivatives of the covariant vectors A_i and A_α are

$$\frac{\delta A_i}{\delta t} = \frac{dA_i}{dt} - \left\{ \begin{matrix} k \\ i\ j \end{matrix} \right\}_g A_k \frac{dx^j}{dt} \tag{10.23}$$

and

$$\frac{\delta A_\alpha}{\delta t} = \frac{dA_\alpha}{dt} - \left\{ \begin{matrix} \gamma \\ \alpha\ \beta \end{matrix} \right\}_a A_\gamma \frac{dx^\beta}{dt} \tag{10.24}$$

Consider a tensor field T^i_α, which is a contravariant vector with respect to a transformation of space coordinates x^i and a covariant vector with respect to a transformation of surface vector u^α.

If T^i_α, is defined over a surface curve C and the parameter along C is t, T^i_α, is a function of t. We introduce a parallel vector field A_i along C, regarded as a space curve, and a parallel vector field B^α along C as a surface curve and form an invariant

$$\phi(t) = T^i_\alpha A_i B^\alpha .$$

Then,

$$\frac{d\phi(t)}{dt} = \frac{dT^i_\alpha}{dt} A_i B^\alpha + T^i_\alpha \frac{dA_i}{dt} B^\alpha + T^i_\alpha A_i \frac{dB^\alpha}{dt}, \tag{10.25}$$

but since $A_i(t)$ and $B^\alpha(t)$ are parallel, then $\frac{\delta A_i}{\delta t} = 0$ *and* $\frac{\delta B^\alpha}{\delta t} = 0$.

From (10.23) and (10.22), we have

$$\frac{dA_i}{dt} = \left\{ \begin{matrix} k \\ i\ j \end{matrix} \right\}_g A_k \frac{dx^j}{dt} \text{ and } \frac{dB^\alpha}{dt} = -\left\{ \begin{matrix} \alpha \\ \beta\ \gamma \end{matrix} \right\}_a B^\beta \frac{du^\gamma}{dt}.$$

(10.25) becomes

$$\frac{d\phi(t)}{dt}=\frac{dT^i_\alpha}{dt}A_i\,B^\alpha+T^i_\alpha\left\{\begin{matrix}k\\ i\;\;j\end{matrix}\right\}_g A_k\frac{dx^j}{dt}B^\alpha-T^i_\alpha A_i\left\{\begin{matrix}\alpha\\ \beta\;\;\gamma\end{matrix}\right\}_a B^\beta\frac{du^\gamma}{dt}$$

$$=\left[\frac{dT^i_\alpha}{dt}+T^j_\alpha\left\{\begin{matrix}i\\ j\;\;k\end{matrix}\right\}_g\frac{dx^k}{dt}-T^i_\delta\left\{\begin{matrix}\delta\\ \alpha\;\;\gamma\end{matrix}\right\}_a\frac{du^\gamma}{dt}\right]A_i\,B^\alpha. \tag{10.26}$$

By the quotient law we conclude that the expression in the bracket is a tensor of the same character as T^i_α. We call this tensor an intrinsic derivative of T^i_α, with respect to t, and write

$$\frac{\delta T^i_\alpha}{\delta t}=\frac{dT^i_\alpha}{dx}+T^j_\alpha\left\{\begin{matrix}i\\ j\;\;k\end{matrix}\right\}_g\frac{dx^k}{dt}-T^i_\delta\left\{\begin{matrix}\delta\\ \alpha\;\;\gamma\end{matrix}\right\}_a\frac{du^\gamma}{dt} \tag{10.27}$$

$$=\left[\frac{\partial T^i_\alpha}{\partial u^\gamma}+T^j_\alpha\left\{\begin{matrix}i\\ j\;\;k\end{matrix}\right\}_g x^k_\gamma-T^i_\delta\left\{\begin{matrix}\delta\\ \alpha\;\;\gamma\end{matrix}\right\}_a\right]\frac{du^\gamma}{dt}$$

$$\left[\text{since } x^k_\gamma=\frac{\partial x^k}{\partial u^\gamma};\frac{\partial x^k}{\partial u^\gamma}\frac{du^\gamma}{dt}=\frac{dx^k}{dt}\right]$$

Since $\frac{du^\gamma}{dt}$ is an arbitrary surface vector, we conclude that

$$T^i_{\alpha,\gamma}=\frac{\partial T^i_\alpha}{\partial u^\gamma}+\left\{\begin{matrix}i\\ \hline j\;\;k\end{matrix}\right\}_g T^j_\alpha\;\;x^k_\gamma-\left\{\begin{matrix}\delta\\ \hline \alpha\;\gamma\end{matrix}\right\}_a T^i_\delta \tag{10.28}$$

and call $T^i_{\alpha,\gamma}$ *a tensor derivative of* T^i_α with respect to u^γ.

The tensor derivative of $T^i_{\alpha,\beta}$ with respect to u^γ is given by

$$T^i_{\alpha\beta,\gamma}=\frac{\partial T^i_{\alpha\beta}}{\partial u^\gamma}+\left\{\begin{matrix}i\\ j\,k\end{matrix}\right\}_g T^j_{\alpha\beta}\,x^k_\gamma-\left\{\begin{matrix}\delta\\ \alpha\;\gamma\end{matrix}\right\}_a T^i_{\delta\beta}\;\;-\left\{\begin{matrix}\delta\\ \beta\;\gamma\end{matrix}\right\}_a T^i_{\alpha\delta} \tag{10.29}$$

If the surface coordinates at any n-point P *of* S are geodesic and the space coordinates are orthogonal Cartesian, we see that at that point the tensor derivatives reduce to the ordinary derivatives. It is concluded that the operations of tensor differentiatio of products and sums follow the usual rules and the tensor derivatives of g_{ij}, $a_{\alpha\beta}$, ϵ_{ijk}, $\epsilon_{\alpha\beta}$ and their associated tensors vanish.

10.5 Second Fundamental Form of a Surface

We begin by calculating the tensor derivative of the tensor x^i_α, representing the components of the surface base vectors a_α. We have

$$x^i_{\alpha,\beta} = \frac{\partial^2 x^i}{\partial u^\alpha \partial u^\beta} + \left\{ \begin{matrix} i \\ j\,k \end{matrix} \right\}_g x^j_\alpha x^k_\beta - \left\{ \begin{matrix} \delta \\ \alpha\,\beta \end{matrix} \right\}_a x^i_\delta \quad (10.30)$$

$$\left[\text{since, } x^i_\alpha = \frac{\partial x^i}{\partial u^\alpha} \right]$$

from which we get $x^i_{\alpha,\beta.} = x^i_{\beta,\alpha.}$ (10.31)

Since the tensor derivative of $a_{\alpha\beta}$ vanishes, $a_{\alpha\beta} = g_{ij} x^i_\alpha x^j_\beta$.
We obtain $g_{ij} x^i_{\alpha,\gamma} x^j_\beta + g_{ij} x^i_\alpha x^j_{\beta,\gamma} = 0,$ (10.32)
interchanging α, β, and γ cyclically,

$$g_{ij} x^i_{\beta,\alpha}, x^j_\gamma + g_{ij} x^i_\beta x^j_{\gamma,\alpha} = 0 \quad (10.33)$$

$$g_{ij} x^i_{\gamma,\beta}, x^j_\alpha + g_{ij} x^i_\alpha x^j_{\alpha,\beta} = 0. \quad (10.34)$$

From (10.32), (10.33), and (10.34) and using (10.31), we get

$$g_{ij} x^i_{\alpha,\beta} x^j_\gamma = 0 \quad (10.35)$$

This is the orthogonality relation which states that $x^i_{\alpha,\beta}$ is a space vector normal to the surface and it is directed along the unit normal $\boldsymbol{n}^i$.

Hence, there exists a set of functions $b_{\alpha\beta}$ such that

$$x^i_{\alpha,\beta} = b_{\alpha\beta} n^i . \tag{10.36}$$

The quantities $b_{\alpha\beta}$ are the components of a symmetric surface tensor and the differential quadratic form

$$\mathcal{B} \equiv b_{\alpha\beta} du^\alpha du^\beta \tag{10.37}$$

is the desired second fundamental form.

10.5.1 Equivalence of Definition of Tensor $b_{\alpha\beta}$

We define

$$b_{\alpha\beta} = -\frac{1}{2}\left(\frac{\partial \boldsymbol{n}}{\partial u^\alpha}\frac{\partial \boldsymbol{r}}{\partial u^\beta} + \frac{\partial \boldsymbol{n}}{\partial u^\beta}\frac{\partial \boldsymbol{r}}{\partial u^\alpha}\right).$$

We note that $\boldsymbol{n}$ and $\boldsymbol{a}_\alpha = \dfrac{\partial \boldsymbol{r}}{\partial u^\alpha}$ are orthogonal, hence

$$\boldsymbol{n}.\frac{\partial \boldsymbol{r}}{\partial u^\alpha} = 0 \ \ and \ \ \boldsymbol{n}.\frac{\partial \boldsymbol{r}}{\partial u^\beta} = 0.$$

Differentiating these two scalar products with respect to u^β *and* u^α, respectively, and adding, we get

$$\left(\frac{\partial \boldsymbol{n}}{\partial u^\alpha}\frac{\partial \boldsymbol{r}}{\partial u^\beta} + \frac{\partial \boldsymbol{n}}{\partial u^\beta}\frac{\partial \boldsymbol{r}}{\partial u^\alpha}\right) + 2n\left(\frac{\partial^2 \boldsymbol{r}}{\partial u^\beta \partial u^\alpha}\right) = 0$$

$$\text{or } \frac{1}{2}\left(\frac{\partial \boldsymbol{n}}{\partial u^\alpha}\frac{\partial \boldsymbol{r}}{\partial u^\beta} + \frac{\partial \boldsymbol{n}}{\partial u^\beta}\frac{\partial \boldsymbol{r}}{\partial u^\alpha}\right) = -\boldsymbol{n}\left(\frac{\partial^2 \boldsymbol{r}}{\partial u^\beta \partial u^\alpha}\right).$$

Hence,

$$b_{\alpha\beta} = \boldsymbol{n}\left(\frac{\partial^2 \boldsymbol{r}}{\partial u^\beta \partial u^\alpha}\right), \tag{10.38}$$

but,

$$\frac{\partial \boldsymbol{r}}{\partial u^\alpha} = \boldsymbol{a}_\alpha = \boldsymbol{b}_i x^i_\alpha .$$

Therefore,

$$\frac{\partial^2 \boldsymbol{r}}{\partial u^\beta \partial u^\alpha} = \boldsymbol{b}_i \frac{\partial x^i_\alpha}{\partial u^\beta} + \frac{\partial \boldsymbol{b}_i}{\partial u^\beta} x^i_\alpha$$

$$= \boldsymbol{b}_i \frac{\partial x^i_\alpha}{\partial u^\beta} + \frac{\partial \boldsymbol{b}_i}{\partial x^\beta} x^i_\alpha x^j_\beta$$

$$= \boldsymbol{b}_i \left(\frac{\partial x^i_\alpha}{\partial u^\beta} + {}_g\left\{ \begin{matrix} i \\ j\ k \end{matrix} \right\} x^j_\alpha x^k_\beta \right).$$

Using (10.30) in the above equations, we have

$$\frac{\partial^2 \boldsymbol{r}}{\partial u^\beta \partial u^\alpha} = \boldsymbol{b}_i \left(x^i_{\alpha,\beta} + {}_a\left\{ \begin{matrix} \delta \\ \alpha\ \beta \end{matrix} \right\} x^i_\delta \right) \tag{10.39}$$

Multiplying (10.39) scalarly by n *and considering* n is orthogonal to $a_\delta = \boldsymbol{b}_i x^i_\delta$ so that $n\ a_\delta = 0$,

therefore
$$\boldsymbol{n}.\frac{\partial^2 \boldsymbol{r}}{\partial u^\beta \partial u^\alpha} = \boldsymbol{n}.\boldsymbol{b}_i x^i_{\alpha,\beta}$$

$$= x^i_{\alpha,\beta} n_i$$

$$= b_{\alpha\beta} \text{ [by (10.36)]}.$$

This establishes the equivalence of the two definitions of the second fundamental form.

Equation (10.36) is known as the *formulas of Gauss.*

10.6 The Integrability Condition

In order to get insight into the significance of the tensor $b_{\alpha\beta}$, let us examine more closely the Gauss formulas

$$x^i_{\alpha,\beta} = b_{\alpha\beta} n^i, \tag{10.40}$$

where $$x^i_{\alpha,\beta} = \frac{\partial^2 y}{\partial u^\alpha \partial u^\delta} + {}_g\left\{ \begin{matrix} i \\ j\,k \end{matrix} \right\} x^j_\alpha x^k_\delta - {}_a\left\{ \begin{matrix} \delta \\ \alpha\,\beta \end{matrix} \right\} x^i_\delta$$

and $$n_i = \frac{1}{2}\epsilon^{\alpha\beta}\epsilon_{ijk}x^j_\alpha x^k_\beta$$

with $$x^j_\alpha = \frac{\partial x^j}{\partial u^\alpha}.$$

If we insert these expressions in Equation (10.40), we obtain a set of second order partial differential equations in which variables x^i are the functions of surface coordinates u^α.

The surface S is defined as $x^i = x^i(u^1,u^2)$ $(i=1,2,3)$, (10.41)
are immersed; they are also functions of

$$a_{\alpha\beta} = g_{ij}\frac{\partial x^i}{\partial u^\alpha}\frac{\partial x^j}{\partial u^\beta} = g_{ij}x^i_\alpha x^j_\beta \text{ and } b_{\alpha\beta}$$

In order for tensors $a_{\alpha\beta}$ *and* $b_{\alpha\beta}$ *to* be related to some surface, it is necessary that x^i satisfies the integrability conditions

$$\frac{\partial^2 x^i_\alpha}{\partial u^\gamma \partial u^\beta} = \frac{\partial^2 x^i_\alpha}{\partial u^\beta \partial u^\gamma}, \tag{10.42}$$

whenever the functions x^j_α are of class C^2, which is equivalent to

$$x^i_{\alpha,\beta\gamma} - x^i_{\alpha,\gamma\beta} = R^\sigma_{\alpha\beta\gamma}x^i_{\delta,} \tag{10.43}$$

where $R^\sigma_{\alpha\beta\gamma}$ is the Riemann tensor of the second kind formed with the coefficients $a_{\alpha\beta}$ of the first fundamental quadratic form. We shall see that integrability conditions (10.42) impose certain restrictions on possible choices of functions $a_{\alpha\beta}$ and $b_{\alpha\beta}$. These restricted conditions are known as the equations of *Gauss and Codazzi.*

Example 10.6.1. Show that $b_{\alpha\beta} = g_{ij}x^i_{\alpha,\beta}n^j = \frac{1}{2}\epsilon^{\gamma\delta}\epsilon_{ijk}\ x^i_{\alpha,\beta}x^j_\gamma x^k_\delta$.

Solution: Here, from Equation (10.10)

$$A^i = x^i_\alpha A^\alpha, B^j = x^j_\beta B^\beta.$$

If θ is the angle between A^i *and* B^j, we can write

$$ABsin\theta = \epsilon_{\alpha\beta} A^\alpha B^\beta.$$

If n is the unit surface normal, then

$$n_i = \frac{\mathbf{A}\times\mathbf{B}}{AB|sin\theta|} = \frac{\epsilon_{ijk}\, x^j_\alpha A^\alpha x^k_\beta B^\beta}{\epsilon_{\alpha\beta}\, A^\alpha B^\beta \,|}$$

or

$$\left(n\epsilon_{\alpha\beta} - \epsilon_{ijk}\, x^j_\alpha x^k_\beta\right) A^\alpha B^\beta = 0.$$

Since A^α *and* B^β are arbitrary, then

$$n_i\epsilon_{\alpha\beta} - \epsilon_{ijk}\, x^j_\alpha x^k_\beta = 0,$$

or

$$n_i\epsilon_{\alpha\beta} = \epsilon_{ijk}\, x^j_\alpha x^k_\beta$$

or

$$n_i\epsilon^{\alpha\beta}\epsilon_{\alpha\beta} = \epsilon^{\alpha\beta}\epsilon_{ijk}\, x^j_\alpha x^k_\beta,$$

or

$$n_i.2 = \epsilon^{\gamma\delta}\epsilon_{ijk}\, x^j_\gamma x^k_\delta$$

$$\therefore\ n_i = \frac{1}{2}\epsilon^{\gamma\delta}\epsilon_{ijk}\, x^j_\gamma x^k_\delta.$$

Using $x^i_{\alpha,\beta} = b_{\alpha\beta}n^i$, we have

$$b_{\alpha\beta} = x^i_{\alpha,\beta}n_i = x^i_{\alpha,\beta}g_{ij}n^j = g_{ij}x^i_{\alpha,\beta}n^j$$

$$\therefore\ b_{\alpha\beta} = x^i_{\alpha,\beta}n_i = x^i_{\alpha,\beta}\left(\frac{1}{2}\epsilon^{\gamma\delta}\epsilon_{ijk}\, x^j_\gamma x^k_\delta\right) = \frac{1}{2}\epsilon^{\gamma\delta}\epsilon_{ijk}\, x^j_\gamma x^k_\delta x^i_{\alpha,\beta}.$$

Combining the above two equations, we get

$$b_{\alpha\beta} = g_{ij}x^i_{\alpha,\beta}n^j = \frac{1}{2}\epsilon^{\gamma\delta}\epsilon_{ijk}\, x^j_\gamma x^k_\delta x^i_{\alpha,\beta}.$$

10.7 Formulás of Weingarten

Let n^i be the unit normal vector to surface S. We begin with the relation

$$g_{ij}n^i n^j = 1$$

and when we form its tensor derivative, we have $g_{ij}n^i_{,\alpha} n^j + g_{ij}n^i n^j_{,\alpha} = 0$

or $$g_{ij}n^i n^j_{,\alpha} = 0. \tag{10.44}$$

This shows that $n^j_{,\alpha}$, considered as a space vector, is orthogonal to the unit normal n^i and hence it lies in the tangent plane to the surface.

Accordingly, it can be represented as a linear form in the base vectors $x^i_{,\alpha}$. $x^i_{,\alpha}$ is also called a surface base vector.

$$n^i_{,\alpha} = c^\beta_\alpha x^i_\beta. \tag{10.45}$$

Since n^i is normal to the surface, we have the orthgonality relation

$$g_{ij}x^i_\alpha \eta^i = 0,$$

whose tensor derivative is

$$g_{ij}x^i_{\alpha,\beta}n^i + g_{ij}x^i_\alpha n^i_{,\beta} = 0, \tag{10.46}$$

but from Gauss's Formula: $$x^i_{\alpha,\beta} = b_{\alpha\beta}n^i, \tag{10.47}$$
so that the substitution from (10.46) and (10.45) in (10.47) yields

$$g_{ij}b_{\alpha\beta}n^i n^j + g_{ij}x^i_\alpha c^\gamma_\beta x^i_\gamma = 0,$$

or $$1.b_{\alpha\beta} + g_{ij}x^i_\alpha x^j_\gamma c^\gamma_\beta = 0, \qquad [\because g_{ij}n^i n^j = 1]$$

or $$b_{\alpha\beta} + g_{ij}x^i_\alpha x^j_\gamma c^\gamma_\beta = 0$$

Since, $a_{\alpha\beta} = g_{ij}x^i_\alpha x^j_\beta$

$$b_{\alpha\beta} + a_{\alpha\gamma}c^\gamma_\beta = 0.$$

We have

$$\therefore\ b_{\alpha\beta} = -a_{\alpha\gamma} c^{\gamma}_{\beta}.$$

Solving this equation for c^{γ}_{β}, we get

$$c^{\gamma}_{\beta} = -a^{\alpha\gamma} b_{\alpha\beta}.,\quad (since,\ c^{\beta}_{\delta} = -a^{\alpha\beta} b_{\alpha\delta})$$

So that Equation (10.45) becomes

$$\begin{aligned} n^{i}_{,\delta} &= c^{\beta}_{\delta} x^{i}_{\beta.} \\ &= -a^{\alpha\beta} b_{\alpha\delta}\ x^{i}_{\beta.} \\ n^{i}_{,\alpha} &= -a^{\beta\gamma} b_{\beta\alpha}\ x^{i}_{\gamma.} \end{aligned} \tag{10.48}$$

These are the *Weingarten formulas.*

We can use these equations in deriving the equations of Gauss and Codazzi.

10.7.1 Third Fundamental Form

If we write

$$c_{\alpha\beta} = g_{ij}\ n^{i}_{,\alpha}\ n^{j}_{,\beta},$$

we see that $c_{\alpha\beta}$ is a symmetric covariant surface tensor of type (0,2) and we call the quadratic form $C \equiv c_{\alpha\beta} du^{\alpha} du^{\beta}$ the third fundamental form of the surface.

Using Weingarten formula (10.48), we get

$$\begin{aligned} C_{\alpha\beta} &= g_{ij}\ n^{i}_{,\alpha}\ n^{j}_{,\beta} = g_{ij}\left(-a^{\delta\gamma} b_{\delta\alpha}\ x^{i}_{\gamma.}\right)\left(-a^{\mu\vartheta} b_{\mu\beta}\ x^{j}_{\vartheta.}\right) = a_{\gamma\vartheta} a^{\delta\gamma} b_{\delta\alpha} a^{\mu\vartheta} b_{\mu\beta} \\ &= a^{\mu\delta} b_{\delta\alpha} b_{\mu\beta}. \end{aligned}$$

This is the relation between three fundamental forms on a surface, but here the third fundamental form is not an actual fundamental form because this can be obtained from first and second fundamental forms.

10.8 Equations of Gauss and Codazzi

Now we will establish Codazzi's Equation using the Weingarten Equation.

We follow the integrability conditions

$$x^i_{\alpha,\beta\gamma} - x^i_{\alpha,\gamma\beta} = R^\sigma_{\alpha\beta\gamma} x^i_\delta. \tag{10.49}$$

Now, from tensor derivative Equation (10.47), we use (10.48) to obtain: differentiating (10.47) and substituting (10.48)

$$x^i_{\alpha,\beta\gamma} = b_{\alpha\beta,\gamma} n^i + b_{\alpha\beta} n^i_{,\gamma}$$

$$= b_{\alpha\beta,\gamma} \,.\, n^i - b_{\alpha\beta}\ a^{\delta\lambda} b_{\delta\gamma} x^i_{\lambda.}$$

Similarly, $$x^i_{\alpha,\gamma\beta} = b_{\alpha\gamma,\beta} \,.\, n^i - b_{\alpha\gamma}\ a^{\delta\lambda} b_{\delta\beta} x^i_{\lambda.} \tag{10.50}$$

Substituting it in (10.49), we get

$$\left(x^i_{\alpha,\beta\gamma} - x^i_{\alpha,\gamma\beta}\right) = \left(b_{\alpha\beta,\gamma} - b_{\alpha\gamma,\beta}\ \right).\ n^i - a^{\delta\lambda}\left(b_{\alpha\beta} b_{\delta\gamma} - b_{\alpha\gamma} b_{\delta\beta}\right) x^i_{\lambda.}$$

Hence,

$$\left(b_{\alpha\beta,\gamma} - b_{\alpha\gamma,\beta}\ \right).\ n^i - a^{\delta\lambda}\left(b_{\alpha\beta} b_{\delta\gamma} - b_{\alpha\gamma} b_{\delta\beta}\right) x^i_{\lambda.} = R^\sigma_{\alpha\beta\gamma} x^i_\delta. \tag{10.51}$$

To obtain the equation of Codazzi, we multiply (10.51) by n_i and since $x^i_\alpha n_i = 0$ (by using $g_{ij} x^i_\alpha \eta.^i = 0$)

$$\left(b_{\alpha\beta,\gamma} - b_{\alpha\gamma,\beta}\ \right).\, n_i\ n^i - a^{\delta\lambda}\left(b_{\alpha\beta} b_{\delta\gamma} - b_{\alpha\gamma} b_{\delta\beta}\right) x^i_{\lambda.} n_i = R^\sigma_{\alpha\beta\gamma} x^i_\delta n_i,$$

we get desire result: $$\left(b_{\alpha\beta,\gamma} - b_{\alpha\gamma,\beta}\ \right) = 0. \tag{10.52}$$

This is known as the *Codazzi Equation,*
which will constitute all integrability conditions of the formula of Weingarten.

To obtain the *equations of Gauss,*
we multiply (10.51) by $g_{ij} x^j_\rho$ and get

$$g_{ij} x^j_\rho \left(b_{\alpha\beta,\gamma} - b_{\alpha\gamma,\beta}\ \right).\ n^i - a^{\delta\lambda}\left(b_{\alpha\beta} b_{\delta\gamma} - b_{\alpha\gamma} b_{\delta\beta}\right) x^i_{\lambda.} g_{ij} x^j_\rho = R^\sigma_{\alpha\beta\gamma} x^i_\delta g_{ij} x^j_\rho.$$

Since $g_{ij}x^j_\rho x^i_\sigma = a_{\sigma p}$ and 1st term be zero, we have

$$a_{\sigma p}R^{\sigma}_{\alpha\beta\gamma} = -a^{\delta\lambda}\left(b_{\alpha\beta}b_{\delta\gamma} - b_{\alpha\gamma}b_{\delta\beta}\right)a_{\lambda\rho}.$$

Since $a^{\delta\lambda}a_{\lambda\rho} = \delta^{\delta}_{\rho}$

$$R_{\rho\alpha\beta\gamma} = b_{\alpha\gamma}b_{\rho\beta} - b_{\alpha\beta}b_{\rho\gamma} \tag{10.53}$$

which is known as the *Gauss Equation.*

Since α,β assumes values 1,2 and $b_{\alpha\beta}=b_{\beta\alpha}$, we see that there are two independent equations of Codazzi and only one independent equation of Gauss.

The independent equations of Codazzi are

$$b_{\alpha\alpha,\beta} - b_{\alpha\beta,\alpha} = 0 \ (\alpha \neq \beta) \ (\text{no sum on } \alpha).$$

We know

$$b_{\alpha\beta,\gamma} = \frac{\partial b_{\alpha\beta}}{\partial u^{\gamma}} - {}_a\begin{Bmatrix}\delta \\ \alpha\ \gamma\end{Bmatrix} b_{\delta\beta} - {}_a\begin{Bmatrix}\delta \\ \beta\ \gamma\end{Bmatrix} b_{\alpha\delta}$$

$$\left[\therefore R_{\alpha\alpha\beta\gamma} = R_{\alpha\beta\gamma\gamma} = 0\ ;\ R_{1212} = R_{2121} = -R_{2112} = -R_{1221}\right].$$

We have

$$b_{\alpha\alpha,\beta} - b_{\alpha\beta,\alpha} = \frac{\partial b_{\alpha\alpha}}{\partial u^{\beta}} - {}_a\begin{Bmatrix}\delta \\ \alpha\ \beta\end{Bmatrix} b_{\delta\alpha} - {}_a\begin{Bmatrix}\delta \\ \alpha\ \beta\end{Bmatrix}$$

$$b_{\alpha\delta} - \frac{\partial b_{\alpha\beta}}{\partial u^{\alpha}} + {}_a\begin{Bmatrix}\delta \\ \alpha\ \alpha\end{Bmatrix} b_{\delta\beta} + {}_a\begin{Bmatrix}\delta \\ \beta\ \alpha\end{Bmatrix} b_{\alpha\delta}$$

Therefore, $$\frac{\partial b_{\alpha\alpha}}{\partial u^{\beta}} - \frac{\partial b_{\alpha\beta}}{\partial u^{\alpha}} - {}_a\begin{Bmatrix}\delta \\ \alpha\ \beta\end{Bmatrix} b_{\delta\alpha} + {}_a\begin{Bmatrix}\delta \\ \alpha\ \alpha\end{Bmatrix} b_{\delta\beta} = 0 \tag{10.54}$$
$\alpha \neq \beta$ no sum on α
and the equation of Gauss,

$$R_{\rho\alpha\beta\gamma} = b_{\alpha\gamma}\, b_{\rho\beta} - b_{\alpha\beta}\, b_{\rho\gamma}.$$

Here, $\rho=1, \alpha=2, \beta=1, \gamma=2,$

therefore, $$R_{1212}=b_{22}\,b_{11}-b_{21}\,b_{12}=b_{22}\,b_{11}-b^2{}_{12} \tag{10.55}$$
[since other $R_{\rho\alpha\beta\gamma}=0$].

This equation relates the coefficients $b_{\alpha\beta}$ and $a_{\alpha\beta}$ in the two fundamental quadratic forms.

10.9 Mean and Total Curvatures of a Surface

We know the total curvature (from 9.39)

$$\kappa=\frac{R_{1212}}{a}\text{, where } a=|a_{\alpha\beta}|.$$

We can write Equation (10.55) in the form

$$\kappa=\frac{R_{1212}}{a}=\frac{b_{22}b_{11}-b^2{}_{12}}{a}=\frac{b}{a}. \tag{10.56}$$

The Gaussian Curvature is equal to the quotient of the determinants of the second and first fundamental quadratic forms.

We can define another important invariant H, given by the formula

$$H\equiv\frac{1}{2}a^{\alpha\beta}b_{\alpha\beta}, \tag{10.57}$$

which is called the *mean curvature of the surface.*

Example 10.9.1. Prove that $C-2HB+\kappa A=0$, where the notations have their usual meaning.

Solution: Here, the third fundamental form is $(C)=c_{\alpha\beta}\,du^\alpha\,du^\beta$,
the second fundamental form is $(B)=b_{\alpha\beta}\,du^\alpha\,du^\beta$,
the first fundamental form is $(A)=a_{\alpha\beta}\,du^\alpha\,du^\beta$

$\kappa=\frac{R_{1212}}{a}$, and the mean curvature is $H=\frac{1}{2}a^{\alpha\beta}\,b_{\alpha\beta}$.

The Gauss formula is $R_{\rho\alpha\beta\gamma}=b_{\alpha\gamma}\,b_{\rho\beta}-b_{\alpha\beta}\,b_{\rho\gamma}$ and $R_{\rho\alpha\beta\gamma}=\kappa\epsilon_{\rho\alpha}\epsilon_{\beta\gamma}$

$$\therefore\quad \kappa\epsilon_{\rho\alpha}\epsilon_{\beta\gamma}=b_{\alpha\gamma}b_{\rho\beta}-b_{\alpha\beta}b_{\rho\gamma}.$$

Multiplying both sides by $a^{\rho\gamma}$, we get

$$\kappa\epsilon_{\rho\alpha}\epsilon_{\beta\gamma}a^{\rho\gamma} = a^{\rho\gamma}b_{\alpha\gamma}b_{\rho\beta} - a^{\rho\gamma}b_{\alpha\beta}b_{\rho\gamma},$$

or
$$-\kappa a_{\alpha\beta} = a^{\rho\gamma}b_{\alpha\gamma}b_{\rho\beta} - a^{\rho\gamma}b_{\alpha\beta}b_{\rho\gamma} \text{ as } -a_{\alpha\beta} = \epsilon_{\rho\alpha}\epsilon_{\beta\gamma}a^{\rho\gamma},$$

or
$$\kappa a_{\alpha\beta} = 2Hb_{\alpha\gamma} - c_{\alpha\beta} \text{ since } c_{\alpha\beta} = a^{\rho\gamma}b_{\alpha\beta}b_{\rho\gamma},$$

or
$$c_{\alpha\beta} - 2Hb_{\alpha\gamma} + \kappa a_{\alpha\beta} = 0,$$

or
$$c_{\alpha\beta}du^{\alpha}du^{\beta} - 2Hb_{\alpha\gamma}du^{\alpha}du^{\beta} + \kappa a_{\alpha\beta}du^{\alpha}du^{\beta} = 0$$

$$\therefore \quad C - 2HB + \kappa A = 0.$$

Example 10.9.2. Show that $c_{\alpha\beta}a^{\alpha\beta} = 4H^2 - 2\kappa$.

Solution: Using the relations from above,

$$\kappa a_{\alpha\beta} = 2Hb_{\alpha\gamma} - c_{\alpha\beta},$$

or
$$\kappa a_{\alpha\beta}\lambda^{\alpha}\lambda^{\beta} = 2Hb_{\alpha\gamma}\lambda^{\alpha}\lambda^{\beta} - c_{\alpha\beta}\lambda^{\alpha}\lambda^{\beta},$$

or
$$\kappa = 2Hb_{\alpha\beta}\lambda^{\alpha}\lambda^{\beta} - c_{\alpha\beta}\lambda^{\alpha}\lambda^{\beta}.$$

Using this relation $\kappa a_{\alpha\beta} = 2Hb_{\alpha\beta} - c_{\alpha\beta}$
and multiplying $a^{\alpha\beta}$, we get

$$\kappa a_{\alpha\beta}a^{\alpha\beta} = 2Hb_{\alpha\beta}a^{\alpha\beta} - c_{\alpha\beta}a^{\alpha\beta}$$

$2\kappa = 2H.2H - c_{\alpha\beta}a^{\alpha\beta}$ as $a_{\alpha\beta}a^{\alpha\beta} = 2$.
Therefore, $c_{\alpha\beta}a^{\alpha\beta} = 4H^2 - 2\kappa$.

Example 10.9.3. Find the Gaussian and mean curvature of the surface

$$x^1\, x^2 = x^3.$$

Solution: Let the parametric representation of the surface be given by

$$x^1 = u^1,\ x^2 = u^2 \text{ and } x^3 = u^1\, u^2$$

$$\therefore\ r = \left(x^1, x^2, x^3\right) = \left(u^1, u^2, u^1 u^2\right)$$

$$a_{\alpha\beta} = \frac{\partial y^i}{\partial u^\alpha}\frac{\partial y^i}{\partial u^\beta}$$

$$\therefore\ a_{11} = \frac{\partial y^i}{\partial u^1}\frac{\partial y^i}{\partial u^1} = \left(\frac{\partial y^1}{\partial u^1}\right)^2 + \left(\frac{\partial y^2}{\partial u^1}\right)^2 + \left(\frac{\partial y^3}{\partial u^1}\right)^2 = 1 + (u^2)^2$$

$$a_{22} = \frac{\partial y^i}{\partial u^2}\frac{\partial y^i}{\partial u^2} = \left(\frac{\partial y^1}{\partial u^2}\right)^2 + \left(\frac{\partial y^2}{\partial u^2}\right)^2 + \left(\frac{\partial y^3}{\partial u^2}\right)^2 = 1 + (u^1)^2$$

$$a_{12} = \frac{\partial y^i}{\partial u^1}\frac{\partial y^i}{\partial u^2} = \frac{\partial y^1}{\partial u^1}\frac{\partial y^1}{\partial u^2} + \frac{\partial y^2}{\partial u^1}\frac{\partial y^2}{\partial u^2} + \frac{\partial y^3}{\partial u^1}\frac{\partial y^3}{\partial u^2} = u^1 u^2.$$

Therefore, $a = \begin{vmatrix} 1+(u^2)^2 & u^1 u^2 \\ u^1 u^2 & 1+(u^1)^2 \end{vmatrix} = 1 + (u^1)^2 + (u^2)^2$

$$a^{11} = \frac{1+(u^1)^2}{1+(u^1)^2+(u^2)^2},\quad a^{22} = \frac{1+(u^2)^2}{1+(u^1)^2+(u^2)^2},\quad a^{12} = \frac{-u^1 u^2}{1+(u^1)^2+(u^2)^2}.$$

Since $r = \left(x^1, x^2, x^3\right) = \left(u^1, u^2, u^1 u^2\right)$,
here we have $\boldsymbol{A} = (1, 0, u^2)$ and $\boldsymbol{B} = (0, 1, u^1)$

The normal vector is $n = \dfrac{A \times B}{|A \times B|} = \dfrac{1}{\sqrt{1+(u^1)^2+(u^2)^2}}\left(-u^2, -u^1, 1\right).$

Now, $b_{\alpha\beta} = -\dfrac{1}{2}\left(\dfrac{\partial n}{\partial u^\alpha}\cdot\dfrac{\partial r}{\partial u^\beta} + \dfrac{\partial n}{\partial u^\beta}\cdot\dfrac{\partial r}{\partial u^\alpha}\right)$

$$b_{11}=-\frac{1}{2}\left(\frac{\partial n}{\partial u^1}\cdot\frac{\partial r}{\partial u^1}+\frac{\partial n}{\partial u^1}\cdot\frac{\partial r}{\partial u^1}\right)=-\frac{\partial n}{\partial u^1}\cdot\frac{\partial r}{\partial u^1}=-\frac{1}{\sqrt{1+(u^1)^2+(u^2)^2}}$$

$$(0,-1,0).(1,0,u^2)=0$$

$$b_{22}=-\frac{1}{2}\left(\frac{\partial n}{\partial u^2}\cdot\frac{\partial r}{\partial u^2}+\frac{\partial n}{\partial u^2}\cdot\frac{\partial r}{\partial u^2}\right)=-\frac{\partial n}{\partial u^2}\cdot\frac{\partial r}{\partial u^2}=-\frac{1}{\sqrt{1+(u^1)^2+(u^2)^2}}$$

$$(-1,0,0).(0,1,u^1)=0$$

$$b_{12}=-\frac{1}{2}\left(\frac{\partial n}{\partial u^1}\cdot\frac{\partial r}{\partial u^2}+\frac{\partial n}{\partial u^2}\cdot\frac{\partial r}{\partial u^1}\right)=-\frac{1}{2}\left(-\frac{1+(u^1)^2+(u^2)^2}{\{1+(u^1)^2+(u^2)^2\}^{\frac{3}{2}}}-\frac{1+(u^1)^2+(u^2)^2}{\{1+(u^1)^2+(u^2)^2\}^{\frac{3}{2}}}\right)$$

$$=\frac{1}{\sqrt{1+(u^1)^2+(u^2)^2}}=b_{21}.$$

$$\text{Therefore, } b=\begin{vmatrix} 0 & \frac{1}{\sqrt{1+(u^1)^2+(u^2)^2}} \\ \frac{1}{\sqrt{1+(u^1)^2+(u^2)^2}} & 0 \end{vmatrix}$$

$$\kappa=\frac{\frac{1}{1+(u^1)^2+(u^2)^2}}{1+(u^1)^2+(u^2)^2}=\frac{1}{\left[1+(u^1)^2+(u^2)^2\right]^2}.$$

The mean curvature H is given by

$$H=\frac{1}{2}a^{\alpha\beta}\,b_{\alpha\beta}=\frac{1}{2}\left(a^{11}\,b_{11}+a^{22}\,b_{22}+2a^{12}\,b_{12}\right)=a^{12}\,b_{12}$$

$$=\frac{-u^1u^2}{1+(u^1)^2+(u^2)^2}\times\frac{1}{\sqrt{1+(u^1)^2+(u^2)^2}}=\frac{-u^1u^2}{\{1+(u^1)^2+(u^2)^2\}^{\frac{3}{2}}}.$$

Example 10.9.4. Show that the given surface

$$x^1 = u^1 \cos u^2,\ x^2 = u^1 sinu^2 \ and \ x^3 = cu^2$$

is a minimal surface.
Solution: The first fundamental form of the surface is

$$a_{\alpha\beta} = \frac{\partial y^i}{\partial u^\alpha}\frac{\partial y^i}{\partial u^\beta}$$

$$\therefore\ a_{11} = \frac{\partial y^i}{\partial u^1}\frac{\partial y^i}{\partial u^1} = \left(\frac{\partial y^1}{\partial u^1}\right)^2 + \left(\frac{\partial y^2}{\partial u^1}\right)^2 + \left(\frac{\partial y^3}{\partial u^1}\right)^2$$
$$= (\cos u^2)^2 + (\sin u^2)^2 = 1$$

$$a_{22} = \frac{\partial y^i}{\partial u^2}\frac{\partial y^i}{\partial u^2} = \left(\frac{\partial y^1}{\partial u^2}\right)^2 + \left(\frac{\partial y^2}{\partial u^2}\right)^2 + \left(\frac{\partial y^3}{\partial u^2}\right)^2$$
$$= (-u^1 sinu^2)^2 + (u^1 cosu^2)^2 + c^2 = (u^1)^2 + c^2$$

$$a_{12} = \frac{\partial y^i}{\partial u^1}\frac{\partial y^i}{\partial u^2} = \frac{\partial y^1}{\partial u^1}\frac{\partial y^1}{\partial u^2} + \frac{\partial y^2}{\partial u^1}\frac{\partial y^2}{\partial u^2} + \frac{\partial y^3}{\partial u^1}\frac{\partial y^3}{\partial u^2}$$
$$= \cos u^2\left(-u^1 \sin u^2\right) + \left(sinu^2\right)\left(u^1 \cos u^2\right) = 0$$

$$\therefore\ a = \begin{vmatrix} 1 & 0 \\ 0 & (u^1)^2 + c^2 \end{vmatrix} = (u^1)^2 + c^2$$

$$a^{11} = \frac{(u^1)^2 + c^2}{(u^1)^2 + c^2} = 1, a^{22} = \frac{1}{(u^1)^2 + c^2}, a^{12} = a^{21} = 0.$$

Here, $r = \left(x^1, x^2, x^3\right) = \left(u^1 \cos u^2, u^1 sinu^2, cu^2\right)$.

We have $A = \left(\frac{\partial x^1}{\partial u^1}, \frac{\partial x^2}{\partial u^1}, \frac{\partial x^3}{\partial u^1}\right) = \left(\cos u^2, sinu^2, 0\right); B = \left(\frac{\partial x^1}{\partial u^2}, \frac{\partial x^2}{\partial u^2}, \frac{\partial x^3}{\partial u^2}\right)$
$= \left(-u^1 \sin u^2, u^1 cosu^2, c\right)$

$$A \times B = \left(c sinu^2, -c \cos u^2, u^1\right)$$

$$n = \frac{1}{\sqrt{(u^1)^2 + c^2}} \left(c sinu^2, -c \cos u^2, u^1\right).$$

Now, $b_{\alpha\beta} = -\frac{1}{2}\left(\frac{\partial n}{\partial u^\alpha} \cdot \frac{\partial r}{\partial u^\beta} + \frac{\partial n}{\partial u^\beta} \cdot \frac{\partial r}{\partial u^\alpha}\right)$

$$b_{11} = -\frac{1}{2}\left(\frac{\partial n}{\partial u^1} \cdot \frac{\partial r}{\partial u^1} + \frac{\partial n}{\partial u^1} \cdot \frac{\partial r}{\partial u^1}\right) = -\frac{\partial n}{\partial u^1} \cdot \frac{\partial r}{\partial u^1} = -\frac{1}{\sqrt{(u^1)^2 + c^2}}(0,0,1).$$

$$\left(\cos u^2, sinu^2, 0\right) = 0$$

$$b_{22} = -\frac{\partial n}{\partial u^2} \cdot \frac{\partial r}{\partial u^2} = -\frac{1}{\sqrt{(u^1)^2 + c^2}}\left(c cosu^2, c \sin u^2, 0\right).$$

$$\left(-u^1 \sin u^2, u^1 cosu^2, c\right) = 0$$

$$b_{12} = -\frac{1}{2}\left(\frac{\partial n}{\partial u^1} \cdot \frac{\partial r}{\partial u^2} + \frac{\partial n}{\partial u^2} \cdot \frac{\partial r}{\partial u^1}\right) = -\frac{1}{2} \frac{1}{\sqrt{(u^1)^2 + c^2}}$$

$$\left[(0,0,1).\left(-u^1 \sin u^2, u^1 cosu^2, c\right) + \left(c cosu^2, c \sin u^2, 0\right).\left(\cos u^2, sinu^2, 0\right)\right]$$

$$= \frac{-c}{\sqrt{(u^1)^2 + c^2}} = b_{21}$$

$$H = \frac{1}{2} a^{\alpha\beta}\, b_{\alpha\beta} = \frac{1}{2}\left(a^{11}\, b_{11} + a^{22}\, b_{22} + 2a^{12}\, b_{12}\right) =$$

$$= \frac{1}{2}\left(1.0 + \frac{1}{(u^1)^2 + c^2}.0 + 2.0.\frac{-c}{\sqrt{(u^1)^2 + c^2}}\right) = 0.$$

Since the mean curvature $H=0$ at every point of the given surface, it is a minimal surface.

10.10 Exercises

1. Show that in the usual notations

$$b_{\alpha\beta} = -\frac{1}{2}\left(\frac{\partial n}{\partial u^{\alpha}}\cdot\frac{\partial r}{\partial u^{\beta}} + \frac{\partial n}{\partial u^{\beta}}\cdot\frac{\partial r}{\partial u^{\alpha}}\right),$$

   where n is the unit normal and r is the position vector of the point on the surface.
2. Find the mean curvature of the surface $r=(u,v,u^2-v^2)$.
3. Calculate the Gaussian and mean curvature of the given surface

$$y^1=u^1 \cos u^2$$
$$y^2=u^1 \sin u^2$$
$$y^3=f(u^2).$$

4. When the equation of surface S, referred to a set of orthogonal Cartesian axes, is taken in the form $y^3 = f(y^1, y^2)$, in parametric form $y^1 = u^1, y^2 = u^2, y^3 = f(u^1, u^2)$, and partial derivatives of $f(y^1, y^2)$ are denoted by $f_{y^1} \equiv p, f_{y^2} \equiv q, f_{y^1y^1} \equiv r, f_{y^1y^2} \equiv s, f_{y^2y^2} \equiv t,$ then the coefficients of $\alpha_{\alpha\beta}$ in $ds^2 = a_{\alpha\beta}du^{\alpha}du^{\beta}$

$$a_{11} = 1+p^2, a_{12} = pq, a_{22} = 1+q^2$$

   and the coefficients of $b_{\alpha\beta}$ of the second fundamental form are

$$b_{11} = \frac{r}{\sqrt{1+p^2+q^2}}, b_{12} = \frac{s}{\sqrt{1+p^2+q^2}}, b_{22} = \frac{t}{\sqrt{1+p^2+q^2}}.$$

   (i) Show these.

   (ii) Compute $\alpha_{\alpha\beta}$ and $b_{\alpha\beta}$ for the surface of the sphere: $y^3 = \sqrt{a^2 - (y^1)^2 - (y^2)^2}$.

11
Curves on a Surface

11.1 Introduction

We assumed in last chapter that if the surface coordinates are u^α and space coordinates x^i, then the equations of a surface are given by

$$S: x^i = x^i\,(u^1, u^2). \tag{11.1}$$

If the coordinates u^α are given as functions of a parameter, then the point represented by these coordinates describes a curve on the surface as the parameter varies. If we shall take the arc length s of the curve as the arc parameter, then the equations of a smooth curve C lying on surface S *is*

$$C: u^\alpha = u^\alpha\,(s). \tag{11.2}$$

If the values of u^α (s) are inserted in (11.1), we obtain the space coordinates x^i of C in the form

$$S: x^i = x^i\,(u^1, u^2),$$

$$S: x^i = x^i\,(u^1, u^2 = x^i\,(u^1(s), u^2(s)) = x^i(s), \tag{11.3}$$

which is the equation of C regarded as a space curve.

The properties of C can then be studied with the aid of the Serret-Frenet formulas by analyzing the rates of the unit tangent vector λ, the unit principal normal μ, and the unit binormal ν. Then, its curvature χ and its torsion τ are connected with these vectors by Serret-Frenet formulas as

Dipankar De. Introduction to Differential Geometry with Tensor Applications, (349–380) © 2022 Scrivener Publishing LLC

$$\frac{\delta\lambda^i}{\delta s} = \chi\mu^i \quad \chi > 0$$
$$\frac{\delta\mu^i}{\delta s} = \tau\nu^i - \chi\lambda^i$$
$$\frac{\delta\nu^i}{\delta s} = -\tau\mu^i.$$

11.2 Curve on a Surface: Theorem of Meusnier

If we regard C as a surface curve, defined by equation (11.2), the components λ^α of unit tangent vector λ are related to the space components λ^i of the same vector by the formulas

$$\lambda^i = \frac{\mathrm{dxi}}{\mathrm{ds}} = \frac{\partial \mathrm{x}^i}{\partial u^\alpha}\frac{\mathrm{d}u^\alpha}{\mathrm{ds}} \equiv \mathrm{x}^i_a \lambda^\alpha, \tag{11.4}$$

$$\text{where } \lambda^\alpha = \frac{\mathrm{d}u^\alpha}{\mathrm{ds}}.$$

The unit surface vector η^α normal to λ^α and χ_g is geodesic curvature of C, as shown in Figure (11.1), so that

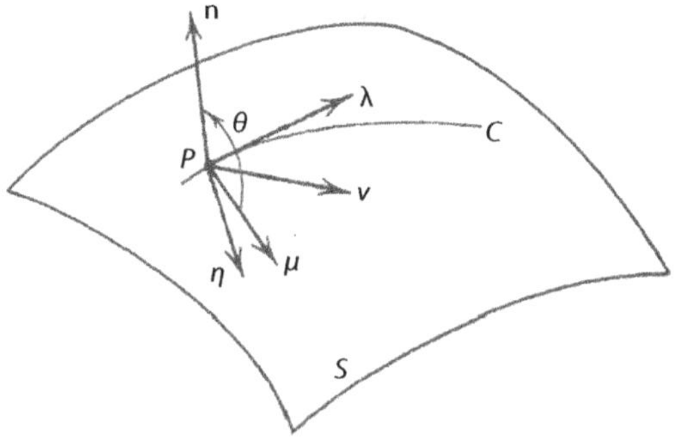

Figure 11.1

$$\frac{\partial \lambda^{\alpha}}{\partial s} = \chi_g \eta^{\alpha}, \tag{11.5}$$

where η^{α} is the unit normal to C in the tangent plane to the surface and χ_g is the geodesic curvature of C.

If we differentiate (11.4) intrinsically with respect to s, we obtain

$$\frac{\delta \lambda^{i}}{\delta s} = x^{i}_{\alpha,\beta} \lambda^{\alpha} \frac{du^{\beta}}{ds} + x^{i}_{\alpha} \frac{\delta \lambda^{\alpha}}{\delta s}.$$

or $x\mu^{i} = x^{i}_{\alpha,\beta}\lambda^{\alpha}\lambda^{\beta} + x^{i}_{\alpha}\chi_{g}\eta^{\alpha}$ (using the Frenet Formula and (11.5))

The space components η^{i} of η are $\eta^{i} = x^{i}_{\alpha}\eta^{\alpha}$ and the Gauss formula is $x^{i}_{\alpha,\beta} = b_{\alpha\beta}n^{i}$.

The equation becomes

$$\chi\mu^{i} = b_{\alpha\beta}\, n^{i}\, \lambda^{\alpha}\lambda^{\beta} + \chi_{g}\, \eta^{i}, \tag{11.6}$$

where n^{i} is the unit normal to the surface S.

Equation (11.6) states that the principal normal μ to C lies in the plane of the vectors n and η. Since n, η and λ are orthogonal, $n \times \eta = \lambda$.

We get $$\epsilon_{ijk}\, n^{j}\, \eta^{k} = \lambda_{i}. \tag{11.7}$$

and since λ is orthogonal to the plane of n *and* μ,

$$\therefore \mu \times n = -\sin\theta.\ \lambda. \tag{11.8}$$

$$\text{or } \mu \times n = \epsilon_{ijk}\, \mu^{j}\, n^{k} = -\sin\theta.\ \lambda$$

(θ is the angle between n *and* μ and $\mu \times n$ is a vector along λ)

On multiplying (11.8) by χ, we get

$$\epsilon_{ijk}\, \mu^{j}\, n^{k}\, \chi = -\chi \sin\theta\, \lambda_{i}.$$

Substituting the value of $\chi\mu^{i}$ from (11.6) in above equation,

$$\epsilon_{ijk}\, n^{k}\, (b_{\alpha\beta}\, n^{j}\, \lambda^{\alpha}\, \lambda^{\beta} + \chi_{g}\, \eta^{j}) = -\chi \sin\theta\, \lambda_{i},$$

but $$\epsilon_{ijk}\, n^j\, n^k = 0 \; and \; \epsilon_{ijk}\, \eta^j\, n^k = -\lambda_i$$

$$\chi_g\,(-\lambda_i) = -\chi sin\theta\, \lambda_i.$$

We have

$$\chi_g = \chi sin\theta. \tag{11.9}$$

On the other hand, if we form the scalar product of both members of Equation (11.6) with n_i and note that

$$n_i.\, \mu^i = cos\theta$$

$$(\text{since, } n^i\, n_i = 1,\; n_i.\, \eta^i = 0),$$

$$n_i.\, \chi\mu_i = n_i.\, b_{\alpha\beta}\, n^i\, \lambda^\alpha\, \lambda^\beta + n^i.\, \chi_g\, \eta^i$$

$$\chi cos\theta = b_{\alpha\beta}\, \lambda^\alpha\, \lambda^\beta. \tag{11.10}$$

The invariant $b_{\alpha\beta}\, \lambda^\alpha\, \lambda^\beta$ in (11.10) has the same value for all curves on S with the same tangent vector $\boldsymbol{\lambda}$ *at P*. In particular, it has this value for the curve formed by the intersection of the normal plane containing $\boldsymbol{n}$ and $\boldsymbol{\lambda}$, but every normal plane section angle θ is either 0 or π radians, so the normal plane section is

$$\chi cos\theta = \chi \text{ or } -\chi.$$

Since the rhs of (11.10) is an invariant, the value of $\chi cos\theta$ for every curve C, tangent to λ is equal to the curvature $\chi_{(n)}$ of the normal section in the direction λ.

The curvature $\chi_{(n)}$ is called the *normal curvature of the surface* S in the direction λ.

We can thus write (11.10) as

$$\chi_{(n)} = b_{\alpha\beta}\, \lambda^\alpha\, \lambda^\beta, \tag{11.11}$$

where $\chi_{(n)} = \chi\cos\theta$.

According to (11.6), we can write

$$\chi\mu^i = \chi_{(n)}\, n^i. + \chi_g\, \eta^i.$$

This equation states that $\chi_{(n)}$ *and* χ_g *are the components of the curvature vector* $\chi\mu^i$ *in the directions of the vectors* n^i *and* η^i.

11.2.1 Theorem of Meusnier

The radius of curvature $R = \frac{1}{\chi}$ of any curve at a given point on the surface is equal to the product of the radius of curvature $R_{(n)} = \frac{1}{\chi_{(n)}}$ of the corresponding normal section at that point by the cosine of the angle between the normal to the surface and the principal normal to the curve.

In symbols, we have $R = \pm R_{(n)} \cos\theta$.

If S is a sphere, every normal section is a great circle of the sphere and if C is any circle drawn on the sphere, then the preceding result becomes obvious from the elementary geometry consideration, as shown in Figure (11.2).

This is known as Meusnier's Theorem.

We know $ds^2 = a_{\alpha\beta} du^\alpha du^\beta$ *and* $\frac{du^\alpha}{ds} = \lambda^\alpha$.

We see that the formula

$$\chi_{(n)} = \frac{b_{\alpha\beta} du^\alpha du^\beta}{a_{\alpha\beta} du^\alpha du^\beta} = \frac{\mathcal{B}}{\mathcal{A}}. \tag{11.12}$$

Result 11.2.1. If the surface is a plane, the normal curvature $\chi_{(n)} = 0$ at all points of the plane, and if it is a sphere, $\chi_{(n)} = \frac{1}{R}$, where R is the radius of the sphere.

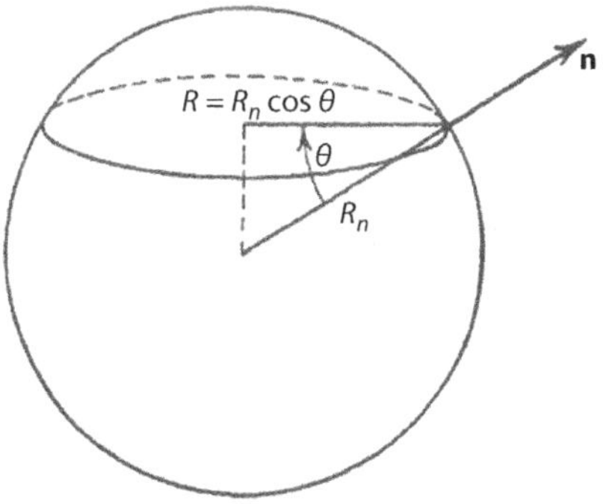

Figure 11.2

Accordingly, we conclude from (11.12) that for the plane $b_{\alpha\beta} = 0$ and for the sphere $b_{\alpha\beta}du^\alpha du^\beta = \frac{1}{R}a_{\alpha\beta}du^\alpha du^\beta$ that $a_{\alpha\beta} = Rb_{\alpha\beta}$ at all points of the sphere.

Therefore, for the sphere $b_{\alpha\beta} = \frac{1}{R}a_{\alpha\beta}$,
It is implied that $b_{\alpha\beta}$ *and* $a_{\alpha\beta}$ are proportional.

Result 11.2.2. Since η^i is perpendicular to λ^i, it lies in the plane containing n^i *and* μ^i. It is tangent to the surface and the angle between μ^i and η^i is $\frac{\pi}{2}-\theta$.

Multiplying (*scalar*) (11.6) by η^i, we get

$$\chi\mu^i = b_{\alpha\beta}n^i\lambda^\alpha\lambda^\beta + \chi_g\eta^i$$

$$\chi\cos\left(\frac{\pi}{2}-\theta\right) = \chi_g \tag{11.13}$$

$$\therefore \quad \chi_g = \chi\sin\theta.$$

Result 11.2.3. From (11.12), we have

$$\chi_{(n)} = \frac{b_{\alpha\beta}du^\alpha du^\beta}{a_{\alpha\beta}du^\alpha du^\beta} = \frac{\mathcal{B}}{\mathcal{A}}.$$

Squaring and adding (11.11) and (11.13), we get

$$(\chi_g)^2 + (\chi_{(n)})^2 = (\chi)^2.$$

Also, we have $\chi\mu^i = \chi_{(n)}\, n^i. + \chi_g\,\eta^i$.
Multiplying it by $g_{ij}\mu^j$,

$$g_{ij}\mu^j\chi\mu^i = g_{ij}\mu^j b_{\alpha\beta}\lambda^\alpha\lambda^\beta n^i. + \chi_g\eta^i g_{ij}\mu^j$$

$$\chi = b_{\alpha\beta}\lambda^\alpha\lambda^\beta\cos\theta. + \chi_g\cos\left(\frac{\pi}{2}-\theta\right)$$

$$= \chi_{(n)}\cos\theta + \chi_g\sin\theta.$$

Example 11.2.1. If θ is the angle between the principal normal and the surface normal, show that

$$\mu^i = cos\theta n^i + sin\theta\eta^i.$$

Solution: From (11.6) we have

$$\chi\mu^i = b_{\alpha\beta}\, n^i\, \lambda^\alpha\, \lambda^\beta + \chi_g\, \eta^i,$$

using (11.13) and (11.10), i.e., $\chi cos\theta = b_{\alpha\beta}\,\lambda^\alpha\,\lambda^\beta$ and $\chi_g = \chi \sin\theta$

$$\chi\mu^i = \chi cos\theta\; n^i + \chi \sin\theta\; \eta^i$$

$$\text{or } \mu^i = cos\theta\; n^i + \sin\theta\; \eta^i.$$

Example 11.2.2. Prove that the normal curvatures in the direction of the coordinate curves are $\frac{b_{11}}{a_{11}}$ and $\frac{b_{22}}{a_{22}}$, respectively.

Solution: We know the unit vector $\lambda^\alpha_{(1)}$ along u^1- curve and $\lambda^\beta_{(2)}$ along u^2- curve are

$\lambda^\alpha_{(1)} = \frac{1}{\sqrt{a_{11}}}\delta^\alpha_{(1)}$ and $\lambda^\beta_{(2)} = \frac{1}{\sqrt{a_{22}}}\delta^\beta_{(2)}$, respectively.

Therefore, (11.11) gives

$$\chi_{(n)} = b_{\alpha\beta}\,\lambda^\alpha_{(1)}\,\lambda^\beta_{(1)}$$

$$\chi^1_{(n)} = b_{\alpha\beta}\frac{1}{\sqrt{a_{11}}}\delta^\alpha_{(1)}\frac{1}{\sqrt{a_{11}}}\delta^\beta_{(1)} = \frac{b_{11}}{a_{11}}$$

$$\chi^2_{(n)} = b_{\alpha\beta}\frac{1}{\sqrt{a_{22}}}\delta^\alpha_{(2)}\frac{1}{\sqrt{a_{22}}}\delta^\beta_{(2)} = \frac{b_{22}}{a_{22}}.$$

Example 11.2.3. If the surface is plane, show that $b_{\alpha\beta} = 0$.

Solution: If the surface is plane, then any normal section at any point of the plane is a straight line and therefore its curvature is zero, i.e., $\chi_{(n)} = 0$

$$\therefore\; \chi_{(n)} = b_{\alpha\beta}\,\lambda^\alpha\,\lambda^\beta = 0$$

or $b_{\alpha\beta}\, du^\alpha\, du^\beta = 0$ for all directions of du^α (here du^α is arbitrary).

Thus, for a plane, $b_{\alpha\beta} = 0$.

Example 11.2.4. If a curve is a geodesic on the surface, prove that it is either a straight line or its principal normal is orthogonal to the surface at every point and conversely.
Solution: From (11.9), $\chi_g = \chi \sin\theta$.
Since the curve is geodesic, $\chi_g = 0$

$$\therefore \chi \sin\theta = 0.$$

Thus, either $\chi = 0$ or $\sin\theta = 0$.
If $\chi = 0$, the curve is a straight line and if $\sin\theta = 0$, then $\theta = 0$ *or* π.
Hence, the principal normal and the surface normal are colinear and the principal normal is orthogonal to the surface at every point.
Conversely it follows immediately.

Example 11.2.5. Find the normal curvature of the right helicoid.
Solution: The parametric representation of right helicoids is given by

$$x^1 = u^1 \cos u^2,\ x^2 = u^1 \sin u^2 \text{ and } x^3 = cu^2$$

$$\text{or } \mathbf{r} = (x^1, x^2, x^3) = (u^1 \cos u^2, u^1 \sin u^2, cu^2).$$

$$\therefore a_{11} = \frac{\partial y^i}{\partial u^1}\frac{\partial y^i}{\partial u^1} = \left(\frac{\partial y^1}{\partial u^1}\right)^2 + \left(\frac{\partial y^2}{\partial u^1}\right)^2 + \left(\frac{\partial y^3}{\partial u^1}\right)^2 = (\cos u^2)^2 + (\sin u^2)^2 = 1$$

$$a_{22} = \frac{\partial y^i}{\partial u^2}\frac{\partial y^i}{\partial u^2} = \left(\frac{\partial y^1}{\partial u^2}\right)^2 + \left(\frac{\partial y^2}{\partial u^2}\right)^2 + \left(\frac{\partial y^3}{\partial u^2}\right)^2 = (-u^1 \sin u^2)^2 + (u^1 \cos u^2)^2 + c^2 = (u^1)^2 + c^2$$

$$a_{12} = \frac{\partial y^i}{\partial u^1}\frac{\partial y^i}{\partial u^2} = \frac{\partial y^1}{\partial u^1}\frac{\partial y^1}{\partial u^2} + \frac{\partial y^2}{\partial u^1}\frac{\partial y^2}{\partial u^2} + \frac{\partial y^3}{\partial u^1}\frac{\partial y^3}{\partial u^2} = \cos u^2(-u^1 \sin u^2)^2 + (\sin u^2)(u^1 \cos u^2) = 0$$

$$\therefore \quad a = \begin{vmatrix} 1 & 0 \\ 0 & (u^1)^2 + c^2 \end{vmatrix} = (u^1)^2 + c^2$$

$$\text{Now, } b_{\alpha\beta} = -\frac{1}{2}\left(\frac{\partial n}{\partial u^{\alpha}}\cdot\frac{\partial r}{\partial u^{\beta}} + \frac{\partial n}{\partial u^{\beta}}\cdot\frac{\partial r}{\partial u^{\alpha}}\right)$$

$$b_{11} = -\frac{1}{2}\left(\frac{\partial n}{\partial u^{1}}\cdot\frac{\partial r}{\partial u^{1}} + \frac{\partial n}{\partial u^{1}}\cdot\frac{\partial r}{\partial u^{1}}\right) = -\frac{\partial n}{\partial u^{1}}\cdot\frac{\partial r}{\partial u^{1}}$$

$$= -\frac{1}{\sqrt{(u^1)^2 + c^2}}(0,0,1).(\cos u^2, sinu^2, 0) = 0$$

$$b_{22} = -\frac{\partial n}{\partial u^{2}}\cdot\frac{\partial r}{\partial u^{2}} = -\frac{1}{\sqrt{(u^1)^2 + c^2}}(ccosu^2, c\sin u^2, 0).$$

$$(-u^1 \sin u^2, u^1 cosu^2, c) = 0$$

$$b_{12} = -\frac{1}{2}\left(\frac{\partial n}{\partial u^{1}}\cdot\frac{\partial r}{\partial u^{2}} + \frac{\partial n}{\partial u^{2}}\cdot\frac{\partial r}{\partial u^{1}}\right)$$

$$= -\frac{1}{2}\frac{1}{\sqrt{(u^1)^2 + c^2}}[(0,0,1).(-u^1 \sin u^2, u^1 cosu^2, c)$$

$$+ (ccosu^2, c\sin u^2, 0).(\cos u^2, sinu^2, 0) = \frac{-c}{\sqrt{(u^1)^2 + c^2}} = b_{21}$$

$$b = |b_{\alpha\beta}| = \begin{vmatrix} b_{11} & b_{12} \\ b_{21} & b_{22} \end{vmatrix} = -b_{12}b_{12} = -\frac{-c^2}{(u^1)^2 + c^2}$$

$$\text{Gaussian Curvature} \quad \kappa = \frac{|b_{\alpha\beta}|}{|a_{\alpha\beta}|} = \frac{\dfrac{-c^2}{(u^1)^2 + c^2}}{(u^1)^2 + c^2} = \frac{-c^2}{[(u^1)^2 + c^2]^2} < 0.$$

It is a negative curvature and has minimal surface since H = 0.

Thus, $\chi_{(1)} + \chi_{(1)} = 2H = 0$ and $\chi_{(1)}.\chi_{(2)} = \kappa = \frac{-c^2}{[(u^1)^2 + c^2]^2}$.

The normal curvature $\chi_{(n)}$ of the given right helicoids is

$$\chi_{(n)} = \frac{b_{\alpha\beta} du^\alpha du^\beta}{a_{\alpha\beta} du^\alpha du^\beta} = \frac{2b_{12} du^1 du^2}{a_{11}(du^1)^2 + a_{22}(du^2)^2}$$

$$= \frac{\dfrac{-2c^2}{\sqrt{(u^1)^2 + c^2}} du^1 du^2}{1.(du^1)^2 + [(u^1)^2 + c^2](du^2)^2}$$

$$= \frac{1}{\sqrt{(u^1)^2 + c^2}} \frac{-2c du^1 du^2}{(du^1)^2 + [(u^1)^2 + c^2](du^2)^2}$$

11.3 The Principal Curvatures of a Surface

We know tangent vector $\lambda^\alpha = \dfrac{du^\alpha}{ds}$ on the surface such that the normal curvature $\chi_{(n)}$ is given by formula

$$\chi_{(n)} = b_{\alpha\beta}\, du^\alpha\, du^\beta, \tag{11.14}$$

assuming an extreme value.

Since vector λ^α is unit a vector, $\chi_{(n)}$ in (11.14) has to be maximized subject to the considering relation

$$a_{\alpha\beta}\, \lambda^\alpha\, \lambda^\beta = 1. \tag{11.15}$$

Following the usual procedure of determining constrained maxima and minima, we deduce that a necessary condition for an extremum is

$$b_{\alpha\beta}\, \lambda^\beta + \Lambda a_{\alpha\beta}\, \lambda^\beta = 0, \tag{11.16}$$

where Λ is the lagrange multiplier. If Equation (11.16) is multiplied by λ^α and account is taken of relations (11.14) and (11.15), it follows at once that $\Lambda = -\chi_{(n)}$. Thus, Equation (11.26) for the determination of directions yielding extreme values of $\chi_{(n)}$ can be written as

$$(b_{\alpha\beta} - \chi_{(n)}\, a_{\alpha\beta})\, \lambda^\beta = 0 \text{ (11.17) } (\alpha = 1, 2, 3)$$

The set of homogenous equation (11.17) will possess nontrivial solutions for λ^{β} if the values of $\chi_{(n)}$ are the roots of the determinal equation

$$|b_{\alpha\beta} - \vartheta a_{\alpha\beta} =| \, 0. \tag{11.18}$$

The quadratic equation (11.18), when written out in expanded form, is

$$\vartheta^2 - (a_{\alpha\beta} b_{\alpha\beta})\vartheta + \frac{b}{a} = 0, \tag{11.19}$$

where $b = | b_{\alpha\beta}|$ and $a = | a_{\alpha\beta}|$.

Since the Gaussian curvature κ is given by $\kappa = \frac{b}{a}$ and the mean curvature $H \equiv \frac{1}{2} a^{\alpha\beta} b_{\alpha\beta,}$,
we see that Equation (11.19) assumes the form

$$\vartheta^2 - 2H\vartheta + \kappa = 0. \tag{11.20}$$

The roots $\vartheta = \chi_{(1)}$ *and* $\vartheta = \chi_{(2)}$ of (11.20) are called the *principal curvatures of the surface* and the directions $\lambda^{\alpha}_{(1)}$ and $\chi^{\alpha}_{(2)}$, corresponding to these extreme values of $\chi_{(n)}$, are the principal directions of the surfaces.

From (11.20) it is clear that the principal curvature $\chi_{(1)}$ and $\chi_{(2)}$ are related to mean and Gaussian curvatures by the formulas:

$$\chi_{(1)} + \chi_{(2)} = 2H$$

$$\chi_{(1)} \cdot \chi_{(2)} = \kappa \tag{11.21}$$

From (11.16), it follows that the principal directions are determined by

$$(b_{\alpha\beta} - \chi_{(1)} a_{\alpha\beta}).\lambda^{\beta}_{(1)} = 0$$

$$(b_{\alpha\beta} - \chi_{(2)} a_{\alpha\beta}).\lambda^{\beta}_{(2)} = 0$$

If the first of these equations is multiplied by $\lambda^{\alpha}_{(2)}$ and the second by $\lambda^{\alpha}_{(1)}$ and the results subtracted, we obtain

$$(\chi_{(1)} - \chi_{(2)}) a_{\alpha\beta} \lambda^{\alpha}_{(1)} \lambda^{\beta}_{(1)} = 0. \tag{11.22}$$

If $\chi_{(1)} \neq \chi_{(2)}$, Equation (11.22) gives us

$$a_{\alpha\beta}\lambda^{\beta}_{(1)}\lambda^{\beta}_{(2)} = 0. \tag{11.23}$$

That is, *the principal directions are orthogonal.* If the extreme values of $\chi_{(n)}$, are equal at a given point, then every direction is a principal direction, and the point is called an *umbillic point.*

Theorem 11.1. At each point of a surface, there exist two mutually orthogonal directions for which the normal curvature attains its extreme values.

11.3.1 Umbillic Point

A point at which principal curvatures of the surface are equal, i.e., $\chi_{(1)} = \chi_{(2)}$, is called an umbillic or naval point. From Equations (11.20) and (11.21),

$$\chi_{(1)} + \chi_{(2)} = 2H$$

$$\chi_{(1)} \cdot \chi_{(2)} = \kappa.$$

Therefore, $$2\,\chi_{(1)} = 2H \textit{ and } (\chi_{(2)})^2 = \kappa$$

$$H^2 = \kappa.$$

From (10.56) and (10.57), $$\left(\frac{1}{2}a^{\alpha\beta}b_{\alpha\beta}\right)^2 = \kappa,$$

or $$(a^{\alpha\beta}b_{\alpha\beta})^2 = 4\frac{\mathrm{b}}{\mathrm{a}},$$

or $$(a^{11}b_{11} + a^{22}b_{22} + a^{12}b_{12} + a^{21}b_{21})^2 = 4\frac{\mathrm{b}}{\mathrm{a}} \quad [\because a_{11} = aa^{22},$$

$$a_{12} = -aa^{21}, a_{22} = aa^{11}],$$

or $$4a(a_{11}b_{12} - a_{12}b_{11})^2 + [a_{11}(a_{11}b_{22} - a_{22}b_{11}) - 2a_{12}(a_{11}b_{12} - a_{12}b_{11})]^2 = 0.$$

Since $a_{ij}\, dx^i\, dx^j$ is positive definite and a is positive,

$$a_{11}\, b_{12} - a_{12}\, b_{11} = a_{11}\, b_{22} - a_{22}\, b_{11} = 0,$$

$$\text{implying } \frac{b_{11}}{a_{11}} = \frac{b_{12}}{a_{12}} = \frac{b_{22}}{a_{22}}. \tag{11.24}$$

Thus, at the umbillic point, we have the above condition (11.24). We know

$$\chi_{(n)} = \frac{b_{\alpha\beta} du^\alpha du^\beta}{a_{\alpha\beta} du^\alpha du^\beta}.$$

Therefore, $\chi_{(n)}$ is independent of the direction of $\dfrac{du^\alpha}{ds}$.

At the umbillic point, the normal curvature is the same in every direction. At all other points where $\chi_{(1)} \neq \chi_{(2)}$, we have

$$a_{\alpha\beta} \lambda^\alpha_{(1)} \lambda^\beta_{(2)} = 0. \tag{11.25}$$

At each point of a surface that is non-umbillic, there exist two mutually orthogonal directions for which the normal curvature attains its extreme values.

11.3.2 Lines of Curvature

A curve on a surface such that the tangent line to it at every point is directed along a principal direction is called a *line of curvature.*

From (11.17), the equations for determination of directions yielding extreme values of $\chi_{(n)}$ *are*

$$(b_{\alpha\beta} - \chi_{(n)}\, a_{\alpha\beta})\lambda^\beta = 0$$

$$\text{or } b_{\alpha\beta}\, \lambda^\beta = \chi_{(n)}\, a_{\alpha\beta} \lambda^\beta.$$

For eliminating $\chi_{(n)}$ from these equations and setting $\lambda_\beta = \dfrac{du^\beta}{ds}$:

$$\alpha = 1\ gives \qquad b_{1\beta}\lambda^{\beta} = \chi_{(n)}a_{1\beta}\lambda^{\beta}$$

$$\alpha = 2\ gives \qquad b_{2\beta}\lambda^{\beta} = \chi_{(n)}a_{2\beta}\lambda^{\beta}.$$

We get

$$\frac{b_{1\beta}\lambda^{\beta}}{b_{2\beta}\lambda^{\beta}} = \frac{a_{1\beta}\lambda^{\beta}}{a_{2\beta}\lambda^{\beta}}$$

$$\frac{b_{1\beta}du^{\beta}}{b_{2\beta}du^{\beta}} = \frac{a_{1\beta}du^{\beta}}{a_{2\beta}du^{\beta}},$$

or

$$\frac{b_{11}du^1 + b_{12}du^2}{b_{21}du^1 + b_{22}du^2} = \frac{a_{11}du^1 + a_{12}du^2}{a_{21}du^1 + a_{22}du^2},$$

or $(\mathrm{b}_{11}a_{12} - b_{12}a_{11})(du^1)^2 + (b_{12}a_{22} - b_{22}a_{12})(du^2)^2 + (b_{11}a_{22}$

$$- b_{22}a_{11})du^1du^2 = 0 \qquad (11.26)$$

Equation (11.26) represents the lines of curvature of the surface.

11.3.3 Asymptotic Lines

The directions on the surface given by

$$b_{\alpha\beta}\lambda^{\alpha}\lambda^{\beta} = 0 \qquad (11.27)$$

are called asymptotic directions and the curves whose tangents are asymptotic directions are called the asymptotic lines of the surface.

Example 11.3.1. Show that a straight line on a surface is an asymptotic line.

Solution: As the curve is a straight line, its curvature is $\kappa = 0$.

From $\chi cos\theta = b_{\alpha\beta}\lambda^{\alpha}\lambda^{\beta}$,

implying that $b_{\alpha\beta}\lambda^{\alpha}\lambda^{\beta} = 0$.

Thus, the straight line is an asymptotic line.

Example 11.3.2. Show that the parametric curves are asymptotic lines if $b_{11} = b_{22} = 0$.

Solution: Let the curves be asymptotic.

We have,

$$b_{\alpha\beta}\lambda^{\alpha}\lambda^{\beta} = 0. \qquad \text{(i)}$$

$$\text{For } u^1-\text{curve}, b_{\alpha\beta}\lambda^{\alpha}_{(1)}\lambda^{\beta}_{(1)} = 0,$$

$$\text{using } \lambda^{\alpha}_{(1)} = \frac{1}{\sqrt{a_{11}}}\delta^{\alpha}_{(1)}, \lambda^{\beta}_{(1)} = \frac{1}{\sqrt{a_{11}}}\delta^{\beta}_{(1)},$$

$$\therefore \quad b_{\alpha\beta}\frac{1}{\sqrt{a_{11}}}\delta^{\alpha}_{(1)}\frac{1}{\sqrt{a_{11}}}\delta^{\beta}_{(1)} = 0,$$

$$\text{or } b_{11}\frac{1}{\sqrt{a_{11}}}\delta^{1}_{(1)}\frac{1}{\sqrt{a_{11}}}\delta^{1}_{(1)} = 0,$$

$$\text{or } \frac{b_{11}}{a_{11}} = 0 \Rightarrow b_{11} = 0.$$

Similarly, for u^2 –curve, $b_{\alpha\beta}\lambda^{\alpha}_{(2)}\lambda^{\beta}_{(2)} = 0.$
From this equation, we get $\frac{b_{22}}{a_{22}} = 0 \Rightarrow b_{22}0.$
Conversely, suppose $b_{11} = b_{22} = 0.$
Using (i), it gives

$$b_{11}\lambda^1\lambda^1 + b_{22}\lambda^2\lambda^2 + b_{21}\lambda^2\lambda^1 + b_{12}\lambda^1\lambda^2 = 0$$

or $b_{12}\lambda^1\lambda^2 = 0$, or $b_{12}\, du^1\, du^2 = 0.$
Here, $du^1\, du^2 = 0$ as $b_{12} \neq 0.$
The curves are parametric.

Example 11.3.3. Show that the coordinate lines are asymptotic lines if $a_{12} = b_{12} = 0.$
Solution: At each point of S where either $b_{\alpha\beta}\, du^\alpha\, du^\beta \neq 0$ or is not proportional to $a_{\alpha\beta}\, du^\alpha\, du^\beta$, then (11.26) specifies two orthogonal directions,

$$\frac{du^2}{du^1} = \phi_\alpha(u^1, u^2), \tag{11.28}$$

which coincide with the direction of the principal curvatures. Each equation of (11.27) determines a family of curves on S covering the surface without gap. These two families of curves are orthogonal and, if they are taken as a parametric net on S, the first fundamental form has the form

$$ds^2 = \bar{a}_{11}(d\bar{u}^1)^2 + \bar{a}_{22}(d\bar{u}^2)^2.$$

According to (11.26), in the coordinate system $\bar{u}^{\alpha}$ takes the form

$$-\bar{b}_{12}\bar{a}_{11}(d\bar{u}^1)^2 + \bar{b}_{12}\bar{a}_{22}(d\bar{u}^2)^2 + (\bar{b}_{11}\bar{a}_{22} - \bar{b}_{22}\bar{a}_{11})d\bar{u}^1 d\bar{u}^2 = 0$$

and its solutions are $\bar{u}^1 = constant, and\ \bar{u}^2 = constant$

If we take $d\bar{u}^1 \neq 0$ and $d\bar{u}^2 = 0$, we see that $\bar{b}_{12} = 0$, since $\bar{a}_{11} = 0$.

Thus, a necessary condition for the net of lines of curvature to be orthogonal is that $a_{12} = b_{12} = 0$.

Consequently, we may always choose coordinates u^1, u^2 on S so that the lines of curvature are the coordinate curves of this system which is allowable at any point of S which is not umbilic.

Conversely, if $a_{12} = b_{12} = 0$, then equation (11.26) has the solutions

$$u^1 = constant,\ u^2 = constant,$$

so that the coordinate lines are the lines of curvature.

Example 11.3.4. Show that the parametric curves are the lines of curvature if $a_{12} = b_{12} = 0$.

Solution: Let us assume that parametric curves are the lines of curvature. Then, they form an orthogonal net and will satisfy (11.26),

$$(b_{11}\,a_{12} - b_{12}\,a_{11})\,(du^1)^2 + (b_{12}\,a_{22} - b_{22}\,a_{12})\,(du^2)^2 + (b_{11}\,a_{22} - b_{22}\,a_{11})\,du^1\,du^2 = 0.$$

For u^1 *curve* $du^2 = 0$.

$$\therefore b_{11}\,a_{12} - b_{12}\,a_{11} = 0 \text{ (since } du^1 \neq 0) \tag{i}$$

and for u^2 *curve* $du^1 = 0$,

$$b_{12}\,a_{22} - b_{22}\,a_{12} = 0, \text{ as } du^2 \neq 0. \tag{ii}$$

Multiplying (i) by a_{22}, and (ii) by a_{11}, and adding these two equations, we get

$$a_{22}\,(b_{11}\,a_{12} - b_{12}\,a_{11}) + a_{11}\,(b_{12}\,a_{22} - b_{22}\,a_{12}) = 0,$$

or $$a_{22}\,b_{11}\,a_{12} - a_{11}\,b_{22}\,a_{12} = 0,$$

or $$a_{12}\,(a_{22}\,b_{11} - a_{11}\,b_{22}) = 0,$$

Implying that

either $$a_{12} = 0 \; or \; a_{22}\,b_{11} - a_{11}\,b_{22} = 0.$$

For the parametric curve, $a_{22}\,b_{11} \neq a_{11}\,b_{22}$ and so $a_{12} = 0$.

The condition $a_{12} = 0$ is that of orthogonality satisfied by all lines of curvature.

Similarly, multiplying (i) by a_{22} and (ii) by a_{11}, and adding these two equations, we get

$$b_{12}\,(a_{22}\,b_{11} - a_{11}\,b_{22}) = 0$$

$$\Rightarrow b_{12} = 0.$$

Conversely, let $a_{12} = b_{12} = 0$.

Equation (11.26) becomes

$$(b_{11}\,a_{22} - b_{22}\,a_{11})\,du^1\,du^2 = 0.$$

This is true if the lines of curvature $b_{11}\,a_{22} \neq b_{22}\,a_{11}$

$$\therefore\; du^1\,du^2 = 0.$$

Hence, when $du^1 \neq 0$ *and* $du^2 = 0 \Rightarrow u^2 = constant.$

When $du^2 \neq 0$ *and* $du^1 = 0 \Rightarrow u^1 = constant,$ the parametric curves are the lines of curvature.

Example 11.3.5. Show that the lines of curvature on a minimal surface form an isometric system.

Solution: If the parametric curves are the lines of curvature, then

$$a_{12} = b_{12} = 0.$$

For a minimal surface, $H = \dfrac{1}{2} a^{\alpha\beta} b_{\alpha\beta} = 0.$

$$\Rightarrow \; a^{\alpha\beta}\, b_{\alpha\beta} = 0$$

$$\text{or } (a^{11}\, b_{11} + a^{12}\, b_{12} + a^{22}\, b_{22} + a^{21}\, b_{21} = 0. \left[a^{11} = \frac{a_{22}}{a}, a^{22} = \frac{a_{11}}{a}, a^{12} = \frac{a_{21}}{a}\right]$$

Here, $a_{12} = b_{12} = 0$

$$\therefore \frac{a_{22}}{a} b_{11} + a^{12}\, b_{12} + \frac{a_{11}}{a} b_{22} + a^{21}\, b_{21} = 0$$

$$\text{or } \frac{1}{a}(a_{22}b_{11} + a_{11}b_{22}) = 0$$

$$\Rightarrow \; b_{11} = b_{22} = 0$$

$$\therefore \; b_{\alpha\beta} = 0.$$

Therefore, the surface is a plane, i.e., the surface is isometric with the Euclidean plane.

Example 11.3.6. Find the principal curvature of the surface defined by

$$x^1 = u^1 \cos u^2, x^2 = u^1 \sin u^2, x^3 = f(u^1).$$

Find the condition that it is a minimal surface.
Solution: The parametric representation of the surface is

$$x^1 = u^1 \cos u^2, x^2 = u^1 \sin u^2, x^3 = f(u^1)$$

$$\text{or } r = (u^1 \cos u^2, u^1 \sin u^2, f(u^1))$$

$$ds^2 = (dy^1)^2 + (dy^2)^2 + (dy^3)^2 = (du^1)^2 + (u^1)^2(du^2)^2 + \left(\frac{\partial f}{\partial u^1}\right)^2 (du^1)^2$$

$$= \left(1 + \left(\frac{\partial f}{\partial u^1}\right)^2\right)(du^1)^2 + (u^1)^2(du^2)^2.$$

The first fundamental magnitudes $a_{\alpha\beta}$ are

$$\therefore a_{11} = 1 + \left(\frac{\partial f}{\partial u^1}\right)^2 = 1 + (f')^2, a_{22} = (u^1)^2, a_{12} = a_{21} = 0$$

$$\therefore a = \begin{vmatrix} a_{11} & 0 \\ 0 & a_{22} \end{vmatrix} = a_{11}a_{22} = (u^1)^2[1 + (f')^2],$$

$$\text{where } f_1 = \frac{\partial f}{\partial u^1} \text{ and } f_{11} = \frac{\partial^2 f}{\partial u^1 \partial u^2}$$

$$a^{11} = \frac{a_{22}}{a_{11}a_{22}} = \frac{1}{a_{11}} = \frac{1}{1 + (\mathrm{f}')^2}, a^{22} = \frac{1}{a_{22}} = \frac{1}{(u^1)^2}, a^{12} = a^{21} = 0.$$

Here, we have $A = \left(\frac{\partial x^1}{\partial u^1}, \frac{\partial x^2}{\partial u^1}, \frac{\partial x^3}{\partial u^1}\right) = (\cos u^2, \sin u^2, f_1)$ and

$$B = \left(\frac{\partial x^1}{\partial u^2}, \frac{\partial x^2}{\partial u^2}, \frac{\partial x^3}{\partial u^2}\right) = (-u^1 \sin u^2, u^1 \cos u^2, 0),$$

$$A \times B = (-f_1 u^1 \cos u^2, -f_1 u^1 \sin u^2, u^1),$$

$$\textit{and } n = \frac{A \times B}{|A \times B|} = \frac{1}{u^1\sqrt{1 + (f_1)^2}}(-f_1 u^1 \cos u^2, -f_1 u^1 \sin u^2, u^1)$$

$$= \frac{1}{\sqrt{1 + (f_1)^2}}(-f_1 \cos u^2, -f_1 \sin u^2, 1).$$

The symmetric covariant tensors $b_{\alpha\beta}$ are

$$b_{\alpha\beta} = -\frac{1}{2}\left(\frac{\partial n}{\partial u^\alpha} \cdot \frac{\partial r}{\partial u^\beta} + \frac{\partial n}{\partial u^\beta} \cdot \frac{\partial r}{\partial u^\alpha}\right)$$

$$b_{11} = -\frac{1}{2}\left(\frac{\partial n}{\partial u^1} \cdot \frac{\partial r}{\partial u^1} + \frac{\partial n}{\partial u^1} \cdot \frac{\partial r}{\partial u^1}\right) = -\frac{\partial n}{\partial u^1} \cdot \frac{\partial r}{\partial u^1} = \frac{f_{11}}{\sqrt{1 + (f_1)^2}}$$

$$b_{22} = -\frac{1}{2}\left(\frac{\partial n}{\partial u^2} \cdot \frac{\partial r}{\partial u^2} + \frac{\partial n}{\partial u^2} \cdot \frac{\partial r}{\partial u^2}\right) = -\frac{\partial n}{\partial u^2} \cdot \frac{\partial r}{\partial u^2} = \frac{u^1 f_1}{\sqrt{1 + (f_1)^2}}$$

$$b_{12} = -\frac{1}{2}\left(\frac{\partial n}{\partial u^1} \cdot \frac{\partial r}{\partial u^2} + \frac{\partial n}{\partial u^2} \cdot \frac{\partial r}{\partial u^1}\right) = 0 = b_{21}.$$

Therefore, $b = \begin{vmatrix} \dfrac{f_{11}}{\sqrt{1+(f_1)^2}} & 0 \\ 0 & \dfrac{u^1 f_1}{\sqrt{1+(f_1)^2}} \end{vmatrix} = \dfrac{u^1 f_1 f_{11}}{1+(f_1)^2}$.

The equation of the principal curvature is

$$\chi_P^2 - (a^{\alpha\beta} b_{\alpha\beta})\chi_P + \frac{b}{a} = 0$$

$$\text{or } \chi_P^2 - \left[\frac{f_{11}}{(1+(f_1)^2)^{\frac{3}{2}}} + \frac{f_1}{u^1\sqrt{1+(f_1)^2}}\right]\chi_P + \frac{u^1 f_1 f_{11}}{u^1[1+(f_1)^2]^2} = 0.$$

Here, $\chi_{(1)} = \dfrac{f_{11}}{(1+(f_1)^2)^{\frac{3}{2}}}$ and $\chi_{(2)} = \dfrac{f_1}{u^1\sqrt{1+(f_1)^2}}$.

If ρ_1 and ρ_2 are the corresponding radii of curvatures, then

$$\rho_1 = \frac{1}{\chi_{(1)}} = \frac{(1+(f_1)^2)^{\frac{3}{2}}}{f_{11}} \text{ and } \rho_2 = \frac{u^1\sqrt{1+(f_1)^2}}{f_1}.$$

The condition for the surface is minimal if $H = 0$,

or $$2H = \chi_{(1)} + \chi_{(2)} = 0,$$

or $$\frac{f_{11}}{(1+(f_1)^2)^{\frac{3}{2}}} + \frac{f_1}{u^1\sqrt{1+(f_1)^2}} = 0$$

$$f_1(1+(f_1)^2) + u^1 f_{11} = 0.$$

Example 11.3.7. Find the principal curvature of the surface defined by

$$x^1 = u^1, x^2 = u^2, x^3 = f(u^1, u^2).$$

Solution: The parametric representation of the surface is

$$x^1 = u^1, x^2 = u^2, x^3 = f(u^1, u^2),$$

or

$$r = (u^1, u^2, f(u^1, u^2)).$$

$$a_{11} = \left(\frac{\partial x^i}{\partial u^1}\right)^2 = 1 + \left(\frac{\partial f}{\partial u^1}\right)^2 = 1 + f_1^2$$

$$a_{22} = \left(\frac{\partial x^i}{\partial u^2}\right)^2 = 1 + \left(\frac{\partial f}{\partial u^2}\right)^2 = 1 + f_2^2$$

$$a_{12} = \frac{\partial x^i}{\partial u^1}\frac{\partial x^i}{\partial u^2} = 1.0 + 0.1 + \frac{\partial f}{\partial u^1}\frac{\partial f}{\partial u^2} = f_1 f_2 = a_{21}$$

$$\therefore a = \begin{vmatrix} a_{11} & a_{12} \\ a_{21} & a_{22} \end{vmatrix} = \begin{vmatrix} 1+f_1^2 & f_1 f_2 \\ f_1 f_2 & 1+f_2^2 \end{vmatrix} = 1 + f_1^2 + f_2^2$$

$$a^{11} = \frac{1+f_2^2}{1+f_1^2+f_2^2}, \quad a^{22} = \frac{1+f_1^2}{1+f_1^2+f_2^2}, a^{12} = \frac{f_1 f_2}{1+f_1^2+f_2^2}$$

Now, for calculating the second fundamental $b_{\alpha\beta}$

let $A = \left(\frac{\partial x^1}{\partial u^1}, \frac{\partial x^2}{\partial u^1}, \frac{\partial x^3}{\partial u^1}\right) = (1, 0, f_1)$ and $B = \left(\frac{\partial x^1}{\partial u^2}, \frac{\partial x^2}{\partial u^2}, \frac{\partial x^3}{\partial u^2}\right) = (1, 0, f_2)$

$$\therefore A \times B = (-f_1, -f_2, 1)$$

$$n = \frac{A \times B}{|A \times B|} = \frac{1}{\sqrt{1+f_1^2+f_2^2}}(-f_1, -\mathrm{f}_2, 1)$$

$$b_{\alpha\beta} = -\frac{1}{2}\left(\frac{\partial n}{\partial u^{\alpha}}\cdot\frac{\partial r}{\partial u^{\beta}} + \frac{\partial n}{\partial u^{\beta}}\cdot\frac{\partial r}{\partial u^{\alpha}}\right)$$

$$b_{11} = -\frac{1}{2}\left(\frac{\partial n}{\partial u^{1}}\cdot\frac{\partial r}{\partial u^{1}} + \frac{\partial n}{\partial u^{1}}\cdot\frac{\partial r}{\partial u^{1}}\right) = -\frac{\partial n}{\partial u^{1}}\cdot\frac{\partial r}{\partial u^{1}}$$

$$= -\left(1,0,f_1\right)[((1+f_1^2+f_2^2)^{\frac{-1}{2}}(-f_{11},-f_{12},0)$$

$$+(-f_1,-f_2,1)\left\{-\frac{1}{2}(1+f_1^2+f_2^2)^{\frac{-3}{2}}2(f_1 f_{11}+f_2+f_{12})\right\}]$$

$$= \frac{f_{11}}{(1+f_1^2+f_2^2)^{\frac{1}{2}}}$$

$$b_{22} = -\frac{1}{2}\left(\frac{\partial n}{\partial u^{2}}\cdot\frac{\partial r}{\partial u^{2}} + \frac{\partial n}{\partial u^{2}}\cdot\frac{\partial r}{\partial u^{2}}\right) = -\frac{\partial n}{\partial u^{2}}\cdot\frac{\partial r}{\partial u^{2}} = \frac{f_{22}}{(1+f_1^2+f_2^2)^{\frac{1}{2}}}$$

$$b_{12} = -\frac{1}{2}\left(\frac{\partial n}{\partial u^{1}}\cdot\frac{\partial r}{\partial u^{2}} + \frac{\partial n}{\partial u^{2}}\cdot\frac{\partial r}{\partial u^{1}}\right) = \frac{f_{12}}{(1+f_1^2+f_2^2)^{\frac{1}{2}}} = b_{21}$$

$$b = \frac{1}{k^2}\begin{vmatrix} f_{11} & f_{12} \\ f_{12} & f_{22}\end{vmatrix} = \frac{f_{11}f_{22}-f_{12}^2}{1+f_1^2+f_2^2} \quad \text{where } k=(1+f_1^2+f_2^2)^{\frac{1}{2}}$$

The equation of principal curvature is

$$\chi_p^2 - (a^{\alpha\beta}b_{\alpha\beta})\chi_p + \frac{b}{a} = 0$$

$$\text{or } \chi_p^2 - \left[(f_{11}+f_{22}) - \frac{f_1^2 f_{11} + f_2^2 f_{22} + 2 f_1 f_2 f_{12}}{1+f_1^2+f_2^2}\right]\chi_p + f_{11}f_{22} - f_{12}^2 = 0.$$

There is a quadratic equation in χ_p which gives the values of $\chi_{(1)}$ and $\chi_{(2)}$.

Example 11.3.8. Show that any curvature on a sphere is a line of curvature.
Solution: The parametric representation of a sphere of radius a is given by

$$x^1 = asinu^1 \cos u^2, x^2 = asinu^1 \, sinu^2 \text{ and } x^3 = acosu^1.$$

Therefore, $r = (asinu^1 \cos u^2, asinu^1 \, sinu^2, acosu^1)$.

The second order tensors $a_{\alpha\beta}$ for the first fundamental form are

$$ds^2 = (dx^1)^2 + (dx^2)^2 + (dx^3)^2 = (acosu^1 \cos u^2 \, du^1 - asinu^1 \, sinu^2 \, du^2)^2 + (acosu^1 \sin u^2 \, du^1 + asinu^1 \, cosu^2 \, du^2)^2 + (-asinu^1 \, du^1)^2$$

$$ds^2 = a^2 (du^1)^2 + a^2 \sin^2 u^1 (du^2)^2$$

$$\therefore a_{11} = a^2, a_{22} = a^2 \sin^2 u^1.$$

$$a_{12} = a_{21} = 0$$

The second fundamental form are given

$$b_{11} = -a, b_{22} = -aSin^2 u^1 \quad b_{12} = b_{21} = 0$$

Since $a^{12} = a^{21} = 0$, then the given curve has the line of curvature.
Also, $\frac{b_{11}}{a_{11}} = \frac{b_{12}}{a_{12}} = \frac{b_{22}}{a_{22}} = -\frac{1}{a}$
That is, $b_{\alpha\beta}$ are proportional to $\alpha_{\alpha\beta}$
Now, $(b_{11} a_{12} - b_{12} a_{11})(du^1)^2 + (b_{11} a_{22} - b_{22} a_{11}) du^1 du^2 + (b_{12} a_{22} - b_{22} a_{12}) (du^2)^2$

$$= (-a \, Sin^2 u^1 + a \, Sin^2 u^1) = 0$$

Implies that any curve on a sphere is a line of curvature

Example 11.3.9. Find the equation of principal curvatures of the given surface (right helicoid)

$$x^1 = u^1 \cos u^2, x^2 = u^1 \, sinu^2 \text{ and } x^3 = cu^2$$

and show that it is a minimal surface.
Solution: The parametric surface is

$$x^1 = u^1 \cos u^2, x^2 = u^1 \, sinu^2 \text{ and } x^3 = cu^2$$

$$\therefore r = (u^1 \cos u^2, u^1 \, sinu^2, cu^2)$$

The first fundamental form of the surface

$$a_{\alpha\beta} = \frac{\partial y^i}{\partial u^\alpha}\frac{\partial y^i}{\partial u^\beta}$$

$$\therefore a_{11} = \frac{\partial y^i}{\partial u^1}\frac{\partial y^i}{\partial u^1} = \left(\frac{\partial y^1}{\partial u^1}\right)^2 + \left(\frac{\partial y^2}{\partial u^1}\right)^2 + \left(\frac{\partial y^3}{\partial u^1}\right)^2$$

$$= (\cos u^2)^2 + (\sin u^2)^2 = 1$$

$$a_{22} = \frac{\partial y^i}{\partial u^2}\frac{\partial y^i}{\partial u^2} = \left(\frac{\partial y^1}{\partial u^2}\right)^2 + \left(\frac{\partial y^2}{\partial u^2}\right)^2 + \left(\frac{\partial y^3}{\partial u^2}\right)^2$$

$$= (-u^1 sinu^2)^2 + (u^1 cosu^2)^2 + c^2 = (u^1)^2 + c^2$$

$$a_{12} = \frac{\partial y^i}{\partial u^1}\frac{\partial y^i}{\partial u^2} = \frac{\partial y^1}{\partial u^1}\frac{\partial y^1}{\partial u^2} + \frac{\partial y^2}{\partial u^1}\frac{\partial y^2}{\partial u^2} + \frac{\partial y^3}{\partial u^1}\frac{\partial y^3}{\partial u^2}$$

$$= \cos u^2(-u^1 \sin u^2) + (\sin u^2) + (u^1 \cos u^2) = 0$$

$$\therefore a = \begin{vmatrix} 1 & 0 \\ 0 & (u^1)^2 + c^2 \end{vmatrix} = (u^1)^2 + c^2$$

$$a^{11} = \frac{(u^1)^2 + c^2}{(u^1)^2 + c^2} = 1, a^{22} = \frac{1}{(u^1)^2 + c^2}, a^{12} = a^{21} = 0.$$

Here, $r = (x^1, x^2, x^3) = (u^1 \cos u^2, u^1\, sinu^2, cu^2)$

$$\text{We have } A = \left(\frac{\partial x^1}{\partial u^1}, \frac{\partial x^2}{\partial u^1}, \frac{\partial x^3}{\partial u^1}\right) = (\cos u^2, sinu^2, 0);$$

$$B = \left(\frac{\partial x^1}{\partial u^2}, \frac{\partial x^2}{\partial u^2}, \frac{\partial x^3}{\partial u^2}\right) = (-u^1 \sin u^2, u^1 cosu^2, c)$$

$$A \times B = (csinu^2, -c\cos u^2, u^1)$$

$$n = \frac{1}{\sqrt{(u^1)^2 + c^2}}(csinu^2, -c\cos u^2, u^1).$$

$$\text{Now, } b_{\alpha\beta} = -\frac{1}{2}\left(\frac{\partial n}{\partial u^\alpha}\cdot\frac{\partial r}{\partial u^\beta} + \frac{\partial n}{\partial u^\beta}\cdot\frac{\partial r}{\partial u^\alpha}\right)$$

$$b_{11} = -\frac{1}{2}\left(\frac{\partial n}{\partial u^1}\cdot\frac{\partial r}{\partial u^1} + \frac{\partial n}{\partial u^1}\cdot\frac{\partial r}{\partial u^1}\right) = -\frac{\partial n}{\partial u^1}\cdot\frac{\partial r}{\partial u^1} =$$

$$-\frac{1}{\sqrt{(u^1)^2+c^2}}(0,0,1).(\cos u^2, sinu^2, 0) = 0$$

$$b_{22} = -\frac{\partial n}{\partial u^2}\cdot\frac{\partial r}{\partial u^2} = -\frac{1}{\sqrt{(u^1)^2+c^2}}(ccosu^2, c\sin u^2, 0).(-u^1\sin u^2,$$

$$u^1cosu^2, c) = 0$$

$$b_{12} = -\frac{1}{2}\left(\frac{\partial n}{\partial u^1}\cdot\frac{\partial r}{\partial u^2} + \frac{\partial n}{\partial u^2}\cdot\frac{\partial r}{\partial u^1}\right)$$

$$= -\frac{1}{2}\frac{1}{\sqrt{(u^1)^2+c^2}}[(0,0,1).(-u^1\sin u^2, u^1cosu^2, c)$$

$$+(ccosu^2, c\sin u^2, 0).(\cos u^2, sinu^2, 0) = \frac{-c}{\sqrt{(u^1)^2+c^2}} = b_{21}$$

$$b = b_{11}b_{22} - b_{12}b_{21} = 0.0 - \frac{-c}{\sqrt{(u^1)^2+c^2}}\cdot\frac{-c}{\sqrt{(u^1)^2+c^2}} = -\frac{c^2}{\sqrt{(u^1)^2+c^2}}$$

$$a = (u^1)^2 + c^2$$

$$H = \frac{1}{2}a^{\alpha\beta}b_{\alpha\beta} = \frac{1}{2}(a^{11}b_{11} + a^{22}b_{22} + 2a^{12}b_{12}) =$$

$$= \frac{1}{2}\left(1.0 + \frac{1}{(u^1)^2+c^2}.0 + 2.0.\frac{-c}{\sqrt{(u^1)^2+c^2}}\right) = 0.$$

Since the mean curvature is H=0 at every point of the given surface, it is a minimal surface.

The equation of principal surface is

$$\chi_p^2 - (a^{\alpha\beta}b_{\alpha\beta})\chi_p + \frac{b}{a} = 0.$$

$$\text{Here, } a^{\alpha\beta}b_{\alpha\beta} = a^{11}b_{11} + a^{22}b_{22} + a^{12}b_{12} + a^{21}b_{21}$$

$$= 1.0 + \frac{1}{(u^1)^2 + c^2}.0 + 0 + 0 = 0$$

$$\text{and } \frac{b}{a} = -\frac{\frac{c^2}{(u^1)^2 + c^2}}{(u^1)^2 + c^2} = -\frac{c^2}{[(u^1)^2 + c^2]^2}.$$

$$\text{The equation is } \chi_P^2 - \frac{c^2}{[(u^1)^2 + c^2]^2} = 0$$

$$\chi_{(1)}\text{'s} = \pm\frac{c}{\sqrt{(u^1)^2 + c^2}}.$$

Hence, $2H = \chi_{(1)} + \chi_{(2)} = 0$.
Therefore, the surface is a minimal surface.

Example 11.3.10. Give a surface of revolution S,

$$y^1 = r\cos\phi.\ y^2 = r\sin\phi,\ y^3 = f(r)$$

with $f(r)$ of class C^2. Prove that the lines of curvature on S are the meridians, $\phi = constant$, and the parallels are $r = constant$.

Solution: Here, $u = (r\cos\phi.,\ r\sin\phi, f(r))$

$$ds^2 = (1 + f_1^2)(dr)^2 + r^2(d\phi)^2$$

$$a_{11} = 1 + f_1^2, a_{22} = r^2, a_{12} = a_{21} = 0$$

$$a = r^2(1 + f_1^2).$$

$$\text{We have } A=\left(\frac{\partial y^1}{\partial r},\frac{\partial y^2}{\partial r},\frac{\partial y^3}{\partial r}\right)=(cos\phi, sin\phi, f_1);$$

$$B=\left(\frac{\partial y^1}{\partial \phi},\frac{\partial y^2}{\partial \phi},\frac{\partial y^3}{\partial \phi}\right)=(-r\sin\phi, rcos\phi,0)$$

$$A\times B=(-rf_1\cos\phi,-rf_1\sin\phi,r)$$

$$n=\frac{A\times B}{|A\times B|}=\frac{1}{\sqrt{1+f_1^2}}(-f_1cos\phi,-f_1sin\phi,1)$$

$$b_{11}=\frac{\partial n}{\partial r}\cdot\frac{\partial u}{\partial r}$$

$$=-(cos\,\phi.,sin\,\phi,f_1)\left[\frac{1}{\sqrt{1+f_1^2}}(-f_{11}cos\phi,-f_{11}sin\phi,0)-\frac{1}{2}2f_1f_{11}\right.$$

$$(1+f_1^2)^{\frac{-3}{2}}(-f_1\cos\phi-f_1\sin\phi,1)$$

$$=\frac{f_{11}}{\sqrt{1+f_1^2}}$$

$$b_{22}=-\frac{\partial n}{\partial \phi}\cdot\frac{\partial u}{\partial \phi}=\frac{rf_1}{\sqrt{1+f_1^2}},b_{12}=b_{21}=0$$

$$b=\begin{vmatrix} b_{11} & b_{12} \\ b_{21} & b_{21}\end{vmatrix}=b_{11},\, b_{22}=\frac{rf_1f_{11}}{1+f_1^2}.$$

The equation of the line of curvature is

$$(b_{11}a_{12}-b_{12}a_{11})(dr)^2+(b_{12}a_{22}-b_{22}a_{12})(d\phi)^2+b_{11}a_{22}-b_{22}a_{11})drd\phi=0$$

or $(b_{11}a_{22}-b_{22}a_{11})du^1du^2=0$

$$\left[\frac{f_{11}}{\sqrt{1+f_1^2}}r^2-\frac{rf_1}{\sqrt{1+f_1^2}}(1+f_1^2)\right]drd\phi=0$$

$\therefore drd\phi=0$ or $r=constant\ or\ \phi=constant.$

11.4 Rodrigue's Formula

A line of curvature is characterized by

$$\frac{\partial n^i}{\partial s} + \chi_{(n)} \frac{dx^i}{ds} 0,$$

where $\chi_{(n)}$ is the principal curvature of the surface.
Proof: From the Weingarten formula, we have

$$n^i_{,\alpha} = -a^{\beta\gamma} b_{\beta\alpha} x^i_{\gamma.},$$

$$\text{or} \quad n^i_{,\alpha} \frac{du^\alpha}{ds} = -a^{\beta\gamma} b_{\beta\alpha} x^i_{\gamma.} \frac{du^\alpha}{ds} \qquad \left[n^i_{,\alpha} \frac{du^\alpha}{ds} = \frac{\partial n^i}{\partial u^\alpha} \frac{du^\alpha}{ds} = \frac{\partial n^i}{\partial s} \right],$$

$$\text{or} \qquad \frac{\partial n^i}{\partial s} = -a^{\beta\gamma} b_{\beta\alpha} x^i_{\gamma.} \lambda^\alpha. \tag{11.29}$$

We know from the line of curvature

$$\chi_{(n)} = \frac{b_{\alpha\beta}}{a_{\alpha\beta}}$$

$$\text{or } b_{\alpha\beta}\lambda^\beta = a_{\alpha\beta}\chi_{(n)}\lambda^\beta \text{ and } b_{\alpha\beta}\lambda^\alpha = a_{\alpha\beta}\chi_{(n)}\lambda^\alpha.$$

$$\text{Now,} \frac{\partial n^i}{\partial s} = -a^{\beta\gamma}(b_{\beta\alpha}\lambda^\alpha)x^i_{\gamma.} = -a^{\beta\gamma}(a_{\alpha\beta}\chi_{(n)}\lambda^\alpha)x^i_{\gamma.}$$

$$= -\delta^\gamma_\alpha \chi_{(n)}\lambda^\alpha x^i_{\gamma.} = -\chi_{(n)}\lambda^\alpha x^i_{\alpha.}$$

$$= -\chi_{(n)} \frac{du^\alpha}{ds} \frac{\partial x^i}{\partial u^\alpha} = -\chi_{(n)} \frac{dx^i}{ds}$$

$$\text{or} \quad \frac{\partial n^i}{\partial s} = -\chi_{(n)} \frac{dx^i}{ds}$$

$$\therefore \frac{\partial n^i}{\partial s} + \chi_{(n)} \frac{dx^i}{ds} = 0.$$

Conversely, let

$$\frac{\partial n^i}{\partial s} + \chi_{(n)}\frac{dx^i}{ds} = 0.$$

Using the Weingarten formula, we have

$$n^i_{,\alpha} = -a^{\beta\gamma} b_{\beta\alpha} x^i_{\gamma.}$$

$$\frac{\partial n^i}{\partial s} = -a^{\beta\gamma} b_{\beta\alpha}\, x^i_{\gamma.} \lambda^\alpha$$

$$\chi_{(n)}\frac{dx^i}{ds} - a^{\beta\gamma} b_{\beta\alpha} x^i_{\gamma.} \lambda^\alpha = 0 \text{ (substituting the above result).}$$

Taking the inner product with $g_{ik}x^k_p$, we get

$$\chi_{(n)} g_{ik} x^k_p \frac{dx^i}{ds} - a^{\beta\gamma} b_{\beta\alpha} g_{ik} x^k_p x^i_{\gamma.} \lambda^\alpha = 0$$

and

$$\chi_{(n)} g_{ki} x^k_p \frac{dx^k}{ds} - a^{\beta\gamma} b_{\beta\alpha} a_{\gamma p} \lambda^\alpha = 0.$$

Interchanging dummy indices i and k in the 1st term, we get

$$\chi_{(n)} g_{ik} x^k_p \frac{\partial x^k}{\partial u^\alpha}\frac{du^\alpha}{ds} - b_{p\alpha}\lambda^\alpha = 0,$$

$$\text{or } -b_{p\alpha}\lambda^\alpha + \chi_{(n)} g_{ik} x^k_p x^k_\alpha \lambda^\alpha = 0,$$

$$\text{or } -b_{p\alpha}\lambda^\alpha + \chi_{(n)} a_{p\alpha}\lambda^\alpha = 0,$$

$$\text{or } (\chi_{(n)} a_{p\alpha} - b_{p\alpha})\lambda^\alpha = 0,$$

which is the equation of the line of curvature.

The Rodrigues formula is characteristic for lines of curvature.

Theorem 11.4.1. (Euler's Theorem on Normal Curvature)
If the lines of curvature are not indeterminant at a given point P on the surface and if θ is the angle between a given direction and a principal direction at P, then the normal curvature is given by the formula

$$\chi_{(n)} = \chi_{(1)} \cos^2 \theta + \chi_{(2)} \sin^2 \theta.$$

Proof: We assume that P is not an umbilic point. If the parametric curves are taken as the lines of curvature, then the principal curvature is given by

$$\chi_{(1)} = \frac{b_{11}}{a_{11}} \text{ and } \chi_{(2)} = \frac{b_{22}}{a_{22}}.$$

Let θ be the angle between a given direction $(\delta u^1, \delta u^2)$ and the principal direction at a given point P. Since the coordinate curves are orthogonal, we have

$$\cos\theta = \sqrt{a_{11}}\frac{du^1}{ds} \text{ and } \sin\theta = \sqrt{a_{22}}\frac{du^2}{ds}.$$

Also, let $\chi_{(1)}, \chi_{(2)}$ be the principal curvatures at P. If the parametric curves are taken as lines of curvature, then the normal curvature at P is given by

$$\begin{aligned}\chi_{(n)} &= \frac{b_{11}(du^1)^2 + b_{22}(du^2)^2}{ds^2} = b_{11}\left(\frac{du^1}{ds}\right)^2 + b_{22}\left(\frac{du^2}{ds}\right)^2 \\ &= b_{11}\left(\frac{\cos\theta}{\sqrt{a_{11}}}\right)^2 + b_{22}\left(\frac{\sin\theta}{\sqrt{a_{22}}}\right)^2 \\ &= \frac{b_{11}}{a_{11}}\cos^2\theta + \frac{b_{22}}{a_{22}}\sin^2\theta \\ &= \chi_{(1)}\cos^2\theta + \chi_{(2)}\sin^2\theta.\end{aligned}$$

This theorem states that the normal curvature corresponding to any direction can be simply represented in terms of the principal curvatures. The total curvature in this case is given by

$$\kappa = \frac{b_{11}}{a_{11}}\frac{b_{22}}{a_{22}} \; and \quad H = \frac{1}{2}\left(\frac{b_{11}}{a_{11}} + \frac{b_{22}}{a_{22}}\right).$$

We conclude that the lines of curvature on a minimal surface form an isometric system.

11.5 Exercises

1. Find the principal curvature of the surface defined by

$$x^1 = u^1 \cos u^2, x^2 = u^1 \sin u^2, x^3 = f(u^2).$$

Hence, find the Gaussian and the mean curvature.

2. Find the principal curvature of the surface defined by

$$x^1 = u^1 \cos u^2, x^2 = u^1 sin u^2 \; and \; x^3 = a \; log\left(u^1 + \sqrt{(u^1)^2 - a^2}\right)$$

3. Find the conditions for the meridians to be lines of curvature of the helicoid

$$x^1 = u^1 \cos u^2, x^2 = u^1 \; sin u^2 \; and \; x^3 = f(u^1) + cu^2.$$

12

Curvature of Surface

12.1 Introduction

We discuss the determination of the shape of the surface S in a neighborhood of any of its points considering arbitrary curves on S. We enable restriction of S and the planes which are orthogonal to the tangent plane to S at a point P. We introduce the normal Curvature $\chi_{(n)}$ and consider the behavior of $\chi_{(n)}$ as a function of direction du^2: du^1 of the tangent plane to S. We also study the Gauss-Bonnet theorem with geometric interpretation of a formula that suggests an alternative definition of Gaussian curvature and introduces a few concepts from geometry of n-dimensional metric manifolds, which are of interest in applications to dynamics and relativity.

12.2 Surface of Positive and Negative Curvature

The first fundamental form is positive definite (since $ds^2 = a_{\alpha\beta}du^\alpha du^\beta$), hence the sign of $\chi_{(n)}$ depends on the second fundamental form only.

A surface at all points of which the Gaussian curvature κ (> 0) is positive, is called a *surface of positive curvature,*

$$\text{i.e., } \kappa = \frac{b}{a} = \frac{b_{22}b_{11} - b^2{}_{12}}{a} \Rightarrow b_{22}b_{11} - b^2{}_{12} > 0 \quad \text{as } a > 0 \textit{ on the surface}$$

and since $\chi_{(n)} = b_{\alpha\beta}\,\lambda^\alpha\,\lambda^\beta$, we see that the principal radii $R_{(n)} = \frac{1}{\chi_{(n)}}$ to all normal sections of the surface have positive curvature and do not differ in sign. It will differ in sign if $\kappa < 0$.

Then, the equation

$$b_{\alpha\beta}\,\lambda^\alpha\,\lambda^\beta = 0 \tag{12.1}$$

defines two directions for which the radii of curvatures are infinite.

Dipankar De. Introduction to Differential Geometry with Tensor Applications, (381–396) © 2022 Scrivener Publishing LLC

A surface at all points of which the Gaussian curvature κ (< 0) is negative is called a *surface of negative curvature.* If κ = 0 at a given point, the directions given by (12.1) coincide and for this direction R is infinite.

From geometrical considerations, it is clear that ellipsoid, biparted hyperboloid, and elliptic paraboloids are surfaces of positive curvature and hyperboloids of one sheet and hyperbolic paraboloids are surfaces of negative curvature.

A Point on S is said to be *elliptic* if the signs of the principal curvatures $\chi_{(1)}, \chi_{(2)}$ are the same. It follows that $\chi_{(n)}$ at an elliptic point does not change for any direction of the normal section. A point is hyperbolic if $\chi_{(1)}, \chi_{(2)}$ have opposite signs. At a hyperbolic point there are two directions for which $\chi_{(n)} = 0$. A point is parabolic if one of the values of $\chi_{(1)}$ or $\chi_{(2)}$ is zero. In special cases, if $\chi_{(1)} = \chi_{(2)}$ and all values of $\chi_{(n)}$ are equal, such points are called *spherical or umbilical.*

In the neighborhood of a spherical point, the surface looks like a sphere and we can prove that if all points of S are spherical, then the surface is a sphere.

Example 12.2.1. Show that the helicoids $y^1 = u^1 cosu^2$

$y^2 = u^1 sinu^2$
$y^3 = au^2$ are a surface of negative curvature.

Solution: Here, $y^1 = u^1 cosu^2$, $y^2 = u^1 sinu^2$ and $y^3 = au^2$

$$\text{or } r = (u^1 cosu^2, u^1 sinu^2, au^2)$$

$$a_{11} = \left(\frac{\partial y^i}{\partial u^1}\right)^2 = 1, a_{22} = \left(\frac{\partial y^i}{\partial u^2}\right)^2 = a^2 + (u^1)^2,$$

$$a_{12} = \frac{\partial y^1}{\partial u^1}\frac{\partial y^1}{\partial u^2} + \frac{\partial y^2}{\partial u^1}\frac{\partial y^2}{\partial u^2} + \frac{\partial y^3}{\partial u^1}\frac{\partial y^3}{\partial u^2} = 0 = a_{21},$$

$$a = a_{11}a_{22} - a_{12}{}^2 = a^2 + (u^1)^2 > 0$$

$$n = \frac{(cosu^2, sinu^2, 0) \times (-u^1 sinu^2, u^1 cosu^2, a)}{\sqrt{(u^1)^2 + a^2}}$$

$$= \frac{1}{\sqrt{(u^1)^2 + a^2}}(asinu^2, -acosu^2, u^1)$$

$$b_{11} = -\frac{\partial n}{\partial u^1}\cdot\frac{\partial r}{\partial u^1} = -\left\{\frac{1}{\sqrt{(u^1)^2+a^2}}(0,0,1) - \frac{1}{2}\frac{2u^1}{[(u^1)^2+a^2]^{\frac{3}{2}}}(asinu^2,-a\cos u^2,u^1)\right\}$$

$$((cosu^2,sinu^2,0)=0$$

$$b_{22} = -\frac{\partial n}{\partial u^2}\cdot\frac{\partial r}{\partial u^2} = 0 \text{ and } b_{12} = -\frac{-a}{(u^1)^2+a^2}$$

$$b = b_{11}b_{22} - b_{12}{}^2 = -\frac{a^2}{(u^1)^2+a^2},$$

$\kappa = \dfrac{|b_{\alpha\beta}|}{|a_{\alpha\beta}|} = -\dfrac{a^2}{[(u^1)^2+a^2]^2} < 0$ and the given surface is a surface of negative curvature.

12.3 Parallel Surfaces

Definition 12.3.1. Let S be a smooth surface defined by equations

$$y^i = y^i(u^1, u^2), (i = 1, 2) \tag{12.2}$$

where the coordinates y^i are orthogonal Cartesian. A surface $\overline{S}$ is determined by the equations

$$\overline{y}^i(u^1,u^2) = y^i(u^1,u^2) + \text{hn}^i(u^1,u^2), \tag{12.3}$$

where n^i is a unit normal to S *and* h is the distance measured along the normal n, called a *parallel surface* to S.

Parallel surfaces figure prominently in the theory of elastic plates and shells, where relations connecting the Gaussian curvatures κ and the mean curvature H *of* S with the corresponding invariants for the surface $\overline{S}$ are important.

12.3.1 Computation of $\overline{a}_{\alpha\beta}$ and $\overline{b}_{\alpha\beta}$

We have

$$a_\alpha = b_i\frac{\partial y^i}{\partial u^\alpha}, \tag{12.4}$$

where the base vector a_α along the curves $u^\alpha = constant$ and base vector b_i along the y^i – *axes.*

We introduce the notations $\frac{\partial y^i}{\partial u^\alpha} = y^i_\alpha$ *and* $\frac{\partial \bar{y}^i}{\partial u^\alpha} = \bar{y}^i_\alpha$.

Differentiating (12.3), we get

$$\bar{y}^i_\alpha = y^i_\alpha + \mathrm{h} n^i_{,\alpha} \tag{12.5}$$

Multiplying n_i on both sides,

$$\bar{y}^i_\alpha n_i = y^i_\alpha n_i + \mathrm{h} n^i_{,\alpha} n_i,$$

but $y^i_\alpha n_i = 0$ for a_α is orthogonal to $\boldsymbol{n}$ and $n^i_{,\alpha} n_i = 0$ since $n^i n_i = 1$
[a_α *lies on curves* $u^\alpha = constant$]

$$\therefore \ \bar{y}^i_\alpha n_i = 0. \tag{12.6}$$

Also, the unit vector $\bar{n}_i$ to $\bar{S}$ is orthogonal to the base vector $\bar{y}^i_\alpha$ on $\bar{S}$, so that

$$\bar{y}^i_\alpha \bar{n}_i = 0. \tag{12.7}$$

From (12.6) and (12.7), it is implied that n_i and $\bar{n}_i$ are collinear and since they are unit vector

$$n_i = \bar{n}_i,$$

the metric coefficients $\bar{a}_{\alpha\beta}$ of S are given by

$$\bar{a}_{\alpha\beta} = \bar{y}^i_\alpha \bar{y}^i_\beta.$$

Using (12.5) in the above equation,

$$\begin{aligned}\bar{a}_{\alpha\beta} &= (y^i_\alpha + \mathrm{h} n^i_{,\alpha})(y^i_\beta + \mathrm{h} n^i_{,\beta}) \\ &= (y^i_\alpha y^i_\beta + \mathrm{h} n^i_{,\alpha} y^i_\beta + \mathrm{h} n^i_{,\beta} y^i_\alpha + \mathrm{h}^2 n^i_{,\alpha} n^i_{,\beta}).\end{aligned}$$

Applying the Weingarten formula in this expression,

$$n^i_{,\alpha} = -a^{\delta\gamma} b_{\delta\alpha} y^i_\gamma$$

$$\bar{a}_{\alpha\beta} = (y^i_\alpha y^i_\beta + \mathrm{h}(-a^{\delta\gamma} b_{\gamma\alpha} x^i_\gamma) y^i_\beta + \mathrm{h}(-a^{\delta\gamma} b_{\gamma\beta} x^i_\gamma) y^i_\alpha + \mathrm{h}^2 n^i_{,\alpha} n^i_{,\beta})$$

$$= a_{\alpha\beta} - 2hb_{\alpha\beta} + \mathrm{h}^2 n^i_{,\alpha} n^i_{,\beta} \quad [\text{since } y^i_\alpha y^i_\beta = a_{\alpha\beta}]$$

$$\bar{a}_{\alpha\beta} = a_{\alpha\beta} - 2hb_{\alpha\beta} + \mathrm{h}^2 n^i_{,\alpha} n^i_{,\beta} \tag{12.8}$$

The last term of the right hand side of (12.8) can be expressed in terms of Gaussian curvature κ and the mean curvature *H*. Also using the Weingarten formula, we get

$$n^i_{,\alpha} n^i_{,\beta} = a^{\delta\gamma} b_{\alpha\delta} y^i_\gamma a^{\lambda\mu} b_{\beta\lambda} y^i_\mu \qquad (\text{Weingarten formula: } n^i_{,\alpha} = -a^{\delta\gamma} b_{\gamma\alpha} x^i_\gamma)$$

$$= a^{\delta\gamma} b_{\alpha\delta} a^{\lambda\mu} b_{\beta\lambda} a_{\gamma\mu}, \quad \text{since} \quad y^i_\gamma y^i_\mu = a_{\gamma\mu}$$

$$n^i_{,\alpha} n^i_{,\beta} = a^{\mu\lambda} b_{\alpha\mu} b_{\beta\lambda}. \tag{12.9}$$

The Gauss Equation is

$$R_{\alpha\beta\gamma\delta} = b_{\beta\delta} b_{\alpha\gamma} - b_{\beta\gamma} b_{\alpha\delta},$$

where $R_{\alpha\beta\gamma\delta} = \kappa \epsilon_{\alpha\beta} \epsilon_{\gamma\delta}$.

Equation (10.53) becomes

$$\kappa \epsilon_{\alpha\beta} \epsilon_{\gamma\delta} = b_{\beta\delta} b_{\alpha\gamma} - b_{\beta\gamma} b_{\alpha\delta}.$$

Multiplying both sides by $a^{\alpha\delta}$, summing on δ

$$\kappa \epsilon_{\alpha\beta} \epsilon_{\gamma\delta} a^{\alpha\delta} = b_{\beta\delta} b_{\alpha\gamma} a^{\alpha\delta} - b_{\beta\gamma} b_{\alpha\delta} a^{\alpha\delta},$$

Using $a^{\alpha\delta} \epsilon_{\alpha\beta} \epsilon_{\gamma\delta} = -a_{\beta\gamma}$

$$-\kappa a_{\beta\gamma} = a^{\alpha\delta} b_{\beta\delta} b_{\alpha\gamma} - 2Hb_{\beta\gamma}, \tag{12.10}$$

since $H = \frac{1}{2} b_{\alpha\delta} a^{\alpha\delta}$

Substituting the right-hand side of the 1st term of (12.10) from (12.9), we get

$$-\kappa a_{\alpha\beta} = n^i_{,\alpha} n^i_{,\beta} - 2Hb_{\alpha\beta} \text{ [changing suffices } \alpha,\beta \textit{ in the place of } \beta \textit{ and } \gamma]$$

$$\text{or } n^i_{,\alpha} n^i_{,\beta} = -\kappa a_{\alpha\beta} + 2Hb_{\alpha\beta}. \tag{12.11}$$

Now we can write (12.8) by substituting the values of (12.11):

$$\bar{a}_{\alpha\beta} = a_{\alpha\beta} - 2hb_{\alpha\beta} + \mathrm{h}^2(-\kappa a_{\alpha\beta} + 2Hb_{\alpha\beta})$$

$$\text{or} \quad \bar{a}_{\alpha\beta} = a_{\alpha\beta}(1-\mathrm{h}^2\kappa) - 2hb_{\alpha\beta}(1-hH). \tag{12.12}$$

The above formula (12.12) enables us to compute $\bar{a}_{\alpha\beta}$ at a given point $\bar{P}(u^1,u^2)$ on $\bar{S}$ with the values of $a_{\alpha\beta}$, $b_{\alpha\beta}$, κ, H at the $P(u^1, u^2)$ on S.

To compute the coefficients of $\bar{b}_{\alpha\beta}$ on $\bar{S}$ and using Equation (10.36),

$$y^i_{\alpha,\beta} = b_{\alpha\beta} n^i \text{ or } b_{\alpha\beta} = y^i_{\alpha,\beta} n_i$$

and we get from (12.5)

$$\bar{y}^i_{\alpha} = y^i_{\alpha} + \mathrm{h} n^i_{,\alpha}.$$

Differentiating $\bar{y}^i_{\alpha,\beta} = y^i_{\alpha,\beta} + \mathrm{h} n^i_{,\alpha\beta}$,

so that
$$\begin{aligned} \bar{b}_{\alpha\beta} &= \bar{y}^i_{\alpha,\beta}\, n_i \\ &= (y^i_{\alpha,\beta} + \mathrm{h} n^i_{,\alpha\beta})\, n_i \\ &= y^i_{\alpha,\beta} n_i + \mathrm{h} n^i_{,\alpha\beta}\, n_i \end{aligned}$$

$$\bar{b}_{\alpha\beta} = b_{\alpha\beta} + \mathrm{h} n^i_{,\alpha\beta}\, n_i. \tag{12.13}$$

Since the coordinates y^i are rectangular Cartesian, $n^i = n_i$, $n^i n^i = 1$ we have $n^i_{,\alpha} n^i = 0$.

On differentiating this orthogonality relation, we find

$$n^i_{,\alpha\beta} n^i = -n^i_{,\alpha} n^i_{,\beta}$$

(12.13) becomes

$$\begin{aligned}\bar{b}_{\alpha\beta} &= b_{\alpha\beta} + \mathrm{h} n^i_{,\alpha\beta}\, n_i \\ &= b_{\alpha\beta} + \mathrm{h}(-n^i_{,\alpha} n^i_{,\beta})\, n_i n_i \\ &= b_{\alpha\beta} - \mathrm{h} n^i_{,\alpha} n^i_{,\beta}.\end{aligned}$$

Using (12.11) in the above equation, i.e., using $n^i_{,\alpha} n^i_{,\beta} = -\kappa a_{\alpha\beta} + 2Hb_{\alpha\beta}$,

$$\begin{aligned}\bar{b}_{\alpha\beta} &= b_{\alpha\beta} - \mathrm{h} n^i_{,\alpha} n^i_{,\beta} \\ &= b_{\alpha\beta} - \mathrm{h}(-\kappa a_{\alpha\beta} + 2Hb_{\alpha\beta}) \\ &= b_{\alpha\beta}(1 - 2Hh) - \kappa a_{\alpha\beta} h\end{aligned}$$

$$\therefore \bar{b}_{\alpha\beta} = b_{\alpha\beta}(1 - 2Hh) - \kappa a_{\alpha\beta} h \tag{12.14}$$

The result of the mentioned monograph by T. Y. Thomas is

$$\bar{\kappa} = \frac{\kappa}{1 + \mathrm{h}^2\kappa - 2\mathrm{hH}}$$
$$\text{and} \qquad \bar{\mathrm{H}} = \frac{\mathrm{H} - \mathrm{h}\kappa}{1 + \mathrm{h}^2\kappa - 2\mathrm{hH}}. \tag{12.15}$$

It follows that when S is a developable surface, then the parallel surfaces $\bar{S}$ are also developable.

12.4 The Gauss-Bonnet Theorem

The classical result, however, relates the integral of the Gaussian curvature evaluated over the area of an arbitrary smooth surface to the line integral of the geodesic curvature computed over the curve that bounds the area. Gauss viewed this result as the most important theorem of the geometry of surfaces at large.

Let D be the region bound by a closed piecewise smooth curve C over surface S, as shown in Figure (12.1). D is homeomorphic to a circular disc. We know that a unit tangent vector λ^α to a surface curve C is related to the unit vector η^α to λ^α by

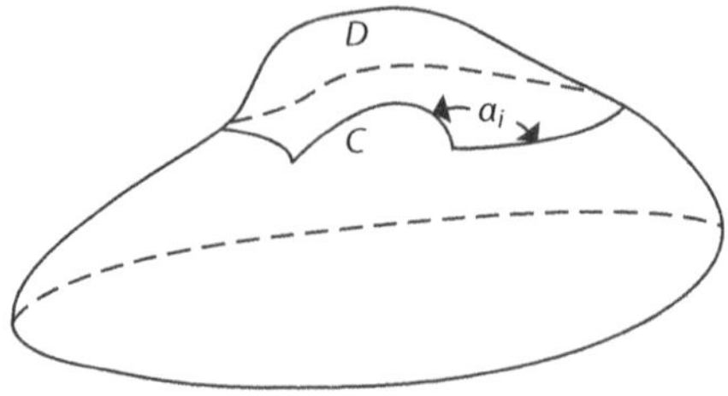

Figure 12.1

$$\frac{\delta\lambda^\alpha}{\delta s} = \chi_g \eta^\alpha. \tag{12.16}$$

where χ_g is geodesic curvature of C. Moreover, if C is a geodesic, then $\chi_g = 0$ at all points of C.

Conversely, since η^α is orthogonal to λ^α, it follows

$$\epsilon_{\alpha\beta}\eta^\alpha\lambda^\beta = 1$$

$$\text{or } \eta^\alpha = \epsilon_{\alpha\beta}\lambda^\beta$$

$$\therefore \chi_g = \frac{\delta\lambda^\alpha}{\delta s}\eta_\alpha = \frac{\delta\lambda^\alpha}{\delta s}\epsilon_{\alpha\beta}\lambda^\beta. \tag{12.17}$$

Now, integration over C is expressed as

$$\int_C \chi_g ds = \int_C \epsilon_{\alpha\beta}\lambda^\beta \frac{\delta\lambda^\alpha}{\delta s} ds \tag{12.18}$$

and when the line integral in the right-hand member of (12.18) is transformed into a surface integral by Green's formula, we get

$$\int_C \chi_g ds = -\int\int_C k\, d\sigma + 2\pi - \Sigma(\pi - \alpha_i), \tag{12.19}$$

where the α_i are interior angles of the contour C and $d\sigma = \sqrt{a}du^1 du^2$ is the element of the surface area of D. If C is smooth, the sum is $\Sigma\,(\pi - \alpha_i) = 0$.

Formula (12.19) gives us the statement of the *Gauss-Bonnet theorem.* With the aid of Green's theorem, we give a geometrical interpretation of (12.19) as an alternative definition of Gaussian curvature.

Consider a sphere S of radius R and a spherical triangle $P_1 P_2 P_3$ on S, as shown in Figure (12.2), formed by the arcs $P_1 P_2$, P_2P_3, $P_1 P_3$ of three great circles. We denote that the interior angles made by them at P_1, P_2, *and* P_3 are α_1, α_2, *and* α_3, respectively and cover the sphere by some coordinate net (u_1, u_2).

If the base vector along the u_1 –coordinate line at P is $\boldsymbol{a}_1$ and $\boldsymbol{A}$ (P_1) is an arbitrary surface vector at P_1, we denote the angle between $\boldsymbol{a}_1$ and $\boldsymbol{A}$ (P_1) as φ. Let θ be the angle between $\boldsymbol{A}$ (P_1) and geodesic arc $P_1 P_2$. When $\boldsymbol{A}$ (P_1) is propogated in a parallel manner along the geodesic triangle $P_1 P_2 P_3$, it assumes the position $\boldsymbol{A}'$ (P_1). Its object is to determine the angle φ' between $\boldsymbol{a}_1$ and $\boldsymbol{A}'$ (P_1).

During the parallel propagation of $\boldsymbol{A}$ (P_1) along $P_1 P_2$, if the angle θ is unchanged and vector $\boldsymbol{A}$ assumes the position $\boldsymbol{A}$ (P_2) with geodesic arc $P_2 P_3$, then

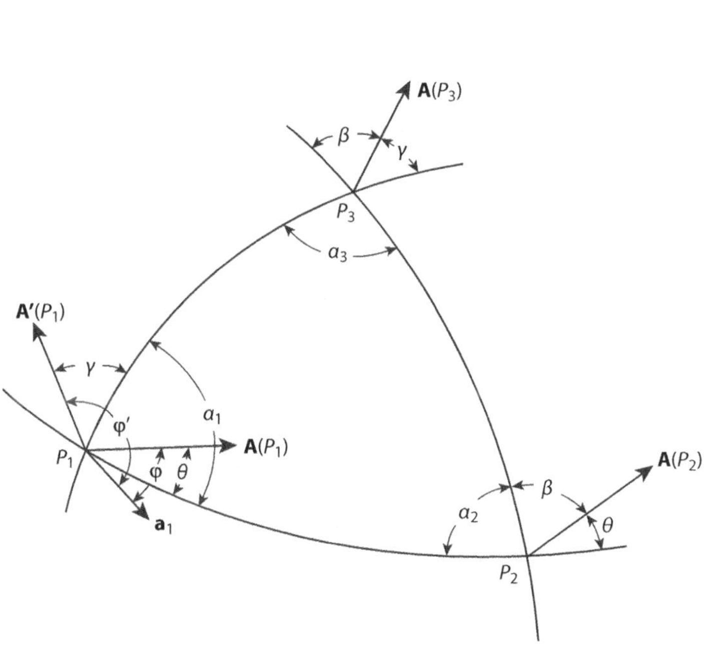

Figure 12.2

$$\beta = \pi - (\alpha_2 + \theta).$$

In the course of parallel propagation of $\boldsymbol{A}$ (P_2) along $P_2\,P_3$, vector $\boldsymbol{A}$ continues making angle β with arc $P_2\,P_3$ *and* assumes position (P_3). Let γ be the angle between $\boldsymbol{A}$ (P_3) and geodesic arc $P_1\,P_3$.

$$\therefore \gamma = \alpha_3 - \beta = \alpha_3 - [\pi - (\alpha_2 + \theta)] = \alpha_3 + \alpha_2 + \theta - \pi$$

On continuing propagations of $\boldsymbol{A}$ along $P_1\,P_3$, vector $\boldsymbol{A}$ maintains angle γ with $P_1\,P_3$ until it reaches point P_1 when it assumes position $\boldsymbol{A}'$ (P_1). Now φ' is made by $\boldsymbol{a}_1$ and $\boldsymbol{A}'$ (P_1) is

$$\begin{aligned}\varphi' &= \gamma + \alpha_1 + \varphi - \theta \\ &= \alpha_3 + \alpha_2 + \theta - \pi + \alpha_1 + \varphi - \theta \\ &= \alpha_1 + \alpha_2 + \alpha_3 + \varphi - \pi.\end{aligned}$$

Now, the angle between $\boldsymbol{A}$ (P_1) and $\boldsymbol{A}'$ (P_1) is $\varphi' - \varphi$

$$\varphi' - \varphi = \alpha_1 + \alpha_2 + \alpha_3 - \pi. \tag{12.20}$$

This difference of angles is called the *spherical excess* of spherical triangle $P_1\,P_2\,P_3$.

If instead of interior angles α_i we introduce the exterior angles $\theta_i = \pi - \alpha_i$, (12.20) becomes

$$\varphi' - \varphi = [3\pi - (\theta_1 + \theta_2 + \theta_3)] - \pi = 2\pi - \Sigma_{i=1}^{n}\theta_i.$$

When vector $\boldsymbol{A}$ is propagated along a geodesic polygon of n sides, entirely similar computations yield for spherical excess of a polygon.

$\varphi' - \varphi = \Sigma_{i=1}^{n}\alpha_i - (n-2)\pi$ (sum of interior angles of polygon of n sides $= (n-2)\pi$) or $\varphi' - \varphi = 2\pi - \Sigma_{i=1}^{n}\theta_i$, if using exterior angles $\theta_i = \pi - \alpha_i$, but it is known for spherical trigonometry that spherical excess of a geodesic polygon is equal to $\frac{\sigma}{R^2}$, where σ is an area of a polygon and R is the radius of a sphere.

$$\therefore\ \varphi' - \varphi = 2\pi - \Sigma_{i=1}^{n}\theta_i = \frac{\sigma}{R^2}$$

Since Gaussian curvature $\kappa = \frac{1}{R^2}$,

$$\kappa = \frac{2\pi - \sum_{i=1}^{n} \theta_i}{\sigma}. \tag{12.21}$$

This formula can be generalized to obtain the Gauss-Bonnet formula (12.19) for the case where C is a geodesic polygon.

Thus, if the region D (Figure 12.12) is subdivided by a geodesic polygon into regions of area $d\sigma$, then

$$\iint_D \mathrm{k}d\,\sigma = 2\pi - \sum_{i=1}^{n} \theta_i, \tag{12.22}$$

which coincides with (12.19) since $\chi_g = 0$ when C is a geodesic polygon.

Formula (12.22) was first obtained by Gauss, which is generalized by Bonnet to yield (12.19), which is directly derived from (12.18) using Green's formula.

The left-hand side of (12.22) $\iint_D \kappa d\,\sigma$ is called the *integral curvature* of D. It turns out that integral curvature is a topological invariant. Two surfaces are said to be topologically equivalent if they can be mapped into one another by a continuous one to one transformation.

It can be shown by using (12.19) that

$\iint_D \kappa d\,\sigma = 4\pi$ for all regular surfaces topologically equivalent to a sphere

and $\iint_D \kappa d\,\sigma = 0$ for all regular surfaces topologically equivalent to a trous.

12.5 The n-Dimensional Manifolds

In this section, we introduce a few concepts from geometry of n-dimensional metric manifolds which are of interest to apply in dynamics and relativity. Here, many of these concepts are direct generalizations of ideas introduced in this chapter related to the study of surfaces imbedded in three-dimensional Euclidean manifolds.

The elements of distance between two neighboring points in n-dimensional manifolds is

$$ds^2 = g_{ij}\, dx^i\, dx^j \ (i, j = 1, 2, \ldots n) \ with \ g_{ij} \neq 0. \tag{12.23}$$

The space is *Euclidean* if there exists a transformation of coordinates x^i such that the transformation of ds^2 is a quadratic form with constant coefficients since every real quadratic form with constant coefficients can be reduced by a real linear transformation to the form

$$ds^2 = \lambda_i (dx^i)^2, \text{ where } \lambda_i = \pm 1. \tag{12.24}$$

This form of (12.24) can be used to define a *Euclidean n-dimensional manifold.*

If, in particular, the form (12.24) is definite, we shall say that the manifold is *purely Euclidean*, but if it is indefinite, the manifold will be called *pseudo-Euclidean.*

A linear manifold determined by a set of n equations

$$C: x^i = x^i (t),\ t_1 \le t \le t_2$$

with suitable differentiability properties will be said to define a curve C in an n-dimensional manifold.

If (12.23) is positive definite, we shall say that the positive number

$$s = \int_{t_1}^{t_2} \sqrt{g_{ij} \frac{dx^i}{dt}\frac{dx^j}{dt}} dt \text{ is the length of the curve C.}$$

If vector $\lambda^i = \dfrac{dx^i}{ds}$ defines the direction of the curve, then

$$g_{ij} \lambda^i \lambda^j = 1 \text{ (from 12.23)},$$

so that λ^i is a unit vector. The length of any vector A^i is given by

$$A = \sqrt{g_{ij} A^i A^j}.$$

If λ^i and μ^i are two unit vectors, we define the cosine of the angle between them as

$$cos\theta = g_{ij} \lambda^i \mu^j. \tag{12.25}$$

Example 12.5.1. The angle between unit tangent vector λ^i and normal vector μ^j is real if the form $g_{ij}\, dx^i\, dx^j$ is positive definite.

Proof: By Cauchy-Schwarz inequality, we can write

$$(g_{ij}\, x^i\, y^j)^2 \le (g_{ij}\, x^i\, x^j)\,(g_{ij}\, y^i\, y^j), \tag{12.26}$$

where form $g_{ij}\, dx^i\, dx^j \ge 0$.

Let $Q\,(x) = g_{ij}\, x^i\, x^j$ be positive definite.

If we replace x^i with $x^i + \lambda y^i$ in it where λ is an arbitrary scalar, we obtain

$$\begin{aligned} Q\,(x + \lambda y) &= g_{ij}\,(x^i + \lambda y^i)\,(x^j + \lambda y^j) \\ &= g_{ij}\,(x^i + \lambda y^i)\,(x^j + \lambda y^j) \\ &= g_{ij}\, x^i\, x^j + 2\lambda x^i\, y^j + g_{ij}\, y^i\, y^j \lambda^2 \\ &= g_{ij}\, x^i\, x^j + 2\lambda\, g_{ij}\, x^i\, y^j + g_{ij}\, y^i\, y^j \lambda^2 \\ &= Q\,(x) + 2Q\,(x, y)\,\lambda + Q\,(y)\,\lambda^2 \end{aligned}$$

$Q\,(x + \lambda y)$ is a quadratic equation of λ.

It is clear that $Q\,(x + \lambda y) \ge 0$.

The sign of equality holds if $x^i + \lambda y^i = 0$.

Hence,

$$f\,(\lambda) \equiv Q\,(x) + 2Q\,(x, y)\,\lambda + Q\,(y)\,\lambda^2 = 0$$

possessing no distinct real roots if

$$4\,[Q\,(x, y)]^2 - 4Q\,(x)\,Q\,(y) \le 0$$

$$\text{or } [Q\,(x, y)]^2 - Q\,(x)\,Q\,(y) \le 0.$$

That is, $(g_{ij}\, x^i\, y^j)^2 \le (g_{ij}\, x^i\, x^j)\,(g_{ij}\, y^i\, y^j)$.

This equation establishes (12.26).

If we now set $x^i = \lambda^i$ and $y^i = \mu^i$, we get

$$\frac{(g_{ij}\lambda^i \mu^i)^2}{(g_{ij}\lambda^i \lambda^j)(g_{ij}\mu^i \mu^j)} \le 1.$$

Since λ^i and μ^i are unit vectors, it is concluded that $(g_{ij}\, \lambda^i\, \mu^i)^2 \leq 1$.
That implies that $cos^2\theta \leq 1$

$$\text{or } -1 \leq cos\theta \leq 1.$$

That is, the angle between λ^i and μ^i are real.

12.6 Hypersurfaces

A set of n – equations

$$x^i = x^i\,(u^1, u^2, \ldots. u^n),\ (i = 1, 2, \ldots.n),\ m \leq n, \tag{12.27}$$

is said to define a *hypersurface* over a neighborhood of variables u^α if (a) the x^i in (12.27) are of class C^2 and (b) the Jacobian matrix $\frac{\partial x^i}{\partial u^\alpha}(\alpha = 1, 2, \ldots m)$ is of rank m at each point of the neighborhood.

We also know that there are $n - m$ linearly independent normals to $V_{m.}$ If we take $n = m + 1$, then $V_{m.}$ is said to be the hypersurface of the enveloping space V_{m+1}.

Now, under these circumstances an m-dimensional hyper surface with a Riemannian metric

$$ds^2 = a_{\alpha\beta}\, du^\alpha\, du^\beta,\ (\alpha, \beta = 1, 2, \ldots.m) \tag{12.28}$$

can be imbedded in an n-dimensional Euclidean manifold with

$$ds^2 = dx^i\, dx^i. \tag{12.29}$$

Now, it requires that

$$a_{\alpha\beta} = \frac{\partial x^i}{\partial u^\alpha}\frac{\partial x^i}{\partial u^\beta}\ and\ (i = 1, 2 \ldots n). \tag{12.30}$$

The set of $\frac{1}{2}m(m+1)$ partial differential equations of (12.30) in n-variables x^i will not be expected to possess a solution unless $n \geq \frac{1}{2}m(m+1)$. Thus, if $m = 2$, $n \geq 3$, *if* $m = 3$, $n \geq 6$, *and so on.*

Thus, it is possible to prove that a neighborhood of R_m can be imbedded in E_n if $n \geq \frac{1}{2}m(m+1)$.

12.7 Exercises

1. Given a surface of revolution S

$$y^1 = r\cos\phi.\ y^2 = r\sin\phi,\ y^3 = f(r)$$

   with $f(r)$ of class, show that the points on a surface of revolution S for which $f_1 f_{11} > 0$ are elliptic, those for which $f_1 f_{11} < 0$ are hyperbolic, and if $f_{11} = 0$, then S is a cone.

2. Find the nature of a point on a unit sphere

$$(x^1)^2 + (x^2)^2 + (x^3)^2 = 1.$$

Part III
ANALYTICAL MECHANICS

13
Classical Mechanics

13.1 Introduction

Classical mechanics originated with the work of Galileo and were developed extensively by Isaac Newton. It deals with the motion of particles in a fixed frame of reference. In order to develop the science of mechanics of a universe consisting of more than two particles, it is necessary to adjoin Newtonian laws with the principle of superposition of effects. The basic premise of Newtonian mechanics is the concept of absolute time measurement between two reference frames at a constant velocity. In this system, other coordinate systems may be used so that the metric remains Euclidean.

13.2 Newtonian Laws of Motion

Newton's Laws of Motion are stated in the following form:

1. Everybody continues in its state of rest or of uniform motion in a straight line until it is compelled by impressed forces to change that state.
2. The change of motion is proportional to impressed motive force and takes place in the direction of the straight line in which that force is impressed.
3. To every action there is always an equal and opposite reaction.

The first law depends on the dynamical concept of force and the kinematical idea of uniform rectilinear motion.

The second law of motion introduces the kinematical concept of motion and the dynamic idea of force. To understand its meaning, it should be noted that Newton uses the term motion in the sense of momentum as the

Dipankar De. Introduction to Differential Geometry with Tensor Applications, (399–446) © 2022 Scrivener Publishing LLC

product of mass and velocity. Thus, the change of motion means "the rate of change of momentum with respect to time."

$$\therefore \boldsymbol{F} = \frac{d(m\boldsymbol{v})}{dt} \tag{13.1}$$

If we postulate the invariance of mass then, (13.1) can be written as

$$\boldsymbol{F} = m\boldsymbol{a}, \text{ where } \boldsymbol{a} = \frac{d\boldsymbol{v}}{dt}. \tag{13.2}$$

If $\boldsymbol{F} = 0$, then $\frac{d(m\boldsymbol{v})}{dt} = 0$ implies that $m\boldsymbol{v}$ = constant.

Hence, $\boldsymbol{v}$ is a constant vector.

Therefore, the first law is a consequence of the second law.

The third law of motion states that accelerations always occur in pairs. In terms of force, we may say that if a force acts on a given body, the body itself exerts an equal and opposite directed force on some other body. Newton called the two aspects of the force action and reaction.

Newtonian Law is sometimes called *inertia* to distinguish it from the Newtonian Law of Gravitation.

This law states that the force of attraction between a pair of particles is proportional to the product of their masses, is inversely proportional to the square of the distance between them, and directed along the line joining the particles,

$$\boldsymbol{F} = k\frac{M_1 M_2}{r^3}\boldsymbol{r}, \tag{13.3}$$

where k is a universal constant and $\boldsymbol{r}$ is a vector directed from mass M_1 *to mass* M_2.

In order to develop the science of mechanics of a universe consisting of more than two bodies, it is necessary to include the principles of superposition of effects and make assumptions regarding the nature of constraints.

13.3 Equations of Motion of Particles

Theorem 13.3.1. The work done in displacing a particle along its trajectory is equal to the change in the kinetic energy of particle.

Proof: Let the position of moving particle P be determined by a vector $\boldsymbol{r}$.

The equation of the path C of the moving particle can be written as

$$C: x^i = x^i\,(t) \tag{13.4}$$

and we call the curve C the *trajectory* of the moving particle.

The velocity of P is a vector $\boldsymbol{v} = \frac{dr}{dt}$, whose components are

$$v^i = \frac{dx^i}{dt} \tag{13.5}$$

and acceleration $a = \frac{d^2\boldsymbol{r}}{dt^2}$ and its components

$$\begin{aligned} a^i &= \frac{\delta v^i}{\delta t} = \frac{dv^i}{dt} + \begin{Bmatrix} i \\ j \quad k \end{Bmatrix} v^j \frac{dx^k}{dt} \\ &= \frac{d^2x^i}{dt^2} + \begin{Bmatrix} i \\ j \quad k \end{Bmatrix} \frac{dx^j}{dt}\frac{dx^k}{dt} \end{aligned} \tag{13.6}$$

Where $\frac{\delta v^i}{\delta t}$ is the intrinsic derivative and $\begin{Bmatrix} i \\ j \quad k \end{Bmatrix}$ the second kind of Christoffel symbol calculated from the metric tensor g_{ij}.

If the mass of a particle is m, then by Newton's Second Law of Motion $F = m\frac{d^2r}{dt^2}$, where its components express

$$\boldsymbol{F}^i = m\frac{\delta v^i}{\delta t} = ma^i. \tag{13.7}$$

The first clear introduction of the idea of the energy of mechanics as a quantity equal to the product of mass and the square of velocity of the particle was made by Huygens in the 17th century. The full use of this idea in the concept of work did not come until the 19th century.

We define that the element of work done by the force $\boldsymbol{F}$ is producing a displacement $\boldsymbol{dr}$ by the invariant $dW = \boldsymbol{F}.\ \boldsymbol{dr}$ and its components of $\boldsymbol{F}$ *and* $\boldsymbol{dr}$ are F^i *and* dx^i respectively.

$$\text{Therefore, } dW = g_{ij}\, F^i.\ dx^j = F_j dx^j, \tag{13.8}$$

where $F_j = g_{ij}\, F^i$.

The work done in displacing a particle along the trajectory C, joining a pair of points P_1 and P_2 is line integral:

$$W = \int dW = \int_{P_1}^{P_2} F_i dx^i. \tag{13.9}$$

Now, by (13.7) we get

$$\begin{aligned} W &= \int_{P_1}^{P_2} g_{ij} F^i dx^i \\ &= \int_{P_1}^{P_2} g_{ij} m \frac{\delta v^i}{\delta t} dx^i \\ W &= \int_{P_1}^{P_2} m g_{ij} \frac{\delta v^i}{\delta t}\frac{dx^j}{dt} dt = \int_{P_1}^{P_2} m g_{ij} \frac{\delta v^i}{\delta t} v^j dt, \end{aligned} \tag{13.10}$$

but $\dfrac{\delta(g_{ij}v^i v^j)}{\delta t} = 2g_{ij}\dfrac{\delta v^i}{\delta t}v^j$ since g_{ij} is an invariant.

$$\frac{\delta(g_{ij}v^i v^j)}{\delta t} = \frac{d(g_{ij}v^i v^j)}{dt}$$

and

$$\frac{d(g_{ij}v^i v^j)}{dt} = 2g_{ij}\frac{\delta v^i}{\delta t}v^j.$$

Using this result in (13.10), we get

$$W = \int_{P_1}^{P_2} \frac{1}{2} m \frac{d(g_{ij} v^i v^j)}{dt} dt = \frac{m}{2} g_{ij} v^i v^j \Big|_{P_1}^{P_2} = T_2 - T_1,$$

where $T = \frac{m}{2} g_{ij} v^i v^j = \frac{1}{2} m v^2 = \textit{Kinetic Energy}$ of a particle.

Thus, we conclude that the work done in displacing a particle along the trajectory is equal to the change in the kinetic energy of the particle.

13.4 Conservative Force Field

The force field F_i is such that the integral (13.9) is independent of the path. Therefore, the integrated $F_i dx^i$ is an exact differential, i.e.,

$$dW = F_i dx^i \tag{13.11}$$

of the work function W. The negative of the work function W is called the force potential or *potential energy.* We get from (13.11),

$$F_i = -\frac{\partial V}{\partial x^i}, \tag{13.12}$$

where the potential energy V is the function of coordinates x^i.

Hence, fields of force are called *conservative* if $F_i = -\frac{\partial V}{\partial x^i}$.

Theorem 13.4.1. A necessary and sufficient condition that a force field F_i, defined in a simple connected region, is conservative if $F_{i,j} = F_{j,i}$.

Proof: Suppose F_i is conservative. Then, $F_i = -\frac{\partial V}{\partial x^i}$.

$$\text{Now, } F_{i,j} = \frac{\partial F_i}{\partial x^j} - \left\{ \begin{matrix} k \\ i \quad j \end{matrix} \right\} F_k$$

$$= \frac{\partial\left(-\frac{\partial V}{\partial x^i}\right)}{\partial x^j} - \left\{ \begin{matrix} k \\ i \quad j \end{matrix} \right\} F_k$$

$$= -\frac{\partial^2 V}{\partial x^j \partial x^i} - \left\{ \begin{matrix} k \\ i \quad j \end{matrix} \right\} F_k$$

$$\text{and } F_{j,i} = \frac{\partial F_j}{\partial x^i} - \left\{ \begin{matrix} k \\ j \quad i \end{matrix} \right\} F_k$$

$$= -\frac{\partial^2 V}{\partial x^i \partial x^j} - \left\{ \begin{matrix} k \\ j \quad i \end{matrix} \right\} F_k.$$

From these two relations, it is implied that

$$F_{i,j} = F_{j,i}.$$

Conversely, let $F_{i,j} = F_{j,i}$.

$$\text{Then, } \frac{\partial F_i}{\partial x^j} - \left\{ \begin{matrix} k \\ i \quad j \end{matrix} \right\} F_k = \frac{\partial F_j}{\partial x^i} - \left\{ \begin{matrix} k \\ j \quad i \end{matrix} \right\} F_k$$

$$\text{or } \frac{\partial F_i}{\partial x^j} = \frac{\partial F_j}{\partial x^i}, \text{ as } \left\{ \begin{matrix} k \\ i \quad j \end{matrix} \right\} \textit{ is symmetric due to } i, j.$$

$$\text{Take } F_i = -\frac{\partial V}{\partial x^i}$$

$$\therefore \frac{\partial F_i}{\partial x^j} = -\frac{\partial^2 V}{\partial x^j \partial x^i} = -\frac{\partial^2 V}{\partial x^i \partial x^j}$$

$$= \frac{\partial}{\partial x^i}\left(-\frac{\partial V}{\partial x^j}\right)$$

$$= \frac{\partial F_j}{\partial x^i}$$

$$\frac{\partial F_i}{\partial x^j} = \frac{\partial F_j}{\partial x^i}$$

It implies that we can take $F_i = -\frac{\partial V}{\partial x^i}$.

Therefore, F_i is conservative.

13.5 Lagrangean Equations of Motion

Consider a particle moving on the curve C: $x^i = x^i(t)$ and curve C is the trajectory of the particle.

At time t, the particle is at point $P(x^i)$.

$$\text{The kinetic energy } T = \frac{1}{2}mv^2$$
$$= \frac{m}{2} g_{ij}\dot{x}^i\dot{x}^j, \tag{13.13}$$

Since $\dot{x}^i = v^i$.

Differentiating (13.13),

$$\text{now } \frac{\partial T}{\partial \dot{x}^i} = mg_{ij}\dot{x}^j$$

$$\text{and } \frac{\partial T}{\partial x^i} = \frac{m}{2}\frac{\partial g_{ij}}{\partial x^i}\dot{x}^i\dot{x}^j$$

$$\text{or } \frac{d}{dt}\left(\frac{\partial T}{\partial \dot{x}^i}\right) = \frac{d}{dt}(mg_{ij}\dot{x}^j)$$

$$= m\left[g_{ij}\ddot{x}^j + \frac{\partial g_{ij}}{\partial x^k}\dot{x}^k\dot{x}^j\right].$$

Subtracting the above two equations, we get

$$\frac{d}{dt}\left(\frac{\partial T}{\partial \dot{x}^i}\right) - \frac{\partial T}{\partial x^i} = m\left[g_{ij}\ddot{x}^j + \frac{\partial g_{ij}}{\partial x^k}\dot{x}^k\dot{x}^j\right] - \frac{m}{2}\frac{\partial g_{ij}}{\partial x^i}\dot{x}^i\dot{x}^j$$

$$= m\left[g_{ij}\ddot{x}^j + \left(\frac{1}{2}\frac{\partial g_{ij}}{\partial x^k} + \frac{1}{2}\frac{\partial g_{ik}}{\partial x^j} - \frac{1}{2}\frac{\partial g_{jk}}{\partial x^i}\right)\dot{x}^j\dot{x}^k\right]$$

$$= m(g_{ij}\ddot{x}^j + [jk,i]\dot{x}^j\dot{x}^k)$$

$$= mg_{il}(\ddot{x}^l + g^{il}[jk,i]\dot{x}^j\dot{x}^k)$$

$$= mg_{il}\left(\ddot{x}^l + \begin{Bmatrix} l \\ j\ k \end{Bmatrix}\dot{x}^j\dot{x}^k\right),$$

Since $a^i = \dfrac{d^2x^i}{dt^2} + \begin{Bmatrix} i \\ j\ k \end{Bmatrix}\dfrac{dx^j}{dt}\dfrac{dx^k}{dt} = \ddot{x}^l + \begin{Bmatrix} l \\ j\ k \end{Bmatrix}\dot{x}^j\dot{x}^k$ (from (13.6)).

Therefore, in parentheses on the right side of the above equation is acceleration and since $mg_{il}a^l = ma_i = F_i$, it is a component of the force field.

Therefore,

$$\frac{d}{dt}\left(\frac{\partial T}{\partial \dot{x}^i}\right) - \frac{\partial T}{\partial x^i} = F_i \tag{13.14}$$

Equation (13.14) is the *Lagrangean Equation of Motion.*

This equation gives the statement of Newton's 2nd Law of Motion in the form of Lagrange.

For a conservative system of forces,

$$F_i = -\frac{\partial V}{\partial x^i}.$$

Equation (13.14) becomes

$$\frac{d}{dt}\left(\frac{\partial T}{\partial \dot{x}^i}\right) - \frac{\partial T}{\partial x^i} = -\frac{\partial V}{\partial x^i}$$

or

$$\frac{d}{dt}\left(\frac{\partial T}{\partial \dot{x}^i}\right) - \frac{\partial (T-V)}{\partial x^i} = -0.$$

Since the potential V is a function of the coordinate x^i alone, let $L = T - V$, and the Lagrangean function

$$\frac{d}{dt}\left(\frac{\partial (T-V)}{\partial \dot{x}^i}\right) - \frac{\partial (T-V)}{\partial x^i} = 0$$

$$\text{or}\ \frac{d}{dt}\left(\frac{\partial L)}{\partial \dot{x}^i}\right) - \frac{\partial L}{\partial x^i} = 0 \qquad (13.15)$$

Equation (13.15) is known as Lagrange's Equation of Motion for conservative, holonomic systems.

Example 13.5.1. Show the covariant components of the acceleration vector in a spherical coordinate system with

$ds^2 = (dx^1)^2 + (x^1 dx^2)^2 + (x^1)^2 \sin^2 x^2\, (dx^3)^2$ are

$$a_1 = x^1 - x^1(\dot{x}^2)^2 - x^1(\dot{x}^3 \sin^2 x^2)^2$$

$$a_2 = \frac{d}{dt}[(x^1)^2 \dot{x}^2] - (x^1)^2 \sin x^2 \cos x^2 (\dot{x}^3)^2$$

$$a_3 = \frac{d}{dt}[(x^1 \sin x^2)^2 \dot{x}^3]$$

Solution: In a spherical coordinate system,

$$ds^2 = (dx^1)^2 + (x^1dx^2)^2 + (x^1)^2\sin^2x^2\,(dx^3)^2.$$

If v is the velocity of the particle,

$$v^2 = \left(\frac{ds}{dt}\right)^2 = \left(\frac{dx^1}{dt}\right)^2 + (x^1)^2\left(\frac{dx^2}{dt}\right)^2 + (x^1)^2\sin^2 x^2\left(\frac{dx^3}{dt}\right)^2$$
$$= (\dot{x}^1)^2 + (x^1)^2(\dot{x}^2)^2 + (x^1)^2\sin^2 x^2(\dot{x}^3)^2.$$

If T is kinetic energy,

$$T = \frac{1}{2}mv^2 = \frac{1}{2}m(\dot{x}^1)^2 + (x^1)^2(\dot{x}^2)^2 + (x^1)^2\sin^2 x^2(\dot{x}^3)^2] \qquad \text{(i)}$$

Using the Lagrangean Equation of Motion,

$$\frac{d}{dt}\left(\frac{\partial T}{\partial \dot{x}^1}\right) - \frac{\partial T}{\partial x^1} = F_i, \textit{ where } F_1 = ma_1.$$

Take $i = 1$,

$$ma_1 = \frac{d}{dt}\left(\frac{\partial T}{\partial \dot{x}^1}\right) - \frac{\partial T}{\partial x^1}.$$

From i, we have

$$ma_1 = \frac{d}{dt}\left[\frac{\partial}{\partial \dot{x}^1}\left\{\frac{1}{2}m[(\dot{x}^1)^2 + (x^1)^2(\dot{x}^2)^2 + (x^1)^2\sin^2 x^2(\dot{x}^3)^2\right\}\right]$$
$$-\frac{\partial}{\partial x^1}\left\{\frac{1}{2}m[(\dot{x}^1)^2 + (x^1)^2(\dot{x}^2)^2 + (x^1)^2\sin^2 x^2(\dot{x}^3)^2\right\}\Big]$$
$$a_1 = \frac{d}{dt}\left[\frac{1}{2}2\dot{x}^1\right] - 2\frac{1}{2}x^1[(\dot{x}^2)^2 + \sin^2 x^2(\dot{x}^3)^2] = \ddot{x}^1 - x^1(\dot{x}^2)^2 - x^1\left(\sin x^2 x^3\right)^2.$$

Take $i = 2$,

$$ma_2 = \left(\frac{d}{dt}\frac{\partial T}{\partial \dot{x}^2}\right) - \frac{\partial T}{\partial x^2}$$

From (i), we have

$$ma_2 = \frac{d}{dt}\left[\frac{\partial}{\partial \dot{x}^2}\left\{\frac{1}{2}m[(\dot{x}^1)^2 + (x^1)^2(\dot{x}^2)^2 + (x^1)^2 \sin^2 x^2 (\dot{x}^3)^2\right\}\right]$$

$$-\frac{\partial}{\partial x^2}\left\{\frac{1}{2}m[(\dot{x}^1)^2 + (x^1)^2(\dot{x}^2)^2 + (x^1)^2 \sin^2 x^2 (\dot{x}^3)^2\right]$$

$$a_2 = \frac{d}{dt}\left[\frac{1}{2}2(x^1)^2 \dot{x}^2\right] - (x^1)^2 \sin x^2 \cos x^2 (\dot{x}^3)^2]$$

$$= \frac{d}{dt}[(x^1)^2 \dot{x}^2] - (x^1)^2 \sin x^2 \cos x^2 (\dot{x}^3)^2$$

Take $i = 3$,

$$ma_3 = \frac{d}{dt}[\frac{\partial}{\partial \dot{x}^3}\{\frac{1}{2}m[(\dot{x}^1)^2 + (x^1)^2(\dot{x}^2)^2 + (x^1)^2 \sin^2 x^2 (\dot{x}^3)^2]$$

$$-\frac{\partial}{\partial x^3}\{\frac{1}{2}m[(\dot{x}^1)^2 + (x^1)^2(\dot{x}^2)^2 + (x^1)^2 \sin^2 x^2 (\dot{x}^3)^2\}]$$

$$a_3 = \frac{d}{dt}[x^3 (x^1)^2 \sin^2 x^2].$$

Example 13.5.2. Use the Lagrangean equations to show that if a particle is not subjected to the action of forces, then its trajectory is given by

$$y^i = a^i t + b^i,$$

where the a^i *and* b^i are constants and the y^i are orthogonal Cartesian coordinates.
Solution: If v is the velocity of a particle, then we know

$$v^2 = g_{ij} \dot{y}^i \dot{y}^j,$$

where y^i are the orthogonal Cartesian coordinates.

$$\text{Since } g_{ij} = 0, \textit{ if } i \neq j$$

$$= 1, \textit{ if } i = j,$$

so $v^2 = (\dot{y}^i)^2$

We know $T = \frac{1}{2}mv^2 = \frac{1}{2}m(\dot{y}^i)^2$.

The Lagrangean Equation of Motion is

$$\frac{d}{dt}\left(\frac{\partial T}{\partial \dot{y}^i}\right) - \frac{\partial T}{\partial y^i} = F_i$$

Since the particle is not subjected to the action of forces, i.e., $F_i = 0$,

therefore, $\frac{d}{dt}\left(\frac{\partial T}{\partial \dot{y}^i}\right) - \frac{\partial T}{\partial y^i} = 0$

$$\text{or } \frac{d}{dt}(m\dot{y}^i) - 0 = 0$$

$$\text{or } \frac{d}{dt}(\dot{y}^1) = 0$$

$$\text{or } \dot{y}^i = constant = a^i$$

$$\therefore y^i = a^i t + b^i.$$

Example 13.5.3. Prove that if a particle moves so that its velocity is constant in magnitude, then its acceleration vector is either orthogonal to the velocity vector or it is zero.

Solution: We know

$g_{ij}v^i v^j = v^2 = constant.$

Taking the intrinsic derivative,

$$\frac{\delta}{\delta t}(g_{ij}v^i v^j) = 0,$$

$$\text{or } g_{ij}\left[\frac{\delta}{\delta t}(v^i)v^j + \frac{\delta}{\delta t}(v^j)v^i\right] = 0,$$

$$\text{or } g_{ij}\frac{\delta}{\delta t}(v^i)v^j + g_{ij}\frac{\delta}{\delta t}(v^j)v^i = 0,$$

$$\text{or } 2g_{ij}\frac{\delta}{\delta t}(v^i)v^j = 0$$

(interchanging *i and j* in 2nd term and $g_{ij} = g_{ji}$ *and s* is symmetric metric)

$$g_{ij}\frac{\delta}{\delta t}(v^i)v^j = 0$$

This shows that the acceleration vector $\frac{\delta}{\delta t}(v^i)$ is either orthogonal to v^j or zero, i.e., $\frac{\delta}{\delta t}(v^i) = 0$.

13.6 Applications of Lagrangean Equations

We consider several examples of the application of Lagrangean Equations for determining the trajectories which include the cases of particles moving on smooth curves and surfaces.

(a) Free-moving particle: If a particle is not subjected to the action of forces, then we have

$$\frac{d}{dt}\left(\frac{\partial T}{\partial \dot{y}^i}\right) - \frac{\partial T}{\partial y^i} = 0 \tag{13.16}$$

and $T = \frac{1}{2}m\dot{y}^i\dot{y}^i$, implying that $\ddot{y}^i = 0$.

Integrating this, we get that $y^i = a^i t + b^i$ represents a straight line.

(b) Simple Pendulum:

Let a pendulum bob of mass m be suspended by a string (Figure 13.1). In spherical coordinates,

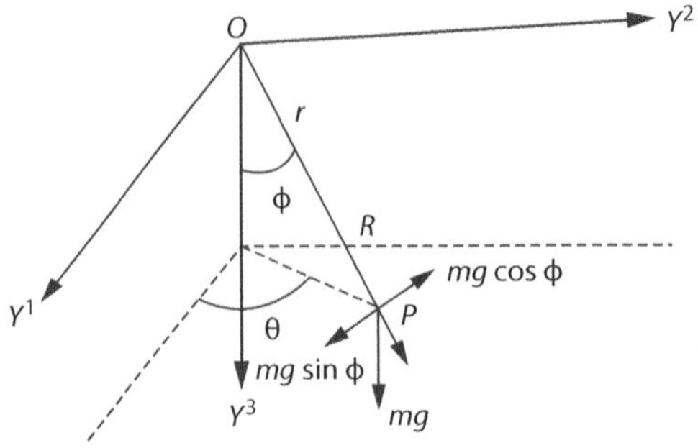

Figure 13.1

$$ds^2 = dr^2 + r^2 d\phi^2 + r^2 sin^2\phi d\theta^2$$

$$\text{and } T = \frac{1}{2}mv^2 = \frac{1}{2}m(\dot{r}^2 + r^2\dot{\phi}^2 + r^2 sin^2\phi\dot{\theta}^2) \qquad \text{(i)}$$

Lagrangean Equation of Motion:

$$\frac{d}{dt}\left(\frac{\partial T}{\partial \dot{x}^i}\right) - \frac{\partial T}{\partial x^i} = F_i$$

Here, $x^1 = r$, $x^2 = \phi$, $x^3 = \theta$.

$$\text{For } x^1 = r, \qquad \frac{d}{dt}\left(\frac{\partial T}{\partial \dot{r}}\right) - \frac{\partial T}{\partial r} = mgcos\phi - R.$$

$$\text{From i,} \qquad \ddot{r} + r(\dot{\phi}^2 + sin^2\phi\dot{\theta}^2) = gcos\phi - \frac{R}{m}. \qquad \text{(ii)}$$

$$\text{For } x^2 = \phi, \qquad r\ddot{\phi} + 2\dot{r}\dot{\phi} - r\dot{\theta}^2 sin\phi cos\phi = -gsin\phi \qquad \text{(iii)}$$

$$\text{and } x^3 = \theta \qquad \frac{d}{dt}(r^2 sin^2\phi\dot{\theta}) = 0. \qquad \text{(iv)}$$

If the motion in one plane is from (ii), (iii), and (iv), by taking $\dot{\theta} = 0$

$$\ddot{r} - r\,\dot{\phi}^2 = gcos\phi - \frac{R}{m}$$

$$r\ddot{\phi} + 2\dot{r}\dot{\phi} = -g sin\phi.$$

If $\dot{r} = 0$, we get $\ddot{\phi} = -\left(\frac{g}{r}\right) sin\phi$ which is the equation of a simple pendulum.

For small angles of oscillation, the vibration is a simple harmonic. For large vibrations, the solution is given in terms of elliptic functions.

(c) Constant Gravitational Field

We take a Cartesian reference frame and Y^3-axis to the normal to the plane of the earth, so the potential V of the constant Gravitational field is = mgy^3 if the positive Y^3-axis is directed upward. Therefore,

$$\ddot{y}^1 = 0, \ddot{y}^2 = 0, \ddot{y}^3 = -g.$$

The equation of the path of trajectory is

$$y^a = a^a t + b^a \ (a = 1, 2)$$

$$y^3 = -\frac{1}{2} g t^2 + at + b.$$

This trajectory is a parabola whose axis is parallel to the Y^3-axis.

(d) Motion of a Particle on a Curve

Let a particle move on a curve

$$x^i = x^i(s) \ (i = 1, 2, 3), \tag{13.17}$$

with s being the arc parameter.

We suppose that C has a continuously turning tangent, so the curve is of C^2.

The components v^i of velocity vector $\boldsymbol{v}$ of the particle are

$$v^i = \frac{dx^i}{dt} = \frac{dx^i}{ds}\frac{ds}{dt} = v\lambda^i, \tag{13.18}$$

Where $\lambda^i = \frac{dx^i}{dt}$ *is the* unit tangent vector to C and $v = \frac{ds}{dt}$ is the magnitude of $\boldsymbol{v}$.

Now acceleration vector $\boldsymbol{a}$, by intrinsically differentiating (13.18), is

$$a^i = \frac{\delta v^i}{\delta t} = \frac{\delta v}{\delta t}\lambda^i + v\frac{\delta \lambda^i}{\delta t} = \frac{dv}{dt}\lambda^i + v\frac{\delta \lambda^i}{\delta t}, \tag{13.19}$$

and since v is a scalar, $\frac{\delta v}{\delta t} = \frac{dv}{dt}$.

Again,

$$\frac{\delta \lambda^i}{\delta t} = \frac{\delta \lambda^i}{\delta s}\frac{ds}{dt} = v\frac{\delta \lambda^i}{\delta s} = v\kappa\mu^i, \tag{13.20}$$

using the Serret-Frenet formula $\frac{\delta \lambda^i}{\delta s} = \kappa\mu^i, \kappa > 0$, where κ is curvature and μ is principal normal unit.

Now, substituting $\frac{\delta \lambda^i}{\delta s}$ of (13.20) in (13.19), we get

$$a^i = \frac{dv}{dt}\lambda^i + v^2\kappa\mu^i, \tag{13.21}$$

which states that an acceleration vector a lies in the osculating plane (λ^i, μ^i). Here, the component in the direction of the principal normal is, $\frac{v^2}{R}$ where is $R = \frac{1}{\kappa}$ the radius curvature of C.

The force

$$\boldsymbol{F}^i = m\boldsymbol{a}^i = m\frac{dv}{dt}\lambda^i + mv^2\kappa\mu^i. \tag{13.22}$$

Since F lies in the osculating plane of the curve, the component of all external forces normal to this plane is zero. This condition enables us to compute that reaction R is normal to C, i.e.,

$$R^i\lambda_i = 0.$$

If $\boldsymbol{R} = 0$, the curve C is called the *natural trajectory* of the particle.

Example 13.6.1. Establish the energy equation $T + V = constant$.
Solution: Let a bead of mass m slide under gravity along a smooth curve C lying in the vertical Y^1Y^2 -plane, as shown in Figure (13.2).

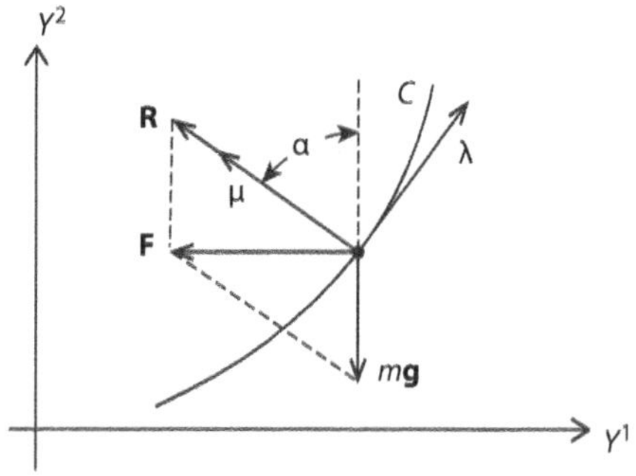

Figure 13.2

The force F acting on m is

$$\boldsymbol{F} = m\boldsymbol{g} + \boldsymbol{R},$$

where R is the pressure exerted by the curve on the particle and mg is the gravitational force. Since the curve is smooth, R is normal to C. If α is the angle between the direction of R and the Y^2-axis, the components of F in the directions of the tangent λ and the principal normal μ are $\boldsymbol{F}_{(\lambda)} = -mgsign\alpha$ and $\boldsymbol{F}_{(\mu)} = -mgcos\alpha + R$

Using (13.22), we get

$$m\frac{dv}{dt} = -mgsign\alpha \text{ and } mv^2\kappa = -mgcos\alpha + R. \tag{13.23}$$

Here, $cosa = \dfrac{dy^1}{ds}, sina = \dfrac{dy^2}{ds}$ and $\dfrac{dv}{dt} = \dfrac{dv}{ds}\dfrac{ds}{dt}$

$$\therefore m\frac{dv}{dt} = mv\frac{dv}{ds} = -mgsign\alpha = -mg\frac{dy^2}{ds}$$

or $mv\dfrac{dv}{ds} = -mg\dfrac{dy^2}{ds}$. Integrating this, we get

$$\frac{1}{2}mv^2 = -mgy^2 + constant \tag{13.24}$$

Since R is zero in the direction of the path, we can write directly from (13.24).

$T + V = Constan$, where V is potential energy.

Example 13.6.2. Let a particle of mass m move under gravity along a smooth cycloid (Figure 13.3)

$$y^1 = a(\theta - sin\theta)$$

$$y^2 = a(1 + cos\theta), 0 \leq \theta \leq 2\pi \quad \text{(i)}$$

Solution: The equation $m\frac{dv}{dt} = -mg\frac{dy^2}{ds}$ (from 13.23)

$$\text{or } m\frac{d^2s}{dt^2} = -mg\frac{dy^2}{ds}. \quad \text{(ii)}$$

Here,

$$s = \int_0^\theta \sqrt{(dy^1)^2 + (dy^2)^2} = \int_0^\theta \sqrt{(dy^1)^2 + (dy^2)^2}d\theta = a\int_0^\theta \sqrt{2(1 - cos\theta)}d\theta$$

$$= 2a\int_0^\theta sin\frac{\theta}{2}d\theta = 4a\left(1 - cos\frac{\theta}{2}\right)$$

$$(dy^1)^2 + (dy^2)^2 = [ad\theta - cos\theta d\theta]^2 + [-asin\theta d\theta]^2$$

$$= a^2(2 - 2cos\theta)d\theta^2),$$

$$\text{or } (s - 4a)^2 = \left(-4acos\frac{\theta}{2}\right)^2 = 16a^2\left(cos\frac{\theta}{2}\right)^2,$$

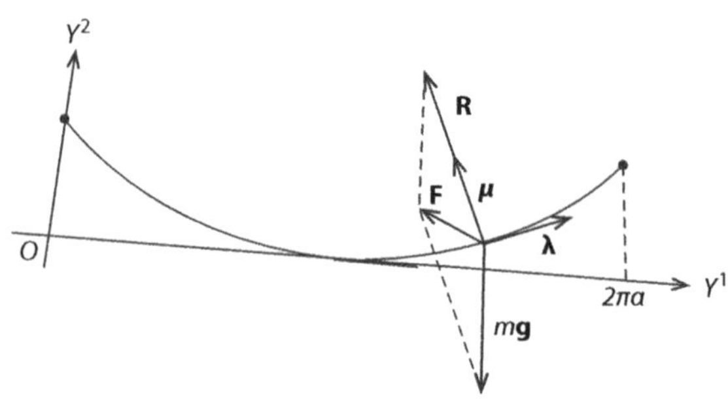

Figure 13.3

$$\text{or } \frac{(s-4a)^2}{8a} = 2a\left(cos\frac{\theta}{2}\right)^2 = a(1+cos\theta) = y^2.$$

From (ii), we get

$$\ddot{s} = -g\frac{dy^2}{ds} = -g\frac{d}{ds}\left(\frac{(s-4a)^2}{8a}\right) = \frac{-g}{8a}2(s-4a) = \frac{-g}{4a}(s-4a)$$

$$\text{or } \ddot{s} + \frac{g}{4a}s = g$$

The general solution is

$$S = C_1 \cos\left(\sqrt{\frac{g}{4a}}t + C_2\right) + 4a. \tag{iii}$$

The integration constants c_1 *and* c_2 are determined by initial conditions of motion on a cycloid.

The period of motion $\left(2\pi / \sqrt{\frac{g}{4a}}\right)$ is independent of amplitude c_1.

This fact first was discovered by Christian Huygens about 300 years ago. Huygens proposed the use of a cycloidal pendulum in the construction of Isochronous clocks.

Using the second equation of (13.23), it is shown that $R = 2mgcos\alpha$.

(d) Motion of a Particle on a Surface

Let the equations of regular surface S be given in parametric form

$$S: x^i = x^i(u^1, u^2)\ (i = 1, 2, 3). \tag{13.25}$$

The force F is the resultant of all external forces acting on the particle, thus including the reaction R of the surface on the particle. When surface is smooth, R is normal to S and represents pressure that constrains the particle to remain on S.

The space component v^i of the velocity v of the particle is related to the surface components v^α by the formula

$$v^i = \frac{dx^i}{dt} = \frac{\partial x^i}{\partial u^a}\frac{du^a}{dt} = x^i_a \dot{u}^a, \qquad (13.26)$$

$$v^i = x^i_a v^a$$

where $v^a = \dot{u}^a$

$$\text{The acceleration } a^i = \frac{\delta v^i}{\delta t},$$

$$\therefore\ a^i = \frac{\delta v^i}{\delta t} = x^i_a \frac{\delta v^a}{\delta t} + v^a \frac{\delta x^i_a}{\delta t} \qquad (13.27)$$

$$= x^i_a a^a + v^a x^i_{\alpha,\beta} v^\beta,$$

where $a^a = \dfrac{\delta v^a}{\delta t.}$

We take the Gauss formula

$x^i_{\alpha,\beta} = b_{\alpha\beta} n^i$, substuting in (13.27), we get

$$a^i = x^i_\alpha a^\alpha + v^\alpha v^\beta b_{\alpha\beta} n^i. \qquad (13.28)$$

Thus, $a^i = x^i_\alpha a^\alpha + b_{\alpha\beta} n^i v^2 \lambda^\alpha \lambda^\beta$ $\left[v^\alpha = \dfrac{du^\alpha}{dt} = \dfrac{\partial u^\alpha}{\partial x}\dfrac{dx}{dt} = v\lambda^\alpha \right]$ since the normal curvature $\kappa_{(n)} = b_{\alpha\beta}\lambda^\alpha\lambda^\beta$

$$a^i = x^i_\alpha a^\alpha + \kappa_{(n)} n^i v^2.$$

Since $F^i = ma^i$, we have

$$F^i = m x^i_\alpha a^\alpha + m\kappa_{(n)} v^2 n^i. \qquad (13.29)$$

The first term of the right hand side of (13.29) is the component of **F** in the tangent plane to S and second term is the component of **F** along the normal n.

The component of F in the direction of the normal n is

$$F^i n_i = m x^i_\alpha a^\alpha n_i + m\kappa_{(n)} v^2 n^i n_i \qquad \text{(from 13.29)}$$

$$= 0 + m\kappa_{(n)} v^2$$

$$F^i n_i = m\kappa_{(n)} v^2, \tag{13.30}$$

since surface vector x^i_α is orthogonal to normal n_i and $n_i n^i = 1$.

The components of F in the plane tangent to S, are given by

$$\begin{aligned} g_{ij} x^j_\gamma F^i &= m g_{ij} x^j_\gamma x^i_\alpha a^\alpha + m\kappa_{(n)} g_{ij} x^j_\gamma v^2 n^i \\ &= m a_{\gamma\alpha} a^\alpha + 0, \end{aligned}$$

Since $g_{ij} x^j_\gamma x^i_\alpha = a_{\gamma\alpha}$ and $g_{ij} x^j_\gamma n^i = 0$ because the surface vector x^j_γ are orthogonal to n_j.

Thus, $x^j_\gamma F_j = m a_\gamma$

and let $F_\gamma \equiv x^j_\gamma F_j$. From this we obtain a pair of Newtonian equations

$$F_\gamma = m a_\gamma \tag{13.31}$$

relating the surface force vector F_γ to the surface acceleration vector a_γ.

It can be written in Lagrangean form,

$$\text{Kinetic Energy } T = \frac{m}{2} v^2 \text{ is}$$

$$T = \frac{m}{2} a_{\alpha\beta} v^\alpha v^\beta = \frac{m}{2} a_{\alpha\beta} \dot{u}^\alpha \dot{u}^\beta .$$

The Lagrangean Equation of motion,

$$\frac{d}{dt}\left(\frac{\partial T}{\partial \dot{u}^\alpha}\right) - \frac{\partial T}{\partial u^\alpha} = F_\alpha, \tag{13.32}$$

Where F_α is defined in (13.31). When the force is conservative in $F_\alpha = -\frac{\partial V}{\partial u^\alpha}$, V is potential energy.

The velocity v^α of the particle along the trajectory is $v^\alpha = v\lambda^\alpha$, hence

$$a^{\alpha}=\frac{\delta v^{\alpha}}{\delta t}=\lambda^{\alpha}\frac{dv}{dt}+v\frac{\delta\lambda^{\alpha}}{\delta t}$$
$$=\lambda^{\alpha}\frac{dv}{dt}+v^{2}\frac{\delta\lambda^{\alpha}}{\delta s}.$$

We know $\frac{\delta\lambda^{\alpha}}{\delta s}=\kappa_{g}\eta^{\alpha}$

Where η^{α} is the unit normal to the trajectory in the tangent plane and κ_g is the geodesic curvature, we can write

$$a^{\alpha}=\frac{\delta v^{\alpha}}{\delta t}=\lambda^{\alpha}\frac{dv}{dt}+v^{2}\kappa_{g}\eta^{\alpha}$$
$$=\lambda^{\alpha}v\frac{dv}{ds}+v^{2}\kappa_{g}\eta^{\alpha}$$

Therefore, $a^{\alpha}=\frac{1}{2}\frac{dv^{2}}{ds}\lambda^{\alpha}+v^{2}\kappa_{g}\eta^{\alpha}$.

It follows from the result of (13.22),

$$F^{\alpha}=\frac{dT}{ds}\lambda^{\alpha}+2T\kappa_{g}\eta^{\alpha} \qquad \text{(taking unit mass)}$$

If the vector F^{α} vanishes identically, then $\frac{dT}{ds}=0$ and $\kappa_g = 0$ along the trajectory.

The first of these equations states that v = constant and if $v \neq 0$, then the trajectory is a geodesic [curve on a surface S if its geodesic curvature is zero].

Example 13.6.3. Let a particle of mass m be constrained to move under gravity on a smooth paraboloid of revolution.

Solution: The equation of a paraboloid of revolution, as illustrated in Figure (13.4), is

$$y^{3}=\frac{1}{4a}[(y^{1})^{3}+(y^{2})^{2}], \quad a=constant \tag{13.33}$$

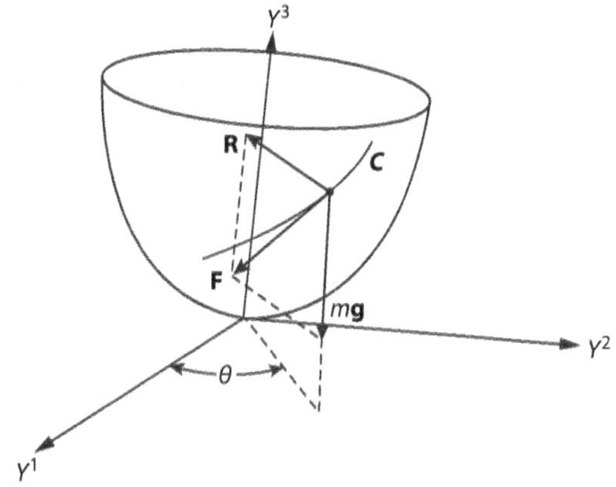

Figure 13.4

Introducing cylindrical coordinates (r, θ, z), the coordinates are

$$y^1 = r\cos\theta,\ y^2 = r\sin\theta,\ and\ y^3 = z.$$

Substitute in (13.33) and we get

$$z = \frac{r^2}{4a} \tag{13.34}$$

and kinetic Energy $T = \frac{1}{2} m\dot{y}^i\dot{y}^i$ becomes

$$T = \frac{1}{2} m \left[\left(1 + \frac{r^2}{4a^2} \right) \dot{r}^2 + r^2\dot{\theta}^2 \right]$$

The potential energy is

$$V = mgy^3 = mg\frac{r^2}{4a}$$

Since the surface is smooth and the reaction $\boldsymbol{R}$ is normal to S',
we use the Lagrangean Equation of Motion

$\frac{d}{dt}\left(\frac{\partial T}{\partial \dot{u}^{\alpha}}\right)-\frac{\partial T}{\partial u^{\alpha}}=F_{\alpha}$, with $F_{\alpha}=-\frac{\partial V}{\partial u^{\alpha}}$ (since components of $\boldsymbol{R}$ in tangent plane to S are zero) and parametrize the surface by setting $u^1 = r$, $u^2 = \theta$.

For $\alpha = 1$

$$\frac{d}{dt}\left(\frac{\partial T}{\partial \dot{u}^{1}}\right)-\frac{\partial T}{\partial u^{1}}=-\frac{\partial V}{\partial u^{1}},$$

$$\text{or } \frac{d}{dt}\left(\frac{\partial T}{\partial \dot{r}}\right)-\frac{\partial T}{\partial r}=-\frac{\partial V}{\partial r}$$

$$\frac{d}{dt}\left[m\dot{r}\left(1+\frac{r^2}{4a^2}\right)\right]-m\frac{r\dot{r}^2}{4a^2}=-mg\frac{r}{2a}$$

$$\ddot{r}\left(1+\frac{r^2}{4a^2}\right)+\frac{r\dot{r}^2}{4a^2}-r\dot{\theta}^2=-g\frac{r}{2a}$$

$$\left(1+\frac{r^2}{4a^2}\right)\ddot{r}+\frac{r\dot{r}^2}{4a^2}-r\dot{\theta}^2=-\frac{gr}{2a}$$

For $\alpha = 2$

$$\frac{d}{dt}(r^2\dot{\theta})=0 \tag{13.35}$$

The 2nd equation of (13.35) gives

$$r^2\dot{\theta}=h, constant \tag{13.36}$$

The 1st equation becomes

$$\left(1+\frac{r^2}{4a^2}\right)\ddot{r}+\frac{r\dot{r}^2}{4a^2}-\frac{h^2}{r^3}=-\frac{gr}{2a}, \tag{13.37}$$

which has a unique solution with initial values.

If the particle is constrained to move on a horizontal circle $r = constant$, then (13.37) becomes

$$\frac{h^2}{r^3} = \frac{gr}{2a} \qquad \Rightarrow h^2 = \frac{gr^4}{2a}$$

and

$$\dot{\theta} = \frac{h}{r^2} \Rightarrow \quad \dot{\theta}^2 = \frac{h^2}{r^4} = \frac{\frac{gr^4}{2a}}{r^4} = \frac{g}{2a},$$

so that the angular velocity $\dot{\theta}$ is independent of the radius of the circle.

If we replace the surface of the paraboloid in this example by the surface of the sphere, we have the problem of a spherical pendulum.

13.7 Himilton's Principle

Consider a particle of mass m in a three-dimensional Euclidean manifold, referring to a curvilinear system of coordinates X. The particle is in motion under the effect of force $\boldsymbol{F}$ and our problem is to determine the trajectory.

$$C: x^i = x^i\,(t),\ (i = 1, 2, 3)\ t_1 \le t \le t_2,$$

where t denotes the time.

Theorem 13.7.1. If a particle is at point P_1 at time t_1 and is at P_2 at a time t_2, then the motion of a particle takes place in such a way that

$$\int_{t_1}^{t_2} (\delta T + F_i \delta x^i) dt = 0, \tag{13.38}$$

where $x^i = x^i\,(t)$ are the coordinates of the particle along a trajectory and $x^i + \delta x^i$ are the coordinates along a varied path beginning at P_1 at time t_1 and ending at P_2 at a time t_2.

Proof: Consider a particle moving on the curve

$$C: x^i = x^i\,(t),\ (i = 1, 2, 3)\ t_1 \le t \le t_2$$

The kinetic energy of the particle, $T = \frac{1}{2} m g_{ij} \dot{x}^i \dot{x}^j$.

Here, T is a function of $(x^i, \dot{x}^i)$. Let C′ be another curve,

joining at t_1 and t_2 to show that if C is

$C' : \bar{x}^i(\epsilon, t) = x^i(t) + \delta x^i(t),$

with $\delta x^i(t) = \epsilon \xi^i(t)$ *and* $\xi^i(t_1) = \xi^i(t_2) = 0,$

then

$$\delta(T) = \frac{\partial T}{\partial \dot{x}^i} \delta \dot{x}^i + \frac{\partial T}{\partial x^i} \delta x^i. \tag{13.39}$$

$$\text{Now, } \int_{t_1}^{t_2} (\delta T + F_i \delta x^i) dt = \int_{t_1}^{t_2} \left(\frac{\partial T}{\partial \dot{x}^i} \delta \dot{x}^i + \frac{\partial T}{\partial x^i} \delta x^i + F_i \delta x^i \right) dt$$

$$= \int_{t_1}^{t_2} \frac{\partial T}{\partial \dot{x}^i} \delta \dot{x}^i dt + \int_{t_1}^{t_2} \frac{\partial T}{\partial x^i} \delta x^i dt + \int_{t_1}^{t_2} F_i \delta x^i dt$$

$$= \int_{t_1}^{t_2} \frac{\partial T}{\partial x^i} \delta x^i dt + \left[\frac{\partial T}{\partial \dot{x}^i} \delta x^i \right]_{t_1}^{t_2} - \int_{t_1}^{t_2} \frac{d}{dt} \left(\frac{\partial T}{\partial \dot{x}^i} \right) \delta x^i dt$$

$$+ \int_{t_1}^{t_2} F_i \delta x^i dt.$$

Since $\delta x^i(t_1) = 0, \delta x^i(t_2) = 0$, then $\left[\frac{\partial T}{\partial \dot{x}^i} \delta x^i \right]_{t_1}^{t_2} = 0$

$$\therefore \int_{t_1}^{t_2} (\delta T + F_i \delta x^i) dt = \int_{t_1}^{t_2} \frac{\partial T}{\partial x^i} \delta x^i dt - \int_{t_1}^{t_2} \frac{d}{dt} \left(\frac{\partial T}{\partial \dot{x}^i} \right) \delta x^i dt + \int_{t_1}^{t_2} F_i \delta x^i dt.$$

$$\Rightarrow \int_{t_1}^{t_2} (\delta T + F_i \delta x^i) dt = \int_{t_1}^{t_2} \left[\frac{\partial T}{\partial x^i} - \frac{d}{dt} \left(\frac{\partial T}{\partial \dot{x}^i} \right) + F_i \right] \delta x^i dt$$

We know the Lagrangean Equation of Motion

$$\frac{d}{dt}\left(\frac{\partial T}{\partial \dot{x}^i}\right)-\frac{\partial T}{\partial x^i}=F$$

$$\text{or}\quad \frac{\partial T}{\partial x^i}-\frac{d}{dt}\left(\frac{\partial T}{\partial \dot{x}^i}\right)=-Fi, \tag{13.40}$$

$$\text{so}\quad \int_{t_1}^{t_2}(\delta T+F_i\delta x^i)dt=0$$

In particular, if this field is conservative, then there exists a potential function $V(x^1, x^2, x^3)$, such that

$$\frac{\partial V}{\partial x^i}=-F_i.$$

From (13.38) we get $\int_{t_1}^{t_2}(\delta T+F_i\delta x^i)dt=0$,

$$\text{or } \int_{t_1}^{t_2}(\delta T-\frac{\partial V}{\partial x^i}\delta x^i)dt=0,$$

or

$$\int_{t_1}^{t_2}(\delta T-\delta V)dt=0 \text{ or } \int_{t_1}^{t_2}\delta(T-V)dt=0. \tag{13.41}$$

We defined the Lagrangean Function $L = T - V$, so that we can write

$$\int_{t_1}^{t_2}\delta L\,dt=0.$$

Since the limits of integration are fixed, we have a concise formulation of Himilton's principle for a conservative field in the form

$$\delta\int_{t_1}^{t_2}L\,dt=0. \tag{13.42}$$

We can state from this result:

In a conservative field of force, a particle moves so that the integral $\int_{t_1}^{t_2}L\,dt$, evaluated along the trajectory $x^i = x^i(t)$, $t_1 \le t \le t_2$ has a stationary value in comparison with its values for all neighboring paths beginning at point P_1 at $t = t_1$ and ending at point P_2 at $t = t_2$.

Equation of Motion (13.15)

$\frac{d}{dt}\left(\frac{\partial L}{\partial \dot{x}^i}\right) - \frac{\partial L}{\partial x^i} = 0$ follows at once from Formulation (13.42).

Theorem 13.7.1. Integral of Energy: The motion of a particle in a conservative field of force is such that the sum of its kinetic and potential energies is a constant.

Consider a particle moving on the curve

$$C: x^i = x^i(t), \qquad t_1 \leq t \leq t_2,$$

where t denotes the time. The kinetic energy T of the particle is given by

$$T = \frac{1}{2} m g_{ij} \dot{x}^i \dot{x}^j,$$

since Kinetic energy is invariant,

$$\begin{aligned}
\frac{dT}{dt} &= \frac{\delta T}{\delta t} = \frac{\delta}{\delta t}\left(\frac{1}{2} m g_{ij} \dot{x}^i \dot{x}^j\right) \\
&= \frac{1}{2} m g_{ij} \frac{\delta}{\delta t}(\dot{x}^i \dot{x}^j) \\
&= \frac{1}{2} m g_{ij} \left(\frac{\delta \dot{x}^i}{\delta t} \dot{x}^j + \frac{\delta \dot{x}^j}{\delta t} \dot{x}^i\right) \qquad (13.43) \\
&= m g_{ij} \frac{\delta \dot{x}^i}{\delta t} \dot{x}^j = m g_{ij} a^i v^j \\
\frac{dT}{dt} &= m a_i v^i,
\end{aligned}$$

where a^i is the acceleration and v^j is the velocity of the particle.

For a conservative field of force, $F_i = -\frac{\partial V}{\partial x^i} = m a^i$, substituting in (13.43), we get

$$\frac{dT}{dt} = ma_i v^i = -v^i \frac{\partial V}{\partial x^i} = -\frac{\partial V}{\partial x^i}\frac{dx^i}{dt}. \tag{13.44}$$

$$\therefore \frac{dT}{dt} = -\frac{dV}{dt}$$

Integrating this, we get $T + V = h$, where h is the constant of integration.

13.8 Principle of Least Action

In 1744, Euler showed that the integral $\int mv\, ds$ has a stationary value along the trajectory of a particle moving in a central field of force. In 1760, Lagrange extended Euler's result by demonstrating that the integral $A = \int_{P_1}^{P_2} \boldsymbol{mv.ds}$ has a stationary value along the trajectories of particles moving in a conservative force field, provided that the constraints are not functions of time. This lead him to formulate the principle of least action. Himilton, attempted to understand Lagrange's formulation of the principle.

Let us consider the integral

$$A = \int_{P_1}^{P_2} \boldsymbol{mv.ds} \tag{13.45}$$

evaluated over path

$$C: x^i = x^i(t), \qquad t_1 \le t \le t_2$$

Where C is the trajectory of the particle of mass m moving in a conservative field of force. Here, we suppose that neither kinetic energy T nor potential energy V is a function of time t. In a three-dimensional space with curvilinear coordinates, the integral (13.45) can be written as

$$\begin{aligned} A &= \int_{P_1}^{P_2} \boldsymbol{mv.ds} \\ &= \int_{P_1}^{P_2} mg_{ij}\frac{dx^i}{dt}dx^j \\ &= \int_{t(P_1)}^{t(P_2)} mg_{ij}\frac{dx^i}{dt}\frac{dx^j}{dt}dt. \end{aligned}$$

Here, Kinetic Energy $T = \frac{1}{2} m g_{ij} \frac{dx^i}{dt} \frac{dx^j}{dt}$

$$\therefore \; A = \int_{t(P_1)}^{t(P_2)} 2T dt \tag{13.46}$$

The integral has a physical meaning only when evaluated over the trajectory C, but its value can be computed along any varied path joining the points P_1 *and* P_2. Let us consider a particular set of admissible paths C' along which the function $T + V$, for each value of parameter t, has the same constant value h. The integral A is called the *action integral.*

The principle of least action states that "of all curves of C' passing through P_1 *and* P_2 in the neighbourhood of the trajectory C, which are traversed at a rate such that, for each C', for every value of t, $T + V = h$, that one for which the action integral A is stationary is the trajectory of the particle."

When stated in the form of the variational equation, the principle reads

$$\int_{t(P_1)}^{t(P_2)} 2T dt = 0, \tag{13.47}$$

with the auxiliary condition

$$T + V - h = 0 \text{ on } C'. \tag{13.48}$$

We construct a function $= 2T + \lambda\phi$, where $\phi = T + V - h$ and determines the solution of the system of four equations

$$\begin{cases} \dfrac{\partial F}{\partial x^i} - \dfrac{d}{dt}\left(\dfrac{\partial F}{\partial \dot{x}^i}\right)\dfrac{\partial F}{\partial \dot{x}^i} = 0, (i = 1,2,3) \\ T + V - h = 0 \end{cases}$$

An investigation of this system shows that $\lambda(t) = -1$ and it follows from this fact that the trajectory C is determined by the solution of the system

$$\frac{d}{dt}\left(\frac{\partial T}{\partial \dot{x}^i}\right) - \frac{\partial T}{\partial x^i} = -\frac{\partial V}{\partial x^i} \qquad (i = 1,2,3). \tag{13.49}$$

These are the Lagrangean equations of motion.

We consider this variational problem with fixed end points by change of variable. Since the kinetic energy

$$T=\frac{1}{2}mg_{ij}\dot{x}^i\dot{x}^j=\frac{1}{2}m\left(\frac{ds}{dt}\right)^2$$

$$\text{or}\ \left(\frac{ds}{dt}\right)^2=\frac{2T}{m}\Rightarrow\frac{ds}{dt}=\sqrt{\frac{2T}{m}} \tag{13.50}$$

$$\therefore dt=\sqrt{\frac{m}{2T}}ds=\sqrt{\frac{m}{2(h-V)}}ds,$$

(13.46) can be written as

$$A=\int_{t(P_1)}^{t(P_2)}2Tdt=\int_{t(P_1)}^{t(P_2)}2(h-V)\sqrt{\frac{m}{2(h-V)}}ds=\int_{t(P_1)}^{t(P_2)}\sqrt{2m(h-V)}ds$$

$$A=\int_{t(P_1)}^{t(P_2)}\sqrt{2m(h-V)}ds \tag{13.51}$$

Since the integrand in the preceding integral is clearly independent of t, we now parametrize our varied path C', so that

$$C\colon x^i=x^i(u)\ and\ u_1\le u\le u_2,\ \text{where}\ P_1\colon x^i(u_1)\ \text{and}\ P_2\colon x^i(u_2)$$

and we write $ds=\sqrt{g_{ij}\acute{x}^i\acute{x}^j}\,du$, where $\acute{x}^i=\frac{dx^i}{du}$.

We can write the action integral (13.51) in the form

$$A=\int_{u_1}^{u_2}\sqrt{2m(h-V)g_{ij}\acute{x}^i\acute{x}^j}\,du. \tag{13.52}$$

Since the limits of integration in (13.52) are fixed, we see that the determination of the trajectory is equivalent to finding the geodesics in a three-dimensional manifold with arc element

$$dS^2=2m(h-V)\,g_{ij}dx^idx^j. \tag{13.53}$$

13.9 Generalized Coordinates

In the solutions of most of the mechanical problems it is more convenient to use some other set of coordinates instead of Cartesian coordinates. For example, in the case of a particle moving on the surface of a sphere, the correct coordinates are spherical coordinates r, θ, ϕ and are only two variable quantities.

Let there be a particle or system of n particles moving under possible constraints. For example, a point mass of the simple pendulum or a rigid body moving along an inclined plane. Then, there will be a minimum number of independent coordinates required to specify the motion of the particle or system of particles. The set of independent coordinates sufficient in number to specify unambiguously the system configuration are called *generalized coordinates* and are denoted by q^1, q^2,......q^n where n is the total number of generalized coordinates or degree of freedom.

In a system where the system is defined by n-generalized coordinates q^1, q^2,......q^n and a number of the degree of freedom that is n, that such system is called *an un-connected holonomic system.*

Let there be N particles composing a system and let $x^i_{(\alpha)}(i=1,2,3),(\alpha=1,2,\ldots N)$ be the positional coordinates of these particles referred to some convenient reference frame in E_3. The system of N free particles is described by $3N$ parameters. If the particles are constrained in some way, there will be certain relations among the coordinates $x^i_{(\alpha)}$ and suppose that there are r such independent relations.

$$f^i(x^1_{(1)},x^2_{(1)},x^3_{(1)};x^1_{(2)},x^2_{(2)},x^3_{(2)};\ldots\ldots x^1_{(N)},x^2_{(N)},x^3_{(N)})=0,(i=1,2,\ldots.r) \tag{13.54}$$

If these r equations of constraints in (13.54) can be solved for some r coordinates in terms of the remaining $3N - r$ coordinates, the latter can be viewed as the independent generalized coordinates q^i. It is more convenient however to assume that each of the $3N$ coordinates is expressed in terms of $3N - r = n$ independent variables q^i and written in $3N$ equations

$$x^i_{(\alpha)} = x^i_{(\alpha)}(q^1,q^2,\ldots\ldots q^n,t), \tag{13.55}$$

where we introduce the time parameter t which may enter in the problem explicitly if one deals with moving constraints. If t does not enter explicitly in Equation (13.55), the dynamic system is called a *natural* system.

The velocity of the particles is given by the differential Equation (13.55) with regards to time. Now,

$$\dot{x}^i_{(\alpha)} = \frac{\partial x^i_{(\alpha)}}{\partial q^j}\dot{q}^j + \frac{\partial x^i_{(\alpha)}}{\partial t}. \tag{13.56}$$

The time derivative $\dot{q}^j$ of generalized coordinates q^j are the *generalized velocities.*

For symmetry reasons, it is desirable to introduce a number of superfluous coordinates q^i, describing the system with the aid of $k > n$ coordinates $q^1, q^2, \ldots\ldots q^k$. In this event, there will exist certain relations of the form

$$f^j(q^1, q^2, \ldots\ldots q^k, t) = 0, \tag{13.57}$$

so that the quantities q^i and $\dot{q}^i$ are no longer independent.

Differentiating it, we get

$$\frac{\partial f^j}{\partial q^i}\dot{q}^i + \frac{\partial f^j}{\partial t} = 0. \tag{13.58}$$

It is clear that they are integrable, so that one can deduce Equation (13.57) and use it to eliminate the superfluous coordinates.

In some problems, the functional relations of the type

$$F^j(q^1, q^2, \ldots\ldots q^k, \dot{q}^1, \dot{q}^2, \ldots\ldots \dot{q}^k, t) = 0 \quad (j = 1, 2 \ldots m) \tag{13.59}$$

arise, which are non-integrable. If the non-integrable relations in (13.59) occur in the problems, we shall say that the given system has $k - m$ degrees of freedom, where m is the number of independent non-integrable relations in (13.59) and k is the number of independent coordinates. The dynamic systems involving non-integrable relations in (13.59) are called non-holonomic to distinguish them from holonomic systems in which the number of degrees of freedom is equal to the number of independent generalized coordinates. In other words, a holonomic system is one in which there are no non-integrable relations involving the generalized velocities.

13.10 Lagrangean Equations in Generalized Coordinates

Let there be a system of particles which requires n independent generalized coordinates or a degree of freedom to specify the states of its particles.

The position vectors x^r are expressed as a function of generalized coordinates q^i, $(i = 1, 2,\ldots n)$ and the time t, i.e.,

$$x^r = x^r(q^1, q^2,..q^n, t)$$

and assumes that the functions $x^r(q, t)$ are of class C^2.

The velocity $\dot{x}^r$ of any point of the body is given by

$$\dot{x}^r = \frac{\partial x^r}{\partial q^j}\frac{\partial q^j}{\partial t} + \frac{\partial x^r}{\partial t}$$

$$= \frac{\partial x^r}{\partial q^j}\dot{q}^j + \frac{\partial x^r}{\partial t}(j=1,2,\ldots n),$$

where $\dot{q}^j$ are generalized velocities.

Consider the relation, with n degree of freedom

$$x^r = x^r\,(q^1, q^2, ..q^n) \tag{13.60}$$

involving n independent parameters q^i.

The velocities $\dot{x}^r$ are given by

$$\dot{x}^r = \frac{\partial x^r}{\partial q^j}\dot{q}^j, \quad (r=1, 2, 3;\, j=1, 2,\ldots n), \tag{13.61}$$

where $\dot{q}^j$ transform under any admissible transformation.

$$\overline{q}^k = \overline{q}^k(q^1,q^2,..q^n);\ (k=1,2,\ldots n) \tag{13.62}$$

in accordance with contravariant law.

The kinetic energy of the system is expressed by

$$T = \frac{1}{2}\Sigma m g_{rs}\dot{x}^r\dot{x}^s, (r, s = 1,2,3), \tag{13.63}$$

where m is the mass of the particle located at the point x^r and g_{rs} are the components of the metric tensor.

Substituting the values of $\dot{x}^r$ from (13.61) in (13.63), we get

$$T = \frac{1}{2}\Sigma m g_{rs}\frac{\partial x^r}{\partial q^i}\frac{\partial x^s}{\partial q^j}\dot{q}^i\dot{q}^j$$

$$= \frac{1}{2}a_{ij}\dot{q}^i\dot{q}^j,$$

$$\text{where } a_{ij} \equiv \Sigma m g_{rs}\frac{\partial x^r}{\partial q^i}\frac{\partial x^s}{\partial q^j}(r, s = 1.2.3), (i, j = 1,2,\ldots n)$$

$$\text{since } T = \frac{1}{2}a_{ij}\dot{q}^i\dot{q}^j \tag{13.64}$$

is invariant and the quantities a_{ij} are symmetric, we conclude that a_{ij} are components of a covariant tensor of rank two with respect to the transformations in (13.62) of generalized coordinates.

Since the kinetic energy T is a positive form in the velocities $\dot{q}^i, |a_{ij}| > 0$, we construct the reciprocal tensor a^{ij}.

Using the Lagrangean Equation of Motion articulated in (13.5), we get

$$\frac{d}{dt}\left(\frac{\partial T}{\partial \dot{q}^i}\right) - \frac{\partial T}{\partial q^i} = a_{il}\left(\ddot{q}^l + \begin{Bmatrix} l \\ j \quad k \end{Bmatrix}\dot{q}^j\dot{q}^k\right). \tag{13.65}$$

Put $\ddot{q}^l + \begin{Bmatrix} l \\ j \quad k \end{Bmatrix}\dot{q}^j\dot{q}^k = Q^l$ in (13.65) and it becomes

$$\frac{d}{dt}\left(\frac{\partial T}{\partial \dot{q}^i}\right) - \frac{\partial T}{\partial q^i} = a_{il}Q^l = Q_i \ (i = 1,2,\ldots n). \tag{13.66}$$

Now, from the relations, $\frac{\partial \dot{x}^r}{\partial \dot{q}^i} = \frac{\partial x^r}{\partial q^i}, \frac{\partial \dot{x}^r}{\partial q^i} = \frac{\partial^2 x^r}{\partial x^i \partial q^j}\dot{q}^j$ *and* $\frac{\partial \dot{x}^r}{\partial q^i} = \frac{d}{dt}\left(\frac{\partial x^r}{\partial q^i}\right)$ and using equations (13.61) and (13.63), now the left-hand side of (13.65) becomes

$$\frac{d}{dt}\left(\frac{\partial T}{\partial \dot{q}^i}\right) - \frac{\partial T}{\partial q^i} = \Sigma m a_r \frac{\partial x^r}{\partial q^i}, \tag{13.67}$$

in which $a_j = g_{ij} a^j$ is the acceleration of point P

Newton's 2nd law gives us

$$m a_r = F_r \tag{13.68}$$

where F_r are the components of F acting on the particle located at point P.

From Equation (13.68), we have

$$\Sigma m a_r \frac{\partial x^r}{\partial q^i} = \Sigma F_r \frac{\partial x^r}{\partial q^i}$$

and Equation (13.67) can be written as

$$\frac{d}{dt}\left(\frac{\partial T}{\partial \dot{q}^i}\right) - \frac{\partial T}{\partial q^i} = \Sigma F_r \frac{\partial x^r}{\partial q^i}. \tag{13.69}$$

Comparing (13.66) with (13.67), we get

$$Q_i = \Sigma m a_r \frac{\partial x^r}{\partial q^i} = \Sigma F_r \frac{\partial x^r}{\partial q^i},$$

where vector Q_i is called a *generalized force.*

The equations

$$\frac{d}{dt}\left(\frac{\partial T}{\partial \dot{q}^i}\right) - \frac{\partial T}{\partial q^i} = Q_i \tag{13.70}$$

are known as Lagrangean equations in generalized coordinates.

The solution of these equations is in the form

$$C: q^i = q^i\,(t).$$

If there exists a function $V(q^1, q^2,..q^n)$ such that the system is said to be conservative, for such systems Equation (13.70) assumes the form

$$\frac{d}{dt}\left(\frac{\partial L}{\partial \dot{q}^i}\right)-\frac{\partial L}{\partial q^i}=0, \tag{13.71}$$

where $L = T - V$ is the kinetic potential.

Since $L(q,\dot{q})$ is a function of both the generalized coordinates and velocities,

$$\frac{dL}{dt}=\left(\frac{\partial L}{\partial \dot{q}^i}\right)\ddot{q}^i+\frac{\partial L}{\partial q^i}\dot{q}^i. \tag{13.72}$$

From (13.71), we get $\frac{\partial L}{\partial q^i}=\frac{d}{dt}\left(\frac{\partial L}{\partial \dot{q}^i}\right)$

Then, Equation (13.72) becomes

$$\frac{dL}{dt}=\left(\frac{\partial L}{\partial \dot{q}^i}\right)\ddot{q}^i+\frac{d}{dt}\left(\frac{\partial L}{\partial \dot{q}^i}\right)\dot{q}^i=\frac{d}{dt}\left(\frac{\partial L}{\partial \dot{q}^i}\dot{q}^i\right). \tag{13.73}$$

Since $L = T - V$ but potential energy is not a function of $\dot{q}^i$,

$$\frac{\partial L}{\partial \dot{q}^i}\dot{q}^i=\frac{\partial T}{\partial \dot{q}^i}\dot{q}^i=2T,$$

since $T=\frac{1}{2}a_{ij}\dot{q}^i\dot{q}^j$.

Thus, Equation (13.73) can be written as

$$\frac{d(L-2T)}{dt}=\frac{d(T+V)}{dt}=0,$$

which implies that $T + V = h(constant)$.

Thus, along the dynamical trajectory, the sum of the kinetic and potential energies is a constant.

Example 13.10.1. A simple pendulum consists of a bob of mass m supported by a light inextensible cord of length l. Suppose that the pendulum is set in vibration in some plane which we take as a Y^1 Y^2-plane, as shown in Figure (13.5).

Solution: The Lagrangean Equations are expressed as

$$\frac{d}{dt}\left(\frac{\partial T}{\partial \dot{q}^i}\right)-\frac{\partial T}{\partial q^i}=Q_i. \tag{i}$$

The Kinetic energy is expressed as

$$T=\frac{1}{2}m\dot{y}^i\dot{y}^i. \tag{ii}$$

The equations of the pendulum are

$$\begin{cases} y^1 = l sin\theta = l sin\dfrac{q}{l} \\ y^2 = l(1-cos\theta)=l\left(1-cos\dfrac{q}{l}\right), \end{cases} \tag{iii}$$

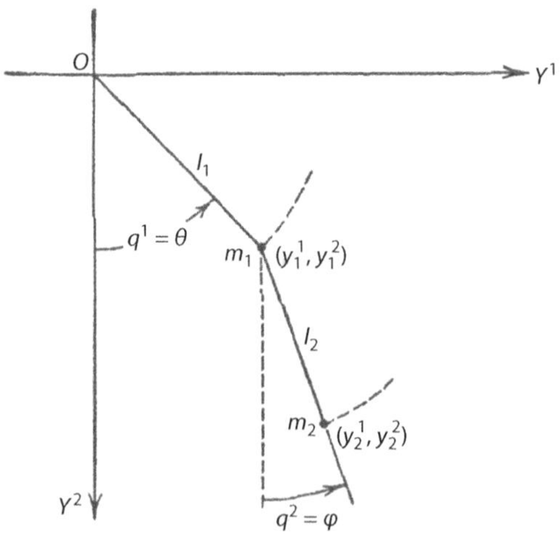

Figure 13.5

where the arc length $q = l\theta$ is a generalized coordinate.

Since

$$\dot{y}^1 = \dot{q}lcos\frac{q}{l}$$

$\dot{y}^2 = \dot{q}lcos\frac{q}{l}$, equation (ii) becomes

$$T = \frac{1}{2}m(\dot{q})^2,$$

The work is expressed as $W_\delta = -mgsin\theta\delta q$

$$= -mgsin\frac{q}{l}\delta q$$

and the generalized force is expressed as $Q = -mgsin\frac{q}{l}$. Thus, Equation (i) yields

$$\ddot{q} + gsin\frac{q}{l} = 0. \tag{iv}$$

Since, for a small displacement, $sin\theta = \theta$,

$$\ddot{q} + k^2 q = 0 \quad \left[\text{since } sin\theta \Rightarrow \theta = \frac{q}{l}\right],$$

where $k^2 = \frac{g}{l}$.

The solution of this equation is

$$q = a\cos(kt + \alpha).$$

13.11 Divergence Theorem, Green's Theorem, Laplacian Operator, and Stoke's Theorem in Tensor Notation

(i) Divergence Theorem

Let F be a vector point function in a closed region V, bound by the regular surface S. Then,

$$\int_V div\boldsymbol{F} = \int_S \boldsymbol{F}.\boldsymbol{n}\, ds, \tag{13.74}$$

where $\boldsymbol{n}$ is unit normal to S.

In orthogonal Cartesian coordinates, the divergence of F is given by

$$div\boldsymbol{F} = \frac{\partial \boldsymbol{F}^i}{\partial \overline{\boldsymbol{x}}^i}. \tag{13.75}$$

If the components of $\boldsymbol{F}$ relative to an arbitrary curvilinear coordinate system X are denoted by F^i, then the covariant derivative of F^i is

$$F^i_{,j} = \frac{\partial F^i}{\partial x^j} + \begin{Bmatrix} i \\ k \quad j \end{Bmatrix} F^k.$$

The invariant $F^i_{,j}$ in Cartesian coordinates represents the divergence of the vector field $\boldsymbol{F}$.

Also,

$$\boldsymbol{F}.\, \boldsymbol{n} = g_{ij} F^i n^j = F^i n_i = g^{ij} F_i n_j = F_i n^i.$$

Then, (13.74) can be written as

$$\int_V div\boldsymbol{F} = \int_S \boldsymbol{F}.\boldsymbol{n}\, ds = \int_S F^i n_i ds. \tag{13.76}$$

(ii) Green's Theorem

Let $u(x^1, x^2, x^3)$ and $v(x^1, x^2, x^3)$ be two scalar functions of class C^2 in V and of class C^1 in the closed region. We denote the gradients of *u and v* by u^i and v^i, respectively.

So, $\nabla u = u^i = \frac{\partial u}{\partial x^i}$ *and* $\nabla v = v^i = \frac{\partial v}{\partial x^i}$.

If we take $F_i = uv_i$,

from the divergence of F^i, we get

$$F^i_{,i} = g^{ij} F_{i,j} = g^{ij}(uv_{i,j} + u_j v_i).$$

We put this in Equation (13.75) and obtain the desired formula

$$\int_V g^{ij}(uv_{i,j} + u_j v_i)dV = \int_S F_i \boldsymbol{n}^i ds = \int_S uv_i \boldsymbol{n}^i dS. \qquad (13.77)$$

The invariant $g^{ij}v_{i,j}$ appearing on the left-hand side of Equation (13.76), when expressed in Cartesian coordinates, is the Laplacian of $v, \frac{\partial^2 v}{\partial y^i \partial y^i}$ and if we denote the Laplacian operator by the symbol ∇^2, we can write $g^{ij} v_{i,j} = \nabla^2 v$ and the inner product of $g^{ij} v_{id} u_j$ can be written as

$$g^{ij} u_j v_i = \nabla u . \nabla v,$$

where we denote ∇ as the operator of the gradient.

From (13.76) we can write

$$\begin{aligned} &\int_V (ug^{ij} v_{i,j} + g^{ij} u_j v_i)dv = \int_S un^i v_i ds, \\ \text{or } &\int_V (\mathrm{u}\nabla^2 \mathrm{v} + \nabla u . \nabla v)dv = \int_S \boldsymbol{un}.\nabla v \, ds, \qquad (13.78) \\ \text{or } &\int_V (\mathrm{u}\nabla^2 \mathrm{v} dv = \int_S \boldsymbol{un}.\nabla v \, ds - \int \nabla u . \nabla v \, dv, \end{aligned}$$

where $\boldsymbol{n}. \nabla v = n^i v_i = \frac{\partial v}{\partial n}$.

Interchanging *u and v* in (13.77) and subtracting the resulting formula from (13.77) yields the symmetrical form of Green's Theorem

$$\int_V (\mathrm{u}\nabla^2 \mathrm{v} - \mathrm{v}\nabla^2 \mathrm{u})dv = \int_S \left(u\frac{\partial v}{\partial n} - v\frac{\partial u}{\partial n} \right) ds. \qquad (13.79)$$

(iii) Expansion Form of the Laplacian Operator

The Laplacian of v *is*

$$g^{ij}\, v_{i,j} = \nabla^2 \mathrm{v}. \tag{13.80}$$

When it is written out in Christoffel symbols associated with curvilinear coordinates x^i, it is

$$\nabla^2 \mathrm{v} = g^{ij}\left(\frac{\partial^2 v}{\partial x^i \partial x^j} - \begin{Bmatrix} k \\ i \quad j \end{Bmatrix} \frac{\partial v}{\partial x^k}\right) \tag{13.81}$$

and as a divergence of vector F^i,

$$F^i_{,i} = \frac{\partial F^i}{\partial x^i} + \begin{Bmatrix} i \\ j \quad i \end{Bmatrix} F^j. \tag{13.82}$$

We know $\begin{Bmatrix} i \\ j \quad i \end{Bmatrix} = \frac{\partial}{\partial x^j} \log\sqrt{g},$ hence divergence $F^i_{,i}$ can be written as

$$\begin{aligned} F^i_{,i} &= \frac{\partial F^i}{\partial x^i} + \left(\frac{\partial}{\partial x^j} \log\sqrt{g}\right) F^j \\ &= \frac{1}{\sqrt{g}} \frac{\partial\left(\sqrt{g} F^i\right.}{\partial x^i}. \end{aligned} \tag{13.83}$$

If we put, $F^i = g^{ij} \dfrac{\partial v}{\partial x^j}$, we get

$$\nabla^2 \mathrm{v} = g^{ij} v_{j,i} = \frac{1}{\sqrt{g}} \frac{\partial\left(\sqrt{g}\, g^{ij} \frac{\partial v}{\partial x^j}\right)}{\partial x}. \tag{13.84}$$

It is the expansion form of the Laplacian operator.

(iv) Stoke's Theorem

Let a portion of regular surface S be bound by a closed regular curve C and $\boldsymbol{F}$ be any vector point function defined on S *and on* C. The theorem of Stoke's states that

$$\int_S \boldsymbol{n.\, curlF}\, ds = \int_C \boldsymbol{F}.\, \boldsymbol{\lambda} ds, \tag{13.85}$$

where λ is the unit tangent vector to C and curl$\boldsymbol{F}$ **is** the vector whose components in orthogonal Cartesian coordinates are determined from

$$\boldsymbol{curlF} = \begin{vmatrix} \boldsymbol{e_1} & \boldsymbol{e_2} & \boldsymbol{e_3} \\ \dfrac{\partial}{\partial y^1} & \dfrac{\partial}{\partial y^2} & \dfrac{\partial}{\partial y^3} \\ F^1 & F^2 & F^3 \end{vmatrix} = \nabla \times \boldsymbol{F}, \tag{13.86}$$

where $\boldsymbol{e}_i$ is the unit base vector in a Cartesian frame.

We consider the covariant derivative $\boldsymbol{F}_{i,\,j}$ of the vector $\boldsymbol{F}_i$ and from a contravariant vector

$$G^i = -\,\epsilon^{ijk}\, \boldsymbol{F}_{j,\,k}, \tag{13.87}$$

we define the vector G to be the curl of $\boldsymbol{F}$.

Since $\boldsymbol{n.\, curlF} = n_i G^i = -\,\epsilon^{ijk}\, \boldsymbol{F}_{j,\,k}\, n_i$

and the components of unit tangent vector λ and $\dfrac{dx^i}{ds}$

Equation (13.84) may be written as

$$-\int_S \epsilon^{ijk} \boldsymbol{F}_{j,k} n_i ds = \int_C F_i \frac{dx^i}{ds} ds. \tag{13.88}$$

The integral $\int_C F_i dx^i$ is called the circulation of $\boldsymbol{F}$ along contour C.

13.12 Hamilton's Canonical Equations

Consider

$$J = \int_{t_1}^{t_2} L(q,\dot{q})dt, \tag{13.89}$$

Where $L = T - V$ is the kinetic potential.

We know $\frac{d}{dt}\left(\frac{\partial L}{\partial \dot{q}^i}\right) - \frac{\partial L}{\partial q^i} = 0$, which is in the form

$$\frac{d}{dt} L_{\dot{q}^i} - L_{q^i} = 0 \, (i = 1, 2 \ldots, n) \tag{13.90}$$

by using the subscript notation for partial derivatives of $L(q,\dot{q})$.

It is convenient to rewrite the system of n Lagrangean Equation (13.90) in the form of an equivalent set of $2n$ first order equations, known as Hamilton's equations.

The function $L(q,\dot{q}) = T(q,\dot{q}) - V(q)$ depends on n generalized coordinates q^i *and n* generalized velocities $\dot{q}^i$. Instead of $\dot{q}^i$, we can introduce a set of n new variables p_i defined by

$$p_i = L_{\dot{q}^i}(q,\dot{q}), \quad (i = 1, 2, \ldots n), \tag{13.91}$$

where (13.91) is solvable for $\dot{q}^i$ in terms of p^i *and* q^i and the condition is that $\left|\frac{\partial L_{\dot{q}^i}}{\partial \dot{q}^i}\right| \neq 0$.

We construct

$$H(p,q) = \dot{q}^i p_i - L(q,\dot{q}). \tag{13.92}$$

Differentiating (13.92) with respect to q^j, we get

$$H_{q^j}(p,q) = \frac{\partial \dot{q}^i}{\partial q^j} p_i - L_{q^j} - L_{\dot{q}^i} \frac{\partial \dot{q}^i}{\partial q^j}$$

and since $p_i = L_{\dot{q}^i}$,

$$H_{q^j}(p,q) = -L_{q^j}. \tag{13.93}$$

Similarly, we have

$$H_{p^j}(p,q)=\dot{q}^j+\frac{\partial \dot{q}^i}{\partial p_j}p_i-L_{\dot{q}^i}\frac{\partial \dot{q}^i}{\partial p_j},$$

which on using (13.91) reduces to

$$H_{p_j}(p,q)=\dot{q}^j, \tag{13.94}$$

but from the Lagrangean Equation in (13.90), we get

$$\frac{d}{dt}L_{\dot{q}^i}=L_{q^i}.$$

From (13.91) and (13.93), we obtain

$$\begin{aligned} &\frac{dp_i}{dt}=\frac{dL_{\dot{q}^i}}{dt}=L_{q^i}=-H_{q^i} \quad (i=1,2,\ldots n) \\ \text{or} \quad &\frac{dp_i}{dt}=-H_{q^i}, \end{aligned} \tag{13.95}$$

which, together with n equation (13.94), yields

$$\frac{dq^i}{dt}=H_{p_i}. \tag{13.96}$$

Equations (13.95) and (13.96) constitute $2n$ first order *Hamilton's Canonical Equations.*

13.12.1 Generalized Momenta

A function $H(p, q)$, *is known as Hamilton's function.*

Here, $H=\dot{q}^i p_i-L(q,\dot{q})=\dot{q}^i L_{\dot{q}^i}-(T-V)=\dot{q}^i\frac{\partial L}{\partial \dot{q}^i}-T+V,$ where $T=\frac{1}{2}a_{ij}\dot{q}^i\dot{q}^j, \frac{\partial T}{\partial \dot{q}^i}=a_{ij}\dot{q}^i$, so that $\dot{q}^i\frac{\partial L}{\partial \dot{q}^i}=\dot{q}^i\frac{\partial (T-V)}{\partial \dot{q}^i}=\dot{q}^i\frac{\partial T}{\partial \dot{q}^i}=\dot{q}^i a_{ij}\dot{q}^j=2T$

$$\therefore H=2T-T+V=T+V.$$

Thus, H is the total energy of the system.

The variables $p_i = L_{\dot{q}^i} = \dfrac{\partial L}{\partial \dot{q}^i} = a_{ij}\dot{q}^i$ are called the generalized momenta and the square of the magnitude of p_i is

$$p^2 = a^{ij} p_i p_i = a^{ij} a_{ik} \dot{q}^k a_{jl} \dot{q}^l = a_{kl} \dot{q}^k \dot{q}^l = 2T.$$

13.13 Exercises

1. Let a particle of mass *m* be constrained to move on the surface of a sphere of radius *a*. Relate the orthogonal Cartesian coordinates y^i to the surface coordinates u^α by the formulas

$$\begin{cases} y^1 = a sinu^1 \cos u^2 \\ y^2 = a sinu^1 a sinu^2 \\ y^3 = a cosu^1. \end{cases}$$

Show that equations $\dfrac{d}{dt}\left(\dfrac{\partial T}{\partial \dot{u}^\alpha}\right) - \dfrac{\partial T}{\partial u^\alpha} = F_\alpha$ becomes

$$\ddot{u}^1 - (\dot{u}^2)^2 sinu^1 \cos u^1 = \frac{F_1}{ma^2}$$

$$\ddot{u}^2 sin^2 u^1 + 2\dot{u}^1 \dot{u}^2 sinu^1 \cos u^1 = \frac{F_2}{ma^2}.$$

Solve these equations for the case when $F^\alpha = 0$, and show that the trajectory is an arc of a great circle and the speed $v = constant$. [*Hints*: The first integral of the second equation is $\dot{u}^2 sin^2 u^1 = constant$. Use this result in the first equation and observe that $v^2 = a^2[(\dot{u}^1)^2 + (\dot{u}^2)^2 sinu^1.)$

2. Find the dynamic equation in spherical coordinates with

$$ds^2 = (dr)^2 + r^2(d\theta)^2 + r^2 sin^2\theta (d\phi)^2.$$

(If $\dot{r} = 0$, we get $\ddot{\theta} = -\left(\dfrac{g}{r}\right) sin\phi$.)

3. Find the dynamic equation in cylindrical coordinates with

$$ds^2 = (dr)^2 + r^2(d\theta)^2 + (dz)^2.$$

4. Show that:
   (a) In plane polar coordinates with $ds^2 = (dr)^2 + r^2 (d\theta)^2$

$$div\, \boldsymbol{F} = \frac{1}{r}\left[\frac{\partial(rFr)}{\partial r} + \frac{\partial F_\theta}{\partial_\theta}\right]$$

$$\nabla^2 \text{v} = \frac{1}{r}\left[\frac{\partial}{\partial r}\left(r\frac{\partial v}{\partial r}\right) + \frac{\partial}{\partial\theta}\left(\frac{1}{r}\frac{\partial v}{\partial\theta}\right)\right],$$

where F_r *and* F_θ are the physical components of the vector F, that is,

$$\boldsymbol{F} = F_r r_1 + F_\theta \theta_1,$$

where r_1 *and* θ_1 are unit vectors.
   (b) In cylindrical coordinates with

$$ds^2 = (dr)^2 + r^2(d\theta)^2 + (dz)^2$$

$$div\, \boldsymbol{F} = \left[\frac{1}{r}\frac{\partial(rF_r)}{\partial r} + \frac{1}{r}\frac{\partial F_\theta}{\partial\theta} + \frac{\partial F_z}{\partial z}\right]$$

$$\nabla^2 \text{v} = \left[\frac{1}{r}\frac{\partial}{\partial r}\left(r\frac{\partial v}{\partial r}\right) + \frac{1}{r^2}\frac{\partial^2 v}{\partial\theta^2} + \frac{\partial^2 V}{\partial z^2}\right],$$

where $\boldsymbol{F} = F_r r_1 + F_\theta \theta_1 + F_z z_1$, and r_1, θ_1, z_1 are unit vectors, so that F_r, F_θ, F_z are the physical components of vector F.

14

Newtonian Law of Gravitations

14.1 Introduction

Sir Isaac Newton was perhaps the first to bring powerful dynamic ideas and necessary mathematics to bear on motion in the solar system. Newton built on the scientific contributions of his predecessors, particularly Kepler and Galileo. The validity of Kepler Laws depends upon the fact that the masses of the planets are very small compared with the mass of the sun. The empirical laws enunciated by Kepler laid Newton, among others, to the conclusion that the force which keeps a planet in its orbit around the sun varies inversely as the square of the distance from the sun to planet. As soon as Newton proved that the gravitational attraction between two homogeneous spheres could be calculated as if the masses of the spheres were concentrated at their centers, the progress of dynamic astronomy was clear and rapid. In this chapter we also study the Gauss theorem and problems of two bodies and restricted three bodies.

14.2 Newtonian Laws of Gravitation

The inverse square law of attraction had its origin in Newton's studies of motion of planetary bodies in what he termed the *eccentric conic sections.* We state this law as follows:

Two material particles attract each other with a force which is directly proportional to the product of their masses and inversely proportional to the square of the distance between them. The line of action of the force is along the line joining the particles.

Dipankar De. Introduction to Differential Geometry with Tensor Applications, (447–468) © 2022 Scrivener Publishing LLC

Thus, the law, in the form of a vector equation is

$$\boldsymbol{F} = \gamma \frac{m_1 m_2}{r_{12}^3} \boldsymbol{r}_{12},$$

where m_1 *and* m_2 are the masses of the particles and $\boldsymbol{r}_{12}$ is the vector from P_1 *and* P_2.

The constant γ depends on the choice of units; in a cgs system, its value 6.664×10^{-8} and its physical dimensions are $M^{-1}L^3T^{-2}$. We usually take it as $\gamma = 1$, so we write

$$\boldsymbol{F} = \frac{m_1 m_2}{r_{12}^3} \boldsymbol{r}_{12}. \tag{14.1}$$

We deal with the particles with continuous distributions of matter and one can subdivide the bodies into infinity. This procedure is for two bodies τ_1 *and* τ_2 leads to the formula

$$\boldsymbol{F} = \int_{\tau_1} \int_{\tau_2} \frac{\rho_1 \rho_2}{r_{12}^3} \boldsymbol{r}_{12} d\tau_1 \tau_2, \tag{14.2}$$

where $d\tau_1$ *and* $d\tau_2$ are the volume elements of bodies τ_1 *and* τ_2, ρ_1 *and* ρ_2, and their density functions and r_{12} is the position vector of $d\tau_2$ relative to $d\tau_1$. We consider that ρ_1 *and* ρ_2 are piecewise, continuous.

It is necessary to verify that the generalized law of gravitation in (14.2) reduces to (14.1) and yields no vanishing couples when the bodies τ_1 *and* τ_2 are allowed to shrink to a point.

We introduce an orthogonal Cartesian reference frame Y and denote the coordinates of points of the bodies τ_1 *and* τ_2 by (y_1^i) and (y_2^i), respectively, as shown in Figure (14.1).

We replace the distributed mass $\rho_1\ \Delta\tau_1$ by the concentrated mass m_1 at $P_1(y_1^1, y_1^2, y_1^3)$ and the mass $\rho_2\ \Delta\tau_2$ by m_2 at $P_1(y_2^1, y_2^2, y_2^3)$.

For components of force,

$$\Delta \boldsymbol{F}^i = \rho_1 \rho_2 \Delta\tau_1 \Delta\tau_2 \frac{y_2^i - y_1^i}{r^3}$$

and for the components of moments, ΔL_i is relative to the origin O,

$$\Delta L_i = e_{ijk} y_1^j \Delta F^k$$
$$= e_{ijk} y_1^j \rho_1 \rho_2 \Delta\tau_1 \Delta\tau_2 \frac{y_2^k - y_1^k}{r^3}$$

Adding these vectorially gives the resultant force

$$F^i = \int_{\tau_1} \int_{\tau_2} \rho_1 \rho_2 \frac{y_2^k - y_1^k}{r^3} d\tau_1 d\tau_2 \tag{14.3}$$

and the resultant moment

$$L_i = \int_{\tau_1} \int_{\tau_2} e_{ijk} y_1^j \rho_1 \rho_2 \frac{y_2^k - y_1^k}{r^3} d\tau_1 d\tau_2. \tag{14.4}$$

We choose the origin O of the coordinate system at P_1, and let τ_1 shrink towards O and τ_2 shrink towards P_2. Since ρ_1 *and* ρ_2 in Equations (14.3) and (14.4) are nonnegative functions, the first mean value theorem for integrals is applicable and we obtain Figure (14.1).

$$F^i = \left[\frac{y_2^k - y_1^k}{r^3}\right] \int_{\tau_1} \int_{\tau_2} \rho_1 \rho_2 d\tau_1 d\tau_2$$

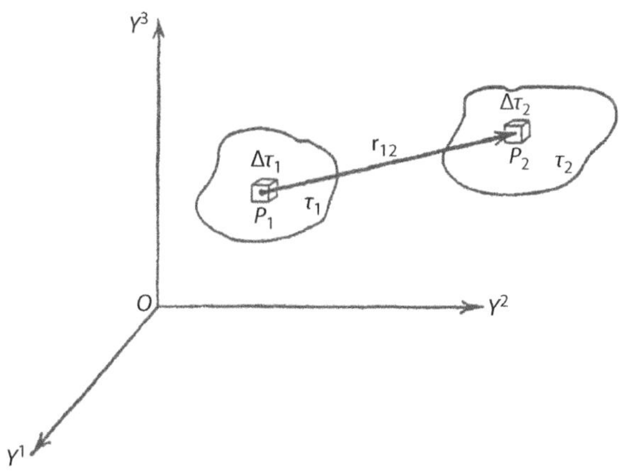

Figure 14.1

and

$$L_i = \left[e_{ijk} y_1^j \frac{y_2^k - y_1^k}{r^3} \right] \int_{\tau_1} \int_{\tau_2} \rho_1 \rho_2 d\tau_1 d\tau_2$$

As dimensions of τ_1 *are* allowed to approach zero, $y_1^i \to 0$ and $L_i \to 0$, whereas the first of the above integrals reduces to

$$F^i = \frac{y_2^i}{r^3} m_1 m_2.$$

This is the Law of Gravitation for two particles at (0, 0, 0) and (y_2^1, y_2^2, y_2^3).

We observed that a material body integrating with a point mass produces no resultant moment ***L***. Moreover, direct calculations show that this is also true when the point mass is replaced by a sphere τ whose density ρ is a continuous function of the radius alone. The resultant force *F*, exterted by the body acting on a point mass $m = \int_\tau \rho d\tau$, is located at the center of the sphere.

Consider a body τ with piecewise continuous density ρ and let $P\ (y^1, y^2, y^3)$ be fixed either within or outside τ. The *gravitational potential* $V\ (P)$ at the point *P* due to τ is defined by the integral

$$V(P) = \int_\tau \frac{\rho(\xi^1, \xi^2, \xi^3)}{r} d\tau(\xi), \tag{14.5}$$

where $r = \sqrt{(y^1 - \xi^1)^2 + (y^2 - \xi^2)^2 + (y^3 - \xi^3)^2}$ is the distance between $P\ (y^1, y^2, y^3)$ and the variable point (ξ^1, ξ^2, ξ^3) is associated with the volume element $d\tau\ (\xi)$ of τ.

If *P* is outside the body, the integral (14.5) is proper.

In particular,

$$\frac{\partial V}{\partial y^i} = -F_i, \tag{14.6}$$

where the F_i are components of gravitational force

$$F_i(P) = \int_\tau \frac{\rho(\xi)}{r} d\tau \tag{14.7}$$

exerted by the body τ on the particle of unit mass at $P(y)$.

If $P(y)$ is within τ, the integral (14.5) is improper since $r = 0$. Although, $V(P)$ is of class C^∞ whenever P is exterior to τ.

14.3 Theorem of Gauss

The integral of the normal component of the gravitational flux computed over a regular surface S containing gravitating masses within it is equal to $4\pi m$, where m is the total mass enclosed by S.

Proof: According to Newton's Law of Gravitation, a particle P and Let θ are the angle between the unit outward normal to n to S and the axis of a cone with its vertex at P. This cone subtends an element of surface $d\sigma$ [Figure (14.2)].

The flux of the gravitational field produced by m is

$$\int_S \mathbf{F}.\mathbf{n} d\sigma = \int_S \frac{m cos\theta}{r^2} \frac{r^2 d\omega,}{cos\theta}$$

where $d\sigma = \dfrac{r^2 d\omega}{cos\theta}$ and $d\omega$ are the solid angle subtended by $d\sigma$.

Thus, we have

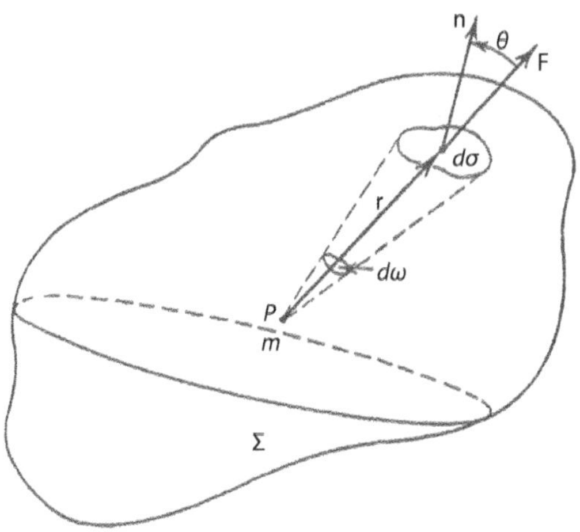

Figure 14.2

$$\int_S \boldsymbol{F.n}d\sigma = \int_S \boldsymbol{m}.d\omega = 4\pi m. \tag{14.8}$$

If there are n discrete particles of masses m_i located within S, then

$$\boldsymbol{F.n} = \sum_{i=1}^{n} \frac{m_i cos\theta}{r^2}$$

and the total flux is

$$\int_S \boldsymbol{F.n}d\sigma = 4\pi \sum_{i=1}^{n} m_i. \tag{14.9}$$

The result of (14.9) can be easily generalized to continuous distributions of matter whenever such distribution is nowhere near the surface S.

The contribution to the flux integration from the mass element $\rho d\tau$, contained within τ, is

$$\int_S \boldsymbol{F.n}d\sigma = \int_S \frac{\rho cos\theta d\tau}{r^2} d\sigma$$

and the contribution from all masses contained easily within S is

$$\int_S \boldsymbol{F.n}d\sigma = \int_S \left(\int_\tau \frac{\rho cos\theta d\tau}{r^2} \right) d\sigma, \tag{14.10}$$

where $\int_\tau$ *denotes* the volume integral over all the bodies interior to S. Since all masses are assumed to be interior to S, r never finishes, so that the integrand in Equation (14.10) is continuous and one can interchange the order of integration to obtain

$$\int_S \boldsymbol{F.n}d\sigma = \int_\tau \rho \left(\int_S \frac{\rho cos\theta d\sigma}{r^2} \right) d\tau,$$

but $\int_S \frac{cos\theta d\sigma}{r^2} = 4\pi$. Since it represents the flux due to a unit mass contained within S,

$$\int_S \boldsymbol{F.n}d\sigma = 4\pi \int_\tau \rho d\tau = 4\pi m,$$

where m denotes the total mass contained within S.

14.4 Poisson's Equation

By the divergence theorem, we have

$$\int_S \boldsymbol{F.n}d\sigma = \int_\tau div\boldsymbol{F}\, d\tau$$

and by Gauss's theorem

$$\int_S \boldsymbol{F.n}d\sigma = 4\pi \int_\tau \rho d\tau. \tag{14.11}$$

Therefore,

$$\int_\tau (div\boldsymbol{F} - 4\pi\rho)d\tau = 0. \tag{14.12}$$

This relation is true for an arbitrary τ and since the integrand in (14.11) is continuous, we conclude that

$$div\boldsymbol{F} - 4\pi\rho = 0 \text{ throughout } \tau. \tag{14.13}$$

In the definition of potential function V, we have

$$F = -\nabla V \text{ and from (14.13),}$$
$$div\boldsymbol{F} = \nabla F = -\nabla^2 V = 4\pi\rho.$$

$$\therefore\ \nabla^2 V = -4\pi\rho \tag{14.14}$$

is the *equation of Poisson.*

If P is not occupied by the mass, then $\rho = 0$ and the potential energy satisfies Laplace's Equation.

$$\nabla^2 V = 0$$

We note in conclusion that the formula

$$V(P) = \int_\tau \frac{\rho d\tau}{r}$$

gives a solution of Equation (14.14) at all points in τ.

14.5 Solution of Poisson's Equation

We know that Green's symmetrical formula is

$$\int_\tau (u\nabla^2 v - v\nabla^2 u)d\tau = \int_S \left(u\frac{\partial v}{\partial n} - v\frac{\partial u}{\partial n} \right) d\sigma \qquad (14.15)$$

where *V* is the volume enclosed by *S* and *u and v* are scalar point functions.

Put $u = \frac{1}{r}$ where r is the distance between the points $P(x^1, x^2, x^3)$ and (ξ^1, ξ^2, ξ^3) *and V* is gravitational potential.

Since $\frac{1}{r}$ has a discontinuity at $(x^i) = (\xi^i)$, we delete $P(x)$ from τ by enclosing it by a sphere σ of radius δ at *P*. Apply Green's formula to region $\tau - \epsilon$, within which $\frac{1}{r}$ and τ possess the desired properties of continuity.

In region $\tau - \epsilon$, $\nabla^2 u = \nabla^2 \frac{1}{r} = 0$ and formula (14.15) yields

$$\int_{\tau-\epsilon} \frac{1}{r}\nabla^2 V d\tau = \int_S \left(\frac{1}{r}\frac{\partial V}{\partial n} - V\frac{\partial \frac{1}{r}}{\partial n} \right) d\sigma + \int_\sigma \left(\frac{1}{r}\frac{\partial V}{\partial n} - V\frac{\partial \frac{1}{r}}{\partial n} \right) d\sigma, \qquad (14.16)$$

where *n* is the unit exterior normal to the surface $S + \sigma$, since on σ the nomal *n* is directed towards *P*, and $\frac{\partial}{\partial n} = -\frac{\partial}{\partial r}$.

Now,

$$\begin{aligned}
\int_\sigma \left(\frac{1}{r}\frac{\partial V}{\partial n} - V\frac{\partial \frac{1}{r}}{\partial n} \right) d\sigma &= \int_\sigma \left(-\frac{1}{r}\frac{\partial V}{\partial r} - V\frac{\partial \frac{1}{r}}{\partial r} \right) d\sigma \\
&= \int_\sigma \left(-\frac{1}{r}\frac{\partial V}{\partial r} - \frac{V}{r^2} \right) r^2 d\omega \\
&= -\int_\sigma \left(r\frac{\partial V}{\partial r} + V \right) d\omega \\
&= -\delta \int_\delta \left(\frac{\partial V}{\partial r} \right)_{r=\delta} d\omega - 4\pi\bar{V}, \qquad (14.17)
\end{aligned}$$

where $\bar{V}$ is the mean value of V over the sphere σ and ω denotes the solid angle.

Let $\delta \to 0$, the right-hand member of (14.17) yields $-4\pi V(P)$ and it follows

$$\int_\sigma \left(\frac{1}{r}\frac{\partial V}{\partial n} - V\frac{\partial \frac{1}{r}}{\partial n} \right) d\sigma = -4\pi V(P).$$

Thus, (14.15) becomes

$$\int_\sigma \frac{1}{r}\nabla^2 V d\tau = \int_\sigma \left(\frac{1}{r}\frac{\partial V}{\partial n} - V\frac{\partial \frac{1}{r}}{\partial n} \right) d\sigma - 4\pi V(P)$$

Since $\epsilon \to 0$, $\int_\epsilon \frac{1}{r}\nabla^2 V d\tau = 0$.

Therefore,

$$V(P) = -\frac{1}{4\pi}\int_\sigma \frac{1}{r}\nabla^2 V d\tau + \frac{1}{4\pi}\int_\sigma \frac{1}{r}\frac{\partial V}{\partial n} d\sigma - \frac{1}{4\pi}\int_\sigma V\frac{\partial \frac{1}{r}}{\partial n} d\sigma. \quad (14.18)$$

This gives the *solution of Poisson's Equation* at the origin.

If V is regular at infinity, i.e., for a sufficiently large value of r, V is such that

$$[V] \le \frac{m}{r} \text{ and } \left[\frac{\partial V}{\partial r}\right] \le \frac{m}{r^2}, \quad (14.19)$$

Where m is constant.

If the integration in Equation (14.18) is extended over all spaces so that $r \to \infty$, then using equation (14.18), (14.19) becomes

$$V(P) = -\frac{1}{4\pi}\int_\infty \nabla^2 d\tau, \quad (14.20)$$

but V is a potential function satisfying Poisson's equation i.e., $\nabla^2 V = -4\pi\rho$.

Hence, from (14.20), we get

$$V(P)=\int_{\infty}\rho\frac{dV}{r}.$$

14.6 The Problem of Two Bodies

Here, we consider a much simpler, very well-known problem in physics: an isolated system of two particles which interact through a central potential. This method is often referred to simply as the *two-body problem.*

The problem of two bodies can be stated as follows: Given a system of two particles interacting in accordance with the law of universal gravitation, what is the trajectory of the system? This problem was solved by Newton in the Principia, Book I. It lies at the basis of all considerations in astronomy. We refer our system to a set of orthogonal Cartesian axes. The coordinates of m_1 and m_2 (at a given instant of time) by (x_1^1, x_1^2, x_1^3) *and* (x_2^1, x_2^2, x_2^3), as shown in Figure (14.3).

We also introduce another Cartesian reference frame Y, moving with m_1 in such a way that m1 is in origin 0 and axis Y^i always remains parallel to axes X^i. The coordinate of mass m_2 relative to the Y – axes are denoted by y^i, = and we have

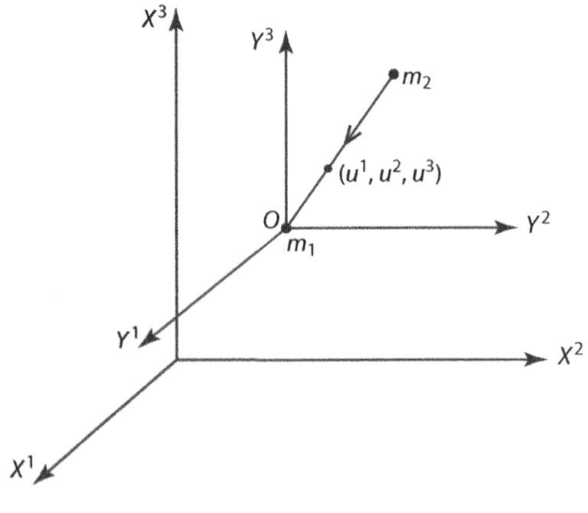

Figure 14.3

$$y^i = x_2^i - x_1^i \quad (i = 1,2,3). \tag{14.21}$$

We choose the coordinates as generalized coordinates, taking them as the center of masses. Thus,

$$u^i = \frac{m_1 x_1^i + m_2 x_2^i}{m_1 + m_2} \quad (i = 1,2,3). \tag{14.22}$$

Our choice of generalized coordinates is as follows:

$$q^1 = y^1, q^2 = y^2, q^3 = y^3, q^4 = u^1, q^5 = u^2, q^6 = u^3.$$

Solving, we get

$$\begin{aligned} x_1^i &= x_2^i - y^i & [m_2 x_2^i &= (m_1 + m_2)u^i - m_1 x_1^i \\ &= u^i + \frac{m_1}{m_2}(u^i - x_1^i) - y^i & x_2^i &= u^i + \frac{m_1}{m_2}(u^i - x_1^i)] \\ &= u^i - \frac{m_2}{m_1 + m_2} y^i \end{aligned}$$

and
$$x_2^i = u^i + \frac{m_1}{m_1 + m_2} y^i. \tag{14.23}$$

This will help us find the positional coordinates x^i in terms of generalized coordinates q^i. Since the magnitude of the force of attraction F is given by $F = \frac{m_1 m_2}{r^2}$, where r is the distance between the particles.

The potential energy V is

$$V = \frac{m_1 m_2}{r} = \frac{m_1 m_2}{\sqrt{(x_1^1 - x_2^1)^2 + (x_1^2 - x_2^2)^2 + (x_1^3 - x_2^3)^2}}$$

V is the function of $y^i = x_2^i = x_1^i$.

We recall the Lagrangean equations

$$\frac{d}{dt}\left(\frac{\partial T}{\partial \dot{q}^i}\right)-\frac{\partial T}{\partial q^i}=-\frac{\partial V}{\partial q^i} \tag{α}$$

For computation

$$\begin{aligned}T&=\frac{1}{2}m_1\dot{x}_1^i\dot{x}_1^i+\frac{1}{2}m_2\dot{x}_2^i\dot{x}_2^i\\&=\frac{1}{2}m_1\left[\dot{u}^i-\frac{m_2}{m_1+m_2}\dot{y}^i\right]^2+\frac{1}{2}m_2\left[\dot{u}^i-\frac{m_1}{m_1+m_2}\dot{y}^i\right]^2\\&=\frac{1}{2}(m_1+m_2)\dot{u}^i\dot{u}^i+\frac{1}{2}\frac{m_1m_2}{m_1+m_2}\dot{y}^i\dot{y}^i,\end{aligned}$$

since $\frac{\partial V}{\partial q^i}=0,\ for\ i=4,5,6,$ (α) can be reduced to

$$\begin{aligned}\frac{m_1m_2}{m_1+m_2}\ddot{y}^1&=-\frac{\partial V}{\partial y^i}\\ \ddot{u}^i&=0 \qquad (i=1,2,3)\end{aligned} \tag{14.24}$$

These differential equations are the motion of our system.

We first note that the motion of mass m_2 relative to m_1 is same as if mass m_1 were fixed and m_2 attracted toward it with a force whose potential is $\frac{m_1+m_2}{m_1}V$.

(14.24) becomes

$$m_2\ddot{y}^i=-\frac{m_1+m_2}{m_1}\frac{\partial V}{\partial y^i}. \tag{14.25}$$

Thus, our problem is reduced to a study of motion under the action of central forces.

The second set of equations of (14.24) states that the center of mass moves in a straight line with constant velocity.

If we carry out the integration in (14.24) under the assumption that m_1 (mass of sun) is much larger than m_2 (mass of earth). If $m_1 >> m_2$, the center of mass u^i will nearly coincide with those of mass m_1.

Thus, $x_1^i=u^i$ and from the second set of equations in (14.24), we conclude that the mass m_1 moves through space with constant velocity.

We need to examine the motion of m_2 relative to m_1.

$$\text{If } m_1 >> m_2, \text{ then } \frac{m_1 + m_2}{m_1} = 1$$

and accordingly (14.25) becomes

$$m_2 \ddot{y}^i = -\frac{\partial V}{\partial y^i}.$$

Let mass m_1 and m_2 be in the plane of Y^1 Y^2, since a force field is a central force so that there is no component of force at right angles to the plane.

Let the polar coordinate of m_2 be (r, θ),

i.e.,
$$y^1 = r cos\theta$$
$$y^2 = r sin\theta$$

$$\text{Kinetic energy}, T = \frac{1}{2} m_2 [(\dot{y}^1)^2 + (\dot{y}^2)^2]$$
$$= \frac{1}{2} m_2 [\dot{r}^2 + r^2 \dot{\theta}^2],$$

and
$$V = -\frac{m_1 m_2}{r}$$

In Lagrangean Equation (α), with $q^1 = r$ and $q^2 = \theta$, we get

$$m_2 \ddot{r} - m_2 r \dot{\theta}^2 = -\frac{m_1 m_2}{r^2}$$

$$\text{and} \quad \frac{d}{dt}(r^2 \dot{\theta}) = 0$$

$$\text{or} \quad \ddot{r} - r\dot{\theta}^2 + \frac{m_1}{r^2} = 0$$

$$r^2 \dot{\theta} = h, constant \tag{14.26}$$

These ordinary differential equations determine the trajectory. The second equation states that sectorial velocity is constant, which is one of Keplar's Laws. The relation $r^2\dot{\theta}=h$ can determine the time required to describe the orbit.

If $h \neq 0$, the trajectory is not a straight line and we have

$$r^2 d\theta = hdt$$

or

$$t=\frac{1}{h}\int_0^\theta r^2 d\theta.$$

We know $\frac{df}{dt}=\frac{df}{d\theta}\frac{d\theta}{dt}$ $\therefore \frac{d}{dt}=\frac{d}{d\theta}\frac{d\theta}{dt}=\frac{d}{d\theta}\left(\frac{h}{r^2}\right)$ and

$\ddot{r}=\frac{d}{dt}\left(\frac{dr}{dt}\right)=\frac{d}{dt}\left(\frac{dr}{d\theta}\frac{d\theta}{dt}\right)=\frac{h}{r^2}\frac{d}{d\theta}\left(\frac{h}{r^2}\frac{dr}{d\theta}\right)$ and using it in (14.26),

$$\frac{h}{r^2}\frac{d}{d\theta}\left(\frac{h}{r^2}\frac{dr}{d\theta}\right)-r\left(\frac{h}{r^2}\right)^2+\frac{m_1}{r^2}=0,$$

Multiplying by r^2,

$$h\frac{d}{d\theta}\left(\frac{h}{r^2}\frac{dr}{d\theta}\right)-r^3\left(\frac{h}{r^2}\right)^2+m_1=0,$$

$$\text{or } h\frac{d}{d\theta}\left(\frac{h}{r^2}\frac{dr}{d\theta}\right)-\frac{h^2}{r}+m_1=0,$$

$$\text{or } \quad \frac{d}{d\theta}\left(\frac{1}{r^2}\frac{dr}{d\theta}\right)-\frac{1}{r}+\frac{m_1}{h^2}=0. \tag{14.27}$$

Change the variable $u=\frac{1}{r}$,

$$\frac{du}{d\theta}=\frac{du}{dr}\frac{dr}{d\theta}=-\frac{1}{r^2}\frac{dr}{d\theta},\ \frac{d^2u}{d\theta^2}=\frac{d}{d\theta}\left(\frac{du}{d\theta}\right)=\frac{d}{d\theta}\left(-\frac{1}{r^2}\frac{dr}{d\theta}\right)=-\frac{d}{d\theta}\left(\frac{1}{r^2}\frac{dr}{d\theta}\right).$$

Now. In (14.27) becomes

$$\frac{d^2u}{d\theta^2}+u=\frac{m_1}{h^2}.$$

The solution of the differential equation is

$$u=\frac{1}{l}[1-e\cos(\theta-\epsilon)]$$

$$\text{or } r=\frac{l}{1-e\cos(\theta-\epsilon)}, \tag{14.28}$$

where $l\equiv\frac{h^2}{m_1}$ and e, ϵ are integration constants.

This is an example of a simple use of Hamilton's Equation.

Example 14.6.1. Consider a particle of mass m moving under the attraction of a central force field with the potential $V(r)$, r being the distance of the particle from the center of attraction.

Solution: We choose the polar coordinates $r = q^1$, $\theta = q^2$ as generalized coordinates, then

$$T=\frac{1}{2}m[r^2+(r\dot{\theta})^2]=\frac{1}{2}a_{ij}\dot{q}^i\dot{q}^j,$$

where $(a_{ij})=\begin{pmatrix} m & 0 \\ 0 & mr^2 \end{pmatrix}$,

but
$$H=T+V=\frac{1}{2}a_{ij}\dot{q}^i\dot{q}^j+V.$$

By (13.95), we get

$$H=\frac{p_1^2}{2m}+\frac{p_2^2}{2mr^2}+V(r).$$

Therefore, $\frac{\partial H}{\partial r}=-\frac{p_2^2}{mr^3}+V'(r), \frac{\partial H}{\partial \theta}=0, \frac{\partial H}{\partial p_1}=\frac{p_1}{m}, \frac{\partial H}{\partial p_2}=\frac{p_2}{mr^2}.$

Hence, Hamilton's Equation in this problem is

$$\frac{dr}{dt}=\frac{p_1}{m},\frac{d\theta}{dt}=\frac{p_2}{mr^2},\frac{dp_1}{dt}=\frac{p_2^2}{mr^3}-V'(r),\frac{dp_2}{dt}=0.$$

From these, we get

$$\frac{d}{dt}(mr^2\dot{\theta})=0,$$

which is Kepler's Law of Planetary Motion.

From these equations, it can be shown that if $V=-\frac{m}{r}$, the orbit is a conic section.

14.7 The Problem of Three Bodies

In general, in the three bodies problem we study the equations of motion of three finite bodies. The problem is restricted in the sense that one of the three masses is taken to be so small that the gravitational effect on the other two masses by the third mass is neglected. Then, the general three body problem reduces to a problem known as a *restricted three body problem.*

The small body is known as an infinitesimal mass and the remaining two massive bodies are known as finite masses. The Earth, Moon, and satellite system constitute a good example of a restricted three body problem.

Let us consider a coordinate system, OXYZ with an origin at O rotating relative to the inertial frame with angular velocity ω about the Z-axis. Without any loss of generality, we can choose the coordinate system such that the X-axis lies along the line joining the finite masses m_1 *and* m_2 with O as a bary center. Let the distance between m_1 *and* m_2 be denoted by ρ, the position of infinitesimal mass m be denoted by (x, y, z), and the radius vector from m *to* m_1 and m *to* m_2 be ρ_1 *and* ρ_2, respectively.

The motions of m_1 *and* m_2 are known. We are only to calculate the motion of m.

Now, the kinetic energy of m is in reference to rotating frame OXYZ.

$$\text{Kinetic Energy, } T = \frac{1}{2}m[(\dot{x} - \omega y)^2 + (\dot{y} + \omega x)^2 + \dot{z}^2]$$

$$= \frac{1}{2}m[\dot{x}^2 + \dot{y}^2 + \dot{z}^2 + 2\omega(x\dot{y} - y\dot{x}) + \omega^2(x^2 + y^2 0] = T_2 + T_1 + T_0$$

$$T = T_2 + T_1 + T_0, \tag{14.29}$$

where

$$T_2 = \frac{1}{2}m(\dot{x}^2 + \dot{y}^2 + \dot{z}^2)$$

$$T_1 = \frac{1}{2}m(2\omega(x\dot{y} - y\dot{x}))$$

$$T_0 = \frac{1}{2}m(\omega^2(x^2 + y^2)) \tag{14.30}$$

$$\text{Potential Energy } V = -m\gamma\left[\frac{m_1}{\rho_1} + \frac{m_2}{\rho_2}\right], \tag{14.31}$$

$$\text{where } \rho_1 = [(x + \alpha)^2 + y^2 + z^2]^{\frac{1}{2}}$$

$$\rho_2 = [(x - b)^2 + y^2 + z^2]^{\frac{1}{2}}, \tag{14.32}$$

where $(-a, 0, 0)$ *and* $(b, 0, 0)$ are coordinates of m_1 and m_2.

The generalized momentum corresponding to the generalized coordinate system is

$$P_x = \frac{\partial T}{\partial \dot{x}} = m(\dot{x} - \omega y)$$

$$P_y = \frac{\partial T}{\partial \dot{y}} = m(\dot{y} - \omega x)$$

$$P_z = \frac{\partial T}{\partial \dot{z}} = m\dot{z}. \tag{14.33}$$

Now, T_2 can be written in the generalized terms of momenta:

$$T_2 = \frac{1}{2}m(\dot{x}^2 + \dot{y}^2 + \dot{z}^2) = \frac{1}{2}m\left[\left(\frac{P_x}{m} + \omega y\right)^2 + \left(\frac{P_y}{m} - \omega x\right)^2 + \left(\frac{P_z}{m}\right)^2\right]$$

$$= \frac{1}{2}m\left[\left(\frac{P_x^2 + P_y^2 + P_z^2}{m^2} + \frac{2\omega}{m}(P_x y - P_y x) + \omega^2(x^2 + y^2)\right)\right]$$

$$T_2 = \frac{P_x^2 + P_y^2 + P_z^2}{2m} + \omega(P_x y - P_y x) + \frac{1}{2}m\omega^2(x^2 + y^2). \tag{14.34}$$

We know the Hamiltonian H is given by

$$H = T_2 - T_0 + V$$

$$= \frac{P_x^2 + P_y^2 + P_z^2}{2m} + \omega(P_x y - P_y x) + \frac{1}{2}m\omega^2(x^2 + y^2)$$

$$- \frac{1}{2}m(\omega^2(x^2 + y^2)) + V$$

$$H = \frac{P_x^2 + P_y^2 + P_z^2}{2m} + \omega(P_x y - P_y x) + V. \tag{14.35}$$

The canonical equation of motion of the infinitesimal mass is given by

$$\dot{x} = \frac{\partial H}{\partial P_x} = \frac{P_x}{m} + \omega y$$

$$\dot{y} = \frac{\partial H}{\partial P_y} = \frac{P_y}{m} - \omega x$$

$$\dot{z} = \frac{\partial H}{\partial P_z} = \frac{P_z}{m}$$

$$\dot{P}_x = -\frac{\partial H}{\partial x} = \omega P_y - \frac{\partial V}{\partial x}$$

$$\dot{P}_y = -\frac{\partial H}{\partial y} = -\omega P_x - \frac{\partial V}{\partial y}$$

$$\dot{P}_z = -\frac{\partial H}{\partial z} = -\frac{\partial V}{\partial z}. \tag{14.36}$$

Equations (14.36) are six first order ordinary differential equations which describe the motion of infinitesimal mass in general terms.

Here, ω will be the function of time. If the finite mass m_1 and m_2 move along a circular orbit, then ω is the angular velocity of m_1 and m_2. Then, H is independent of time. Therefore, there exists a Jacobi integral.

Equations (14.36) are six differential equations of first order. They can be put in the form of three equations of the second order. We have from (14.33)

$$P_x = m(\dot{x} - \omega y)$$

$$\therefore \; \dot{P}_x = m(\ddot{x} - \omega \dot{y})$$

Similarly, $$\dot{P}_y = m(\ddot{y} - \omega \dot{x})$$

and $$\dot{P}_z = m\ddot{z}.$$

Substituting these values in the last three equations of (14.36), we get

$$m(\ddot{x} - \omega \dot{y}) = m\omega(\dot{y} + \omega x) - \frac{\partial V}{\partial x},$$

or $$(\ddot{x} - \omega \dot{y}) = \omega(\dot{y} + \omega x) - \frac{1}{m}\frac{\partial V}{\partial x},$$

or $$\ddot{x} - 2\omega \dot{y} - \omega^2 x = -\frac{1}{m}\frac{\partial V}{\partial x}.$$

Similarly, $$\ddot{y} + 2\omega \dot{x} - \omega^2 y = -\frac{1}{m}\frac{\partial V}{\partial y}$$

and $$\ddot{z} = -\frac{1}{m}\frac{\partial V}{\partial y}. \tag{14.37}$$

Now, let us introduce the modified potential energy, U:

$$\begin{aligned} U &= -V + T_0 \\ &= -m\gamma\left[\frac{m_1}{\rho_1} + \frac{m_2}{\rho_2}\right] + \frac{1}{2}m\omega^2(x^2 + y^2) \end{aligned} \tag{14.38}$$

$$\frac{\partial U}{\partial x} = m\omega^2 x \text{ or } \omega^2 x = \frac{1}{m}\frac{\partial U}{\partial x}.$$

Equation (14.37) can be written as

$$\begin{aligned}\ddot{x} - 2\omega\dot{y} &= \frac{1}{m}\frac{\partial U}{\partial x}\\ \ddot{y} + 2\omega\dot{x} &= \frac{1}{m}\frac{\partial U}{\partial y}\\ \ddot{z} &= -\frac{1}{m}\frac{\partial U}{\partial y}.\end{aligned} \tag{14.39}$$

The Jacobi integral can be obtained from Equation (14.39) and multiplying the above equation, $\dot{x}, \dot{y}, \dot{z}$, respectively, and adding

$$\dot{x}\ddot{x} + \dot{y}\ddot{y} + \dot{z}\ddot{z} = \frac{1}{m}\left(\frac{\partial U}{\partial x}\frac{dx}{dt} + \frac{\partial U}{\partial y}\frac{dy}{dt} + \frac{\partial U}{\partial z}\frac{dz}{dt}\right) = \frac{1}{m}\frac{\partial U}{\partial t},$$

integrating

$$\frac{1}{2}(\dot{x}^2 + \dot{y}^2 + \dot{z}^2) = \frac{1}{m}U + h,$$

$$\text{or } \frac{1}{2}(\dot{x}^2 + \dot{y}^2 + \dot{z}^2) - \frac{1}{m}\left[-m\gamma\left[\frac{m_1}{\rho_1} + \frac{m_2}{\rho_2}\right] + \frac{1}{2}m\omega^2(x^2 + y^2)\right] = h,$$

$$\text{or } \frac{1}{2}(\dot{x}^2 + \dot{y}^2 + \dot{z}^2) - \gamma\left[\frac{m_1}{\rho_1} + \frac{m_2}{\rho_2}\right] - \frac{1}{2}\omega^2(x^2 + y^2) = h. \tag{14.40}$$

This is known as the Jacobi Integral.

14.8 Exercises

1. If a particle of mass m is constrained to move on a smooth surface, show that the system of Hamilton's equation is

$$\frac{du^{\alpha}}{dt}=\frac{\partial H}{\partial p^{\alpha}},\ \frac{dp^{\alpha}}{dt}=-\frac{\partial H}{\partial u^{\alpha}},\quad (\alpha=1,2)\ with\ p_{\alpha}=ma_{\alpha\beta}\dot{u}^{\beta}$$

$$and\ H=\frac{1}{2}ma^{\alpha\beta}p_{\alpha}p_{\beta}+V.$$

2. Show that along the dynamic trajectory $\frac{dH}{dt}=0$ that $H=constant$ is an integral of Hamilton's equations.

3. If $T=\frac{1}{2}(\dot{q})^2$ and $V = kq^2$, k > 0, show that $H=\frac{p^2}{2m}+m\omega^2q^2/2$, where $\omega^2=\frac{k}{m}$. Deduce that $q=\sqrt{\frac{2h}{m\omega^2}}\sin(\omega t+\alpha)$.

Appendix A: Answers to Even-Numbered Exercises

Exercise 1.9

1. (a) a^i (b) $(x^i)^2$ (c) n (d) n (f) $\delta_i^i = n$
2. (b) $(a_1^1 + a_1^2 + \ldots + a_1^n)x^1 + \ldots(a_1^1 + a_n^2 + \ldots + a_n^n)x^n$

   (c) $\frac{\partial}{\partial x^1}\left(\sqrt{g}b^1\right) + \ldots\ldots + \frac{\partial}{\partial x^1}\left(\sqrt{g}b^n\right)$
5. (a) a_{ik}

   (b) $a_{ijp}x^i x^j + a_{ipk}x^i x^k + a_{pjk}x^k x^j$
7. (a) $u^1v^1 + \ldots + u^n v^n + u^1v^2 + \ldots\ldots + u^1v^n + \ldots + u^n v^1 + \ldots\ldots + u^n v^n$

   (b) $a_{ijk}\ x^i x^j\ \ x^k$
8. (a) A_j^{kl} (b) B^{ik} (c) $a^i a_i$ (d) δ_i^i

Exercise 3.9

1. 4, $g^{ij} = \begin{pmatrix} 2 & 3 & \frac{-3}{2} \\ 3 & 5 & \frac{-5}{2} \\ \frac{-3}{2} & \frac{-5}{2} & \frac{3}{2} \end{pmatrix}$

2. $ds^2 = (dy^1)^2 + (y^1)^2(dy^2)^2 + (dy^3)^2$

4. $\begin{pmatrix} 1 & 0 & 0 \\ 0 & -2 & -4 \\ 0 & -4 & 3 \end{pmatrix}, \frac{1}{-22}\begin{pmatrix} 1 & 0 & 0 \\ 0 & 3 & 4 \\ 0 & 4 & 2 \end{pmatrix}$

8. $\frac{1}{sin^2\alpha}\begin{pmatrix} 1 & cos\alpha \\ cos\alpha & 1 \end{pmatrix}$

Exercise 4.6

5. (i) $[12,2]=2x^1,[22,1]=-x^1,\ \left\{ \begin{matrix} 2 \\ 1\ 2 \end{matrix} \right\}=\frac{1}{(x^1)^2},\left\{ \begin{matrix} 3 \\ 1\ 3 \end{matrix} \right\}=$

$\frac{1}{x^1},\left\{ \begin{matrix} 3 \\ 2\ 3 \end{matrix} \right\}=cotx^2,\left\{ \begin{matrix} 1 \\ 2\ 2 \end{matrix} \right\}=-x^1,\left\{ \begin{matrix} 1 \\ 3\ 3 \end{matrix} \right\}=$

$-x^1sinx^2,\left\{ \begin{matrix} 2 \\ 3\ 3 \end{matrix} \right\}=-sinx^2cosx^2$

(ii) $[12,2]=\frac{1}{2}\frac{\partial f}{\partial x^1},,[22,1]=-\frac{1}{2}\frac{\partial f}{\partial x^1},\{22,2]=\frac{1}{2}\frac{\partial f}{\partial x^1},\left\{ \begin{matrix} 2 \\ 1\ 2 \end{matrix} \right\}=$

$\frac{1}{2f}\frac{\partial f}{\partial x^1},\left\{ \begin{matrix} 1 \\ 2\ 2 \end{matrix} \right\}=-\frac{1}{2}\frac{\partial f}{\partial x^1},\left\{ \begin{matrix} 2 \\ 2\ 2 \end{matrix} \right\}=\frac{1}{2f}\frac{\partial f}{\partial x^1},$

4. (i) $\left\{ \begin{matrix} i \\ j\ k \end{matrix} \right\}=0$ (ii) $\left\{ \begin{matrix} i \\ j\ k \end{matrix} \right\}=-x^1,\left\{ \begin{matrix} 2 \\ 1\ 2 \end{matrix} \right\}=\left\{ \begin{matrix} 2 \\ 2\ 1 \end{matrix} \right\}=\frac{1}{x^1}$

(iii) $\left\{ \begin{matrix} 1 \\ 2\ 2 \end{matrix} \right\}=-x^1,\left\{ \begin{matrix} 2 \\ 1\ 2 \end{matrix} \right\}=\frac{1}{x^1},\left\{ \begin{matrix} 3 \\ 3\ 3 \end{matrix} \right\}=-x^1(sinx^2)^2$

$\left\{ \begin{matrix} 3 \\ 1\ 3 \end{matrix} \right\}=\frac{1}{x^1},\left\{ \begin{matrix} 2 \\ 3\ 3 \end{matrix} \right\}=sinx^2\cos x^2$

$\left\{ \begin{matrix} 3 \\ 2\ 3 \end{matrix} \right\}=\left\{ \begin{matrix} 3 \\ 3\ 2 \end{matrix} \right\}=\cot x^2$

11. $\left(-\phi cot\theta,-\frac{2}{r}\phi,\frac{4}{r}\right)$

Exercise 5.7

5. $R_{ij,k}-R_{ik,j}$ 7. $a=-\frac{1}{2}$

Exercise 8.8

5. $\chi = \frac{1}{a}$, $\mu^i = (-1,0,0,)$ *and* $\tau = 0; \nu^i = (0,0,1)$

Exercise 9.11

1. d $\sigma = a^2 sinu\ dudv$

2. $$\frac{d^2u^1}{ds^2} - u^1\left(\frac{du^2}{ds}\right)^2 = 0, \frac{d^2u^2}{ds^2} + \frac{2}{u^1}\frac{du^1}{ds}\frac{du^2}{ds} = 0$$

Exercise 10.10

2. $$\frac{4(v^2 - u^2)}{(1+4v^2+4u^2)^{\frac{3}{2}}}$$

3. $$\kappa = -\frac{(u^1 f_2)^2}{[(u^1)^2 + f_1^{\,2}]^2}, \mathrm{H} = \frac{-\mathrm{f}_1}{2[(u^1)^2 + f_1^{\,2}]^{\frac{3}{2}}}$$

Exercise 11.5

1. $$\mathrm{X}_{\mathrm{p}}^2 + \left[\frac{f_1}{\{(u^1)^2 + f_1^{\,2}\}^{\frac{3}{2}}}\right]\mathrm{X}_{\mathrm{p}} - \frac{(u^1 f_2)^2}{\{(u^1)^2 + f_1^{\,2}\}^2} = 0,\ \kappa = -\frac{(u^1 f_2)^2}{\{(u^1)^2 + f_1^{\,2}\}^2},$$

$$H = \frac{1}{2}\frac{-f_1}{\{(u^1)^2 + f_1^{\,2}\}^{\frac{3}{2}}}$$

2. $\mathrm{X}_{(1)} = \frac{-\mathrm{a}}{(u^1)^2}, \mathrm{X}_{(2)} = \frac{\mathrm{a}}{(u^1)^2}$

3. $[u^1 f_1 f_2 + 1 + f_1^2 = 0]$

Exercise 12.5

2. $\kappa > o$, *elliptic*

References

Arthur E. Fischer, *An Introduction to Conformal Ricci Flow,* Classical and Quantum Gravity, Institute of Physics Publishing (Online, 2004).

Bennett Chow, Peng Lu, Lie Ni, *Hamilton's Ricci Flow*, Vol. I (Preliminary version, 2005).

Chaki, M.C., *Tensor Calculus* (Calcutta Publishers, India, 1987).

De, U.C., *Differential Geometry of Curves and Surfaces in E^3*, (Anamaya Publishers, New Delhi, 2007).

Eisenhart, L.P., *An Introduction to Differential Geometry,* (Princeton University Press, 1940)

Eisenhart, L.P. *Riemannian Spaces of Class Greater than Unity*, Annals of Math. (vol. 38).

Goreux S.J., *Introduction to Tensor Calculus* (New Central Book Agency, India, 1984).

Ishwar, B., *Basics of Dynamical Systems*, (DSTA=2018, ISM, India, 2018).

Levi-Civita, T., *The Absolute Differential Calculus* (Blackie & Sons, London, 1927).

Millman, R.S. and Parker, G.D., *Elements of Differential Geometry*, (Academic Press, 1965).

Nayak, P., *Mechanics: Newtonian, Classical, Relativistic*, (Asian Books, New Delhi, 2008).

Resnik, R., *Introduction of Special Relativity*, (John Wiley, New York, 1968).

Sokolnikoff, I.S. *Tensor Analysis theory and Applications to Geometry and Mechanics of Continua* (John Wiley & Sons, Inc, New York, 1951).

Sokolnikoff, I.S., *Mathematical Theory of Elasticity* (New York, 1946).

Weatherburn, C.E., *Riemannian Geometry and the Tensor Calculus*, (Cambridge University Press, 1938).

Weatherburn, C.E., *Differential Geometry of Three Dimensions*, (Cambridge University Press, London, 1966).

Willmore, T.J., *An Introduction to Differential and Riemannian Geometry*, (Oxford University Press, UK, 1965).

Index

Also of Interest

Check out these other titles from Scrivener Publishing

SIMULATION AND ANALYSIS OF MATHEMATICAL METHODS IN REAL TIME ENGINEERING APPLICATIONS, Edited by T. Ananth Kumar, E. Golden Julie, Y. Harold Robinson, and S. M. Jaisakthi, ISBN: 9781119785378. Written and edited by a group of international experts in the field, this exciting new volume covers the state of the art of real time application of computer science using mathematics. *NOW AVAILABLE!*

MATHEMATICS IN COMPUTATIONAL SCIENCE AND ENGINEERING, Edited by Ramakant Bhardwaj, Jyoti Mishra, and Satyendra Narayan, ISBN: 9781119777151. This groundbreaking new volume, written by industry experts, is a must-have for engineers, scientists, and students across all engineering disciplines working in mathematics and computational science who want to stay abreast with the most current and provocative new trends in the industry. *DUE OUT IN SPRING 2022!*

i-Smooth Analysis: Theory and Applications, by A.V. Kim, ISBN 9781118998366. A totally new direction in mathematics, this revolutionary new study introduces a new class of invariant derivatives of functions and establishes relations with other derivatives, such as the Sobolev generalized derivative and the generalized derivative of the distribution theory. *NOW AVAILABLE!*

Open Ended Problems: A Future Chemical Engineering Approach, by J. Patrick Abulencia and Louis Theodore, ISBN 9781118946046. Although the primary market is chemical engineers, the book covers all engineering areas so those from all disciplines will find this book useful. *NOW AVAILABLE!*

Printed and bound by CPI Group (UK) Ltd, Croydon, CR0 4YY
17/09/2026
14964247-0008